2 Springer Series in Solid-State Sciences

Edited by M. Cardona, P. Fulde, and H.-J. Queisser

Springer Series in Solid-State Sciences

Editors: M. Cardona P. Fulde K. von Klitzing H.-J. Queisser

Volumes 1–39 are listed on the back inside cover

40 **Semiconductor Physics** An Introduction
3rd Edition By K. Seeger

41 **The LMTO Method**
Muffin-Tin Orbitals and Electronic Structure
By H.L. Skriver

42 **Crystal Optics with Spatial Dispersion, and Excitons**
By V.M. Agranovich and V.L. Ginzburg

43 **Resonant Nonlinear Interactions of Light with Matter**
By V.S. Butylkin, A.E. Kaplan, Yu. G. Khronopulo, and E.I. Yakubovich

44 **Elastic Media with Microstructure II**
Three-Dimensional Models
By I.A. Kunin

45 **Electronic Properties of Doped Semiconductors**
By B.I. Shklovskii and A.L. Efros

46 **Topological Disorder in Condensed Matter**
Editors: F. Yonezawa and T. Ninomiya

47 **Statics and Dynamics of Nonlinear Systems**
Editors: G. Benedek, H. Bilz, and R. Zeyher

48 **Magnetic Phase Transitions**
Editors: M. Ausloos and R.J. Elliott

49 **Organic Molecular Aggregates,** Electronic Excitation and Interaction Processes
Editors: P. Reineker, H. Haken, and H.C. Wolf

50 **Multiple Diffraction of X-Rays in Crystals**
By Shih-Lin Chang

51 **Phonon Scattering in Condensed Matter**
Editors: W. Eisenmenger, K. Laßmann, and S. Döttinger

52 **Superconductivity in Magnetic and Exotic Materials**
Editors: T. Matsubara and A. Kotani

53 **Two-Dimensional Systems, Heterostructures, and Superlattices**
Editors: G. Bauer, F. Kuchar, and H. Heinrich

54 **Magnetic Excitations and Fluctuations**
Editors: S. Lovesey, U. Balucani, F. Borsa, and V. Tognetti

55 **The Theory of Magnetism II**
Thermodynamics and Statistical Mechanics
By D.C. Mattis

56 **Spin Fluctuations in Itinerant Electron Magnetism** By T. Moriya

57 **Polycrystalline Semiconductors,**
Physical Properties and Applications
Editor: G. Harbeke

58 **The Recursion Method and Its Applications**
Editors: D. Pettifor and D. Weaire

59 **Dynamical Processes and Ordering on Solid Surfaces**
Editors: A. Yoshimori and M. Tsukada

60 **Excitonic Processes in Solids**
By M. Ueta, H. Kanzaki, K. Kobayashi, Y. Toyozawa, and E. Hanamura

61 **Localization, Interaction, and Transport Phenomena**
Editors: B. Kramer, G. Bergmann, and Y. Bruynseraede

62 **Theory of Heavy Fermions and Valence Fluctuations**
Editors: T. Kasuya and T. Saso

63 **Electronic Properties of Polymers and Related Compounds**
Editors: H. Kuzmany, M. Mehring, and S. Roth

64 **Symmetries in Physics** Group Theory Applied to Physical Problems
By W. Ludwig and C. Falter

65 **Phonons: Theory and Experiments II**
Experiments and Interpretation of Experimental Results By P. Brüesch

66 **Phonons: Theory and Experiments III**
Phenomena Related to Phonons
By P. Brüesch

67 **Two-Dimensional Systems: Physics and New Devices**
Editors: G. Bauer, F. Kuchar, and H. Heinrich

68 **Phonon Scattering in Condensed Matter V**
Editors: A.C. Anderson and J.P. Wolfe

69 **Nonlinearity in Condensed Matter**
Editors: A.R. Bishop, D.K. Campbell, P. Kumar and S.E. Trullinger

70 **From Hamiltonians to Phase Diagrams**
The Electronic and Statistical-Mechanical Theory of sp-Bonded Metals and Alloys
By J. Hafner

71 **High Magnetic Fields in Semiconductor Physics**
Editor: G. Landwehr

72 **One-Dimensional Conductors**
By S. Kagoshima, H. Nagasawa, and T. Sambongi

73 **Quantum Solid-State Physics**
Editors: S.V. Vonsovsky and M.I. Katsnelson

74 **Quantum Monte Carlo Methods** in Equilibrium and Nonequilibrium Systems
Editor: M. Suzuki

75 **Pseudo Potential for the Electronic Structure of Semiconductors**
By M.L. Cohen and J.R. Chelikowsky

76 **Electronic Properties of Conjugated Polymers**
Editors: H. Kuzmany, M. Mehring, and S. Roth

O. Madelung

Introduction to
Solid-State Theory

Translated by B. C. Taylor

With 144 Figures

Springer-Verlag Berlin Heidelberg New York
London Paris Tokyo

Professor Dr. Otfried Madelung

Fachbereich Physik, Universität Marburg, Mainzer Gasse 33,
D-3550 Marburg, Fed. Rep. of Germany

Translator:

Dr. B. C. Taylor

59 Charles Way, Malvern, Worcs. WR 14 2NB, Great Britain

Series Editors:

Professor Dr., Dr. h. c. Manuel Cardona
Professor Dr., Dr. h. c. Peter Fulde
Professor Dr. Klaus von Klitzing
Professor Dr. Hans-Joachim Queisser

Max-Planck-Institut für Festkörperforschung, Heisenbergstrasse 1
D-7000 Stuttgart 80, Fed. Rep. of Germany

1st Edition 1978
2nd Printing 1981

ISBN 3-540-08516-5 Springer-Verlag Berlin Heidelberg New York
ISBN 0-387-08516-5 Springer-Verlag New York Berlin Heidelberg

Library of Congress Cataloging in Publication Data. Madelung, Otfried. Introduction to solid-state theory. (Springer series in solid-state sciences; v. 2). Revised and partly rewritten translation of Festkörpertheorie. Bibliography: p. Includes index. 1. Solid state physics. I. Title. II. Series. QC 176. M 2313 1978 530.4'1
77-26263

Printing: Weihert-Druck GmbH, Darmstadt
Binding: J. Schäffer GmbH & Co. KG., 6718 Grünstadt
2153/3150-54321

Preface to the Second Printing

The necessity for a second printing two years after the publication of the first edition gave the opportunity for some improvements and corrections of the text and for redrawing or substituting several figures. Furthermore, the bibliography has been brought up-to-date. I am grateful to the Springer-Verlag for meeting all my wishes.

Marburg, January 1981 *Otfried Madelung*

From the Preface to the First Edition

This book is intended for graduate students of physics, materials science, and electrical engineering as a textbook in solid-state theory. In addition, it should provide the theoretical background needed by research physicists in solid-state physics and in the solid-state areas of electrical engineering.

The content of this book and the level of presentation are determined by the needs of its intended audience. The field of solid-state physics has grown so large that some selection of topics has to be made. In a book on solid-state *physics* it is still possible to survey the full range of solid-state phenomena and to connect them by a qualitative presentation of theoretical concepts. However, in a text introducing solid-state *theory*, a presentation of *all* theoretical concepts and methods seemed inappropriate. For this reason I have tried to develop the fundamentals of solid-state theory starting from a single unifying point of view—the description by *delocalized* (extended) and *localized states* and by *elementary excitations*. The development of solid-state theory within the last ten years has shown that by a systematic introduction of those concepts large parts of the theory can be described in a unified way. At the same time this form of description gives a "pictorial" formulation of many elementary processes in solids which facilitates their understanding.

Admittedly the attempt to present solid-state physics under one unifying aspect has its shortcomings. Not all parts of solid-state theory fit naturally into this frame. But the limitations imposed by such organization of the book seemed to me justified for several reasons. First, because there are only a few topics which do not fit into this type of description, the range covered is representative of the predominant part of solid-state theory. Secondly, the manner of description chosen seems especially suited for those areas of solid-state physics which are dominant in the application to solid-state electronics. Finally, since so many valuable textbooks and monographs on solid-state theory are available, a new book should intend to complement them rather than to compete.

I have tried to offer a general framework of solid-state theory which the reader can fill in from the more specialized material provided by monographs, review articles, and original papers. In this book, some fields are described in detail and some fields are treated more briefly. Topics which have been covered by comprehensive monographs are in some cases presented here only from the viewpoint of elementary excitations. Thus, spin-waves are emphasized in the chapter on magnetism. The electron-electron interaction by exchange of virtual phonons is the central topic in the chapter on superconductivity, whereas other important aspects of this field are only mentioned briefly. In every case, however, I have tried to inform the reader as completely as possible about additional available literature.

It was not my intention to write a book on solid-state theory for the prospective solid-state *theorist*. I therefore intentionally refrained from using the abstract methods of quantum field theory, important as they are in many-body problems. The general use of these methods seems to me inappropriate for the broad audience to whom this book is directed. On the other hand, some prior knowledge of elementary quantum mechanics as well as of the most important solid-state phenomena is required and assumed. Because of the close connection of all fields of solid-state physics, from the basic theoretical concepts to the technical applications, I have made use of SI-units (Système International) throughout, in contrast to most other textbooks in this field. To each chapter some problems are added. Most of them are not intended to train the reader in theoretical methods but to direct his attention to applications and additional questions which arise from the respective sections. Many of the problems have already been discussed in other monographs or review articles. I have indicated such sources in the Bibliography.

Marburg, April 1978 *Otfried Madelung*

Contents

3. Elementary Excitations

4. Electron-Phonon Interaction: Transport Phenomena

1. Fundamentals

1.1 Introduction

Solids are composed of atoms held together by chemical bonds. Solid-state physics is therefore concerned with those physical properties which are the collective properties of this atomic arrangement. The characteristic properties of free atoms do, of course, determine the nature of the solid they make up, but, when embedded in a crystal lattice, these properties are greatly influenced by the surroundings. Electrical conductivity, ferromagnetism, specific heat, and phase transitions are, moreover, examples of concepts which can be defined for the solid but not for an individual atom. A theoretical description of the properties of solids must therefore use methods appropriate to many-body systems.

The characteristic feature of all solids (as with all *condensed matter*) is their order, i.e., the correlation in the positions of neighbouring atoms. This can be *short-range order* and restricted to a more or less limited volume surrounding an atom. Short-range order can diminish with increasing distance, as in amorphous semiconductors, or it can be restricted to microcrystals which are connected one to another in disjointed fashion. However the majority of all solids has *long-range order*, i.e., a regular *lattice* extending over considerable distances. The great number of structures able to satisfy lattice geometry and bonding criteria is one of the main reasons for the abundance of different solid-state phenomena.

Real crystals always show departures from an ideal structure. Every solid is of finite extent, so crystals are bounded by *surfaces* or *inner boundaries*. This is a trivial observation, but one which is important for many physical phenomena. *Lattice defects*, the presence of impurity atoms, dislocations, and local disturbances of lattice periodicity can never be completely absent in any real crystal.

Even the thermal motion of the lattice atoms constitutes a departure from strict periodicity. The periodic lattice is formed not by the atoms themselves but by their equilibrium positions. The atoms remain at these positions permanently only at the temperature of absolute zero, i.e., when the crystal is in its *ground state*. Departures from this ground state lead to deviations from order. At normal temperatures, however, the deviations are mostly so small that order remains the distinctive feature of a crystal.

Problems in solid-state physics can be related to two basic questions:

1) What is the ground state of a given solid? Why is it stable? What sort of forces hold the atoms in the lattice together?

2) How does the solid behave under external influences?

The first group of questions is characterized by concepts like crystal structure, chemical bonding, cohesion, and binding energy. This group appears at first to take precedence over the second which is concerned with the effect of external influences, but in fact the questions in the first group can only be answered *through* the answers to the second. *For every experiment means intervention, and a disturbance of the ground state.* Only by examining the consequences of such interventions, for example the effects produced by application of an electric field, or a temperature gradient, or by exposure to light, is it possible to also determine the properties of the solid in its ground state.

The phenomena of interest are characterized by the experimental tools available. These are

1) *Electric fields.* The object under investigation is charge transport, i.e., electric current. The phenomenological division of solids into metals, semiconductors, and insulators follows from these investigations, as does the division into electronic and ionic conductors, depending on the mechanism of electrical conduction. Superconductivity also belongs to this topic.

2) *Magnetic fields.* The various types of magnetism—dia- and para-, ferro-, antiferro-, and ferrimagnetism—are phenomena which a solid, depending upon its structure, shows in a magnetic field. A magnetic field is often used as an additional means to increase the variety of the observed effects. An example might be magnetic field as a parameter in investigating charge transport under the influence of an electric field. In this way more information and a greater insight into the characteristics of solids is obtained.

3) *Temperature gradients* lead to the transfer of thermal energy from hot to cold areas. Energy transport is possible along with charge transport.

4) *Optical phenomena*, absorption, reflection, and dispersion of photons provide information on the interaction between electromagnetic waves and solids.

5) *Electrons*, *neutrons*, and other corpuscular rays can be used to probe solid-state characteristics.

6) One can also obtain useful information about crystals by deliberately disturbing the crystal lattice, e.g., by doping with *impurity atoms*, or by producing *lattice disorder* or *dislocations*.

This list of experimental possibilities could be extended. But only the most important ones need to be mentioned here.

It is not possible to devise a single theoretical model to account for all these phenomena. The many-body system of the solid is too complicated for that. Appropriate, simplified models are deduced for particular areas of interest. Any true solid-state theory must, however, aim to bring these individual aspects together under some unifying *concepts*. There are several ways of achieving this.

The concept which has come increasingly to the fore in recent years is that of *elementary excitations*. This concept can be explained as follows:

As is clear from the above, the solid under investigation is usually in an excited state. The energy producing the excitation can be thermal, it can be imposed externally, or it can come from a deliberate disturbance of the lattice structure. It can be fed to various subsystems of the solid. It can be taken up by the valence electrons or by the lattice, it can appear as kinetic energy in the lattice ions, or it can reside in the coupled spins of the lattice ions.

Even for a very weak, local excitation the energy supplied does not usually remain localized at a single lattice particle. There are interactions between the lattice particles (ions and electrons) and these serve to distribute energy from one particle to the others.

From the mechanics of a system of point masses we know how to describe complex modes of oscillation in simple terms. For a system with s degrees of freedom, one introduces s new generalized coordinates (normal coordinates) in such a way that the Hamiltonian—for small oscillations a positive definite quadratic function—is diagonalized. That is, the complex equations of motion are split in normal coordinates into s independent equations representing the motions of free oscillators. In this approach, excited states close to the ground state can be described by the excitation of just a few of these free oscillators. This method of description is used in lattice dynamics to describe the (small) oscillations of lattice ions about their equilibrium positions. The complex, collective oscillation of the lattice is divided into a number of independent normal modes. These normal modes are quantized, and the associated quanta are called *phonons*. Phonons are an example of an *elementary excitation*. In many ways they are equivalent to *photons*, the elementary excitations of the electromagnetic field.

Besides these *collective excitations* there is a second example of how collective interactions in a many-particle system can formally be greatly simplified. If a charged particle moves through a "gas" of similarly charged particles, it will repel the other particles from its path. This can be described *formally* by a model in which no interactions occur; instead the particle is accompanied by a compensating cloud of charges of opposite sign. The interaction, i.e., the effect of the other particles on the one observed, is in fact replaced by the inertia of the charge-cloud that the particle has to carry with it. Here again we have replaced a system of interacting particles by an equivalent system of noninteracting particles, in which the dynamics of the original particles is now replaced by the (different) dynamics of new *quasi-particles*. These quasi-particles are a further example of elementary excitations.

In characterizing the behaviour of solids we find many opportunities to introduce such elementary excitations. Similarly to the phonons or quanta of the lattice vibrations, *plasmons* are introduced to describe collective oscillations of the valence electrons in metals. The spin system of lattice atoms can be represented in terms of spin waves, with *magnons* as the associated quanta.

A further example are the *excitons* used to describe excitations of the valence electrons in semiconductors.

The definition of a quasi-particle is not clear cut. Electrons can be subjected to a variety of interactions as they move through a crystal. Depending on the extent to which these interactions are included in the electron dynamics, i.e., on the approximation used, the electron can appear as a different quasi-particle (free electron, Hartree-Fock-electron, Bloch-electron, screened electron). This is an often overlooked fact and can on occasion give rise to misunderstandings.

To a first approximation, elementary excitations of a similar type are independent. But in higher approximations mutual interactions have to be taken into account. However even in the latter case the concept of elementary interactions can still remain useful. Only interactions which are weak compared to the original interaction have to be taken into account, and these can often be dealt with by perturbation theory.

We shall return to these questions in Sections 3.1.1 and 3.1.5, where we take a closer look at the concept of quasi-particles.

We may often be able to neglect interactions between excitations of a given kind, but interactions between different kinds are always important. It is these interactions which mainly account for the rich variety of solid-state phenomena. Even the setting up of an equilibrium state in the solid demands an interaction, i.e., an energy exchange between the various excitations.

Within the limits of this concept we can now ask the questions: What elementary excitations arise if a given solid is subjected to a small external disturbance? What energy do the quanta of the collective excitations and the quasi-particles have? What interactions should be considered? And finally: How are the elementary excitations affected by external forces? The answers to these questions then give us the answers to the questions on the physical properties of the solid and on its behaviour in an experiment.

Collective excitations like phonons are excitations of the entire solid. A phonon has a definite wave vector and a definite energy, whereas its location is completely undetermined. The same is true for the quasi-particles whose energy and wave vectors are precisely given. The description of excited states of a solid in terms of such *extended* or *delocalized* states is possible only if the solid can be considered as an infinite undisturbed medium. It has the advantage that the elementary processes leading to the excitation can be simply described: By external inputs of energy and momentum, by exchange of energy and momentum between different subsystems, quasi-particles change their state, and quanta of collective excitations are absorbed and emitted. The adoption of the concept of elementary excitations makes both the "pictorial" interpretation of the elementary processes and the mathematical formulation of many problems in solid-state theory relatively simple.

As with every theoretical concept, the concept of elementary excitations has only limited validity and applicability.

First it is clear that the concept is only reasonable for small deviations from the ground state, for when the number of collective excitations and quasi-particles becomes large, when the coupling between them becomes too strong, we again burden the theoretical picture with the many details from which we wanted to free ourselves by this very concept. One set of problems which can therefore certainly not be handled by this model is that of phase transitions.

In addition we must recognize that the description in terms of *extended states* is only one limiting case for the description of physical phenomena in an infinite undistorted solid. The description can also start from *localized states*, e.g., from states concentrated at individual lattice sites. Depending on the nature of the solid but also on the physical question posed, one or other limiting case will be chosen. We shall discuss these alternatives in more detail in Section 8.1.

In a *distorted lattice* a description based on localized states or at least a combination of both limiting cases will always be necessary. Localized point defects, impurity lattice atoms, or other imperfections lead to localized states in addition to the extended states of the host lattice. The catalogue of questions started above can then be extended by questions like: What isolated imperfections are possible in a given crystal? What localized states do occur? What interactions do they have with each other and with the elementary excitations? The answers to these provide the answer to the question on the effect of lattice imperfections on the physical properties of solids.

If the distortion of the lattice is very large, the concept of extended states is only of limited value and a description based on localized states becomes more important. This is the case for alloys and amorphous solids, for example.

The way we plan to present the material in this book is the following. First of all we shall attempt as far as possible to use the extended state description. This means we shall limit our attention to the perfect, infinite crystal and its physical properties. Within this theoretical framework an important role is played by the *one-electron approximation*. As long as we can neglect the interaction between the electrons in the solid or as soon as we can introduce noninteracting quasi-electrons, the description of many solid-state phenomena reduces to describing the behaviour of individual electrons under external influences. The fundamentals of the one-electron approximation will be discussed in Chapter 2.

In Chapter 3 we consider the various elementary excitations which form an important part of the description of solid-state phenomena. We shall also take a closer look at the theoretical basis for the concept of elementary excitations.

The following chapters are then dedicated, respectively, to important interactions between various elementary excitations. Each interaction leads to an important branch of solid-state physics: transport phenomena, superconductivity, optics, thermal properties. At first all these areas will be looked at from the point of view of the particular interaction. The content of the chapters will not, however, be limited to this.

In Chapter 8 we shall change over from extended states to localized states. We shall then have the opportunity to introduce important concepts of the theory of the chemical bond, i.e., the theory which attemps to understand the properties of a solid in terms of the properties of the atoms making up its lattice.

In the last two chapters, localized states will be of increasing interest to us. Localized states associated with point imperfections in an otherwise perfect crystal lattice, and localized surface states call for a description which combines concepts of both limiting cases. The disordered lattice, dealt with in the final chapter, requires new theoretical methods of description.

Finally another comment on the *mathematical methods* of solid-state theory. Two properties of the solid state are of special importance—the solid as a *many-particle system* and the *symmetries of the crystal lattice*. The latter is very important in reducing mathematical complexity. Much information can be gained from considerations of symmetry alone, whithout having to solve the Schrödinger equation quantitatively. We shall point to these possibilities in many places. However, within the scope of this book it is not possible to bring in the tools of group theory in extenso. For an introduction to these methods the reader is referred to [43–49].

The many-body aspect of all the problems demands various mathematical aids. Quantum statistics (Fermi and Bose statistics) provide the energy distributions of the noninteracting elementary excitations. The occupation number representation proves very useful for the quantum mechanical formulation of the interaction processes. This representation is explained in more detail in the Appendix. The methods of quantum field theory (diagrams, Green's functions, scattering theory, density matrix, etc.) are increasingly being used to handle interactions in many-body systems, particularly in disordered systems. A book aimed at a wide readership cannot however make too much use of these abstract modern methods. We shall only use these methods to an extent that will be within the grasp of every reader who has studied conventional quantum mechanics for a single term. For further information we refer to [36–42, 107.7].

1.2 The Basic Hamiltonian

The Schrödinger equation is the starting point for all quantitative calculations of solid-state properties. We begin by setting up the Hamiltonian for the entire problem. It is made up of the kinetic energy of all particles in the solid and of their interaction energies. The solid is comprised of two groups of electrons—*valence electrons* which contribute to chemical bonding and *core electrons* which are tightly bound in the closed shells of the lattice ions and which scarcely influence the properties of the solid. Consequently, we usually consider the *valence electrons* and the *lattice ions* as independent constituents of the solid. It is not, however, always possible to make such a clear distinction, and herein lies the first of our approximations. The justification and the limits of this approximation will be discussed in Section 2.2.16.

The Hamiltonian, then, consists of the kinetic energy of all valence electrons (in future we shall omit the prefix "valence") and all ions, the energy associated with all the interactions between these particles, and eventually the energy associated with interactions with external fields

$$H = H_{\text{el}} + H_{\text{ion}} + H_{\text{el-ion}} + H_{\text{ex}}. \tag{1.1}$$

For the time being we shall neglect the last term. For the electron part we write

$$H_{\text{el}} = H_{\text{el,kin}} + H_{\text{el-el}} = \sum_k \frac{p_k^2}{2m} + \frac{1}{8\pi\varepsilon_0} \sum_{kk'}{}' \frac{e^2}{|\boldsymbol{r}_k - \boldsymbol{r}_{k'}|} \tag{1.2}$$

in which we have inserted a Coulomb term for the interaction. The sums are over all the electron indices, excluding of course $k = k'$ for the interaction term. $\boldsymbol{p}_k$, $\boldsymbol{r}_k$, and m are the momentum, position, and mass of an electron of index k.

For the ion part of the Hamiltonian we correspondingly write

$$H_{\text{ion}} = H_{\text{ion,kin}} + H_{\text{ion-ion}} = \sum_i \frac{P_i^2}{2M_i} + \frac{1}{2} \sum_{ii'}{}' V_{\text{ion}}(\boldsymbol{R}_i - \boldsymbol{R}_{i'}) \tag{1.3}$$

where we have identified ion parameters with capital letters.

We have left open here the explicit form of the ion-ion interaction and have only assumed that it can be described as a sum over two-particle interactions, each dependent only on the difference in the ion coordinates $\boldsymbol{R}_i$.

Correspondingly, for the electron-ion interaction we put

$$H_{\text{el-ion}} = \sum_{k,i} V_{\text{el-ion}}(\boldsymbol{r}_k - \boldsymbol{R}_i)\,. \tag{1.4}$$

It is useful to introduce a further division at this stage. A feature of the crystals is their symmetry, a symmetry resulting from the periodic arrangement of the ions in the lattice. However it is the equilibrium positions about which the ions vibrate which show strong periodicity, rather than their actual positions at any instant. We therefore divide the ion-ion interaction into two components, one describing the interaction when the ions are in their equilibrium position, the other, a correction to account for the vibrations of the lattice

$$H_{\text{ion-ion}} = H^0_{\text{ion-ion}} + H_{\text{ph}}\,, \tag{1.5}$$

$$H_{\text{el-ion}} = H^0_{\text{el-ion}} + H_{\text{el-ph}}\,. \tag{1.6}$$

The index ph for the lattice vibration term is pointing to the phonons, which we shall make use of later to describe the vibrations.

Equations (1.1–6) provide the foundation for the quantum mechanical treatment of most solid-state properties. The next step is the transition from the Hamiltonian *function* to the Hamiltonian *operator*. In the coordinate representation the Hamiltonian operator then depends on the coordinates of all electrons and ions. Correspondingly the wave function on which H operates becomes a function of all these coordinates.

This form of the Hamiltonian allows only limited account to be taken of the electron spin (cf. the next section). However for most of the problems to be discussed here, the nonrelativistic Schrödinger equation without spin-orbit coupling terms will prove to be adequate.

It is not possible to rigorously solve the quantum mechanical problem. Approximations have to be made. In solid-state theory, two simplifications are commonly made. In solving a given problem, individual terms of the Hamiltonian may be neglected, or only partially considered, or handled subsequently as perturbations. This simplified problem is then further simplified using the symmetry of the lattice. The particular approximations made will depend on the questions we are seeking to answer and on the nature of the solid under examination.

There is a problem in completely neglecting individual terms of the Hamiltonian, since all interactions—being always Coulomb interactions—are equally strong. For this reason the *electron component* $H_{\rm el}$ in (1.2), for example, cannot be considered alone, for it describes an electron gas whose charge is not compensated by ions as in the solid. As a first approximation, (1.2) has to be extended at least to include a constant space charge ρ_+, representing the average charge of the ions, and the interaction of the electrons with this space charge. If we include these two additions in a single term H_+, we can write the Hamiltonian for this approximation as

$$H_{\rm el} = -\sum_k \frac{\hbar^2}{2m}\nabla_k^2 + \frac{1}{8\pi\varepsilon_0}\sum_{kk'}{}' \frac{e^2}{|\boldsymbol{r}_k - \boldsymbol{r}_{k'}|} + H_+ \,. \tag{1.7}$$

The electron gas is considered here to be embedded in a constant positive background. This model is often described as "*jellium*". Lattice symmetry retreats completely into the background while the properties of the electron gas, particularly the electron-electron interaction, are to the fore. Many of the properties of metals can be described by means of this approximation. The problem of the free electron gas, with and without electron-electron interaction, will be taken up in Chapters 2 and 3.

The model can be refined by replacing the uniform distribution of space charge by a distribution in which the ions are taken to be fixed in their equilibrium positions $\boldsymbol{R}_i^0$. The electron-ion interaction is then described by the term $H^0_{\rm el-ion}$ from (1.6). In this case lattice symmetry is taken into account. The Hamiltonian, however, then becomes too complicated to deal with without making further approximations. This will be discussed in Section 2.2.

Starting from (1.3) and (1.5), we can handle the *movement of the ions* in a similar way, ending up with an equation which corresponds to (1.7). In this case we introduce a constant negative space charge ρ_- to take account of the neglected electrons and a term to represent their interaction with the ions. Combining both additions into a single term H_- we have

$$H_{\text{ion}} = -\sum_i \frac{\hbar^2}{2M_i} \nabla_i^2 + \frac{1}{2} \sum_{ii'}' V_{\text{ion}}(\boldsymbol{R}_i - \boldsymbol{R}_{i'}) + H_- . \tag{1.8}$$

The second term can be subdivided according to (1.5). This Hamiltonian forms the basis for the study of lattice dynamics which is treated in detail in Section 3.3.

The two terms H_+ and H_- in (1.7) and (1.8) exactly cancel each other. In (1.1) we are then (apart from H_{ex}) left with only $H_{\text{el-ion}}$, the term which couples electron and ion motion. If, following (1.6), we separate the interaction of the electrons with the static ion lattice from $H_{\text{el-ion}}$ and add this component to H_{el}, the only remaining coupling between electrons and ions is the *electron-phonon interaction* $H_{\text{el-ph}}$. This coupling can be dealt with in most cases by perturbation theory, following the solution of (1.7) and (1.8). We shall return to this topic in Chapter 4.

We have been able, through the above, to divide the total solid-state problem into two components—the movement of the electrons in a stationary lattice and the movement of the ions in a uniform space charge of electrons. This decoupling of the total system really requires rigorous justification. The so-called *adiabatic approximation* (Born-Oppenheimer method) is often used to provide such justification. It is based on the following argument: Electrons and ions have very different masses. The ions can respond only slowly to a change in the electron configuration, while the electrons respond adiabatically to a change in the positions of the ions. As far as the motion of the electrons is concerned, it is only the instantaneous configuration of the ions which is of interest. As a first approximation we can therefore adopt for the electrons a Schrödinger equation of the form

$$(H_{\text{el}} + H_{\text{el-ion}})\psi = E_{\text{el}}\psi \tag{1.9}$$

in which the coordinates of the ions are kept fixed. The wave function depends only on the coordinates of the electrons. The coordinates of the ions appear in the equation as parameters. As the starting point for a solution of the total problem we now use the product

$$\Psi = \psi(\boldsymbol{r}_1 \dots \boldsymbol{r}_N; \boldsymbol{R}_1 \dots \boldsymbol{R}_{N'})\varphi(\boldsymbol{R}_1 \dots \boldsymbol{R}_{N'}) \tag{1.10}$$

where the ψ are solutions of (1.9) and N, N' denote the number of electrons and ions, respectively. Putting this into the Schrodinger equation with (1.1) as the Hamiltonian we have

$$H\Psi = (H_{\mathrm{el}} + H_{\mathrm{ion}} + H_{\mathrm{el-ion}})\psi\varphi$$
$$= \psi(H_{\mathrm{ion}} + E_{\mathrm{el}})\varphi - \sum_i \frac{\hbar^2}{2M_i}(\varphi\nabla_i^2\psi + 2\nabla_i\varphi\cdot\nabla_i\psi). \tag{1.11}$$

If the last term here were absent, (1.10) would be a separation ansatz which effectively decouples electron and ion motion. For the movement of the ions an equation of the form

$$(H_{\mathrm{ion}} + E_{\mathrm{el}})\varphi = E\varphi \tag{1.12}$$

would follow, in which E_{el} still depends on the ion positions and therefore provides a contribution from the electrons to the potential energy of the ions.

Equation (1.12) is a Schrödinger equation which involves only the coordinates of the ions. It therefore describes the *motion of the ions*. To describe the *motion of the electrons*, we replace in (1.9) the instantaneous positions of the ions by their mean positions, i.e., we replace $H_{\mathrm{el-ion}}$ by $H^0_{\mathrm{el-ion}}$.

The last term in (1.11) couples the electron and ion systems. One can show that it provides only a small contribution to the total energy of the system in the state Ψ. This cannot, however, be taken as a proof that this term only describes a weak interaction which can subsequently be treated by perturbation theory.

The justification for the ansatz (1.10) is also doubtful. The Schrödinger equation (1.9) has as its solution not one eigenfunction ψ but a complete set of eigenfunctions ψ_n. The ansatz (1.10) should therefore be written as an expansion in terms of this set of eigenfunctions. The restriction to a single wave function neglects all electron transitions induced by the ion motion.

These remarks are intended solely to show that this basic approximation involves problems which call for a more exact analysis. It would not be appropriate to go further into these problems in this introductory chapter. We refer the reader to *Haug* [14] and *Ziman* [35] for a more detailed proof and critical discussion of the adiabatic approximation.

1.3 The Hartree-Fock Approximation

We turn our attention now to the motion of the electrons, as described by (1.7). We consider an *electron gas* which is embedded in a homogeneous, positively charged medium (jellium model) or in a rigid lattice of positively charged ions. The difficulty with the solution of this problem lies in the interaction between the individual electrons. In the absence of this interaction, the many-body problem would decouple into one-body problems which describe the movement of an electron in a given potential field. In view of the obvious advantage of such a *one-electron approximation*, we ask whether this problem could not be reduced to a one-body problem while including at least parts of the electron-electron interaction. Such a reduction is achieved by the *Hartree-Fock approximation*, to which we now turn.

We start with the Hamiltonian

$$H = -\sum_k \frac{\hbar^2}{2m}\nabla_k^2 + \sum_k V(\boldsymbol{r}_k) + \frac{1}{8\pi\varepsilon_0}\sum_{kk'}{}' \frac{e^2}{|\boldsymbol{r}_k - \boldsymbol{r}_{k'}|} = \sum_k H_k + \sum_{kk'}{}' H_{kk'} \,. \tag{1.13}$$

Here we have expressed the interaction $H^0_{\text{el-ion}}$ as $\sum_k V(\boldsymbol{r}_k)$ with $V(\boldsymbol{r}_k) = \sum_i V(\boldsymbol{r}_k - \boldsymbol{R}_i^0)$ according to (1.4).

In (1.13) the first two terms are sums over single-particle operators. The solution would be simple if, in accordance with the above remarks, we could neglect the (relatively strong) electron-electron interaction. The Schrödinger equation $\sum_k H_k\Phi = E\Phi$ could then be separated by putting

$$\Phi(\boldsymbol{r}_1 \ldots \boldsymbol{r}_N) = \varphi_1(\boldsymbol{r}_1)\varphi_2(\boldsymbol{r}_2) \ldots \varphi_N(\boldsymbol{r}_N) \,. \tag{1.14}$$

With $E = \sum_k E_k$, it reduces to one-electron equations $H_k\varphi_k(\boldsymbol{r}_k) = E_k\varphi_k(\boldsymbol{r}_k)$. The term $H_{kk'}$ in (1.13) prevents this possibility, depending as it does on the coordinates of two particles. In spite of this, the ansatz (1.14) allows us to derive an approximate solution which contains parts of the electron-electron interaction.

We start by inserting the function (1.14) into the Schrödinger equation $H\Phi = E\Phi$, with H from (1.13), and calculate the expectation value of the energy $E = \langle\Phi|H|\Phi\rangle$. Since H reduces to a sum of single-particle operators H_k and two-particle operators $H_{kk'}$, the matrix elements become products of integrals $\langle\varphi_k|H_k|\varphi_k\rangle$ or $\langle\varphi_k\varphi_{k'}|H_{kk'}|\varphi_k\varphi_{k'}\rangle$ and integrals $\langle\varphi_j|\varphi_j\rangle$ $(j \neq k, k')$. If we assume that the φ_k are normalized, the latter become unity and we are left with

$$E = \langle\Phi|H|\Phi\rangle = \sum_k \langle\varphi_k|H_k|\varphi_k\rangle + \frac{e^2}{8\pi\varepsilon_0}\sum_{kk'}{}' \left\langle \varphi_k\varphi_{k'} \left| \frac{1}{|\boldsymbol{r}_k - \boldsymbol{r}_{k'}|} \right| \varphi_k\varphi_{k'} \right\rangle . \tag{1.15}$$

This is only the expectation value of the energy for arbitrarily given φ_k. In line with the variation principle, those φ_k which minimize E represent the best set of functions for the ground state, *within the limits of the ansatz* (1.14). We therefore vary (1.15) for any φ_k^* or φ_k and equate the variation to zero. We add the normalization conditions (multiplied with Lagrange parameters E_k) to (1.15) and carry out the variation

$$\delta[E - \sum_k E_k(\langle\varphi_k|\varphi_k\rangle - 1)] = 0 \,. \tag{1.16}$$

We then get

$$\langle\delta\varphi_j|H_j|\varphi_j\rangle + \frac{e^2}{4\pi\varepsilon_0}\sum_{k(\neq j)}\left\langle\delta\varphi_j\varphi_k\left|\frac{1}{|\boldsymbol{r}_k-\boldsymbol{r}_j|}\right|\varphi_j\varphi_k\right\rangle - E_j\langle\delta\varphi_j|\varphi_j\rangle$$
$$= \langle\delta\varphi_j|H_j + \frac{e^2}{4\pi\varepsilon_0}\sum_{k(\neq j)}\left\langle\varphi_k\left|\frac{1}{|\boldsymbol{r}_k-\boldsymbol{r}_j|}\right|\varphi_k\right\rangle - E_j|\varphi_j\rangle = 0\,. \quad (1.17)$$

Since this equation must be valid regardless of the variation $\delta\varphi_j^*$, it follows that the φ_j are defined by the equation

$$\left[-\frac{\hbar^2}{2m}\nabla^2 + V(\boldsymbol{r}) + \frac{e^2}{4\pi\varepsilon_0}\sum_{k(\neq j)}\int\frac{|\varphi_k(\boldsymbol{r}')|^2}{|\boldsymbol{r}-\boldsymbol{r}'|}d\tau'\right]\varphi_j(\boldsymbol{r}) = E_j\varphi_j(\boldsymbol{r})\,. \quad (1.18)$$

Here we are representing the positions of the jth and the kth electrons by $\boldsymbol{r}$ and $\boldsymbol{r}'$, respectively.

Equation (1.18) is a single-particle Schrödinger equation, the *Hartree equation*. It describes an electron (j) at location $\boldsymbol{r}$ in the potential field $V(\boldsymbol{r})$ of the lattice ions, and in the Coulomb potential of an average distribution of all other electrons ($k \neq j$). The Lagrange parameters E_k have the meaning of one-electron energies. We return to this below. We shall also put off further discussion of the equation until later.

We rather extend the expression (1.14) now by applying the Pauli principle. We note that there are $N!$ possible ways of distributing N electrons at the N positions $\boldsymbol{r}_1 \ldots \boldsymbol{r}_N$. In view of the indistinguishability of the electrons, each possibility is equally likely. We choose wave functions $\varphi_j(\boldsymbol{q}_k)$ for the jth electron with coordinates $\boldsymbol{q}_k$ (spatial coordinates $\boldsymbol{r}_k$, and spin coordinate). We use a sum of $N!$ terms of the type described by (1.14), in which all possible electron permutations occur. The individual terms are given plus and minus signs so that Φ changes its sign when two electrons are interchanged. Furthermore we choose the φ_j to be orthogonal, which we can do without any loss of generality. We can write such a wave function in the form of a determinant (*Slater determinant*)

$$\Phi = (N!)^{-1/2}\begin{vmatrix}\varphi_1(\boldsymbol{q}_1) \ldots \varphi_N(\boldsymbol{q}_1)\\ \vdots \qquad\qquad \vdots\\ \varphi_1(\boldsymbol{q}_N) \ldots \varphi_N(\boldsymbol{q}_N)\end{vmatrix}, \quad (1.19)$$

where the factor in front of the determinant is added for normalization purposes. This form can be seen to satisfy the Pauli principle. If two electrons are interchanged, two columns of the determinant are interchanged and Φ changes sign. Again if two electrons have the same coordinates, two columns are identical and Φ vanishes.

With the wave function (1.19), we can once again calculate the expectation value $E = \langle\Phi|H|\Phi\rangle$. It is

$$E = \sum_k \int \varphi_k^*(\boldsymbol{q}_1)H_k\varphi_k(\boldsymbol{q}_1)d\tau_1 + \frac{e^2}{8\pi\varepsilon_0}\sum_{kk'}{}' \int \frac{|\varphi_k(\boldsymbol{q}_1)|^2|\varphi_{k'}(\boldsymbol{q}_2)|^2}{|\boldsymbol{r}_1 - \boldsymbol{r}_2|}d\tau_1 d\tau_2$$

$$- \frac{e^2}{8\pi\varepsilon_0}\sum_{kk'}{}' \int \frac{\varphi_k^*(\boldsymbol{q}_1)\varphi_k(\boldsymbol{q}_2)\varphi_{k'}^*(\boldsymbol{q}_2)\varphi_{k'}(\boldsymbol{q}_1)}{|\boldsymbol{r}_1 - \boldsymbol{r}_2|}d\tau_1 d\tau_2 \,. \tag{1.20}$$

The integration here includes a summation over the spin variables.

Compared with (1.15), an extra term appears. Taking account of the additional condition of orthogonality, the variation equation (1.16) becomes

$$\delta[E - \sum_{kk'} \lambda_{kk'}(\langle\varphi_k|\varphi_{k'}\rangle - \delta_{kk'})] = 0 \tag{1.21}$$

and in a manner similar to (1.17–18) this leads to

$$\left[-\frac{\hbar^2}{2m}\nabla_1^2 + V(\boldsymbol{r}_1)\right]\varphi_k(\boldsymbol{q}_1) + \frac{e^2}{4\pi\varepsilon_0}\sum_{k'}\int\frac{|\varphi_{k'}(\boldsymbol{q}_2)|^2}{|\boldsymbol{r}_1 - \boldsymbol{r}_2|}d\tau_2\varphi_k(\boldsymbol{q}_1)$$

$$-\frac{e^2}{4\pi\varepsilon_0}\sum_{k'}\int\frac{\varphi_{k'}^*(\boldsymbol{q}_2)\varphi_k(\boldsymbol{q}_2)}{|\boldsymbol{r}_1 - \boldsymbol{r}_2|}d\tau_2\varphi_{k'}(\boldsymbol{q}_1) = \sum_{k'}\lambda_{kk'}\varphi_{k'}(\boldsymbol{q}_1)\,. \tag{1.22}$$

It may readily be verified that the Hamiltonian on the left-hand side of (1.22) is hermitian. Then by introducing a transformation $\varphi_i' = \sum_k u_{ik}\varphi_k$, with an appropriately chosen unitary matrix u_{ik}, the matrix $\lambda_{kk'}$ can be diagonalized: $\lambda_{kk'}' = E_k\delta_{kk'}$. If we write φ instead of φ' for the new wave functions, the left-hand side of (1.22) remains unchanged whereas the right-hand side becomes $E_k\varphi_k(\boldsymbol{q}_1)$. We note further that in the absence of spin-orbit coupling, every wave function can be written as the product of a space function and a spin function. In the last term on the left-hand side of (1.22), that leaves us with just a summation over electrons with the same spin, because the orthogonality of the spin functions causes the other spin terms to disappear. Taking this into account, the spin does not explicitly appear further and we can replace the $\boldsymbol{q}_k$ with just the space vectors $\boldsymbol{r}_k$. Finally, as in (1.18) we use $\boldsymbol{r}$ for the coordinates of the electron under observation and $\boldsymbol{r}'$ for the integration variable, and find,

$$\left[-\frac{\hbar^2}{2m}\nabla^2 + V(\boldsymbol{r})\right]\varphi_j(\boldsymbol{r}) + \frac{e^2}{4\pi\varepsilon_0}\sum_{k(\neq j)}\int\frac{|\varphi_k(\boldsymbol{r}')|^2}{|\boldsymbol{r} - \boldsymbol{r}'|}d\tau'\varphi_j(\boldsymbol{r})$$

$$-\frac{e^2}{4\pi\varepsilon_0}\sum_{\substack{k(\neq j)\\ \text{spin}\parallel}}\int\frac{\varphi_k^*(\boldsymbol{r}')\varphi_j(\boldsymbol{r}')}{|\boldsymbol{r} - \boldsymbol{r}'|}d\tau'\varphi_k(\boldsymbol{r}) = E_j\varphi_j(\boldsymbol{r})\,. \tag{1.23}$$

This is the *Hartree-Fock equation.*

We add here some considerations which elucidate the meaning of the quantities E_k, which have so far been introduced formally only as Lagrange parameters. We inquire about the change of energy of the electron system when we remove one of the N electrons (e.g., the ith) from it. To do this we assume that, in view of the large number of electrons in the system, the removal of the ith electron will not alter the other φ_k $(k \neq i)$. The change of energy is then given by

$$\Delta E = \langle \Phi' | H | \Phi' \rangle - \langle \Phi | H | \Phi \rangle \tag{1.24}$$

with H from (1.13) and a wave function Φ' which is obtained from Φ by crossing out the ith row and column in the determinant (1.19). In carrying out the subtraction in (1.24), the only terms which remain are those from (1.20) in which k or k' is equal to i. This leads to

$$\begin{aligned} -\Delta E = &\int \varphi_i^*(\boldsymbol{q}_1) H_i \varphi_i(\boldsymbol{q}_1) d\tau_1 + \frac{e^2}{4\pi\varepsilon_0} \sum_{k(\neq i)} \frac{|\varphi_i(\boldsymbol{q}_1)|^2 |\varphi_k(\boldsymbol{q}_2)|^2}{|\boldsymbol{r}_1 - \boldsymbol{r}_2|} d\tau_1 d\tau_2 \\ &- \frac{e^2}{4\pi\varepsilon_0} \sum_{k(\neq i)} \int \frac{\varphi_i^*(\boldsymbol{q}_1)\varphi_i(\boldsymbol{q}_2)\varphi_k^*(\boldsymbol{q}_2)\varphi_k(\boldsymbol{q}_1)}{|\boldsymbol{r}_1 - \boldsymbol{r}_2|} d\tau_1 d\tau_2 = E_i \,. \end{aligned} \tag{1.25}$$

So E_i has the meaning of the energy parameter in a one-electron Schrödinger equation; $-E_i$ is the energy required to remove an electron from the system. Or expressing it another way: The energy needed to transfer an electron from state i to state k is $E_k - E_i$. This statement is known as *Koopmans theorem.* It is only correct if the assumption above holds.

We return now to the discussion of the Hartree equation (1.18) and the Hartree-Fock equation (1.23). While the Hartree equation was easy to interpret, the newly added third term on the left-hand side of (1.23) has no classical analogue. It is called the *exchange interaction.*

To compare (1.18) and (1.23), we rewrite the integral in the third term of (1.18) as follows:

$$\sum_{k(\neq j)} \int \frac{|\varphi_k(\boldsymbol{r}')|^2}{|\boldsymbol{r} - \boldsymbol{r}'|} d\tau' = \sum_k \int \frac{|\varphi_k(\boldsymbol{r}')|^2}{|\boldsymbol{r} - \boldsymbol{r}'|} d\tau' - \int \frac{|\varphi_j(\boldsymbol{r}')|^2}{|\boldsymbol{r} - \boldsymbol{r}'|} d\tau' . \tag{1.26}$$

In the Hartree equation we therefore subtract from the interaction of the observed electron with all electrons (including itself) that part which represents the interaction of the electron with its own space-charge cloud

$$-\frac{e}{4\pi\varepsilon_0} \int \frac{\rho_j^{\mathrm{H}}(\boldsymbol{r}')}{|\boldsymbol{r} - \boldsymbol{r}'|} d\tau' \varphi_j(\boldsymbol{r}), \quad \rho_j^{\mathrm{H}}(\boldsymbol{r}) = -e|\varphi_j(\boldsymbol{r})|^2 . \tag{1.27}$$

That ρ_j^{H} just represents one-electron charge follows from

$$\int \rho_j^{\mathrm{H}} d\tau = -e .$$

In (1.23) we can include the terms $k = j$ in both sums since these additions cancel themselves. The last term on the left-hand side of the Hartree-Fock equation then corresponds to the final term on the right-hand side of (1.26) and we can correspondingly write

$$\frac{e^2}{4\pi\varepsilon_0} \sum_{\substack{k \\ \text{spin} \parallel}} \int \frac{\varphi_k^*(\boldsymbol{r}')\varphi_j(\boldsymbol{r}')}{|\boldsymbol{r} - \boldsymbol{r}'|} d\tau' \varphi_k(\boldsymbol{r}) = -\frac{e}{4\pi\varepsilon_0} \int \frac{\rho_j^{\text{HF}}(\boldsymbol{r}, \boldsymbol{r}')}{|\boldsymbol{r} - \boldsymbol{r}'|} d\tau' \varphi_j(\boldsymbol{r}) \tag{1.28}$$

with

$$\rho_j^{\text{HF}} = -e \sum_{\substack{k \\ \text{spin} \parallel}} \frac{\varphi_k^*(\boldsymbol{r}')\varphi_j(\boldsymbol{r}')\varphi_j^*(\boldsymbol{r})\varphi_k(\boldsymbol{r})}{\varphi_j^*(\boldsymbol{r})\varphi_j(\boldsymbol{r})} . \tag{1.29}$$

In place of the charge density ρ^{H} (1.27), we thus have the *exchange charge density* ρ^{HF}. This also represents a charge $-e$, as can readily be seen by integration over $\boldsymbol{r}'$.

The important difference between ρ^{H} and ρ^{HF} is that whereas ρ^{H} is distributed over the whole crystal in the same way as the charge density of the $N - 1$ other electrons, ρ^{HF} also depends on $\boldsymbol{r}$, that is, on the location of the particular electron considered. Hence the electron described by (1.23) interacts with a charge distribution which is dependent on the electron position. By the Pauli principle the motion of the electrons of the same spin is correlated!

The spatial distribution of the exchange charge density is difficult to give in the general case. It can, however, be calculated for the special case of free electrons, and reveals typical features of the phenomenon of exchange interaction. Since we shall be dealing with the free electron gas model in both the following chapters, we postpone further discussion until Section 3.1.3.

From (1.27, 28) and using the abbreviation $\rho = \sum_k \rho_k^{\text{H}}$, the Hartree-Fock equation becomes

$$\left[-\frac{\hbar^2}{2m}\nabla^2 + V(\boldsymbol{r}) - \frac{e}{4\pi\varepsilon_0} \int \frac{\rho(\boldsymbol{r}') - \rho_j^{\text{HF}}(\boldsymbol{r}, \boldsymbol{r}')}{|\boldsymbol{r} - \boldsymbol{r}'|} d\tau' \right] \varphi_j(\boldsymbol{r}) = E_j \varphi_j(\boldsymbol{r}) . \tag{1.30}$$

The difficulties in solving this equation stem most of all from the fact that in the interaction term ρ depends on the φ_j, the solutions we are trying to find. An iterative method has therefore to be adopted to solve the equation. This involves putting initially trial functions for the φ_j into the ρ and solving the equation to obtain better estimates of the φ_j. The new φ_j are in turn put into ρ and the cycle is repeated until consistent solutions are attained (self-consistent field approximation).

A further difficulty appears because the third term in (1.30) depends on j, so that there is a different Hartree-Fock equation for each electron. We can get round this by using Slater's approximation which averages the ρ_j^{HF} over all j

$$\bar{\rho}^{\mathrm{HF}} = \frac{\sum_j \varphi_j^*(\boldsymbol{r})\varphi_j(\boldsymbol{r})\rho_j^{\mathrm{HF}}(\mathbf{r}, \boldsymbol{r}')}{\sum_k \varphi_k^*(\boldsymbol{r})\varphi_k(\boldsymbol{r})} = -e\frac{\sum_{jk||} \varphi_k^*(\boldsymbol{r}')\varphi_j(\boldsymbol{r}')\varphi_j^*(\boldsymbol{r})\varphi_k(\boldsymbol{r})}{\sum_k \varphi_k^*(\boldsymbol{r})\varphi_k(\boldsymbol{r})}. \tag{1.31}$$

This averaged charge density is then put into (1.30)

$$\left[-\frac{\hbar^2}{2m}\nabla^2 + V(\boldsymbol{r}) - \frac{e}{4\pi\varepsilon_0}\int\frac{\rho(\boldsymbol{r}') - \bar{\rho}^{\mathrm{HF}}(\boldsymbol{r}, \boldsymbol{r}')}{|\boldsymbol{r} - \boldsymbol{r}'|}\,d\tau'\right]\varphi_j(\boldsymbol{r}) = E_j\varphi_j(\boldsymbol{r}). \tag{1.32}$$

The interaction term is now only a function of $\boldsymbol{r}$, which can be combined with the second term to represent a local potential field which is equally valid for all electrons.

We have thus reached our goal of splitting up the Schrödinger equation for the many-electron problem into one-electron wave equations. The Schrödinger equation for the *one-electron approximation* (1.32) contains, through the third term on the left-hand side, important parts of the electron-electron interaction. We shall explore this equation in detail in Section 2.2. First, however, we turn to the jellium model which we shall examine in the absence of electron-electron interaction in Chapter 2.1, and with electron-electron interaction in Chapter 3.1.

2. The One-Electron Approximation

2.1 The Electron Gas Without Interaction

2.1.1 Introduction

The simplest approximation for the description of an electron gas is to neglect all interactions, the Coulomb interaction of the electrons with each other and the interaction of the electrons with the positive background (or the ion lattice). Every electron is then independent of every other and subject only to external forces.

In spite of its gross simplifications the *model of the interaction-free electron gas* can explain many phenomena. We shall not find out why this is so until Section 2.2.8. It will emerge that the interaction of the electrons with the periodic lattice [including the average electron-electron interaction of the Hartree-Fock approximation (1.32)] can be taken account of in many cases by introducing an effective mass tensor or even a scalar *effective mass* m^*. The problem of an electron moving under the simultaneous influence of external forces and the lattice potential can then be simplified to an equivalent problem where "quasi-electrons", distinguishable from free electrons only by their different mass m^*, move only under the influence of the external forces.

We are not yet in a position to define the limits of validity of this model (see Secs. 2.2.11 and 2.2.12). Let us, however, mention two examples in which this approximation is valid: conduction electrons in *monovalent metals* and in many *semiconductors*. The effective mass values for electrons in metals are somewhat greater than the free electron mass (Na: 1.22 m, Li: 2.3 m); for semiconductors they can be very much less than m (InSb: 0.01 m). There is another aspect in which the electron gas is fundamentally different in both cases. In semiconductors the concentration of electrons is mostly so low that they behave like a gas of non-interacting particles which obey classical Boltzmann statistics. The electron gas in metals however is "degenerate". In view of the high concentration of electrons we have to apply the Pauli principle and, with it, *Fermi statistics* to determine how the electrons occupy the states available to them.

The three following sections are devoted to the discussion of *eigenvalues* and *eigenfunctions* of the noninteracting electron gas, and to the *energy distribution* of electrons in the *ground and excited states*. At the same time these sections serve to bring together a collection of basic concepts which will be used time and again in later chapters (**k**-space, density of states, Fermi sphere, pair excitations, etc.).

The subsequent sections treat the *dynamics* of the electron gas, that is the movement of electrons under external forces. We shall deal separately with the effect of an electric field and a magnetic field. The dia- and paramagnetism of a free electron gas will be the object of our attention in Section 2.1.7.

In a few places it will be possible to compare theory with experiment (specific heat of the electron gas, de Haas-van Alphen effect). In later chapters we shall also return to this model, in discussing transport properties, for example (Drude-Sommerfeld theory of electrical conduction, Wiedemann-Franz law), and optical phenomena (absorption of free carriers, cyclotron resonance).

2.1.2 The Energy States

If we neglect all interactions in the Hamiltonian (1.7), we are left with only the kinetic energy operator. The Schrödinger equation becomes

$$-\frac{\hbar^2}{2m}\sum_j \nabla_j^2 \Phi = E\Phi. \tag{2.1}$$

The wave functions Φ are appropriate combinations of one-electron wave functions, that is, either products (1.14) of such functions or Slater determinants (1.19). Since the Hamiltonian does not contain the spin explicitly, the one-electron functions can be expressed as the product of a space function $\varphi_j(\boldsymbol{r})$ and a spin function. If also we write the energy E in (2.1) as a sum over one-electron energies E_j, then (2.1) splits into one-electron equations

$$-\frac{\hbar^2}{2m}\nabla_j^2 \varphi_j(\boldsymbol{r}) = E_j \varphi_j(\boldsymbol{r}) \tag{2.2}$$

in which only the space functions $\varphi_j(\boldsymbol{r})$ appear. Later we shall reintroduce the spin by way of the Pauli principle.

Since we shall now be concerning ourselves only with (2.2), we can omit the index j. Thus in future, in contrast to (2.1), E will be the one-electron energy. A solution of (2.2) is

$$\varphi(\boldsymbol{r}) = \mathrm{e}^{i\boldsymbol{k}\cdot\boldsymbol{r}}, \qquad E = \frac{\hbar^2 k^2}{2m}. \tag{2.3}$$

Here the φ have not yet been normalized. It is advantageous to restrict the electron gas to a fixed volume V_g, a cube with sides L_x, L_y, and L_z. We then select for the boundary conditions the Born-von Karman *periodic boundary conditions*: $\varphi(x + L_x, y, z) = \varphi(x, y + L_y, z) = \varphi(x, y, z + L_z) = \varphi(x, y, z)$. These boundary conditions simplify the mathematics considerably without (for sufficiently large L_i) influencing the physics of the problem.

The normalized wave function then turns out to be

$$\varphi(\boldsymbol{r}) = \frac{1}{\sqrt{V_g}} e^{i\boldsymbol{k}\cdot\boldsymbol{r}} \tag{2.4}$$

and for the components of $\boldsymbol{k}$ we find from the boundary conditions

$$k_i = \frac{2\pi}{L_i} n_i \qquad (i = x, y, z;\ n_i \text{ integer}). \tag{2.5}$$

The k_i [which in (2.3) only had the meaning of separation constants] can thus be interpreted as quantum numbers which, along with the spin, define the state of the electron.

From (2.5) we see that we can represent the $\boldsymbol{k}$-vectors by a *point lattice* in a $\boldsymbol{k}$-space. Every $\boldsymbol{k}$-point in such a lattice has two one-particle states with opposite spins associated with it.

Again from (2.5) we find that each pair of states occupies a volume $(2\pi)^3/V_g$. In each volume element $d\tau_{\boldsymbol{k}}$ of $\boldsymbol{k}$-space, therefore, there are $2V_g/(2\pi)^3$ states. We divide this number by the volume V_g to define the *density of states* $g(\boldsymbol{k})$

$$g(\boldsymbol{k})d\tau_{\boldsymbol{k}} = \frac{2}{(2\pi)^3} d\tau_{\boldsymbol{k}}. \tag{2.6}$$

The number of states in a given energy range (E, dE) is equal to the volume enclosed between the spherical shells $E + dE = \text{constant}$ and $E = \text{constant}$, multiplied by the density of states in $\boldsymbol{k}$-space. From this we obtain the *density of states on the energy scale*

$$g(E)dE = \frac{1}{2\pi^2}\left(\frac{2m}{\hbar^2}\right)^{3/2} E^{1/2} dE. \tag{2.7}$$

Each state—defined by the three components of the $\boldsymbol{k}$-vector and the spin quantum number—can, according to the Pauli principle, be occupied by only one electron.

We now consider an electron gas with N electrons. The state of lowest energy (*ground state*) will then correspond to the $N/2$ points in $\boldsymbol{k}$-space having the lowest energy, each being occupied by two electrons (Fig. 2.1). In $\boldsymbol{k}$-space these points fill a sphere of radius k_F (*Fermi sphere*). k_F is defined by the condition

$$4\pi \int_0^{k_F} g(\boldsymbol{k}) k^2 dk = \frac{N}{V_g} = n. \tag{2.8}$$

Here $n = N/V_g$ is the electron concentration. We then have the following expression for the Fermi radius k_F and for the energy of electrons at the surface

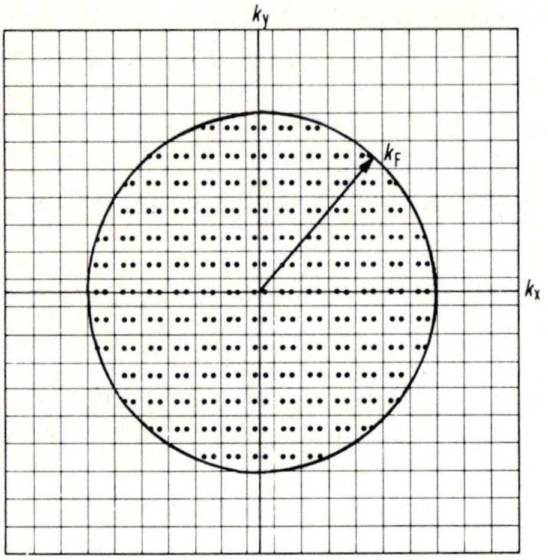

Fig. 2.1. Fermi sphere with radius k_F in $\boldsymbol{k}$-space. Each volume element of size $(2\pi)^3/V_g$ contains two states which, inside the Fermi sphere at $T=0$, are occupied by electrons with opposite spin.

of the Fermi sphere (*Fermi energy*):

$$k_F = (3\pi^2 n)^{1/3}, \qquad E_F = \frac{\hbar^2}{2m}(3\pi^2 n)^{2/3} . \tag{2.9}$$

2.1.3 Excited States

The ground state of the electron gas is the Fermi sphere completely filled with electrons. Every electron in a state $\boldsymbol{k}$, σ within the Fermi sphere is associated with another electron $-\boldsymbol{k}$, $-\sigma$. The momentum $\hbar\boldsymbol{k}$ and spin σ of the electrons just cancel each other. In its ground state the electron gas has therefore no net momentum and no net spin.

Excited states result when energy and momentum are supplied to the electron system. Individual electrons are transferred to higher states outside the Fermi sphere. We shall now look more closely at such individual processes. Let the initial state of the electron be $\boldsymbol{k}_0$, σ and the final state $\boldsymbol{k}$, σ. A momentum $\hbar\boldsymbol{\kappa} = \hbar(\boldsymbol{k} - \boldsymbol{k}_0)$ and energy $E(\boldsymbol{\kappa}) = (\hbar^2/2m)(k^2 - k_0^2)$ are needed for the transition. The momentum supplied to the system can be identified after the transition as the sum of the momentum of the excited electron $\boldsymbol{k}$, σ and that of the now uncompensated electron $-\boldsymbol{k}_0$, $-\sigma$ inside the Fermi sphere. After each individual excitation we have therefore to consider *two* electrons—one outside and one inside the Fermi sphere. These are the only electrons which are relevant to momentum balance. Things are not so simple, however, when energy balance is considered. For the combined energy of the two electrons, $(\hbar^2/2m)(k^2 + k_0^2)$ is not equal to the energy supplied $(\hbar^2/2m)(k^2 - k_0^2)$. We shall now look at the problem in a slightly different way.

We regard the ground state of the electron gas as the "*vacuum state*" of the system, ignoring in this description all the electrons in the filled Fermi sphere.

A single excitation then results in the production of an *electron* outside the Fermi sphere and a *hole* (an unoccupied state) inside the sphere. We interpret the hole as a *quasi-particle*, whose properties have to be characterized more precisely.

Energy balance gives the first pointer towards a quantitative formulation of this concept. The *excitation energy* is

$$E - E_0 = \sum E(\boldsymbol{k})n_{\boldsymbol{k}\sigma} - \sum_{k<k_F} E(\boldsymbol{k}) . \tag{2.10}$$

Here the $n_{\boldsymbol{k}\sigma}$ represent the occupation numbers of the states (0 or 1). The summation is in each case over all states $\boldsymbol{k}$, σ. By rearrangement we have

$$\begin{aligned} E - E_0 &= \sum_{k>k_F} E(\boldsymbol{k})n_{\boldsymbol{k}\sigma} - \sum_{k<k_F} E(\boldsymbol{k})(1 - n_{\boldsymbol{k}\sigma}) \\ &= \sum_{k>k_F} \varepsilon(\boldsymbol{k})n_{\boldsymbol{k}\sigma} + \sum_{k<k_F} |\varepsilon(\boldsymbol{k})|(1 - n_{\boldsymbol{k}\sigma}) , \end{aligned} \tag{2.11}$$

with $\varepsilon(\boldsymbol{k}) = E(\boldsymbol{k}) - E_F$. The excitation energy can thus be written as the sum of the energies of all electrons and holes, these energies being defined as the (positive) energy differences between the energy $E(\boldsymbol{k})$ of the particle and the Fermi energy. *Electron-hole pairs* are thus created at the Fermi surface. The excitation energy is needed to raise electrons above this surface and to draw holes into the Fermi sphere.

Momentum balance gives us

$$\boldsymbol{P} = \sum \hbar\boldsymbol{k}n_{\boldsymbol{k}\sigma} = \sum_{k>k_F} \hbar\boldsymbol{k}n_{\boldsymbol{k}\sigma} + \sum_{k<k_F} (-\hbar\boldsymbol{k})(1 - n_{\boldsymbol{k}\sigma}) . \tag{2.12}$$

Thus every hole in a state $\boldsymbol{k}$, σ contributes a momentum $-\hbar\boldsymbol{k}$ to the momentum balance.

The energy and momentum of the electron and the hole of an electron-hole pair are not uniquely related. For each momentum $\hbar\boldsymbol{\kappa}$ there is a range of possible energies (limited by the Pauli principle). From Fig. 2.2 we have to distinguish two cases. For $\kappa < 2k_F$, not every electron can be excited out of the Fermi sphere, i.e., not all possible states with $\boldsymbol{k}_0 + \boldsymbol{\kappa}$ are unoccupied. For $\kappa > 2k_F$, every $\boldsymbol{k}_0$ can be an initial state. The possible final states however must lie at least the finite energy difference $(\hbar^2/2m)[(\kappa - k_F)^2 - k_F^2]$ above the ground state. For a given κ the maximum energy which can be transferred is in both cases $(\hbar^2/2m)[(\kappa + k_F)^2 - k_F^2]$. Between these two limiting values all initial energies are possible (Fig. 2.3).

We shall now look at the two possible descriptions for the excitations of an electron gas (transfer of electrons from within the Fermi sphere to higher states and production of electron-hole pairs above the vacuum state "filled Fermi sphere") in the context of the occupation number representation which

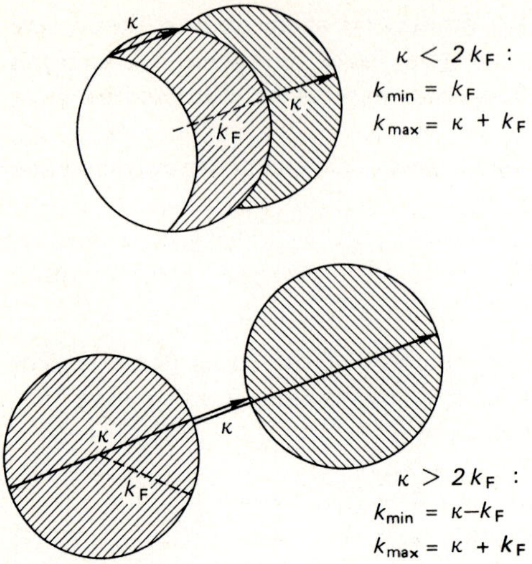

Fig. 2.2. Electron transitions out of the Fermi sphere for a given transferred momentum $\boldsymbol{\kappa}$. Regions inside the Fermi sphere, from which transitions can be made, and regions outside the sphere, into which transitions take place, are shown hatched.

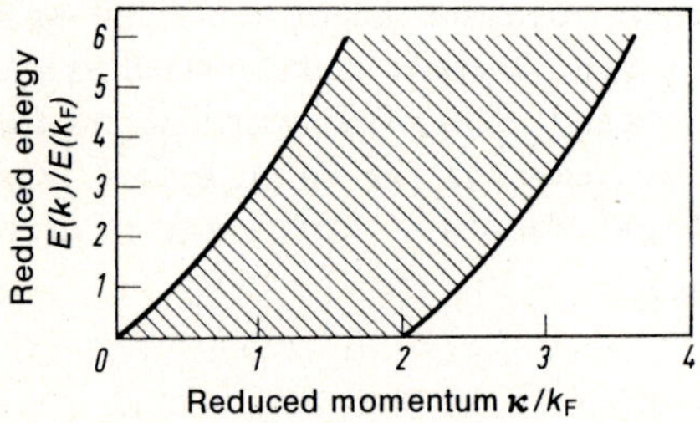

Fig. 2.3. Energy-momentum diagram for electron-hole pair excitations of the electron gas. There is no unique relationship between energy and momentum (hatched region).

is outlined in the Appendix. Besides familiarizing ourselves with methods which will be important for later applications, we shall gain a deeper insight into the concept of electrons and holes as quasi-particles of the electron gas.

From (2.1), (2.3), and (A.31), the Hamiltonian for the free electron gas becomes

$$H = \sum_i \left(-\frac{\hbar^2}{2m} \nabla_i^2 \right) = \sum_{\boldsymbol{k}\sigma} E(\boldsymbol{k}) c_{\boldsymbol{k}\sigma}^+ c_{\boldsymbol{k}\sigma}, \qquad E(\boldsymbol{k}) = \frac{\hbar^2 k^2}{2m}. \tag{2.13}$$

According to the discussion in the Appendix, the operator $c_{\boldsymbol{k}\sigma}^+$ creates an electron in the state $\boldsymbol{k}$, σ; the operator $c_{\boldsymbol{k}\sigma}$ annihilates it. $c_{\boldsymbol{k}\sigma}^+ c_{\boldsymbol{k}\sigma}$ is the particle number operator with the eigenvalues $n_{\boldsymbol{k}\sigma}$. Introduction of the operator for the total number of electrons $N_{\text{op}} = \sum_{\boldsymbol{k}\sigma} c_{\boldsymbol{k}\sigma}^+ c_{\boldsymbol{k}\sigma}$ allows (2.13) to be written as

$$H - E_F N_{\text{op}} = \sum_{\boldsymbol{k}\sigma} \varepsilon(\boldsymbol{k}) c_{\boldsymbol{k}\sigma}^+ c_{\boldsymbol{k}\sigma}. \tag{2.14}$$

This form gives us a better starting point for describing excited states by electron-hole pairs, for in this method of description we are dealing with a system with a *variable number of particles*. In statistical mechanics, for systems with variable particle number, the free energy $F(V, T, N)$ is replaced by the thermodynamic potential $\Omega(V, T, \mu) = F - \mu N$. Here μ is the chemical potential. Likewise the Hamiltonian H is replaced by the operator $H - \mu N_{\text{op}}$. We shall return to this in the following section. There we will also show that E_F is the chemical potential of the electron gas (at $T = 0$).

We rewrite (2.14) now for electron-hole pairs. We shall only use the operators $c^+_{k\sigma}$ and $c_{k\sigma}$ outside the Fermi sphere. Inside it we define equivalent operators $d^+_{k\sigma}$ and $d_{k\sigma}$ for the creation and annihilation of *holes*

$$\begin{aligned} d^+_{k\sigma} &= c_{k\sigma}\text{: creation of a hole = annihilation of an electron} \\ d_{k\sigma} &= c^+_{k\sigma}\text{: annihilation of a hole = creation of an electron.} \end{aligned} \tag{2.15}$$

The commutation relations (A.19), (A.20), and (A.22) are valid for the d^+, d as for the c^+, c. In particular we also have: $1 - c^+_{k\sigma}c_{k\sigma} = c_{k\sigma}c^+_{k\sigma} = d^+_{k\sigma}d_{k\sigma}$. With this rearrangement (2.14) becomes

$$H - E_F N_{\text{op}} = \sum_{k<k_F} \varepsilon(\boldsymbol{k}) + \sum_{k>k_F} \varepsilon(\boldsymbol{k}) c^+_{k\sigma}c_{k\sigma} + \sum_{k<k_F} |\varepsilon(\boldsymbol{k})| d^+_{k\sigma}d_{k\sigma}\,. \tag{2.16}$$

The right-hand side here is made up of the energy of the filled Fermi sphere (referred to E_F as zero energy) and the Hamiltonians of the electrons outside and the holes inside the Fermi sphere.

For many applications a further simplification is helpful. The concept of the hole presents us with the confusing situation that a hole in the state $\boldsymbol{k}, \sigma$ has a momentum $-\hbar\boldsymbol{k}$ while an electron has a momentum $\hbar\boldsymbol{k}$. This asymmetry can be avoided by defining new *quasi-particles* which have momentum $\hbar\boldsymbol{k}$ inside and outside the Fermi sphere. When we note that the creation of a particle with $+\hbar\boldsymbol{k}$ is achieved by the operator $c^+_{k\sigma}$ when outside the Fermi sphere and by $c_{-k,-\sigma}$ when inside the Fermi sphere, it is only a small step to define the following operators:

$$\begin{aligned} \alpha^+_{k\sigma} &= u_k c^+_{k\sigma} + v_{-k} c_{-k,-\sigma} \\ \alpha_{k\sigma} &= u_k c_{k\sigma} + v_{-k} c^+_{-k,-\sigma} \end{aligned}$$

with (2.17)

$$\begin{aligned} u_k &= 1,\ v_k = 0 \qquad \text{for} \quad k > k_F \\ u_k &= 0,\ v_k = 1 \qquad \text{for} \quad k < k_F\,. \end{aligned}$$

For these operators to be adopted as creators and annihilators of Fermi particles, they have to satisfy the commutation relations (A.19), (A.20), and (A.22). This is realized if

$$u_k^2 + v_k^2 = 1\,, \qquad u_k = u_{-k}\,, \qquad v_k = -v_{-k}\,. \tag{2.18}$$

From (2.17) we have first that

$$c_{k\sigma}^{+}c_{k\sigma} = u_k^2\alpha_{k\sigma}^{+}\alpha_{k\sigma} + v_k^2\alpha_{-k,-\sigma}\alpha_{-k,-\sigma}^{+} + u_k v_k(\alpha_{k\sigma}^{+}\alpha_{-k,-\sigma}^{+} + \alpha_{-k,-\sigma}\alpha_{k\sigma}) \quad (2.19)$$

and from that

$$\begin{aligned} H - E_F N_{\text{op}} &= \sum_{k\sigma} \varepsilon(k) c_{k\sigma}^{+} c_{k\sigma} = \sum_{k\sigma} \varepsilon(k)[u_k^2 \alpha_{k\sigma}^{+}\alpha_{k\sigma} + v_k^2(1 - \alpha_{-k,-\sigma}^{+}\alpha_{-k,-\sigma})] \\ &= \sum_{k<k_F} \varepsilon(k) + \sum_{k\sigma} |\varepsilon(k)| \alpha_{k\sigma}^{+}\alpha_{k\sigma}. \end{aligned} \quad (2.20)$$

The second summation on the right-hand side is over *all* states $\boldsymbol{k}, \sigma$. Eq. (2.20) has greater symmetry than (2.16). We shall return to this method of analysis (Bogoliubov-Valatin transformation) when we discuss the theory of superconductivity in Section 5.3.

2.1.4 The Fermi Distribution

We now consider the electron gas at a temperature above absolute zero. The gas is no longer in its ground state: states above k_F will be occupied. In *equilibrium* the probability of a given energy state being occupied will be a function of temperature alone. This occupation probability, called the *Fermi distribution*, can be calculated in various ways. Here we shall outline one of the ways, which will introduce us to concepts we shall be needing later. For a deeper discussion of these concepts we refer the reader to any standard text on statistical mechanics.

The starting point is the probability that in a closed system a subsystem occupies a quantum state E_n. If the subsystem is in thermal contact with its surroundings, this probability is the *Gibbs distribution*

$$w_n = Z^{-1} \exp\left(-\frac{E_n}{k_B T}\right). \quad (2.21)$$

Since $\sum_n w_n = 1$, the constant Z is given by $\sum_n \exp(-E_n/k_B T)$. It is called the *canonical partition function.* Z is connected with the free energy by $Z = \exp(-F/k_B T)$. Using (2.21), we can determine statistical mean values for all quantities which characterize the subsystem

$$\bar{f} = \sum_n f_n w_n = Z^{-1} \sum_n f_n \exp\left(-\frac{E_n}{k_B T}\right). \quad (2.22)$$

If the E_n are eigenvalues of a Schrödinger equation $H|n\rangle = E_n|n\rangle$, then the statistical mean values can also be determined through the *statistical operator* ρ

$$\rho = Z^{-1} \exp\left(-\frac{H}{k_B T}\right),$$

$$Z = \sum_n \left\langle n \middle| \exp\left(-\frac{H}{k_B T}\right) \middle| n \right\rangle = \text{Trace}\left\{\exp\left(-\frac{H}{k_B T}\right)\right\} \tag{2.23}$$

$$\bar{f} = \sum_n \langle n | f\rho | n \rangle = \text{Trace}\,\{f\rho\}\,. \tag{2.24}$$

If there is also a particle exchange with the surroundings (subsystem with *variable particle number*), (2.21) has to be replaced by the probability that the subsystem will be in the nth state and contain N particles.

Instead of (2.21) we then have

$$w_{nN} = Z_N^{-1} \exp\left(-\frac{E_{nN} - \mu N}{k_B T}\right), \qquad Z_N = \sum_{nN} \exp\left(-\frac{E_{nN} - \mu N}{k_B T}\right). \tag{2.25}$$

μ is the chemical potential. Z_N—the *grand partition function*—is connected with the thermodynamic potential $\Omega = F - \mu N$ in the same way as Z with the free energy: $Z_N = \exp(-\Omega/k_B T)$. The statistical operator becomes

$$\rho_N = Z_N^{-1} \exp\left(-\frac{H - \mu N_{\text{op}}}{k_B T}\right). \tag{2.26}$$

The Hamiltonian in (2.23) is replaced by the operator $H - \mu N_{\text{op}}$. We have mentioned this in the previous section.

We shall now apply this formalism to the free electron gas. We shall consider all electrons in a quantum state $\boldsymbol{k}$, σ as a subsystem. For simplicity we shall omit the index σ below. We have therefore only one $E_n = E(\boldsymbol{k})$. The particle number is equal to $n_{\boldsymbol{k}} = 0$ or 1. The energy of the subsystem is $E_{nN} = n_{\boldsymbol{k}} E(\boldsymbol{k})$.

The partition function then becomes

$$Z_{\boldsymbol{k}} = \sum_{n_{\boldsymbol{k}}} \exp\left(-\frac{E(\boldsymbol{k}) - \mu}{k_B T} n_{\boldsymbol{k}}\right) = 1 + \exp\left(-\frac{E(\boldsymbol{k}) - \mu}{k_B T}\right) \tag{2.27}$$

and the statistical operator (in the occupation number representation)

$$\rho_{\boldsymbol{k}} = Z_{\boldsymbol{k}}^{-1} \exp\left(-\frac{E(\boldsymbol{k}) - \mu}{k_B T} c_{\boldsymbol{k}}^{+} c_{\boldsymbol{k}}\right). \tag{2.28}$$

From this we have the average number of particles in state $\boldsymbol{k}$ as

$$\bar{n}_{\boldsymbol{k}} = \text{Trace}\,\{n_{\boldsymbol{k}} \rho_{\boldsymbol{k}}\} = \sum_{n_{\boldsymbol{k}}} \langle n_{\boldsymbol{k}} | n_{\boldsymbol{k}} \rho_{\boldsymbol{k}} | n_{\boldsymbol{k}} \rangle = \langle 1 | \rho_{\boldsymbol{k}} | 1 \rangle$$

$$= Z_{\boldsymbol{k}}^{-1} \exp\left(-\frac{E(\boldsymbol{k}) - \mu}{k_B T}\right) = \frac{1}{1 + \exp\left(\dfrac{E(\boldsymbol{k}) - \mu}{k_B T}\right)}. \tag{2.29}$$

Since $\bar{n}_k$ is limited to between 0 and 1, this average number is identical with the occupation probability of a quantum state $\boldsymbol{k}$. Eq. (2.29) thus represents the desired Fermi distribution. We could have obtained the same result from the thermodynamic potential $\Omega_k = -k_B T \ln Z_k$ using the relationship $\bar{n}_k = -(\partial\Omega_k/\partial\mu)$. From (2.29) we also find that $\bar{n}_k$, the Fermi distribution, is an eigenvalue of the statistical operator.

The Fermi distribution is a function of energy $E(\boldsymbol{k})$ and temperature T.

$$f(E, T) = \left\{1 + \exp\left(\frac{E(\boldsymbol{k}) - \mu}{k_B T}\right)\right\}^{-1}. \tag{2.30}$$

From this and from the density of states (2.7), we find for the electron concentration, i.e., the number of electrons with energies in the range (E, dE) referred to the volume V_g

$$n(E)\,dE = f(E)g(E)\,dE. \tag{2.31}$$

The Fermi distribution is shown in Fig. 2.4 for $T = 0$ and for a temperature $T \neq 0$. For $T = 0$, μ is equal to the energy E_F of the Fermi surface. The Fermi energy is thus identical to the chemical potential of the electron gas at $T = 0$.

For $T \neq 0$, μ corresponds to the energy at which the occupation probability is exactly 1/2. The chemical potential μ is weakly dependent on temperature. Fig. 2.4 also shows the electron concentration for $T = 0$ and for a temperature $T \neq 0$.

By integrating $n(E)$ over all energies the total electron concentration $n = N/V_g$ follows as a function of μ and T. Since N is known, we can determine

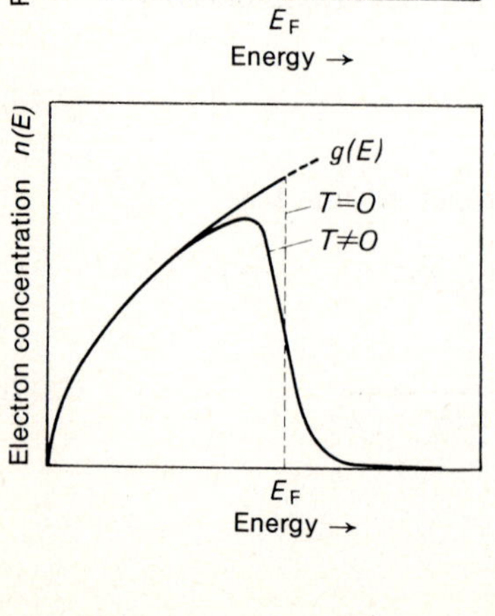

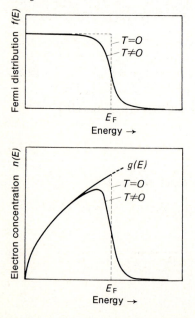

Fig. 2.4. Fermi distribution $f(E)$ and electron concentration $n(E) = g(E)f(E)$ for $T = 0$ and for a temperature $T \neq 0$.

from this the value of the chemical potential μ at a given temperature. From (2.31) and (2.7) we have

$$n = n_0 \frac{2}{\sqrt{\pi}} F\left(\frac{\mu}{k_B T}\right) \tag{2.32}$$

with

$$n_0 = 2\left(\frac{m k_B T}{2\pi\hbar^2}\right)^{3/2} \tag{2.33}$$

and

$$F(x) = \int_0^\infty \frac{y^{1/2}}{1 + \exp(y - x)}\, dy\,. \tag{2.34}$$

$F(x)$ is the so-called *Fermi integral.* Approximations are

$$\begin{aligned} F(x) &\approx \frac{\sqrt{\pi}}{2} e^x \qquad \text{for} \quad x < 0 \quad (\text{for } x < -4, \text{ error } < 2\%) \\ F(x) &\approx \frac{2}{3} x^{3/2} \qquad \text{for} \quad x > 0\,. \end{aligned} \tag{2.35}$$

For nondegeneracy ($x < 0$), $n = n_0 \exp(-\mu/k_B T)$ and the electron gas behaves like a gas of classical particles. In the limit of strong degeneracy ($x > 0$), n becomes proportional to $\mu^{3/2}$. The approximation (2.35) leads here exactly to (2.9).

An integration over the energies of all occupied states gives the total kinetic energy of the electron gas. For the special case of $T = 0$, the mean energy of an electron is found by integration over all states from $E = 0$ to $E = E_F$ to be given by $\bar{E} = (3/5)E_F$. When $T \neq 0$, the integral, like the Fermi integral, can be solved only by approximation. For low temperatures (strong degeneracy) the first term of a power series is sufficient

$$\bar{E} = \frac{3}{5} E_F \left[1 + \frac{5\pi^2}{12}\left(\frac{k_B T}{E_F}\right)^2 + \cdots\right]. \tag{2.36}$$

Here E_F is again the value of μ at $T = 0$. From (2.36) we find for the *specific heat* of the electron gas (per electron)

$$c = \frac{d\bar{E}}{dT} = \frac{\pi^2 k_B}{2} \frac{k_B T}{E_F}\,. \tag{2.37}$$

According to classical statistics (nondegenerate electron gas) every electron would contribute $(3/2)k_B$ to the specific heat. Eq. (2.37), however, shows that for strong degeneracy only a fraction of about k_BT/E_F contributes. This is understandable since for a small increase in temperature only those electrons within about k_BT of E_F are able to find empty states to move into when their energy is increased.

In metals E_F has a value of several eV, so the fraction k_BT/E_F is very small. One of the first successes of the Sommerfeld free electron theory of metals was the prediction that in metals the specific heat of electrons is very much less than classical statistics would suggest.

The linear temperature dependence of specific heat, revealed in (2.37), is also confirmed by experiment. Small deviations are attributed to the effective mass m^* of the metal electrons, a concept which we introduced earlier in this chapter. If we use (2.9) to replace E_F in (2.37), it can be seen that c depends linearly on this mass.

2.1.5 Free Electrons in an Electric Field

From (2.3) we find the expectation value for the momentum of a free electron in state $E(\boldsymbol{k})$ to be given by the de Broglie relation

$$\boldsymbol{p} = \frac{\hbar}{\mathrm{i}}\langle\varphi|\nabla|\varphi\rangle = \hbar\boldsymbol{k}\,. \tag{2.38}$$

The particle property "momentum" is linearly related to the wave property "wave vector".

If the electron momentum is altered according to $\dot{\boldsymbol{p}} = -e\boldsymbol{E}$ by acceleration in a uniform and time-independent electric field, the $\boldsymbol{k}$-vector correspondingly becomes time dependent

$$\dot{\boldsymbol{k}} = -\frac{e}{\hbar}\boldsymbol{E}\,, \qquad \boldsymbol{k}(t) = \boldsymbol{k}(0) - \frac{e}{\hbar}\boldsymbol{E}t \tag{2.39}$$

or, put in another way: Under the action of the electric field the electron moves through states with different $\boldsymbol{k}$-vectors.

We can obtain the result (2.39) from the Schrödinger equation for the electron

$$\left(-\frac{\hbar^2}{2m}\nabla^2 + e\boldsymbol{E}\cdot\boldsymbol{r}\right)\varphi(\boldsymbol{r},t) = \mathrm{i}\hbar\dot{\varphi}(\boldsymbol{r},t) \tag{2.40}$$

by setting

$$\varphi(\boldsymbol{r},t) = A\exp\left(\mathrm{i}\left\{\boldsymbol{k}(t)\cdot\boldsymbol{r} - \frac{1}{\hbar}\int_0^t E(\boldsymbol{k}(\tau))d\tau\right\}\right). \tag{2.41}$$

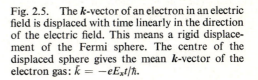

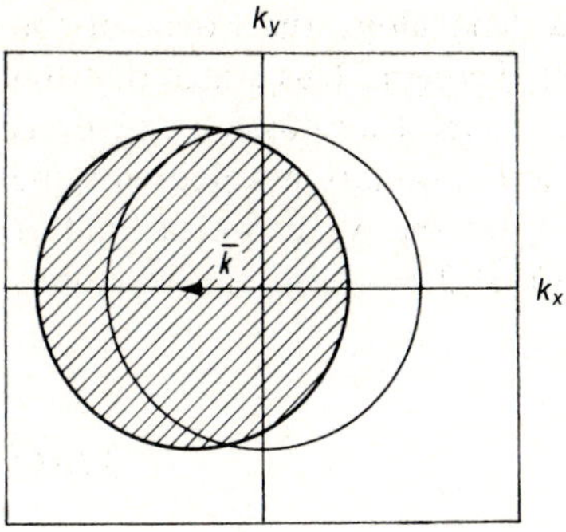

Fig. 2.5. The $\boldsymbol{k}$-vector of an electron in an electric field is displaced with time linearly in the direction of the electric field. This means a rigid displacement of the Fermi sphere. The centre of the displaced sphere gives the mean $\boldsymbol{k}$-vector of the electron gas: $\bar{k} = -eE_x t/\hbar$.

Inserting (2.41) into (2.40) gives

$$E(\boldsymbol{k}(t)) = \frac{\hbar^2 k^2(t)}{2m} \quad \text{with the time-dependent } \boldsymbol{k}\text{-vector (2.39)}. \tag{2.42}$$

If we carry this result over to an electron gas which fills a Fermi sphere in $\boldsymbol{k}$-space, we represent the acceleration of the electron gas in the electric field by a rigid displacement of the sphere in the direction of the $\boldsymbol{k}$-component parallel to the electric field (Fig. 2.5).

This type of representation is useful when we want to describe the energy and momentum balance of the electron gas in an electric field. To describe the movement of a *single* electron in a (not necessarily uniform) electric field, we have to start from a *wave packet*, that is, from a superposition of plane waves. For this we use the general ansatz

$$\psi(\boldsymbol{r}, t) = \sum_{\boldsymbol{k}} c(\boldsymbol{k}, t)\varphi(\boldsymbol{k}, \boldsymbol{r}) \tag{2.43}$$

where the wave packet can be restricted to small volumes in real space and in $\boldsymbol{k}$-space by appropriate choice of the amplitudes $c(\boldsymbol{k}, t)$.

The movement of the electron under the action of an electric field $\boldsymbol{E} = -\nabla\phi$ (ϕ = electrostatic potential) is described by the movement of the centre of gravity of the wave packet (2.43). For this we have to put (2.43) into the time-dependent Schrödinger equation

$$\left(-\frac{\hbar^2}{2m}\nabla^2 - e\phi\right)\psi(\boldsymbol{r}, t) = \mathrm{i}\hbar\dot{\psi}(\boldsymbol{r}, t). \tag{2.44}$$

We can then obtain the change in position and momentum of the centre of gravity from the quantum mechanical equations of motion

$$\dot{\boldsymbol{r}} = \frac{1}{\mathrm{i}\hbar}[\boldsymbol{r}, H], \qquad \dot{\boldsymbol{p}} = \frac{1}{\mathrm{i}\hbar}[\boldsymbol{p}, H]. \tag{2.45}$$

As long as there is little change in the electric field along the extent of the wave packet, we can make use of the Ehrenfest theorem. This states that the expectation values of (2.45) are formally identical to the Hamiltonian equations of motion of classical mechanics. If $\boldsymbol{r}$ and $\boldsymbol{p}$ are expectation values for the position and momentum of the centre of gravity of the wave packet, and H is the classical Hamiltonian $H = E_0 - e\phi = (\hbar^2k^2/2m) - e\phi$, the equations of motion we seek are

$$\dot{\boldsymbol{r}} = \nabla_{\boldsymbol{p}} H = \nabla_{\boldsymbol{p}} E_0 = \frac{1}{\hbar}\nabla_{\boldsymbol{k}} E_0 \tag{2.46}$$

$$\dot{\boldsymbol{p}} = -\nabla_{\boldsymbol{r}} H = e\nabla_{\boldsymbol{r}}\phi = -e\boldsymbol{E}\,. \tag{2.47}$$

Eq. (2.46) is the well known relation for the *group velocity* of a wave packet.

2.1.6 Free Electrons in a Magnetic Field

We now consider the dynamics of an electron gas in the presence of a constant magnetic field $\boldsymbol{B} = (0, 0, B)$. In the equations which follow we shall represent this field by a vector potential $\boldsymbol{A} = (0, Bx, O)$. With this choice of gauge, the magnetic field therefore possesses a z-component only and its vector potential a y-component only.

We proceed in two steps. The first is an extension of the results of the previous section. There we described the motion of an electron in an electric field by a time-dependent $\boldsymbol{k}$-vector (2.39). We now ask how this picture should be extended if a magnetic field is also applied. We shall assume that in the presence of this magnetic field the electron is still described at any moment by three quantum numbers k_i, i.e., by a quantum state in $\boldsymbol{k}$-space.

The second step is to prove these assumptions by solving the Schrödinger equation for an electron in a magnetic field. We shall find that in coupled electric and magnetic fields the description used above is still a useful illustration of electron motion but that its applicability is limited to small magnetic fields.

The classical Hamiltonian for an electron in a magnetic field is

$$H = \frac{1}{2m}(\boldsymbol{p} + e\boldsymbol{A})^2 \equiv \frac{1}{2m}\boldsymbol{P}^2 \quad \text{with} \quad \boldsymbol{P} = \boldsymbol{p} + e\boldsymbol{A} \tag{2.48}$$

and the equations of motion corresponding to (2.46) and (2.47) are

$$\dot{\boldsymbol{r}} = \nabla_{\boldsymbol{p}} H = \frac{\boldsymbol{P}}{m} \tag{2.49}$$

$$\dot{\boldsymbol{p}} = -\nabla_{\boldsymbol{r}} H = -e\dot{\boldsymbol{r}} \times \boldsymbol{B} - e\dot{\boldsymbol{A}} \tag{2.50}$$

or

$$\dot{\boldsymbol{P}} = -e\dot{\boldsymbol{r}} \times \boldsymbol{B} . \tag{2.51}$$

In the derivation of (2.50) we have used the relation $\dot{\boldsymbol{r}} \times \boldsymbol{B}|_y = \dot{x}B = \dot{A}_y$ and $\dot{A}_x = \dot{A}_z = 0$.

From these equations we can readily establish that the electrons execute orbits in the x-y-plane while their motion in the z-direction (the direction of the magnetic field) is not affected. From (2.50) we have

$$\dot{p}_x = -\frac{eB}{m}(p_y + eBx), \quad \dot{p}_y = 0\,(p_y \equiv \hbar k_y), \quad \dot{p}_z = 0\,(p_z \equiv \hbar k_z) \tag{2.52}$$

and further

$$\ddot{x} = -\omega_c\left(\frac{\hbar k_y}{m} + \omega_c x\right), \qquad \ddot{y} = \omega_c \dot{x}, \qquad \ddot{z} = 0, \tag{2.53}$$

$$x = x_0 + \cos\omega_c t, \qquad y = y_0 + \sin\omega_c t, \qquad z = z_0 + \frac{\hbar k_z}{m}t \tag{2.54}$$

with $x_0 = -\hbar k_y/m\omega_c$ and $\omega_c = eB/m$.

In (2.48) we have combined momentum and vector potential into a vector $\boldsymbol{P}$. If we describe the motion of the electron in $\boldsymbol{P}$-space, we find from (2.51) and (2.54) that here too the electrons execute orbits on surfaces of constant energy in the plane perpendicular to $\boldsymbol{B}$. Difficulties arise in transferring this into $\boldsymbol{k}$-space description. If we go from the classical Hamiltonian (2.48) to the Hamiltonian operator, $\boldsymbol{P}$ becomes an operator whose components do *not commute*. As a result the components of $\boldsymbol{P}/\hbar$ (which correspond to the components of $\boldsymbol{k}$ in the absence of a magnetic field) cannot really serve as the axes of a (classical) space in the description of electron motion. In small magnetic fields we can set these considerations aside, and equate the $\boldsymbol{P}/\hbar$-space with the $\boldsymbol{k}$-space of the zero magnetic field case. From (2.51) and (2.47) we then have the following law describing the behaviour of the $\boldsymbol{k}$-vector of the electron:

$$\hbar\dot{\boldsymbol{k}} = -eE - e\boldsymbol{v} \times \boldsymbol{B} . \tag{2.55}$$

This corresponds exactly to the classical statement that the acceleration of an electron (change of classical momentum) is proportional to the Lorentz force of the coupled electric and magnetic fields.

We now consider the exact solution of the Schrödinger equation

$$\frac{P^2}{2m}\psi = \frac{1}{2m}(-i\hbar\nabla + e\boldsymbol{A})^2\psi = E\psi . \tag{2.56}$$

This equation differs from the corresponding equation for the zero magnetic field case in that the x-coordinate explicitly appears in the vector potential. Again we set up a separation ansatz in which the y- and z-components are unchanged from the zero field case. For the x-component we take an ansatz function $\varphi(x)$

$$\psi = \exp\left[\mathrm{i}(k_y y + k_z z)\right]\varphi(x)\,. \tag{2.57}$$

From (2.57) and (2.56) we arrive at the following equation for $\varphi(x)$:

$$-\frac{\hbar^2}{2m}\varphi'' + \frac{m\omega_c^2}{2}(x - x_0)^2\varphi = \left(E - \frac{\hbar^2 k_z^2}{2m}\right)\varphi\,. \tag{2.58}$$

This equation is formally identical to the Schrödinger equation of the one-dimensional oscillator (centred at x_0). The eigenvalues are clearly given by

$$E_\nu = \frac{\hbar^2 k_z^2}{2m} + \left(\nu + \frac{1}{2}\right)\hbar\omega_c\,, \qquad \nu = 0, 1, 2, \ldots\,. \tag{2.59}$$

This is exactly the result we would expect. The energy of the electron is made up of the kinetic energy of its undisturbed motion in the z-direction, together with the quantized energy of the oscillatory motion in the plane orthogonal to the field direction. The quantum energy is determined by the angular frequency ω_c (*cyclotron resonance frequency*).

Equation (2.59) gives only that part of the electron energy which comes from its orbital motion. To this we must add the spin contribution. This contribution is $\pm(g/2)\mu_B B$, the sign depending on the direction of the spin. μ_B is the Bohr magneton $\mu_B = e\hbar/2m = \hbar\omega_c/2B$ and g is the g-factor which, for free electrons, takes the value 2 but which for $m^* \neq m$ can deviate greatly from this value. Eq. (2.59) can then be written

$$E_\pm = \frac{\hbar^2 k_z^2}{2m} + (2\nu + 1)\mu_B B \pm \frac{g}{2}\mu_B B\,. \tag{2.60}$$

In what follows we shall again omit the spin. It can readily be included at any time.

We shall now look briefly at how the density of states changes in a magnetic field. The wave function (2.57) depends on k_y and k_z, and through $\varphi(x)$ again on k_y and on ν. For given k_z and ν we can choose any k_y; the state therefore is degenerate. Since we have to assume that x_0 lies in the volume $V_g(-L_x/2 < x_0 < L_x/2)$, k_y is limited to the range $-(m\omega_c L_x/2\hbar)$ to $+(m\omega_c L_x/2\hbar)$. Since states occur on the k_y-axis at intervals of $2\pi/L_y$ (neglecting spin), the y-component of $\boldsymbol{k}$ can take up $(L_y/2)(m\omega_c L_x/\hbar)$ different values. Since, furthermore, in the range dk_z the z-component of $\boldsymbol{k}$ can take up $(L_z/2\pi)dk_z$ different values, we have for the density of states (dividing by $V_g = L_x L_y L_z$ and adding a factor 2 to take account of spin)

$$g(\nu, k_z)dk_z = \frac{2}{(2\pi)^2}\frac{m\omega_c}{\hbar}dk_z\,. \tag{2.61}$$

Rewriting in terms of E rather than k_z, we get for the density of states in a subband with index ν

$$g(E, \nu)dE = \frac{2}{(2\pi)^2}\frac{\hbar\omega_c}{2}\left(\frac{2m}{\hbar^2}\right)^{3/2}\left(E - \left(\nu + \frac{1}{2}\right)\hbar\omega_c\right)^{-1/2} dE \tag{2.62}$$

and for the total density of states, by summation over all subbands which begin "below" the energy E

$$g(E)dE = \sum_{\nu=0}^{\nu'} g(\nu, E)dE\,. \tag{2.63}$$

It can easily be shown that (2.63) sums the same states as in the zero field case, the states being only redistributed on the energy scale. To show this let B tend to zero. The subbands of different quantum number then move more and more closely together and the summation in (2.63) can be replaced by an integral. If we carry out this integration and then put in the limiting case $B = 0$, we arrive exactly at (2.7).

This redistribution of the states is often pictorially illustrated by a combination of (2.59) and the $\boldsymbol{k}$-space ($\boldsymbol{P}/\hbar$-space) description. While in the absence of a magnetic field the states are uniformly distributed in $\boldsymbol{k}$-space, when a magnetic field is present they lie on concentric cylinders, on each of which the energy relationship (2.59) is satisfied. Figs. 2.6–8 show the various aspects of this description of the quantized motion of free electrons in $\boldsymbol{k}$-space.

For small magnetic fields we can neglect the redistribution of the states according to (2.60). The approximation of (2.55) then applies.

2.1.7 Dia- and Paramagnetism of Free Electrons, the de Haas-van Alphen Effect

We obtain the magnetization of the free electron gas in a magnetic field by differentiating the free energy with respect to the magnetic field: $M = -dF/dB$. From Section 2.1.4 the free energy is

$$F = N\mu + \Omega = N\mu - k_B T \sum_{\boldsymbol{k}} \ln Z_{\boldsymbol{k}}\,. \tag{2.64}$$

If we insert $Z_{\boldsymbol{k}}$ from (2.27) and remember from (2.60) that the energy is different for the two spin directions, we find for the magnetization density m

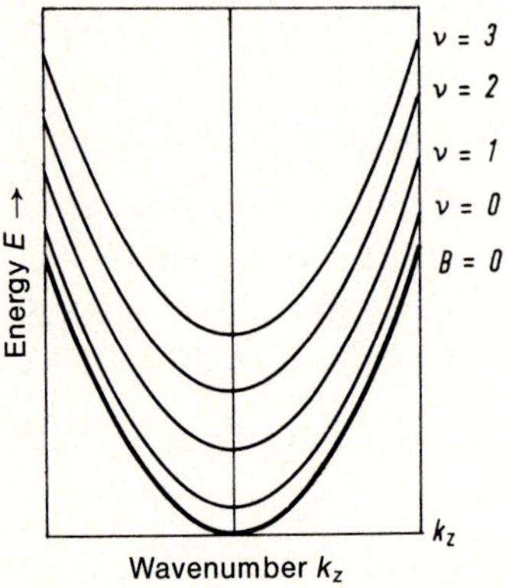

Fig. 2.6. One-dimensional magnetic subbands for free electrons, according to (2.59).

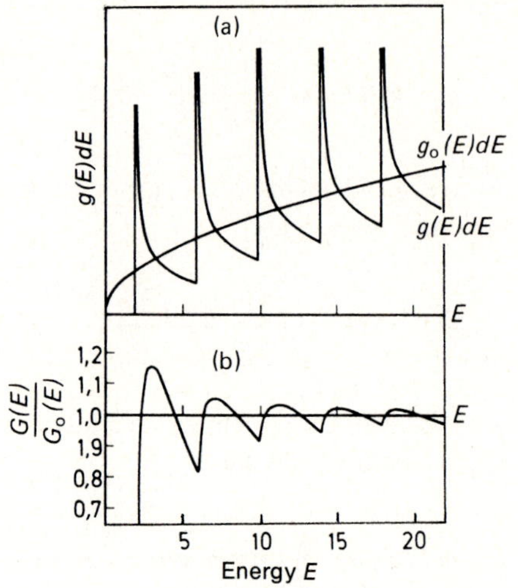

Fig. 2.7. (a) Energy dependence of the density of states for $\boldsymbol{B} = 0$ and $\boldsymbol{B} \neq 0$ ($\hbar\omega_c = 4$). (b) Ratio of the number of states below a certain energy E in a magnetic field $G(E)$ to the number in the absence of a magnetic field $G_0(E)$.

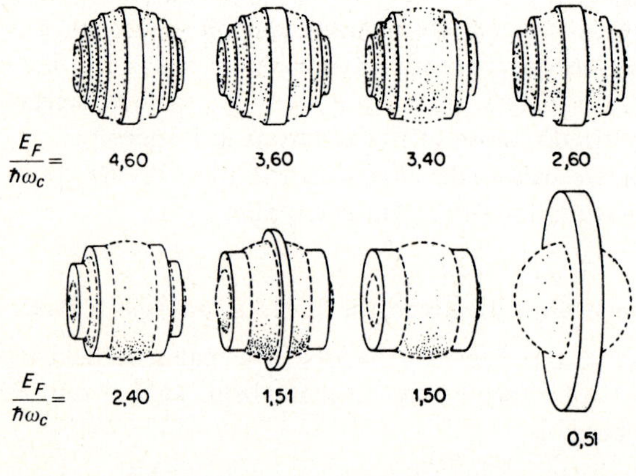

Fig. 2.8. The states, distributed homogeneously in $\boldsymbol{k}$-space in the absence of a magnetic field, lie on concentric cylinders in the presence of a field. The cylinder surfaces shown in the figure together contain the same number of states as the Fermi sphere, also shown. For small fields the shape of the Fermi sphere is approximately maintained; for large fields gross deviations occur (from E. N. Adams, T. Holstein: unpublished).

$$m = \frac{M}{V_g} = -\frac{d}{dB}\left\{n\mu - k_BT\int_0^\infty g(E_+)\ln\left[\exp\left(\frac{\mu - E}{k_BT}\right) + 1\right]dE_+ \right.$$
$$\left. - k_BT\int_0^\infty g(E_-)\ln\left[\exp\left(\frac{\mu - E}{k_BT}\right) + 1\right]dE_-\right\} \quad (2.65)$$

with $E_\pm = E \pm \mu_B B$.

Here we have also assumed that the energy levels defined by (2.60) lie so close together that the summation over the states can be replaced by an integration in energy. The evaluation of this integral is rather complicated. For expediency one limits consideration to low temperatures and takes only the first term in an expansion similar to that in (2.36). It follows then that (derivation in *Wilson* [34], for example)

$$m = \frac{3n\mu_B^2 B}{2E_F}\left[1 - \frac{1}{3} + \frac{\pi k_B T}{\mu_B B}\left(\frac{E_F}{\mu_B B}\right)^{1/2}\sum_{\nu=1}^{\infty}\frac{(-1)^\nu}{\sqrt{\nu}}\cos(\pi\nu)\frac{\sin\left(\frac{\pi}{4} - \frac{\pi\nu E_F}{\mu_B B}\right)}{\sinh\frac{\pi^2\nu k_B T}{\mu_B B}}\right]. \quad (2.66)$$

E_F is again here the value of μ at $T = 0$. The brackets in (2.66) contain three terms. The first orginates from the spin term of the energy. We would have arrived at this term in exactly the same form if we had neglected the orbital quantization, i.e., if we had inserted into (2.65) the density of states and the energy of free electrons in the absence of a magnetic field. This contribution is positive (*Pauli spin-paramagnetism*, Fig. 2.9).

The second term is negative and makes exactly one-third the contribution of the first. In low magnetic fields it represents the orbital quantization of the electrons (*Landau diamagnetism*).

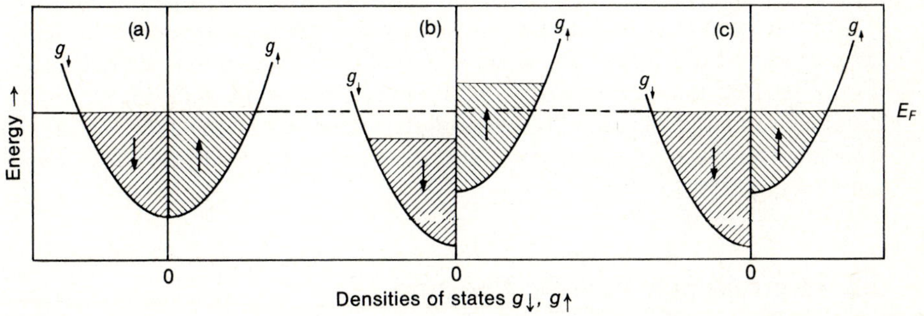

Fig. 2.9a-c. Pauli spin paramagnetism. In a magnetic field a contribution $\pm\mu_B B$ is added to the kinetic energy of the electrons, depending on their spin direction. Below the Fermi energy E_F there are then more states with one spin direction than with the opposite. In the figure the density of states is shown separately for the two spin directions. The occupied states are indicated by hatching, (a) no magnetic field, (b) shift of the states in a magnetic field, (c) occupied states in a magnetic field after reaching equilibrium.

The third term is an additional diamagnetic term which becomes important in high magnetic fields. As a result of the sinh in the denominator the series converges so quickly that often only the first term needs to be considered. This part has oscillatory character. The magnetization (and hence the magnetic susceptibility) is periodic in 1/B with temperature-independent period $2\mu_B/E_F$. These oscillations can only be observed in high magnetic fields and at low temperature (*de Haas-van Alphen effect*).

The origin of the de Haas-van Alphen effect is easy to see. Let $E_F \gg \mu_B B$ (or, equivalently, $E_F \gg \hbar\omega_c$). The electron gas then occupies states in numerous magnetic subbands. At low temperatures all states in the density of states of Fig. 2.7 are occupied, up to a limiting energy E_F which might lie between the threshold energies of the ν-th and the ν–1-th subband. As the magnetic field increases, the energy and the number of states in each subband increase, according to (2.62). The threshold energy likewise increases. Since the total number of electrons is given, there is a continuous rearrangement of the electrons with increasing magnetic field. When the threshold energy $E = (\nu + 1/2)\hbar\omega_c$ grows from a value below E_F to a value above E_F, the electrons in the ν-th band fall back into states in the ν–1-th band. The total energy decreases. With increasing magnetic field it rises again until the next threshold energy $E_{\nu-1}$ exceeds the value E_F. As a result E_F itself becomes (weakly) periodic. The separation between the individual threshold energies is $\hbar\omega_c$. The condition $\hbar\omega_c = 2\mu_B B \gg k_B T$ is therefore important; otherwise the electron distribution in the region of E_F is so widely spread that the oscillations are smoothed out. When the condition $E_F \gg \hbar\omega_c$ is no longer fulfilled, all the electrons are in the lowest subband and the oscillations cease (*quantum limit*).

Equation (2.66) can be extended to the case where the electrons have an effective mass m^*, different from the true electron mass. The Pauli paramagnetism term acquires a factor $(m/m^*)^2$. The paramagnetic and diamagnetic terms are no longer in the ratio 3: 1. In the third term a factor m^*/m appears on the cosine argument. Finally the m in μ_B is replaced everywhere by m^*.

All the terms in (2.66) can be compared with experimental results. In many cases there is good agreement when a suitable effective mass is adopted. We do not go here into a comparison with experiment. As we shall see later, the de Haas-van Alphen effect is particularly suited to establishing *deviations* from the free electron model. We shall therefore postpone further discussion until Section 2.2.11.

2.2 Electrons in a Periodic Potential

2.2.1 Introduction

The basis of this chapter is the *one-electron approximation* of (1.32). This equation describes an electron in a *periodic potential* consisting of the potential of the lattice ions and the mean Coulomb- and exchange-interaction of the Hartree-Fock approximation.

Lattice symmetry and its influence on the form of the eigenvalues and the eigenfunctions of the Schrödinger equation (1.32) will play a significant part in our discussion. An important result which will emerge is the division of the energy spectrum of one-electron states into bands separated by zones in which there are no allowed states (*energy band structure*).

One can understand the origin of these bands from two limiting cases. In bringing *free atoms* together to form a crystal, the discrete levels of these atoms split up into groups of levels which then form an energy band. Or: by the influence of the lattice potential the continuous energy spectrum of a *free electron gas* is broken at certain characteristic energies since electrons with these energies (and corresponding momenta) on their passage through the crystal suffer Bragg reflections from the lattice. Both types of description—starting from tightly bound or completely free electrons—meet in the *band model* of solids. Following on from the treatment of the free electron gas in Section 2.1 we shall concentrate mainly on the second possibility. After a discussion in Section 2.2.2 of the most important symmetries in the crystal lattice and the formulation of the Schrodinger equation for the band model (Sec. 2.2.3), we shall go on to consider the characteristics of a "free" electron gas which is subjected to Bragg reflections (Sec. 2.2.4). In doing so we shall introduce important concepts like the reciprocal lattice and its Brillouin zones. A general treatment of the periodicity of the lattice and the translation invariance of the Hamiltonian then leads to the definition of the band structure function $E_n(\boldsymbol{k})$ and to the possible representations of this function in $\boldsymbol{k}$-space. We use these results in Sections 2.2.6 and 2.2.7 to derive the band structure function by two different approximations. With this understanding of the meaning of the band model behind us, we go on in Section 2.2.8 to examine the general properties of the function $E_n(\boldsymbol{k})$. We shall see that the solutions of the Schrödinger equation for an electron in a periodic potential describe quasi-particles (*crystal* or *Bloch electrons*). The effect of the periodic potential is incorporated into the characteristics of these quasi-particles. We discuss the dynamics of the crystal electron (among other topics) in Section 2.2.9.

Up to this point we have been concerned only with a single electron in a periodic potential. We therefore turn to the problem of the statistics of all (valence) electrons in solids. We fill up (as with the free electron gas in Sec. 2.1) the energy states of the one-electron approximation with all the electrons, in accordance with Fermi statistics. The density of states $g(E)dE$ needed here is derived in Section 2.2.10. In the two sections which follow we illustrate, with examples, the band structures of metals, insulators, and semiconductors. In the following sections we engage in a deeper discussion of band structure. There we shall also employ some of the tools of group theory. The chapter ends with an examination of the important concept of pseudopotentials.

The energy band model is outlined in many textbooks and monographs. In addition to the presentations in *Kittel* [16], *Harrison* [13], and *Ziman* [35], we would particularly recommend the books by *Brillouin* [53], *Callaway* [55], *Harrison* [58], and *Jones* [60]. There are also good treatments in [18, 65, 101.1, 101.13]. We shall refer to special literature on particular topics in the course of Sections 2.2.11, 2.2.12, 2.2.14, and 2.2.16.

2.2.2 The Symmetries of the Crystal Lattice

A crystal is characterized by its regular, perodically repeated structure. The smallest unit of this structure is called the *primitive cell.* Identical primitive cells are joined together, filling the entire volume of the crystal and giving rise to the periodicity of the crystal lattice.

A consequence of this periodicity is that the lattice is invariant to *translations* over distances which are integer multiples of the lattice period. This is only true, of course, for an ideal, infinite crystal or for a crystal which has been subjected to periodic boundary conditions (see Sec. 2.1.2). We shall assume this to be so in what follows.

The lattice symmetry operations referred to as *primitive translations* (or *lattice translations*) can be written as

$$\boldsymbol{R}_n = n_1\boldsymbol{a}_1 + n_2\boldsymbol{a}_2 + n_3\boldsymbol{a}_3\,, \tag{2.67}$$

where the n_i are integers. The $\boldsymbol{a}_i$ are (noncoplanar) *basis vectors* which connect a point in the lattice—e.g., the centre of a primitive cell—with three equivalent points.

The set of all $\boldsymbol{R}_n$ leads to all the equivalent points in the lattice. The $\boldsymbol{R}_n$ form the *point lattice* of the crystal.

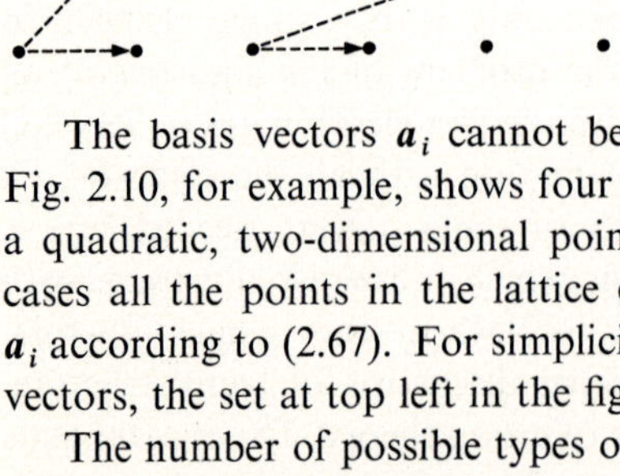

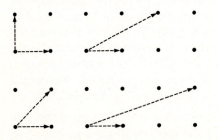

Fig. 2.10. Possible definitions of the two basis vectors in a two-dimensional quadratic point lattice.

The basis vectors $\boldsymbol{a}_i$ cannot be uniquely selected for a given point lattice. Fig. 2.10, for example, shows four possible choices for the two $\boldsymbol{a}_i$ which define a quadratic, two-dimensional point lattice. In these (and innumerable other) cases all the points in the lattice can be reached by combination of the two $\boldsymbol{a}_i$ according to (2.67). For simplicity one chooses a set of the shortest possible vectors, the set at top left in the figure for example.

The number of possible types of point lattices is limited. In two dimensions there are five different types of point lattices. They are defined by the lengths of the vectors $\boldsymbol{a}_1$ and $\boldsymbol{a}_2$, and by the angle between these vectors. One can readily show that only the following possibilities lead to different lattices: $a_1 \neq a_2$, any angle; $a_1 = a_2$, any angle; $a_1 \neq a_2$, right angle; $a_1 = a_2$, right angle; $a_1 = a_2$, angle of 60°. In three dimensions there are fourteen different types of point lattices.

The parallelepipeds with sides defined by the $\boldsymbol{a}_i$ form the primitive cells of the lattice. Each parallelepiped has a lattice point at each of its eight corners, and each corner is common to eight parallelepipeds. Each cell therefore accounts for one lattice point. Later we shall prefer to make use of a unit cell which is constructed in a different manner. For this one puts a lattice point in the middle of the new cell to be constructed. One then determines the bounding surfaces of the new cell by connecting the lattice point with all its neighbours and erecting perpendicular planes in the middle of the connecting lines. By means of this construction, made plain in Fig. 2.11, one encompasses all positions which are nearer to the lattice point considered than to all the others. Each of these *Wigner-Seitz cells* thus contains one lattice point and therefore has the same volume as the parallelepipeds defined by the $\boldsymbol{a}_i$. Fig. 2.12 shows the Wigner-Seitz cells for the four most important point lattices. Wigner-Seitz cells have the property that they are invariant to all lattice symmetry operations (rotations, reflections, and inversion of coordinates) which leave the mid-point of the cell fixed and the lattice invariant.

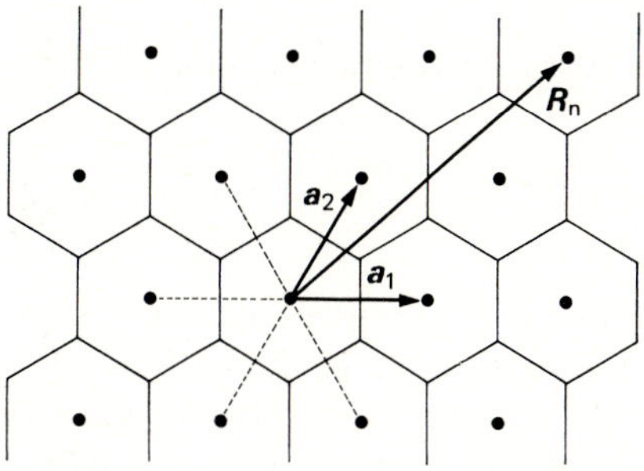

Fig. 2.11. Wigner-Seitz cells of a two-dimensional hexagonal point lattice.

The symmetries of the Wigner-Seitz cell need not be retained in a real crystal. The arrangement of the atoms in the Wigner-Seitz cell, the *basis*, can rather limit these symmetries.

All symmetry operations which leave an ideal, infinite crystal invariant belong to the *space group* of the crystal. Along with the primitive translations (2.67), the space group contains rotations and combinations of rotations and reflections (improper rotations) about given lattice points and axes, coordinate inversion, and also screw-rotations and glide reflections. The last two symmetry operations are combinations of (improper) rotations with (nonprimitive!) translations which by themselves do not leave the lattice invariant.

All these symmetry operations can be described by an orthogonal transformation of the form

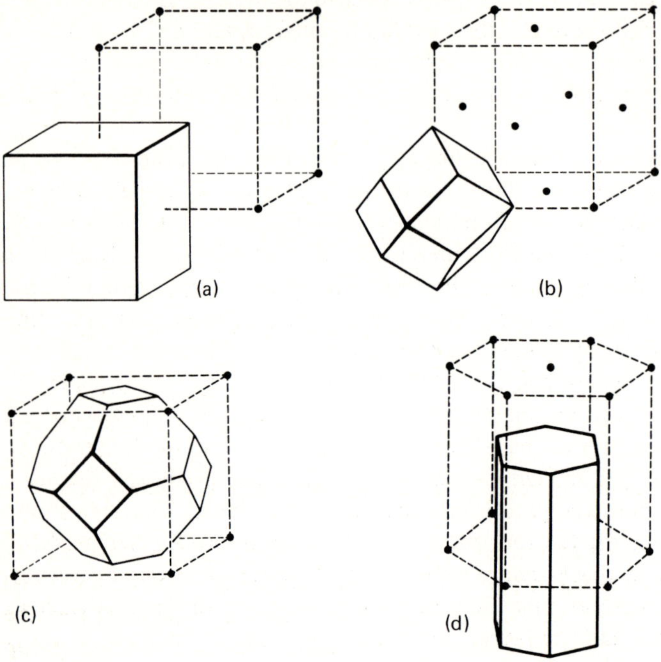

Fig. 2.12a-d. Wigner-Seitz cells of (a) a simple cubic, (b) face-centred cubic, (c) body-centred cubic, and (d) hexagonal point lattice.

$$\boldsymbol{r}' = \alpha \boldsymbol{r} + \boldsymbol{a} \equiv \{\alpha|\boldsymbol{a}\}\boldsymbol{r} \tag{2.68}$$

where α is a (proper or improper) rotation and $\boldsymbol{a}$ is a translation. In the symbolism on the right-hand side, primitive translations are represented by $\{E|\boldsymbol{R}_n\}$ (E = "unit operation", "rotation" by zero degrees), symmetry operations where no translations are involved by $\{\alpha|\boldsymbol{O}\}$, screw rotations and glide reflections by $\{\alpha|\boldsymbol{a}\}$ with $\boldsymbol{a} \neq \boldsymbol{R}_n$.

The elements of the space group form a group in the mathematical sense. Since we shall often make use of the group concept, we introduce here the following definitions:

A *group* is a (finite or infinite) set of elements for which an operation (called multiplication) is given so that the product AB of two elements A and B is well defined. The following *axioms* must be fulfilled:

1) The product of two elements is itself an element of the group: $AB = C$. In general $AB \neq BA$.
2) The multiplication is associative: $A(BC) = (AB)C$.
3) There exists a unit element defined by $AE = A$.
4) For each element A there exists a reciprocal element A^{-1} so that $AA^{-1} = E$.

The operations of the space group obviously satisfy these axioms. The unit element is $\{E|\boldsymbol{O}\}$. From (2.68) we further have for the reciprocal element and for the product of two elements

$$\{\alpha|\boldsymbol{a}\}^{-1} = \{\alpha^{-1}|-\alpha^{-1}\boldsymbol{a}\} \tag{2.69}$$

$$\{\alpha|\boldsymbol{a}\}\{\beta|\boldsymbol{b}\} = \{\alpha\beta|\alpha\boldsymbol{b} + \boldsymbol{a}\}\,. \tag{2.70}$$

All the primitive translations $\{E|\boldsymbol{R}_n\}$ constitute a group, i.e., they fulfil the group axioms. The *translation group* of the $\{E|\boldsymbol{R}_n\}$ is a *subgroup* of the space group. The result of two successive translations is independent of the order in which they are performed; the result of two rotations, however, can depend on the order. The translation group is therefore commutative, the space group in general is not.

In an infinite crystal there is an infinite number of elements in the translation group. If however the crystal is limited to a volume V_g with periodic boundary conditions, the translation group becomes finite and has the same number of elements as the volume V_g has Wigner-Seitz cells. We limit our considerations in the following to this case.

All the rotations α also form a group, the *point group* of the crystal. This is not necessarily a subgroup of the space group since in the latter individual α can occur only coupled with nonprimitive translations. In spite of this the point group has an important meaning. The operations of the point group leave the point lattice (and therefore the Wigner-Seitz cells) invariant. That is, all $\alpha\boldsymbol{R}_n$ as well as the $\boldsymbol{R}_n$ are primitive translations. This follows directly from the first group axiom which says that every product of elements of a group is also a member of the group.

$$\{\alpha|\boldsymbol{a}\}\{E|\boldsymbol{R}_n\}\{\alpha|\boldsymbol{a}\}^{-1} = \{E|\alpha\boldsymbol{R}_n\}\,. \tag{2.71}$$

The condition that a point group must leave the associated point lattice (and the Wigner-Seitz cell) invariant means that it is the symmetry group, or at least a subgroup of the symmetry group of the point lattice. This restricts the number of possible point groups. In two dimensions there are ten point groups (one without a symmetry element, one each with 2-, 3-, 4-, and 6-fold axes of rotation, and five further ones each additionally having a reflection line). In three dimensions there are 32 point groups, which determine the crystal classes. A point group can be coupled with a corresponding point lattice. The point group, the point lattice, and the nonprimitive translations coupled with the elements of the point group completely determine the space group.

From the ten point groups and five point lattices in two dimensions, 17 space groups can be constructed. From the 32 point groups and 14 point lattices in three-dimensional space, a total of 230 space groups can be formed. The particular space group a given crystal has will depend on its translational symmetry and on the arrangement of the atoms in the Wigner-Seitz cell, i.e., on its basis.

A space group is called *symmorphic* if it contains a complete point group as a subgroup. It does not have any nonprimitive translations. Every element $\{\alpha|\boldsymbol{a}\} = \{\alpha|\boldsymbol{R}_n\}$ can be reduced to a (proper or improper) rotation $\{\alpha|\boldsymbol{O}\}$ and a primitive translation $\{E|\boldsymbol{R}_n\}$. Real lattices whose basis does not restrict the symmetry of the Wigner-Seitz cell are called *Bravais lattices*. They are symmorphic. There are 14 Bravais lattices which are identical to the point lattices discussed above.

2.2.3 The Schrödinger Equation for Electrons in a Periodic Potential

We now return to (1.32) which we write in the form

$$H\psi(\boldsymbol{r}) \equiv \left[-\frac{\hbar^2}{2m}\nabla^2 + V(\boldsymbol{r})\right]\psi(\boldsymbol{r}) = E\psi(\boldsymbol{r})\,. \tag{2.72}$$

Here $V(\boldsymbol{r})$ includes the averaged interaction term of the Hartree-Fock approximation, i.e., it already includes parts of the electron-electron interaction as well as the potential of the ions.

In the approximation in which the ions rest in their fixed equilibrium position, the lattice part of the potential is clearly invariant to all operations of the space group. This is also true for the interaction term of the Hartree-Fock approximation.

We thus assume that for $V(\boldsymbol{r})$ in (2.72) and therefore for the whole Hamiltonian

$$V(\{\alpha|\boldsymbol{a}\}\boldsymbol{r}) = V(\boldsymbol{r})\,, \qquad H(\{\alpha|\boldsymbol{a}\}\boldsymbol{r}) = H(\boldsymbol{r})\,. \tag{2.73}$$

The symmetry properties of the Hamiltonian give us information on the structure of the possible solutions (eigenfunctions and eigenvalues) of (2.72). In Sections 2.2.5 and 2.2.13 we shall consider separately the conclusions to be drawn from translational invariance, which will give us the foundations of the band model, and those to be drawn from invariance to the other symmetry operations of the space group.

2.2.4 The Reciprocal Lattice, Bragg Reflections

Because electrons have wavelike properties, electron diffraction at the periodically arranged crystal lattice is possible. This phenomenon is known as *Bragg reflection*. Its discussion will lead us to the important concept of Brillouin zones in $\boldsymbol{k}$-space.

To assist this discussion it is appropriate to introduce the concept of the *reciprocal lattice*. This we take to be the point lattice described by the vectors

$$\boldsymbol{K}_m = m_1\boldsymbol{b}_1 + m_2\boldsymbol{b}_2 + m_3\boldsymbol{b}_3 \tag{2.74}$$

where the m_i are integers and the $\boldsymbol{b}_i$ are related to the $\boldsymbol{a}_i$ of (2.67) by the following relationships:

$$\boldsymbol{a}_i \cdot \boldsymbol{b}_j = 2\pi\delta_{ij} \qquad i, j = 1, 2, 3. \tag{2.75}$$

A vector $\boldsymbol{b}_i$ is therefore at right angles to both vectors $\boldsymbol{a}_j$ and $\boldsymbol{a}_k$ (i, j, k = 1, 2, 3 and cyclic). Thus $\boldsymbol{b}_i = c\boldsymbol{a}_j \times \boldsymbol{a}_k$ and from $\boldsymbol{a}_i \cdot \boldsymbol{b}_i = c\boldsymbol{a}_i \cdot (\boldsymbol{a}_j \times \boldsymbol{a}_k) = 2\pi$ it follows that

$$\boldsymbol{b}_i = 2\pi \frac{\boldsymbol{a}_j \times \boldsymbol{a}_k}{\boldsymbol{a}_i \cdot (\boldsymbol{a}_j \times \boldsymbol{a}_k)}, \qquad \boldsymbol{a}_i = 2\pi \frac{\boldsymbol{b}_j \times \boldsymbol{b}_k}{\boldsymbol{b}_i \cdot (\boldsymbol{b}_j \times \boldsymbol{b}_k)}. \tag{2.76}$$

The two lattices, the point lattice of the $\boldsymbol{R}_n$ and that of the $\boldsymbol{K}_m$, are reciprocal.

Every point lattice has a reciprocal lattice associated with it according to (2.74) and (2.75). The parallelepipeds described by the $\boldsymbol{b}_i$ are primitive cells of the reciprocal lattice. Correspondingly Wigner-Seitz cells in a reciprocal lattice may be constructed. They are called *Brillouin zones*.

One can readily show that, apart from a scale factor, the Brillouin zone of a simple cubic lattice is identical to its Wigner-Seitz cell (Fig. 2.12a). The same is true for the hexagonal lattice (Fig. 2.12d). On the other hand, the shape of the Brillouin zone of the face-centred cubic lattice is like the Wigner-Seitz cell of the body-centred cubic lattice, and vice versa (Figs. 2.12b, c).

From (2.67), (2.74), and (2.75) we find that

$$\boldsymbol{R}_n \cdot \boldsymbol{K}_m = 2\pi(n_1 m_1 + n_2 m_2 + n_3 m_3) = 2\pi N \tag{2.77}$$

where N is an integer. All $\boldsymbol{R}_n$ which, for a given $\boldsymbol{K}_m$, satisfy this equation lie in a plane normal to $\boldsymbol{K}_m$.

Thus every vector $\boldsymbol{K}_m$ (i.e., the three integers m_1, m_2, m_3) can be used to indicate a set of lattice planes of the $\boldsymbol{R}_n$-lattice. Since only the direction of the $\boldsymbol{K}_m$ matters, the m_i are only defined to a common factor. If this factor is chosen so that the m_i have the smallest integer values possible, $\boldsymbol{K}_m$ is (for the given direction) the shortest primitive translation in the reciprocal lattice and the m_i are known as *Miller indices*. A lattice plane with Miller indices (m_1, m_2, m_3) will be intersected by axes in directions $\boldsymbol{a}_1, \boldsymbol{a}_2, \boldsymbol{a}_3$ at distances of $N\boldsymbol{a}_1/m_1$, $N\boldsymbol{a}_2/m_2$, $N\boldsymbol{a}_3/m_3$ (or multiples thereof). Negative Miller indices are written simply as $\bar{m}_i$ instead of $-m_i$. The lattice planes perpendicular to the axes of a cubic crystal (and therefore to the axes of a cartesian coordinate system) then have the indices (100), (010), (001), $(\bar{1}00)$, $(0\bar{1}0)$, $(00\bar{1})$. The set of all symmetrically equivalent lattice planes will be given in braces. Thus all equivalent lattice planes perpendicular to the cubic axes are represented by {100}.

In a similar way, directions in a crystal can be indicated by three indices. One resolves a vector in the desired direction into its three components along

$\boldsymbol{a}_1$, $\boldsymbol{a}_2$, $\boldsymbol{a}_3$. The smallest three integers having the same relationship to one another as these three components are the indices required. They are written in brackets: $[hkl]$. In many crystals—but not in all—these indices are the same as the Miller indices of the set of lattice planes perpendicular to this direction.

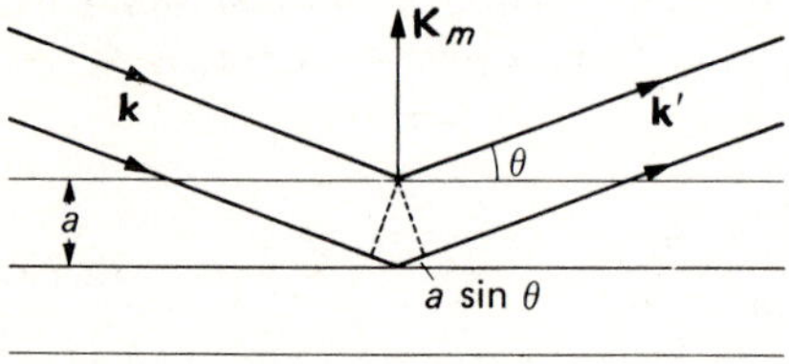

Fig. 2.13. From the left a plane wave arrives, which is reflected from a set of lattice planes defined by the direction of the normal $\boldsymbol{K}_m$ (*Bragg reflection*).

Making use of Fig. 2.13, we shall now consider the origin of Bragg reflections. Let an electron wave propagate in the direction $\boldsymbol{k}$. It will be reflected from the set of lattice planes defined by the vector $\boldsymbol{K}_m$ if the path length of two parallel rays reflected from adjacent planes is a multiple of the de Broglie wavelength $\lambda = 2\pi/k$. From the construction of Fig. 2.13 this is so if

$$2a \sin\theta = N\lambda\,, \tag{2.78}$$

where N is an integer and a the distance between adjacent planes. It can easily be seen from Fig. 2.13 that this condition is equivalent to

$$\boldsymbol{k}' = \boldsymbol{k} + \boldsymbol{K}_m \tag{2.79}$$

or, since $k = k'$

$$k^2 = (\boldsymbol{k} + \boldsymbol{K}_m)^2\,. \tag{2.80}$$

Electrons whose wave vector satisfies (2.80) may be scattered by Bragg reflection. The amount of scattering depends on the scattering cross section which in turn is proportional to the square of the transition matrix element $\langle \boldsymbol{k}'|V|\boldsymbol{k}\rangle = \langle \boldsymbol{k} + \boldsymbol{K}_m|V|\boldsymbol{k}\rangle$. When one inserts plane waves for the wave functions, the matrix element becomes proportional to the Fourier component $V(\boldsymbol{K}_m)$ of the potential. We shall use this result in Section 2.2.6.

There is a very simple construction which illustrates (2.80). It is also important for what follows below. We set up the reciprocal lattice of the $\boldsymbol{K}_m$ in the space associated with the vector $\boldsymbol{k}$ (*$\boldsymbol{k}$-space*). If, starting from one lattice point as the origin, we now draw connecting lines to all other $\boldsymbol{K}_m$, the planes erected at the midpoints of these lines exactly satisfy (2.80).

This construction has been carried out in Fig. 2.14 for the two-dimensional hexagonal point lattice of Fig. 2.11. We note that the net of Bragg reflections becomes increasingly dense with increasing k. All reflections occur first for $\boldsymbol{k}$-values on the boundary of the Brillouin zone defined above.

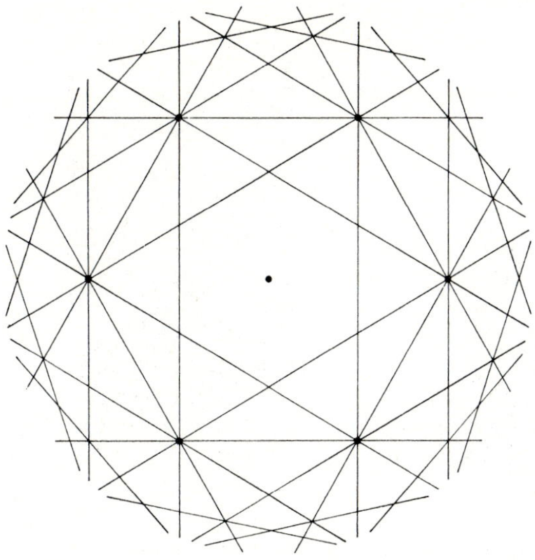

Fig. 2.14. Brillouin zones for the two-dimensional hexagonal point lattice.

The regions encompassed by the planes satisfying (2.80) are also called Brillouin zones, albeit in a somewhat different sense. Considering Fig. 2.14 we see that the six triangles bordering on the inner hexagon have together the same area as the hexagon. By displacing them by appropriately chosen $\boldsymbol{K}_m$ we can make them coincide with the inner hexagon ("*reduction* to the first Brillouin zone"). The bounding triangles to the outside also have the same area. To reduce them to the first zone, we must first divide them into twelve right-angled triangles. Again by suitable choice of $\boldsymbol{K}_m$ these can be displaced to lie exactly on the top of the inner hexagon, etc. These areas (volumes in three dimensions) which can be reduced together to the (first) Brillouin zone are called the 2nd, 3rd, 4th. . . . . Brillouin zones.

In Fig. 2.15 we apply these results to the reduction of the energy of free electrons in a two-dimensional hexagonal point lattice of vanishing strength. Since $V = 0$, the electron energy is given by $E = \hbar^2(k^2 + k^2)/2m$. This energy surface is shown in Fig. 2.15a, divided into parts in accordance with Fig. 2.14. These parts are reduced in Fig. 2.15b to the first Brillouin zone and separated from one another in energy for clarity of representation. We see that every reduced surface is a continuous function $E(\boldsymbol{k})$, albeit with discontinuous derivatives on inner boundary lines. In the next sections we shall identify each of these surfaces as an *energy band* of the band structure.

2.2.5 Consequences of Translational Invariance

Before extending the results of Fig. 2.15 to the case of a nonzero potential $V(\boldsymbol{r})$, we must look more closely at the nature of the solutions of the Schrödinger

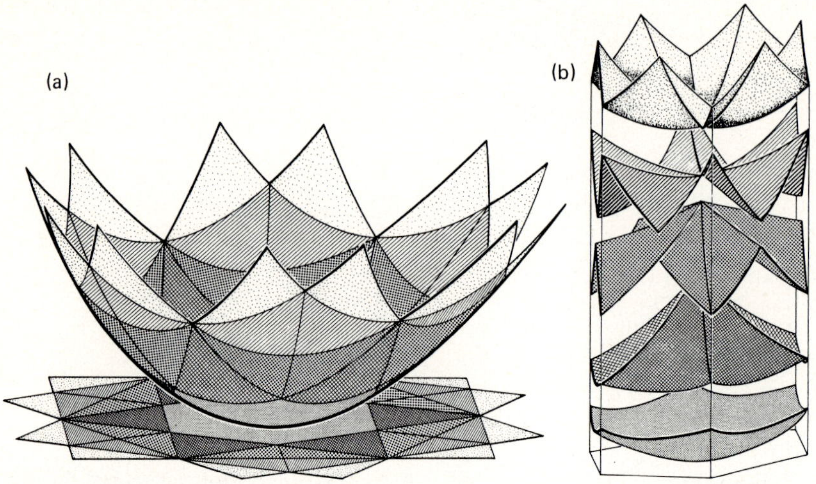

Fig. 2.15. (a) Energy paraboloid for free electrons ($E \sim k^2$) above the $\boldsymbol{k}$-plane for the two-dimensional hexagonal point lattice. Energy paraboloid and $\boldsymbol{k}$-plane are divided into Brillouin zones (or parts of zones) in accordance with Fig. 2.14. (b) Reduction of the paraboloid to the first Brillouin zone.

equation (2.72). The first important results can be deduced from the fact that the Hamiltonian in this equation is invariant under primitive translations.

For a quantitative formulation of this invariance we associate an operator $T_{\boldsymbol{R}_l}$ with every primitive translation $\boldsymbol{R}_l$ through the equation

$$T_{\boldsymbol{R}_l} f(\boldsymbol{r}) = f(\boldsymbol{r} + \boldsymbol{R}_l) . \tag{2.81}$$

The $T_{\boldsymbol{R}_l}$ thus operate on functions of $\boldsymbol{r}$ in such a way that they replace the space vector $\boldsymbol{r}$ in the argument by $\boldsymbol{r} + \boldsymbol{R}_l$.

From (2.73) H is invariant to all $T_{\boldsymbol{R}_l}$. Applying the operator $T_{\boldsymbol{R}_l}$ to (2.72) yields

$$T_{\boldsymbol{R}_l}(H\psi_n) = T_{\boldsymbol{R}_l}(E_n\psi_n) \qquad \text{hence} \quad H(T_{\boldsymbol{R}_l}\psi_n) = E_n(T_{\boldsymbol{R}_l}\psi_n) . \tag{2.82}$$

All $T_{\boldsymbol{R}_l}\psi_n$ are simultaneously with ψ_n eigenfunctions to the same eigenvalue E_n. If E_n is *nondegenerate*, i.e., it has only one eigenfunction ψ_n, $T_{\boldsymbol{R}_l}\psi_n$ must apart from some factor be equal to ψ_n. Since furthermore $|T_{\boldsymbol{R}_l}\psi_n|^2$ must be equal to $|\psi_n|^2$, this factor must have the absolute value 1:

$$T_{\boldsymbol{R}_l}\psi_n = \lambda^{(l)}\psi_n \qquad \text{with} \quad |\lambda^{(l)}|^2 = 1 . \tag{2.83}$$

From (2.83) we see that the $\lambda^{(l)}$ are eigenvalues of the operator $T_{\boldsymbol{R}_l}$. Since $|\lambda^{(l)}|^2 = 1$, $\lambda^{(l)}$ can be put in the form $\exp(\mathrm{i}\alpha_l)$. Since from $\boldsymbol{R}_l + \boldsymbol{R}_m = \boldsymbol{R}_p$

we also have $T_{R_l}T_{R_m} = T_{R_p}$ and $\exp[i(\alpha_l + \alpha_m)] = \exp(i\alpha_p)$, it seems appropriate to write the α_l as the product of a vector $\boldsymbol{k}$ common to all α and its associated $\boldsymbol{R}_l$

$$\lambda^{(l)} = \exp(i\boldsymbol{k}\cdot\boldsymbol{R}_l)\,. \tag{2.84}$$

Here $\boldsymbol{k}$ is not yet connected with the free electron wave vector.

If E_n is *f-fold degenerate*, i.e., the same eigenvalue E_n has f orthogonal eigenfunctions $\psi_{n\kappa}$, the function produced by operating on a $\psi_{n\kappa}$ with T_{R_l} can be represented by a linear combination of all $\psi_{n\kappa}$

$$T_{R_l}\psi_{n\kappa} = \sum_{\kappa'=1}^{f} \lambda^{(l)}_{\kappa\kappa'}\psi_{n\kappa'}\,. \tag{2.85}$$

Every operator T_{R_l} of the translation group is associated with a matrix $\lambda^{(l)}_{\kappa\kappa'}$ through (2.85). These matrices clearly satisfy the same multiplication rules as the T_{R_l}

$$T_{R_l}T_{R_m} = T_{R_p} \to \sum_{\kappa'=1}^{f} \lambda^{(l)}_{\kappa\kappa'}\lambda^{(m)}_{\kappa'\kappa''} = \lambda^{(p)}_{\kappa\kappa''}\,. \tag{2.86}$$

The $\lambda_{\kappa\kappa'}$ therefore also form a group, which is called an *f-dimensional representation of the translation group* to the basis of the $\psi_{n\kappa}$.

Instead of the f eigenfunctions $\psi_{n\kappa}$, by linear combination we can produce a new set of f orthogonal eigenfunctions which form the basis of a new *equivalent* matrix representation. From group theory it is known that among all the equivalent representations of an abelian group—such as the translation group—one can always find one with matrices in diagonal form:

$$\Lambda^{(l)}_{\kappa\kappa'} = \Lambda^{(l)}_{\kappa\kappa}\delta_{\kappa\kappa'}\,. \tag{2.87}$$

Eq. (2.85) then becomes

$$T_{R_l}\psi_{n\kappa} = \sum_{\kappa'} \Lambda^{(l)}_{\kappa\kappa'}\psi_{n\kappa'} = \sum_{\kappa'} \Lambda^{(l)}_{\kappa\kappa}\delta_{\kappa\kappa'}\psi_{n\kappa'} = \Lambda^{(l)}_{\kappa\kappa}\psi_{n\kappa}\,. \tag{2.88}$$

From considerations similar to those following (2.83), we find $|\Lambda^{(l)}_{\kappa\kappa}|^2 = 1$ and $\Lambda^{(l)}_{\kappa\kappa} = \exp(i\boldsymbol{k}_\kappa\cdot\boldsymbol{R}_l)$. Thus for every ψ there is always a $\boldsymbol{k}$, so that ψ, being an eigenfunction of T_{R_l}, has the eigenvalue $\exp(i\boldsymbol{k}\cdot\boldsymbol{R}_l)$. ψ is therefore *classified* by this $\boldsymbol{k}$: $\psi = \psi(\boldsymbol{k}, \boldsymbol{r})$.

We combine (2.83) and (2.88) in the *Bloch theorem*: The nondegenerate solutions of the Schrödinger equation (2.72) and suitably chosen linear combinations of the degenerate solutions are at the same time eigenfunctions

$\psi_n(\boldsymbol{k}, \boldsymbol{r})$ of the translational operator $T_{\boldsymbol{R}_l}$ with the eigenvalues $\exp(\mathrm{i}\boldsymbol{k}\cdot\boldsymbol{R}_l)$

$$T_{\boldsymbol{R}_l}\psi_n(\boldsymbol{k}, \boldsymbol{r}) = \exp(\mathrm{i}\boldsymbol{k}\cdot\boldsymbol{R}_l)\psi_n(\boldsymbol{k}, \boldsymbol{r})\,. \tag{2.89}$$

Since the $\psi_n(\boldsymbol{k}, \boldsymbol{r})$ are at the same time the eigenfunctions of the Hamiltonian, the eigenvalues E_n also depend on $\boldsymbol{k}$

$$E_n = E_n(\boldsymbol{k})\,. \tag{2.90}$$

For degenerate E_n we have furthermore $E_n(\boldsymbol{k}_\kappa) = E_n(\boldsymbol{k}_{\kappa'})$.

From Bloch's theorem it follows that

$$\psi_n(\boldsymbol{k}, \boldsymbol{r} + \boldsymbol{R}_l) = \exp(\mathrm{i}\boldsymbol{k}\cdot\boldsymbol{R}_l)\psi_n(\boldsymbol{k}, \boldsymbol{r})\,. \tag{2.91}$$

If we insert the expression

$$\psi_n(\boldsymbol{k}, \boldsymbol{r}) = \exp(\mathrm{i}\boldsymbol{k}\cdot\boldsymbol{r})u_n(\boldsymbol{k}, \boldsymbol{r}) \tag{2.92}$$

in (2.91), we have for the left-hand side $\exp(i\boldsymbol{k}\cdot(\boldsymbol{r} + \boldsymbol{R}_l))u_n(\boldsymbol{k}, \boldsymbol{r} + \boldsymbol{R}_l)$ and for the right-hand side $\exp(\mathrm{i}\boldsymbol{k}\cdot(\boldsymbol{r} + \boldsymbol{R}_l))u_n(\boldsymbol{k}, \boldsymbol{r})$. We have therefore

$$u_n(\boldsymbol{k}, \boldsymbol{r} + \boldsymbol{R}_l) = u_n(\boldsymbol{k}, \boldsymbol{r})\,. \tag{2.93}$$

$u_n(\boldsymbol{k}, \boldsymbol{r})$ has the periodicity of the lattice. The $\psi_n(\boldsymbol{k}, \boldsymbol{r})$ are called *Bloch functions.* The crystal electrons which they describe are correspondingly called *Bloch electrons.*

The form of the eigenfunctions gives the first clue to the physical meaning of the $\boldsymbol{k}$. If we put u = const, ψ is given by $\psi = c\exp(\mathrm{i}\boldsymbol{k}\cdot\boldsymbol{r})$. The electron in this case behaves as a free particle and is represented by a plane wave with a wave vector $\boldsymbol{k}$. Thus in going over to the free electron limit the vector $\boldsymbol{k}$ of (2.84) becomes identical to the wave vector $\boldsymbol{k}$. If we transfer this meaning to the crystal case, then (2.92) corresponds to the statement that the Bloch electron is represented by a plane wave which is modulated by the periodicity of the lattice.

Before exploring this interpretation further, we shall draw further additional conclusions from Bloch's theorem.

We establish a reciprocal lattice $\boldsymbol{K}_m$ in the space of the vector $\boldsymbol{k}$ ($\boldsymbol{k}$-space) according to Section 2.2.4. From (2.77) $\boldsymbol{K}_m\cdot\boldsymbol{R}_l$ is an integer multiple of 2π. Operating on $\psi(\boldsymbol{k} + \boldsymbol{K}_m, \boldsymbol{r})$ with $T_{\boldsymbol{R}_l}$ then gives

$$\begin{aligned} T_{\boldsymbol{R}_l}\psi(\boldsymbol{k} + \boldsymbol{K}_m, \boldsymbol{r}) &= \exp(\mathrm{i}(\boldsymbol{k} + \boldsymbol{K}_m)\cdot\boldsymbol{R}_l)\psi(\boldsymbol{k} + \boldsymbol{K}_m, \boldsymbol{r}) \\ &= \exp(\mathrm{i}\boldsymbol{k}\cdot\boldsymbol{R}_l)\psi(\boldsymbol{k} + \boldsymbol{K}_m, \boldsymbol{r})\,. \end{aligned} \tag{2.94}$$

Through T_{R_l} not one $\boldsymbol{k}$, but all $\boldsymbol{k}' = \boldsymbol{k} + \boldsymbol{K}_m$, are associated with a $\psi(\boldsymbol{r})$. All these points in $\boldsymbol{k}$-space are equivalent:

$$\psi_n(\boldsymbol{k}, \boldsymbol{r}) = \psi_n(\boldsymbol{k} + \boldsymbol{K}_m, \boldsymbol{r}) . \tag{2.95}$$

We can interpret the result (2.95) as follows. In classifying the solutions of the Schrödinger equation in terms of the vector $\boldsymbol{k}$, the values of $\boldsymbol{k}$ from one Brillouin zone of $\boldsymbol{k}$-space are sufficient. The function $E_n(\boldsymbol{k})$ can correspondingly be limited to the (first) Brillouin zone (Fig. 2.16a). This representation of the function $E_n(\boldsymbol{k})$ in $\boldsymbol{k}$-space is known as the *reduced zone scheme*. In the Brillouin zone $E_n(\boldsymbol{k})$ then gives for every fixed vector $\boldsymbol{k}$ a discrete energy spectrum ($n = 1, 2, 3 \ldots$). For a given n, $E_n(\boldsymbol{k})$ within the Brillouin zone is a continuous and (apart from degenerate points) differentiable function of $\boldsymbol{k}$. It is called an *energy band*. All bands together, i.e., the complete function $E_n(\boldsymbol{k})$, is accordingly referred to as the *band structure*. The fact that $E_n(\boldsymbol{k})$ is continuous and differentiable will be proved in Section 2.2.8.

In view of the equivalence of $\boldsymbol{k}$ with all $\boldsymbol{k} + \boldsymbol{K}_m$ we can also regard the function $E_n(\boldsymbol{k})$ as a periodic (and because of the index n multivalued) function in $\boldsymbol{k}$-space. Zones with the shape of the first Brillouin zone are connected periodically one to another (Fig. 2.16b). This type of representation is called the *repeated zone scheme*.

Finally, starting from the repeated zone scheme we can give $E_n(\boldsymbol{k})$ a unique

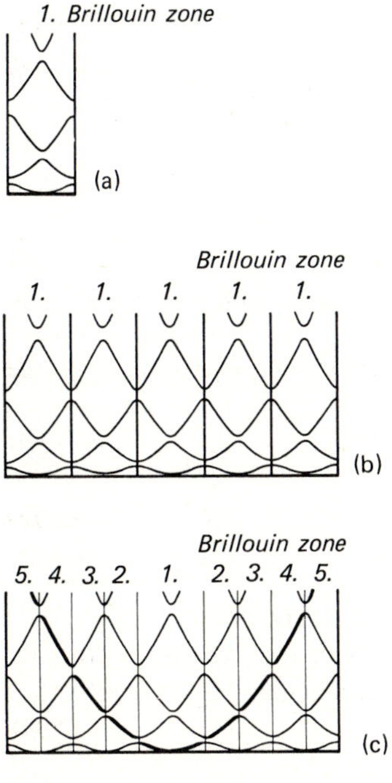

Fig. 2.16a-c. The different representations of band structure in $\boldsymbol{k}$-space, as exemplified by a simple one-dimensional band structure: (a) reduced zone scheme, (b) repeated zone scheme, (c) extended zone scheme.

meaning by dividing the $\boldsymbol{k}$-space in accordance with Fig. 2.14 into the 1st, 2nd, 3rd ... Brillouin zones, and in the mth zone considering only the portion $E_m(\boldsymbol{k})$. This is the *extended zone scheme* (Fig. 2.16c).

In the following sections we shall make frequent use of these three possible representations of the band structure of a solid.

2.2.6 Nearly Free Electron Approximation

In the last section we saw that the two representations in Fig. 2.15 of the function $E(\boldsymbol{k}) = \hbar^2k^2/2m$ for free electrons in a crystal lattice are two possible ways to describe the same physical facts. In Fig. 2.15a the *nonreduced* $\boldsymbol{k}$*-vector* is used; the energy is therefore presented in the extended zone scheme. In Fig. 2.15b every $\boldsymbol{k}$-vector from Fig. 2.15a is shortened by a suitable $\boldsymbol{K}_m$ so that it lies in the first Brillouin zone. This is the reduced zone representation with the *reduced* $\boldsymbol{k}$*-vector*. Additionally there is the possibility of a repeated zone scheme in which all points $\boldsymbol{k} + \boldsymbol{K}_m$ in $\boldsymbol{k}$-space are considered as physically equivalent. Fig. 2.15 can be extended according to this scheme by erecting energy paraboloids at every point $\boldsymbol{K}_m$ (and not merely at $\boldsymbol{K}_m = 0$). These paraboloids then intersect precisely where Bragg reflections occur, and the portions extending through the first Brillouin zone form the energy bands of the reduced zone scheme. At each paraboloid intersection (i.e., where Bragg reflections occur) the function $E_n(\boldsymbol{k})$ for free electrons is degenerate. We now want to see if these degeneracies are removed by some perturbation. Here the perturbation is the nonzero lattice potential $V(\boldsymbol{r})$. In answering this question it will be sufficient to take the perturbation as small. We therefore can apply the usual methods of perturbation theory.

$V(\boldsymbol{r})$ is a function with the periodicity of the lattice. We can expand it in the following Fourier series:

$$V(\boldsymbol{r}) = \sum_{m(\neq 0)} V(\boldsymbol{K}_m) \exp(\mathrm{i}\boldsymbol{K}_m \cdot \boldsymbol{r}) . \tag{2.96}$$

The term with $m = 0$ is the mean value of the potential, which we omit here as unimportant for our further discussion.

Correspondingly we express the periodic factor $u(\boldsymbol{k}, \boldsymbol{r})$ of the Bloch function in terms of a Fourier series

$$\psi(\boldsymbol{k}, \boldsymbol{r}) = \frac{1}{\sqrt{V_g}} \exp(\mathrm{i}\boldsymbol{k} \cdot \boldsymbol{r}) \sum_m u(\boldsymbol{K}_m) \exp(\mathrm{i}\boldsymbol{K}_m \cdot \boldsymbol{r}) . \tag{2.97}$$

Here the term with $m = 0$ represents the undisturbed plane wave.

As long as the potential is weak, the other terms of the expansion are only small perturbations. We therefore take (for the moment): $u(0) \approx 1$, all other $u(\boldsymbol{K}_m)$ small compared with $u(0)$.

Inserting (2.96) and (2.97) into the Schrödinger equation (2.72), we find

$$\frac{1}{\sqrt{V_g}}\sum_m\left(\frac{\hbar^2}{2m}(\boldsymbol{k}+\boldsymbol{K}_m)^2 - E(\boldsymbol{k}) + \sum_l V(\boldsymbol{K}_l)\exp(\mathrm{i}\boldsymbol{K}_l\cdot\boldsymbol{r})\right)u(\boldsymbol{K}_m) \times \exp(\mathrm{i}(\boldsymbol{k}+\boldsymbol{K}_m)\cdot\boldsymbol{r}) = 0\,. \tag{2.98}$$

Multiplication by $(1/\sqrt{V_g})\exp(-\mathrm{i}(\boldsymbol{k}+\boldsymbol{K}_m)\cdot\boldsymbol{r})$ and integration over V_g gives, since $(1/V_g)\int\exp(\mathrm{i}\boldsymbol{K}\cdot\boldsymbol{r})d\tau = \delta_{\boldsymbol{K},0}$

$$\left(\frac{\hbar^2}{2m}(\boldsymbol{k}+\boldsymbol{K}_n)^2 - E(\boldsymbol{k})\right)u(\boldsymbol{K}_n) + \sum_m V(\boldsymbol{K}_n-\boldsymbol{K}_m)u(\boldsymbol{K}_m) = 0\,. \tag{2.99}$$

As a first approximation we equate $E(\boldsymbol{k})$ to the unperturbed solution $\hbar^2k^2/2m$ and take from the summation only the term with $u(0)$. We then have

$$u(\boldsymbol{K}_n) \approx -\frac{V(\boldsymbol{K}_n)}{\dfrac{\hbar^2}{2m}[(\boldsymbol{k}+\boldsymbol{K}_n)^2 - k^2]}\,. \tag{2.100}$$

For small $u(\boldsymbol{K}_m)$, (2.100) gives only small perturbations of the wave function and therefore the energy. The $u(\boldsymbol{K}_n)$ become large however when $(\boldsymbol{k}+\boldsymbol{K}_n)^2 \approx k^2$ which corresponds from (2.79) to the the regions near the Bragg reflections.

If we put $k^2 = (\boldsymbol{k}+\boldsymbol{K}_p)^2$ with a given $\boldsymbol{K}_p$, i.e., if $\boldsymbol{k}$ fulfills the Bragg condition exactly, then in (2.99) the coefficients $u(0)$ and $u(\boldsymbol{K}_p)$ are large, and we have from (2.99) the two equations

$$\begin{aligned}&\left[\frac{\hbar^2}{2m}k^2 - E(k)\right]u(0) + V(-\boldsymbol{K}_p)u(\boldsymbol{K}_p) = 0\\ &\left[\frac{\hbar^2}{2m}(\boldsymbol{k}+\boldsymbol{K}_p)^2 - E(\boldsymbol{k})\right]u(\boldsymbol{K}_p) + V(\boldsymbol{K}_p)u(0) = 0\,.\end{aligned} \tag{2.101}$$

Solving for $E(\boldsymbol{k})$ gives, since $V(-\boldsymbol{K}_p) = V^*(\boldsymbol{K}_p)$ and $k^2 = (\boldsymbol{k}+\boldsymbol{K}_p)^2$,

$$E = \frac{\hbar^2k^2}{2m} \pm |V(\boldsymbol{K}_p)|\,. \tag{2.102}$$

Thus at all $\boldsymbol{k}$-vectors satisfying the Bragg conditions, the energy splits in proportion to the corresponding Fourier component of the potential. In particular this means that $E_n(\boldsymbol{k})$ splits up at the surface of the Brillouin zone if $V \neq 0$. Therefore bands may be separated by energy ranges in which there are no states (*energy gaps*).

This is shown in Figs. 2.17 and 2.18. Fig. 2.17 shows the same reduced band structure as Fig. 2.15b except that now at all points where the bands for free

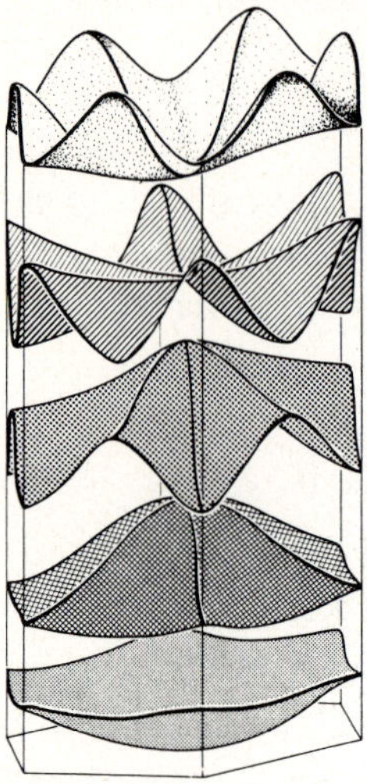

Fig. 2.17. The bands of Fig. 2.15b "smoothed" by the effect of a weak lattice potential.

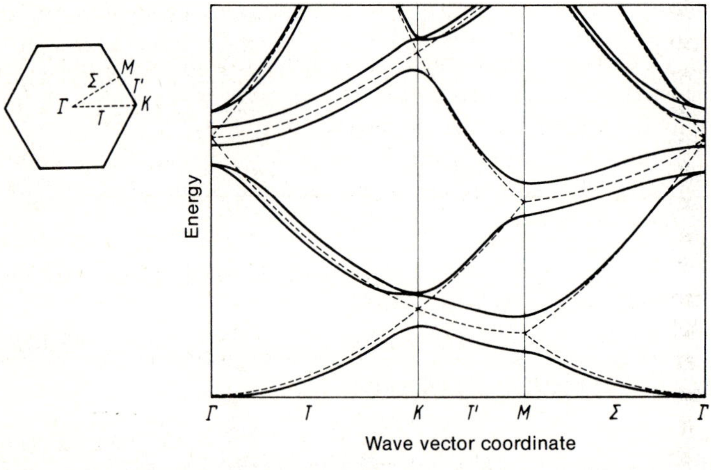

Fig. 2.18. The energy spectrum of Figs. 2.17 and 2.15b (dashed) along lines of symmetry in the Brillouin zone of the hexagonal lattice. Nomenclature of the lines of symmetry is explained at top left.

electrons were degenerate, that degeneracy has been removed by the periodic potential. At the same time the "kinks" in the bands are smoothed out and $E_n(\boldsymbol{k})$ thus becomes a smooth function there as well. Fig. 2.18 shows a section through the Brillouin zone of the previous illustration from the midpoint (Γ) to a corner of the hexagon (K), along one side to its midpoint (M), and back to Γ. The individual parabola segments are intersections of different paraboloids (centred on different $\boldsymbol{K}_m$) in the repeated zone scheme.

2.2.7 Wannier Functions, LCAO Approximation

According to (2.95), the Bloch functions are periodic in $\boldsymbol{k}$-space. They can therefore be represented as a Fourier series

$$\psi_m(\boldsymbol{k}, \boldsymbol{r}) = \frac{1}{\sqrt{N}} \sum_n a_m(\boldsymbol{R}_n, \boldsymbol{r}) \exp(\mathrm{i}\boldsymbol{k} \cdot \boldsymbol{R}_n) . \tag{2.103}$$

The functions $a_m(\boldsymbol{R}_n, \boldsymbol{r})$ in the expansion are called *Wannier functions.* They have a few interesting properties. By inverting (2.103) we have

$$a_m(\boldsymbol{R}_n, \boldsymbol{r}) = \frac{1}{\sqrt{N}} \sum_{\boldsymbol{k}} \exp(-\mathrm{i}\boldsymbol{k} \cdot \boldsymbol{R}_n)\psi_m(\boldsymbol{k}, \boldsymbol{r}) = \frac{1}{\sqrt{N}} \sum_{\boldsymbol{k}} \exp(\mathrm{i}\boldsymbol{k} \cdot (\boldsymbol{r} - \boldsymbol{R}_n))u_m(\boldsymbol{k}, \boldsymbol{r}) . \tag{2.104}$$

The summation is over all $\boldsymbol{k}$-vectors in a Brillouin zone. In view of the lattice periodicity of $u_m(\boldsymbol{k}, \boldsymbol{r})$, the a_m are clearly dependent only on the difference $\boldsymbol{r} - \boldsymbol{R}_n$. Every Wannier function is therefore centred on the midpoint of a Wigner-Seitz cell.

Wannier functions for different bands (index m) and different $\boldsymbol{R}_n$ are orthogonal. This follows from

$$\begin{aligned} \int a_m^*(\boldsymbol{r} - \boldsymbol{R}_n)a_{m'}(\boldsymbol{r} - \boldsymbol{R}_{n'})d\tau \\ &= \frac{1}{N} \sum_{\boldsymbol{k}\boldsymbol{k}'} \exp(\mathrm{i}(\boldsymbol{k} \cdot \boldsymbol{R}_n - \boldsymbol{k}' \cdot \boldsymbol{R}_{n'})) \int \psi_m^*(\boldsymbol{k}, \boldsymbol{r})\psi_{m'}(\boldsymbol{k}', \boldsymbol{r})d\tau \\ &= \frac{1}{N} \sum_{\boldsymbol{k}} \exp[\mathrm{i}\boldsymbol{k} \cdot (\boldsymbol{R}_n - \boldsymbol{R}_{n'})]\delta_{mm'} = \delta_{nn'}\delta_{mm'} . \end{aligned} \tag{2.105}$$

Bloch functions are plane waves modulated with the periodicity of the lattice. The probability for an electron described by $\psi_m(\boldsymbol{k}, \boldsymbol{r})$ to be at a given point is the same for all equivalent points in the lattice. Bloch states are therefore *extended states* in the sense of our discussion in Section 1.1. *Wannier functions*, on the other hand, are *localized* about individual points in the lattice. In characterizing the electronic properties of solids, both systems of description are possible. In the following sections of this chapter we shall use only Bloch functions. In later chapters, when a localized description will suit our purpose better, we shall use Wannier functions.

First of all we want to discuss (2.103) further using a simplified model. We consider a crystal with a Bravais lattice. There will be one atom at the midpoint of each Wigner-Seitz cell. We further assume cubic symmetry (six nearest neighbours in the directions of the cubic axes).

With increasing atomic separation the Wannier functions $a_m(\boldsymbol{r} - \boldsymbol{R}_n)$ can be approximated with increasing accuracy by the atomic functions $\psi^{\text{at}}(\boldsymbol{r} - \boldsymbol{R}_n)$. These ψ^{at} are the solutions of the Schrödinger equation for the free atom

$$\left[-\frac{\hbar^2}{2m}\nabla^2 + V^{\text{at}}(\boldsymbol{r} - \boldsymbol{R}_n)\right]\psi_m^{\text{at}}(\boldsymbol{r} - \boldsymbol{R}_n) = E_{\text{m}}^{\text{at}}\psi_m^{\text{at}}(\boldsymbol{r} - \boldsymbol{R}_n)\,. \tag{2.106}$$

If we use the ψ^{at} as approximations for the Wannier functions of (2.103) and put this into the Schrödinger equation (2.72), we have

$$\left[-\frac{\hbar^2}{2m}\nabla^2 + V(\boldsymbol{r}) - E_m(\boldsymbol{k})\right]\psi_m(\boldsymbol{k}, \boldsymbol{r})$$
$$= \frac{1}{\sqrt{N}}\sum_n \exp(\mathrm{i}\boldsymbol{k}\cdot\boldsymbol{R}_n)[V(\boldsymbol{r}) - V^{\text{at}}(\boldsymbol{r} - \boldsymbol{R}_n) + E_m^{\text{at}} - E_m(\boldsymbol{k})]\psi_m^{\text{at}}(\boldsymbol{r} - \boldsymbol{R}_n) = 0 \tag{2.107}$$

or by multiplication with $\psi_m^{\text{at}*}(\boldsymbol{r})$ and integration

$$[E_m(\boldsymbol{k}) - E_m^{\text{at}}]\sum_n \exp(\mathrm{i}\boldsymbol{k}\cdot\boldsymbol{R}_n)\int\psi_m^{\text{at}*}(\boldsymbol{r})\psi_m^{\text{at}}(\boldsymbol{r} - \boldsymbol{R}_n)d\tau$$
$$= \sum_n \exp(\mathrm{i}\boldsymbol{k}\cdot\boldsymbol{R}_n)\int\psi_m^{\text{at}}(\boldsymbol{r})[V(\boldsymbol{r}) - V^{\text{at}}(\boldsymbol{r} - \boldsymbol{R}_n)]\psi_m^{\text{at}}(\boldsymbol{r} - \boldsymbol{R}_n)d\tau\,. \tag{2.108}$$

Our approximation is certainly justified only if the wave functions of neighbouring atoms overlap weakly. On the left-hand side of (2.108) the term with $n = 0$ is therefore dominant; on the right-hand side, from the terms with $n \neq 0$, only the contributions from the nearest neighbours will be relevant. If we restrict ourselves to these, we have

$$E_m(\boldsymbol{k}) = E_m^{\text{at}} + \int\psi_m^{\text{at}}(\boldsymbol{r})[V(\boldsymbol{r}) - V^{\text{at}}(\boldsymbol{r})]\psi_m^{\text{at}}(\boldsymbol{r})d\tau$$
$$+ \sum_{n.n.}\exp(\mathrm{i}\boldsymbol{k}\cdot\boldsymbol{R}_n)\int\psi_m^{\text{at}*}(\boldsymbol{r})[V(\boldsymbol{r}) - V^{\text{at}}(\boldsymbol{r} - \boldsymbol{R}_n)]\psi_m^{\text{at}}(\boldsymbol{r} - \boldsymbol{R}_n)d\tau\,. \tag{2.109}$$

With the cubic symmetry, the integrals in all the sum terms on the right-hand side are equal. The summation can then be further simplified by taking the nearest neighbours in pairs. If a is the lattice constant, (2.109) can be written with further abbreviations which are readily understood in the form

$$E_m(\boldsymbol{k}) = E_m^{\text{at}} + C_m + 2A_m(\cos k_x a + \cos k_y a + \cos k_z a)\,. \tag{2.110}$$

We interpret this result as follows: For weak interaction between nearest neighbours in a lattice, each energy level for the free atom E_m^{at} is replaced by a *band* of width $12A_m$ (since each cos-term has a range of values $-2A_m$ to

$+2A_m$). The centre of gravity of the band is displaced from E_m^{at} by an amount which depends on the difference between the actual potential at the lattice position and the potential of the free atom. We have thus explained the origin of energy bands in our simple model as a result of the interaction between neighbouring atoms in the solid. Just as the eigenfrequencies of two identical pendula split when we couple them, so the energy states of free atoms split into bands when we bring the atoms together to form a lattice.

The expression (2.103) with atomic orbitals instead of Wannier functions is the starting point of an *approximation method* for the calculation of energy bands in solids. Since the expression represents a linear combination of atomic orbitals, the method is referred to as the *LCAO method.* It will of course have to be extended much further than in our simple example. Possible degeneracies in the atomic levels have to be considered. Atomic orbitals with different quantum numbers have to be combined when a band which does not correspond to a single level of the free atom is to be described (cf. Sec. 2.2.11). For a lattice with a basis (i.e., more than one atom in the Wigner-Seitz cell), atomic orbitals of the different atoms in the basis have to be combined in a suitable manner.

In discussing chemical bonding in Chapter 8 we shall look at the LCAO method from a different standpoint. There we shall also look more closely at the "Coulomb" and "exchange" integrals C_m and A_m which arose in (2.110).

In the last two sections we have determined the characteristic feature of the band model—the succession of allowed and forbidden energy intervals—from two opposite points of views. The results, however, as given in Fig. 2.17 and 2.18 or by (2.110), are only qualitative. We can consider neither the lattice potential to be a small perturbation nor the interaction between lattice atoms to be limited to nearest neighbours. Real band structures will therefore have many features which cannot be derived from the two limiting approximations. In the next sections we shall, therefore, seek general properties of the function $E_n(\boldsymbol{k})$ and consequently of the band structure of a given solid.

2.2.8 General Properties of the Function $E_n(\boldsymbol{k})$

First we shall prove that the function $E_n(\boldsymbol{k})$ is continuous and differentiable in the Brillouin zone. In so doing we shall at the same time obtain some general statements about the properties of the quasi-particle "crystal-electron".

To prove continuity we expand the Bloch function $\psi_n(\boldsymbol{k}, \boldsymbol{r})$ at a point $\boldsymbol{k} + \boldsymbol{\kappa}$ close to $\boldsymbol{k}$. We use the orthogonal system $\chi_n(\boldsymbol{k} + \boldsymbol{\kappa}, \boldsymbol{r}) = \exp(\mathrm{i}\boldsymbol{\kappa}\cdot\boldsymbol{r})\psi_n(\boldsymbol{k}, \boldsymbol{r})$ for the expansion functions. That the χ_n are orthogonal follows from

$$\begin{aligned}\int \chi_{n'}^*(\boldsymbol{k} + \boldsymbol{\kappa}')\chi_n(\boldsymbol{k} + \boldsymbol{\kappa}'')d\tau &= \int \exp(\mathrm{i}(\boldsymbol{\kappa}'' - \boldsymbol{\kappa}')\cdot\boldsymbol{r})\psi_{n'}^*(\boldsymbol{k}, \boldsymbol{r})\psi_n(\boldsymbol{k}, \boldsymbol{r})d\tau \\ &= \int \exp(\mathrm{i}(\boldsymbol{\kappa}'' - \boldsymbol{\kappa}')\cdot\boldsymbol{r})u_{n'}^*(\boldsymbol{k}, \boldsymbol{r})u_n(\boldsymbol{k}, \boldsymbol{r})d\tau\,. \end{aligned} \qquad (2.111)$$

Now $u_{n'}^*, u_n$ has the periodicity of the lattice and can therefore be represented as a Fourier series. Eq. (2.111) can then be transformed into

$$(2.111) = \sum_m A_m^{n'n} \int \exp(\mathrm{i}(\boldsymbol{\kappa}'' - \boldsymbol{\kappa}' - \boldsymbol{K}_m)\cdot\boldsymbol{r})d\tau = V_g \sum_m A_m^{n'n} \delta_{\boldsymbol{\kappa}'',\boldsymbol{\kappa}'+\boldsymbol{K}_m}. \quad (2.112)$$

Furthermore $\boldsymbol{\kappa}'$ and $\boldsymbol{\kappa}''$ are small compared to $\boldsymbol{K}_m$ ($m \neq 0$), i.e., in the summation only the term $m = 0$ remains, and $A_m^{n'n} = 1/V_g \int u_{n'}^* u_n \exp(\mathrm{i}\boldsymbol{K}_m\cdot\boldsymbol{r})d\tau$. Thus

$$(2.111) = V_g A_0^{n'n} \delta_{\boldsymbol{\kappa}''\boldsymbol{\kappa}'} = \delta_{\boldsymbol{\kappa}''\boldsymbol{\kappa}'} \int u_{n'}^* u_n d\tau = \delta_{\boldsymbol{\kappa}''\boldsymbol{\kappa}'} \int \psi_{n'}^* \psi_n d\tau = \delta_{\boldsymbol{\kappa}''\boldsymbol{\kappa}'} \delta_{n'n}. \quad (2.113)$$

The χ_n can therefore be used for the expansion

$$\psi_n(\boldsymbol{k} + \boldsymbol{\kappa}, \boldsymbol{r}) = \sum_j B_{nj}(\boldsymbol{k} + \boldsymbol{\kappa})\chi_j(\boldsymbol{k} + \boldsymbol{\kappa}, \boldsymbol{r}) = \exp(\mathrm{i}\boldsymbol{\kappa}\cdot\boldsymbol{r}) \sum_j B_{nj}\psi_j(\boldsymbol{k}, \boldsymbol{r}). \quad (2.114)$$

$E_n(\boldsymbol{k})$ may not be degenerate in $\boldsymbol{k}$. Operating on (2.114) with the Hamiltonian, we have on the one side

$$\begin{aligned} H\psi_n(\boldsymbol{k} + \boldsymbol{\kappa}, \boldsymbol{r}) &= \left[-\frac{\hbar^2}{2m}\nabla^2 + V(\boldsymbol{r})\right] \exp(\mathrm{i}\boldsymbol{\kappa}\cdot\boldsymbol{r}) \sum_j B_{nj}\psi_j(\boldsymbol{k}, \boldsymbol{r}) \\ &= \exp(\mathrm{i}\boldsymbol{\kappa}\cdot\boldsymbol{r}) \sum_j B_{nj}\left[-\frac{\hbar^2}{2m}\nabla^2 + V(\boldsymbol{r}) + \frac{\hbar^2}{\mathrm{i}m}\boldsymbol{\kappa}\cdot\nabla + \frac{\hbar^2\kappa^2}{2m}\right]\psi_j(\boldsymbol{k}, \boldsymbol{r}) \\ &= \exp(\mathrm{i}\boldsymbol{\kappa}\cdot\boldsymbol{r}) \sum_j B_{nj}\left(E_j(\boldsymbol{k}) + \frac{\hbar^2}{\mathrm{i}m}\boldsymbol{\kappa}\cdot\nabla + \frac{\hbar^2\kappa^2}{2m}\right)\psi_j(\boldsymbol{k}, \boldsymbol{r}); \end{aligned} \quad (2.115)$$

on the other

$$= E_n(\boldsymbol{k} + \boldsymbol{\kappa}) \exp(\mathrm{i}\boldsymbol{\kappa}\cdot\boldsymbol{r}) \sum_j B_{nj}\psi_j(\boldsymbol{k}, \boldsymbol{r}). \quad (2.116)$$

Multiplication by $\exp(-\mathrm{i}\boldsymbol{\kappa}\cdot\boldsymbol{r})\psi_{j'}^*(\boldsymbol{k}, \boldsymbol{r})$ and integration over V_g gives for the right-hand sides of (2.115) and (2.116)

$$E_n(\boldsymbol{k} + \boldsymbol{\kappa})B_{nj'} = \left[E_{j'}(\boldsymbol{k}) + \frac{\hbar^2\kappa^2}{2m}\right]B_{nj'} + \sum_j B_{nj}\frac{\hbar}{m}\boldsymbol{\kappa}\cdot\boldsymbol{p}_{j'j} \quad (2.117)$$

for all j'. Here $\boldsymbol{p}_{j'j}$ is the matrix element

$$\boldsymbol{p}_{j'j} = \frac{\hbar}{\mathrm{i}} \int \psi_{j'}^*(\boldsymbol{k}, \boldsymbol{r})\nabla\psi_j(\boldsymbol{k}, \boldsymbol{r})d\tau. \quad (2.118)$$

Solving the set of equations by the usual perturbation techniques (expansion in a series of increasing powers of $\boldsymbol{\kappa}$) then gives

$$E_n(\boldsymbol{k} + \boldsymbol{\kappa}) = E_n(\boldsymbol{k}) + \frac{\hbar}{m}\boldsymbol{\kappa}\cdot\boldsymbol{p}_{nn} + \frac{\hbar^2\kappa^2}{2m} + \frac{\hbar^2}{m^2}\sum_{\substack{j\\(\neq n)}}\frac{|\boldsymbol{\kappa}\cdot\boldsymbol{p}_{nj}|^2}{E_n(\boldsymbol{k}) - E_j(\boldsymbol{k})}. \qquad (2.119)$$

The limit $\boldsymbol{\kappa} \to 0$ shows first the continuity of the energy at all points in the Brillouin zone for which $E_n(\boldsymbol{k})$ is not degenerate. Secondly it gives

$$\boldsymbol{p}_{nn} = \langle\boldsymbol{p}_n\rangle = \frac{m}{\hbar}\nabla_{\boldsymbol{k}}E. \qquad (2.120)$$

$\boldsymbol{p}_{nn}$ is a diagonal element of the momentum operator, and therefore the expectation value of the momentum of an electron with energy $E_n(\boldsymbol{k})$.

From the second derivative of energy with respect to $\boldsymbol{k}$ we find from (2.119), after going to the limit $\boldsymbol{\kappa} \to 0$,

$$\frac{1}{\hbar^2}\frac{\partial^2 E}{\partial k_\alpha \partial k_\beta} = \frac{1}{m}\delta_{\alpha\beta} + \frac{1}{m^2}\sum_{\substack{j\\(\neq n)}}\frac{p_{nj}^\alpha p_{jn}^\beta + p_{nj}^\beta p_{jn}^\alpha}{E_n(\boldsymbol{k}) - E_j(\boldsymbol{k})}. \qquad (2.121)$$

The interpretation of this equation is simple when we realize that the electron described by the Schrödinger equation is a *quasi-particle* which already has the interaction with the fixed lattice incorporated into its properties. Such a "crystal electron" sees only external forces and the "forces" of lattice vibrations, to which it responds differently than a free electron would.

We list together the corresponding equations for the properties of free electrons and of the quasi-electrons discussed in this section.

	Free electron	Crystal electron
Wave function	$\exp(i\boldsymbol{k}\cdot\boldsymbol{r})$	Bloch function (2.92)
Eigenvalue	$\hbar^2k^2/2m$	Band structure $E_n(\boldsymbol{k})$
Expectation value of momentum	$\hbar\boldsymbol{k}$	(2.120)
$\frac{1}{\hbar^2}\frac{\partial^2 E}{\partial k_\alpha \partial k_\beta}$	$\frac{1}{m}\delta_{\alpha\beta}$	(2.121)

Equation (2.120) is the expression for the *group velocity* of a wave packet, as we also found for free electrons in (2.46). At the same time, however, (2.120) also replaces the de Broglie relation $\boldsymbol{p} = \hbar\boldsymbol{k}$ $(p = h/\lambda)$. For a crystal electron $\boldsymbol{k}$ is therefore no longer equal to the expectation value of the momentum. $\hbar\boldsymbol{k}$ is consequently often referred to as the *crystal momentum* or *pseudomomentum*.

Equation (2.121) shows the different dynamical behaviour of a free electron and a crystal electron. Just as the first derivative of energy with respect to wave number provides the velocity of an electron in a given state, the second derivative provides information about the change of this state. For a free electron the second derivative is the reciprocal of its mass. For the Bloch electron—which already incorporates the effect of the lattice—we have instead of $1/m$ a complicated expression. In the next section we shall see that in an electric field the Bloch electron behaves as though its mass were given by (2.121). The right-hand side of (2.121) has the character of a tensor. It is therefore designated as the *tensor of the effective mass*.

We end this section with two remarks. In Section 2.2.2 we introduced periodic boundary conditions to make the translation group finite. The number of different $\boldsymbol{R}_l$ is then equal to the number of lattice points in the volume V_g. Two primitive translations $\boldsymbol{R}_l$ and $\boldsymbol{R}_l + N_i\boldsymbol{a}_i$ are identical. However that means that

$$\psi(\boldsymbol{k}, \boldsymbol{r} + N_i\boldsymbol{a}_i) = \psi_n(\boldsymbol{k}, \boldsymbol{r}) \tag{2.122}$$

is valid. Therefore with (2.92)

$$\exp(\mathrm{i}\boldsymbol{k}\cdot(\boldsymbol{r} + N_i\boldsymbol{a}_i))u_n(\boldsymbol{k}, \boldsymbol{r}) = \mathrm{e}^{\mathrm{i}\boldsymbol{k}\cdot\boldsymbol{r}}u_n(\boldsymbol{k}, \boldsymbol{r}) \tag{2.123}$$

or

$$\exp(\mathrm{i}N_i\boldsymbol{k}\cdot\boldsymbol{a}_i) = 1\,. \tag{2.124}$$

If we represent $\boldsymbol{k}$ as a vector in the reciprocal lattice, $\boldsymbol{k} = \sum_i \kappa_i\boldsymbol{b}_i$, then

$$\exp(2\pi\mathrm{i}(N_1\kappa_1 + N_2\kappa_2 + N_3\kappa_3)) = 1\,. \tag{2.125}$$

This is fulfilled if the κ_i are restricted to the values

$$\kappa_i = \frac{n_i}{N_i}, \qquad n_i = 1, \ldots, N_i\,. \tag{2.126}$$

Hence there are $N = N_1N_2N_3$ different sets of $\{\kappa_1\kappa_2\kappa_3\}$ and thus N different "allowed" vectors $\boldsymbol{k}$! The vector $\boldsymbol{k}$ can take only discrete values in $\boldsymbol{k}$-space, and only in the limit where V_g tends to infinity is $E_n(\boldsymbol{k})$ a continuous function of $\boldsymbol{k}$. Since V_g can, however, be chosen as large as we like, $E_n(\boldsymbol{k})$ can also be approximated as closely as we like by a continuous function of a variable $\boldsymbol{k}$. In spite of this it is important that $\boldsymbol{k}$ is restricted to the N values

$$\boldsymbol{k} = \sum_i \frac{n_i}{N_i}\boldsymbol{b}_i\,. \tag{2.127}$$

This result is identical with the result (2.5) for free electrons. Because of (2.127) and on account of (2.75), we have for a component k_j: $k_j = (n_j/N_j)b_j = (n_j/N_j)(2\pi/a_j) = (2\pi/L_j)n_j = (2.5)$. In particular, (2.6) for the density of states $g(\boldsymbol{k})$ in $\boldsymbol{k}$-space is also valid here. We shall return to the density of states in Section 2.2.10.

Let us add a further important result. From the Bloch theorem we have

$$T_{\boldsymbol{R}_l}\psi_n^*(\boldsymbol{k}, \boldsymbol{r}) = \exp(-i\boldsymbol{k}\cdot\boldsymbol{R}_l)\psi_n^*(\boldsymbol{k}, \boldsymbol{r}) \tag{2.128}$$

just as

$$T_{\boldsymbol{R}_l}\psi_n(-\boldsymbol{k}, \boldsymbol{r}) = \exp(-i\boldsymbol{k}\cdot\boldsymbol{R}_l)\psi_n(-\boldsymbol{k}, \boldsymbol{r}) . \tag{2.129}$$

Since Bloch's theorem defines the $\boldsymbol{k}$-dependence of the wave function, $\psi^*(\boldsymbol{k}, \boldsymbol{r})$ is identical to $\psi(-\boldsymbol{k}, \boldsymbol{r})$. Since the Hamiltonian is real ($H = H^*$), $\psi^*(\boldsymbol{k}, \boldsymbol{r})$ is degenerate with $\psi(\boldsymbol{k}, \boldsymbol{r})$. Hence $\psi(-\boldsymbol{k}, \boldsymbol{r})$ is also degenerate with $\psi(\boldsymbol{k}, \boldsymbol{r})$, from which we have

$$E(\boldsymbol{k}) = E(-\boldsymbol{k}) . \tag{2.130}$$

This important statement is known as *Kramers theorem.*

2.2.9 Dynamics of Crystal Electrons

In a similar way to our approach to the free electron case in Sections 2.1.5 and 2.1.6, we now consider the Schrödinger equation for the crystal electron in a constant electric field $\boldsymbol{E}$ and a magnetic field $\boldsymbol{B}$. We shall proceed in three steps: We consider first the behaviour of a single electron in a Bloch state under the action of the fields, then the behaviour of all electrons in a band, and finally the dynamics of a wave packet.

Corresponding to (2.40) and (2.48) the Hamiltonian is

$$H = \frac{1}{2m}(\boldsymbol{p} + e\boldsymbol{A})^2 + V(\boldsymbol{r}) + e\boldsymbol{E}\cdot\boldsymbol{r} . \tag{2.131}$$

A Bloch function given at $t = 0$ by $\psi = \psi_n(\boldsymbol{k}_0, \boldsymbol{r})$ changes over a small interval of time dt according to

$$\psi(\boldsymbol{k}, \boldsymbol{r}, dt) = \exp\left(-\frac{i}{\hbar}Hdt\right)\psi_n(\boldsymbol{k}_0, \boldsymbol{r}) = \left(1 - \frac{i}{\hbar}Hdt\right)\psi_n(\boldsymbol{k}_0, \boldsymbol{r}) . \tag{2.132}$$

We ask first whether this wave function is still a Bloch state or not. To do this, we operate on ψ with the translation operator $T_{\boldsymbol{R}}$

$$T_{\boldsymbol{R}}\psi = T_{\boldsymbol{R}}\left(1 - \frac{i}{\hbar}Hdt\right)\psi_n(\boldsymbol{k}_0, \boldsymbol{r}) = \left(1 - \frac{i}{\hbar}Hdt\right)T_{\boldsymbol{R}}\psi_n - \frac{i}{\hbar}dt[T_{\boldsymbol{R}}, H]\psi_n . \tag{2.133}$$

After a few intermediate steps we find for the commutator

$$[T_{\boldsymbol{R}}, H] = \left(\xi \cdot \boldsymbol{R} + \frac{e^2 B^2}{2m} R_x^2\right) T_{\boldsymbol{R}} \quad \text{with} \quad \xi = e\boldsymbol{v} \times \boldsymbol{B} + e\boldsymbol{E} + e\dot{\boldsymbol{A}}\,. \tag{2.134}$$

Remaining in the weak magnetic field approximation, we can neglect the last term on the right-hand side of (2.134) and obtain

$$T_{\boldsymbol{R}}\psi = \left[1 - \frac{\mathrm{i}}{\hbar}(H + \xi \cdot \boldsymbol{R})dt\right] T_{\boldsymbol{R}}\psi_n = (\cdots) \exp(\mathrm{i}\boldsymbol{k}_0 \cdot \boldsymbol{R})\psi_n = \exp(\mathrm{i}\boldsymbol{k} \cdot \boldsymbol{R})\psi \tag{2.135}$$

with

$$\hbar\boldsymbol{k} = \hbar\boldsymbol{k}_0 - \xi\, dt \quad \text{or} \quad \hbar\dot{\boldsymbol{k}} = -\xi\,. \tag{2.136}$$

If we neglect the difference between $\boldsymbol{k}$-space and $(\boldsymbol{k} + (e/\hbar)\boldsymbol{A})$-space, as between $\boldsymbol{k}$-space and $\boldsymbol{P}/\hbar$-space in Section 2.1.6, we find here as in (2.55)

$$\hbar\dot{\boldsymbol{k}} = -e(\boldsymbol{E} + \boldsymbol{v} \times \boldsymbol{B})\,. \tag{2.137}$$

Thus, within the limits of the weak magnetic field approximation (and for all electric fields), we find that a Bloch electron traverses Bloch states such that the change in its $\boldsymbol{k}$-vector (more accurately its $[\boldsymbol{k} + (e/\hbar)\boldsymbol{A}]$-vector) with time is proportional to the Lorentz force.

We can therefore carry over all the results we found for free electrons. For acceleration under the influence of external fields the $\boldsymbol{k}$-vector of the electrons traverses a quasi-continuous succession of states in $\boldsymbol{k}$-space. In a magnetic field only, the $\boldsymbol{k}$-vector remains on a surface of constant energy. Owing to the complicated nature of the energy surfaces in the Brillouin zone, these $\boldsymbol{k}$-vector "orbits" are not circular. Consequently the electron orbit in real space is also much more complex. We shall return to this in Section 2.2.11.

The important difference between the dynamics of a Bloch electron and a free electron lies in the fact that $\boldsymbol{k}$-space is now divided into Brillouin zones and within these zones the dependence of energy on wave vector is given by the band structure function $E_n(\boldsymbol{k})$. We shall examine this now in a simple one-dimensional example—the dynamics of an electron in a band with the structure illustrated in Fig. 2.19, under the influence of an electric field.

Let the band have its minimum at $\boldsymbol{k} = 0$ (point A). In the direction of the $\boldsymbol{k}$-vector component parallel to the electric field there are band maxima at the points B and B' at the edges of the Brillouin zone. A further band, with minima C and C', lies above the one considered.

Before the electric field is applied, the electron in state $\boldsymbol{k} = 0$ has the velocity $\boldsymbol{v} = (1/\hbar)\nabla_{\boldsymbol{k}}E = 0$. When the field is applied, the $\boldsymbol{k}$-vector of the electron

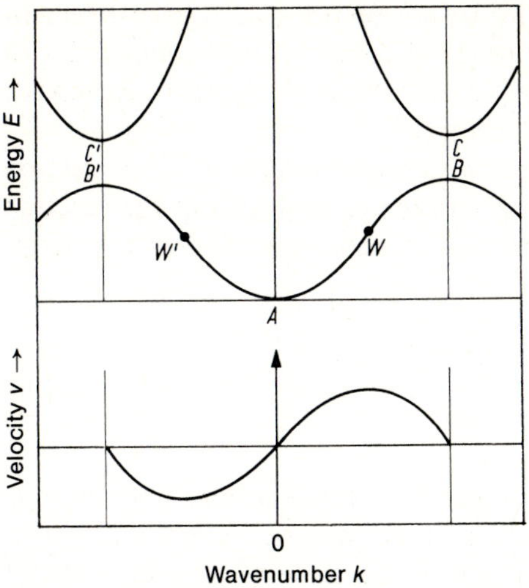

Fig. 2.19. Dynamics of an electron in an energy band. Above: A simple energy band with minimum at A, points of inflection at W and W', and maxima at B and B'. The lower part of a second band with minima at C and C' is also shown. The points B and B' are equivalent, likewise C and C'. Below: Velocity of an electron in the lower band as a function of k (schematic).

runs through states with increasing $\boldsymbol{k}$ and finally reaches the point B. Along this path its velocity increases up to the point of inflection W and then decreases to become zero again at B. This decrease in velocity in spite of acceleration by the electric field means of course that in this region the lattice forces restrain the electron and in so doing overcompensate the acceleration. Finally at B the lattice forces completely prohibit the motion of the electron (Bragg reflection). This effect is described by the effective mass, which as the second derivative of energy with respect to $\boldsymbol{k}$ is negative above the point of inflection. A negative mass means just such a decelerating effect in an accelerating field.

When the electron $\boldsymbol{k}$-vector has reached B it leaves the Brillouin zone and enters the next zone of the repeated zone scheme used here. In view of the periodicity of the function $E(\boldsymbol{k})$, we can then describe the electron by an equivalent point which moves from B' towards A. We can therefore describe the change of the $\boldsymbol{k}$-vector with time as a periodic process in which the $\boldsymbol{k}$-point moves from B' through A to B, then reappears at B' and executes the cycle again, etc. The electron energy then oscillates between $E(A)$ and $E(B) = E(B')$. Associated with this there is an oscillation in real space since the velocity ($\sim dE/dk$) is periodically changing sign.

Such an oscillation is of course never seen since a crystal electron interacts not only with external forces but also with lattice vibrations. These remove energy and momentum from the electron in interaction processes (emission of phonons, Chap. 4) so that between two such processes the electron is never able to traverse more than small portions of the $\boldsymbol{k}$-axis of Fig. 2.19.

We shall now consider the behaviour of *all* electrons in a given band. In this we distinguish three cases: the number of electrons is small compared with the number of states available in the band; all states in the band are occupied; and only a few states in the band are unoccupied.

In the first of these cases the similarity with the behaviour of a free electron gas is apparent. The electrons are in states near the minimum of the function $E_n(\boldsymbol{k})$. If we can approximate $E_n(\boldsymbol{k})$ near the minimum by the first two terms of an expansion

$$E_n(\boldsymbol{k}) = E_{\text{min}} + \alpha k^2 + \cdots = E_{\text{min}} + \frac{\hbar^2 k^2}{2m^*} + \cdots, \tag{2.138}$$

the electrons at the bottom of the band differ from free electrons only through their different mass m^*. Under the action of the electric field the $\boldsymbol{k}$-vectors of all the electrons in $\boldsymbol{k}$-space are displaced in the same way as are those of the free electron gas (Fig. 2.5).

In the second case above, a fully occupied band, the displacement of all electrons in $\boldsymbol{k}$-space changes nothing. Every electron whose $\boldsymbol{k}$-vector crosses the Brillouin zone boundary is then described by a reduced $\boldsymbol{k}$-vector which enters the zone "from the other side". All the states in the band remain occupied. From the Kramers theorem (2.130), every occupied state $\boldsymbol{k}$ will have a corresponding occupied state $-\boldsymbol{k}$; the mean velocity of all electrons remains zero. *A completely filled band makes no contribution to the electric current.*

This result also means that the sum over all velocities $\boldsymbol{v}(\boldsymbol{k})$ associated with the states in a band is zero. In a given band let the states $\{\boldsymbol{k}_1\}$ be occupied and the states $\{\boldsymbol{k}_2\}$ be empty. It then follows that

$$\boldsymbol{I} = -e \sum_{\boldsymbol{k}_1} \boldsymbol{v}(\boldsymbol{k}) = +e \sum_{\boldsymbol{k}_2} \boldsymbol{v}(\boldsymbol{k}). \tag{2.139}$$

The current carried by electrons in the states $\{\boldsymbol{k}_1\}$ is thus the same as that which would be carried by fictitious positively charged particles occupying the states $\{\boldsymbol{k}_2\}$. We therefore find here the possibility—similar to Section 2.1.3—of representing the behaviour of an ensemble of electrons (here in a band, for free electrons in the Fermi sphere) in terms of the behaviour of fictitious *holes* which take up the unoccupied states. A *positive charge* has clearly to be associated with these holes.

This possibility is important in the case of a band which has only a few empty states. These may be found at the top of the band. Analogous to (2.138), we can expand the energy in this region as

$$E_n(k) = E_{\text{max}} + \frac{\hbar^2 k^2}{2m^*} = E_{\text{max}} - \frac{\hbar^2 k^2}{2|m^*|}. \tag{2.140}$$

Here we have used the fact that the effective mass is negative in the vicinity of the band maximum. If we consider an unoccupied state (a hole), we find that its contribution to the current is $I = +e\boldsymbol{v}(\boldsymbol{k})$. By differentiation we have

$$\dot{I} = +e\dot{\boldsymbol{v}}(\boldsymbol{k}) = e\frac{d}{dt}\left(\frac{1}{\hbar}\nabla_{\boldsymbol{k}}E\right) = -\frac{e}{|m^*|}\hbar\dot{\boldsymbol{k}} = \frac{e^2}{|m^*|}(\boldsymbol{E} + \boldsymbol{v} \times \boldsymbol{B}). \tag{2.141}$$

This is precisely the law which governs the acceleration of a *positively charged particle with a positive mass* $|m^*|$! Thus if we want to describe the dynamics of the electrons in a nearly filled band by the behaviour of a few holes, we must accord these quasi-particles a positive (effective) mass as well as a positive charge.

There is an important difference between the holes introduced in Section 2.1.3 as the unoccupied states inside the Fermi sphere and the holes in states of a band. Under the action of an electric field, the whole Fermi sphere moves in the first case, and with it the holes. In the second case, on the other hand, only the holes in the Brillouin zone (fixed in $\boldsymbol{k}$-space) are displaced.

Pair excitation in the band model means the transition of an electron from a state in one band into a band of higher energy. A finite threshold energy is clearly needed to create an *electron-hole pair*. We shall return to this in detail in various future sections.

We go on now to consider the dynamics of a wave packet of Bloch functions in a given band structure. The function $E_n(\boldsymbol{k})$ is periodic in $\boldsymbol{k}$-space and can be expanded in a Fourier series

$$E_n(\boldsymbol{k}) = \sum_m E_{nm} \exp(\mathrm{i}\boldsymbol{R}_m \cdot \boldsymbol{k}). \tag{2.142}$$

If we formally construct an operator $E_n(-\mathrm{i}\nabla)$ by replacing all $\boldsymbol{k}$ in $E_n(\boldsymbol{k})$ by $-\mathrm{i}\nabla$, we find that it has the following properties:

$$\begin{aligned} E_n(-\mathrm{i}\nabla)\psi_n(\boldsymbol{k}, \boldsymbol{r}) &= \sum_m E_{nm} \exp(\boldsymbol{R}_m \cdot \nabla)\psi_n(\boldsymbol{k}, \boldsymbol{r}) \\ &= \sum_m E_{nm}\left[1 + \boldsymbol{R}_m \cdot \nabla + \frac{1}{2}(\boldsymbol{R}_m \cdot \nabla)^2 + \cdots\right]\psi_n(\boldsymbol{k}, \boldsymbol{r}) \\ &= \sum_m E_{nm}\psi_n(\boldsymbol{k}, \boldsymbol{r} + \boldsymbol{R}_m) = \sum_m E_{nm} \exp(\mathrm{i}\boldsymbol{R}_m \cdot \boldsymbol{k})\psi_n(\boldsymbol{k}, \boldsymbol{r}) \\ &= E_n(\boldsymbol{k})\psi_n(\boldsymbol{k}, \boldsymbol{r}). \end{aligned} \tag{2.143}$$

The Bloch functions $\psi_n(\boldsymbol{k}, \boldsymbol{r})$ are therefore eigenfunctions of the operator $E_n(-\mathrm{i}\nabla)$ with eigenvalues $E_n(\boldsymbol{k})$.

We now examine the time-dependent Schrödinger equation

$$\left[-\frac{\hbar^2}{2m}\nabla^2 + V(\boldsymbol{r}) - e\phi\right]\psi = -\frac{\hbar}{\mathrm{i}}\dot{\psi}. \tag{2.144}$$

We represent the electron described by this equation as a wave packet constructed from all the Bloch states of all bands

$$\psi = \sum_{n,k} c_n(\boldsymbol{k}, t)\psi_n(\boldsymbol{k}, \boldsymbol{r}) . \tag{2.145}$$

We then have

$$\sum_{n,k} c_n(\boldsymbol{k}, t)\left[-\frac{\hbar^2}{2m}\nabla^2 + V(\boldsymbol{r}) - e\phi\right]\psi_n(\boldsymbol{k}, \boldsymbol{r}) = -\frac{\hbar}{\mathrm{i}}\dot{\psi}$$
$$= \sum_{n,k} c_n(\boldsymbol{k}, t)[E_n(\boldsymbol{k}) - e\phi]\psi_n(\boldsymbol{k}, \boldsymbol{r}) = \sum_{n,k} c_n(\boldsymbol{k}, t)[E_n(-\mathrm{i}\nabla) - e\phi]\psi_n(\boldsymbol{k}, \boldsymbol{r}) . \tag{2.146}$$

This equation can be rearranged further if we make the restrictive assumption that the electric field is too weak to induce transitions from one band to another. The electron should therefore remain always in one band (index n), and this band should not be degenerate with another band. It is then sufficient to construct the wave packet from the states of the band concerned, and the summation in (2.145) and (2.146) is only over all $\boldsymbol{k}$ of the band n. In the last equation of (2.146), however, the operator in the brackets does not depend on $\boldsymbol{k}$ so that it can be taken in front of the summation over $\boldsymbol{k}$. We therefore have

$$[E_n(-\mathrm{i}\nabla) - e\phi]\sum_{k} c_n(\boldsymbol{k}, t)\psi_n(\boldsymbol{k}, \boldsymbol{r}) = [E_n(-\mathrm{i}\nabla) - e\phi]\psi = -\frac{\hbar}{\mathrm{i}}\dot{\psi} . \tag{2.147}$$

With this we have arrived at a new Schrödinger equation which differs from (2.144) in that the periodic potential $V(\boldsymbol{r})$ no longer appears explicitly! Instead of the kinetic energy operator applying to free electrons, we have a new equivalent Hamiltonian $E_n(-\mathrm{i}\nabla)$. This reveals exactly the quasi-particle properties of the crystal electron. The periodic potential is incorporated in the properties of the electron; the wave packet behaves in an electric field as though it were a free particle with charge $-e$ and with a dispersion relationship between energy and wave vector given by $E_n(\boldsymbol{k})$. $E_n(\boldsymbol{k})$ thus appears in place of the relationship $E = \hbar^2k^2/2m$ for free electrons, and the second derivative of the function $E_n(\boldsymbol{k})$ [(2.121)] appears in place of the reciprocal mass of the free electron. We shall continue the discussion of (2.147) in Section 2.2.12.

In the derivation of (2.147) we restricted ourselves to an electric field. We can also derive a similar equation for simultaneously applied electric and magnetic fields. Again we must neglect band-to-band transitions and restrict ourselves to small magnetic fields. As the magnetic field increases, splitting of the energy bands into magnetic subbands will occur just as with free electrons. The relationships here however are so complicated that we choose not to examine them more closely.

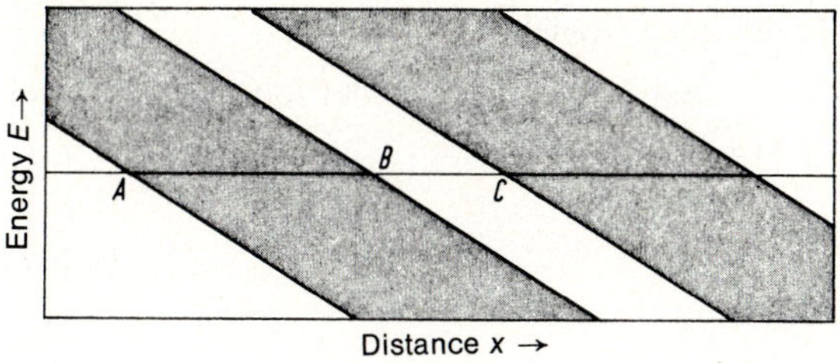

Fig. 2.20. If we add the electrostatic energy of a constant electric field to the energy of the band model, the energy bands in the *E-x* diagram appear inclined. An electron (wave packet) moving from *A* to *B* will be reflected back into the band, or it can pass into the next higher band by tunnel effect.

We close this section with a comment on the approximation (2.147) by which we were able to describe the electron wave function solely in terms of Bloch states from a single band. The addition of Bloch functions from other bands means the possibility of an electron transition into another band under the action of the electric field. This phenomenon is known as *internal field emission* or *Zener effect*. To understand it we consider Figs. 2.19 and 2.20. In Fig. 2.20 the electrostatic energy of the electron in the electric field is added to its energy according to the band model. Since the electrostatic energy varies with distance, the energy axis is extended by a position coordinate. In this representation the bands appear inclined. The spatial oscillation connected with the oscillation in energy between E_A and E_B is represented in Fig. 2.20 as a periodic motion between points *A* and *B*. When the electron reaches *B* it is reflected back to *A*. From the representation in Fig. 2.20, however, we can see that there is also the possibility that the electron *tunnels* through the energy gap between the bands and reaches *C*. Corresponding to this in Fig. 2.19 would be the possibility that an electron reaching *B*, instead of jumping back to *B'*, would make a transition to *C*. To calculate the probability of a transition from *B* to *C*, i.e., the penetrability of the potential barrier between *B* and *C*, we need to know the wave function on both sides. Approximations can be made to evaluate this effect in an algebraic way but we shall not treat them in detail. The probability of tunneling is certainly dependent on the distance *BC*. *BC* increases with increasing width of the energy gap and with decreasing electric field (weaker tilting of the bands in Fig. 2.20). The transition probability therefore takes the form

$$w(B \to C) \propto \exp\left(-cE_G^{3/2}/|\boldsymbol{E}|\right) \tag{2.148}$$

where the constant c contains the effective masses at the two band edges.

2.2.10 The Density of States in the Band Model

At the end of Section 2.2.8 we saw that in the band model too the density of states (per band) in $\boldsymbol{k}$-space is given by

$$g(\boldsymbol{k})d\tau_{\boldsymbol{k}} = \frac{2}{(2\pi)^3}d\tau_{\boldsymbol{k}}\,. \tag{2.149}$$

More important is the density of states on the energy scale, i.e., the number of states in a given energy interval, referred to the volume V_g. We find this in the following way.

We consider an energy band with a minimum at $\boldsymbol{k} = 0$. The proof is easily modified if the band has minima at other locations in $\boldsymbol{k}$-space. The number of states with energy smaller than a given energy E_0 is equal to $g(\boldsymbol{k})$ times the volume of the region of the Brillouin zone enclosed by the surface $E = E_0$

$$G(E_0) = \int_{E_{\min}}^{E_0} g(E)dE = \int_{E_0} g(\boldsymbol{k})d\tau_{\boldsymbol{k}}\,. \tag{2.150}$$

The integration over $\boldsymbol{k}$ can be divided into an integration in energy and a surface integral over surfaces of constant energy (of a band n!)

$$G_n(E_0) = \frac{2}{(2\pi)^3}\int_{E_{\min}}^{E_0}\int_{E_n=\mathrm{const}} \frac{dE_n df}{|\nabla_{\boldsymbol{k}}E_n(\boldsymbol{k})|}\,. \tag{2.151}$$

By comparing (2.150) and (2.151), we have the density of states sought

$$g_n(E)dE = \left[\frac{2}{(2\pi)^3}\int_{E_n=\mathrm{const}} \frac{df}{|\nabla_{\boldsymbol{k}}E_n(\boldsymbol{k})|}\right]dE\,. \tag{2.152}$$

$g_n(E)$ is thus given when the band structure $E_n(\boldsymbol{k})$ of the energy band is known.

Energy bands can *overlap*. This means that the highest energy states of a band n can lie above the lowest energy states of the following band $n + 1$. In this case the states at the same energy add together. The total density of states follows from (2.152) by summation over n.

Figure 2.21 shows this schematically. Three bands follow one another in energy. Bands A and B overlap, while C is separated from B by an energy range without any allowed state (*energy gap*). Other examples of densities of states appear in Figs. 2.29 and 2.32.

With the help of the density of states and the Fermi distribution (2.30), we can determine the distribution $n(E)dE$ of the electrons in a solid. At $T = 0$, all states below the limiting energy E_F are occupied, all above it are empty. For $T \neq 0$, the boundary between occupied and empty states is blurred.

All the bands which represent the closed electron shells of the individual atoms (deep-lying bands) are always fully occupied with electrons. The next

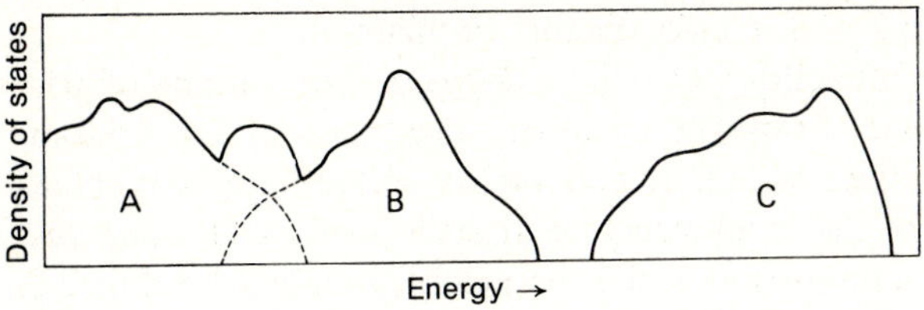

Fig. 2.21. Density of states (schematic) with overlapping bands and with a band separated from the others by an energy gap.

higher-lying band contains the valence electrons (*valence band*). For the ground state ($T = 0$) we distinguish two cases.

a) The valence band is completely filled by the valence electrons. An energy gap lies between the highest state in the valence band and the lowest state of the next higher *conduction band.*
b) The valence band (or a group of overlapping valence bands) is not filled. The first empty state lies immediately above the highest occupied state.

We can then make the following statements about the electrical behaviour of the solid: Since a completely filled band makes no contribution to conduction, the solid in case a) behaves like an *insulator*, while in case b) if shows the characteristics of a *metal.* Between these limits lies the *semiconductor.* It belongs to case a), but, in contrast to insulators, the energy gap between valence and conduction band is so small that thermal excitations can produce electron transitions between valence and conduction band. Or, put another way, the width of the region, set by the Fermi distribution, in which occupied and unoccupied states occur in appreciable numbers alongside each other is greater than the width of the energy gap between valence and conduction band.

In *metals* the Fermi energy lies in a single band or in several overlapping bands. At $T = 0$, the surface $E = E_F$ in the Brillouin zone will therefore separate occupied and unoccupied states (*Fermi surface*). The *shape of the Fermi surface* will determine the properties of the metal at all temperatures (Sec. 2.2.11)

In *semiconductors*, where there are only a few electrons in the conduction band and only a few holes in the valence band, we shall be interested in the band structure near the *extrema of the conduction and valence bands.* (Sec. 2.2.12).

Our main interest in *insulators* will be optical phenomena whereby electrons from one band are raised into a higher one. To describe these processes (which are of course also important in semiconductors) we shall need detailed knowledge of the structure of several bands (see Chap. 6).

2.2.11 The Band Structure of Metals, Fermi Surfaces

We have just recognized the importance of band structure in determining the energy distribution of the electrons in a solid and their behaviour in external fields. In this and the next section we want to give examples of the form of the function $E_n(\boldsymbol{k})$ for metals and semiconductors (insulators).

$E_n(\boldsymbol{k})$ is a many-valued function periodic in $\boldsymbol{k}$-space. The volume of periodicity is the Brillouin zone. From the symmetry properties of the Brillouin zone, certain points and lines within it and on its surface are particularly important. A knowledge of the band structure at such points and along such lines is often sufficient for a discussion of the characteristics of a solid. Fig. 2.22 shows their nomenclature for the Brillouin zones of the four most important point lattices.

In the discussion in Section 2.2.6 we explained the origin of band structure and the succession of allowed and forbidden regions of energy in terms of Bragg reflections which cut out regions from the continuous energy spectrum of free electrons. Another qualitative explanation (Sec. 2.2.7) started with the discrete energy levels of the free atom and interpreted the bands by the splitting up of atomic levels through interactions in the crystal lattice. In this interpreta-

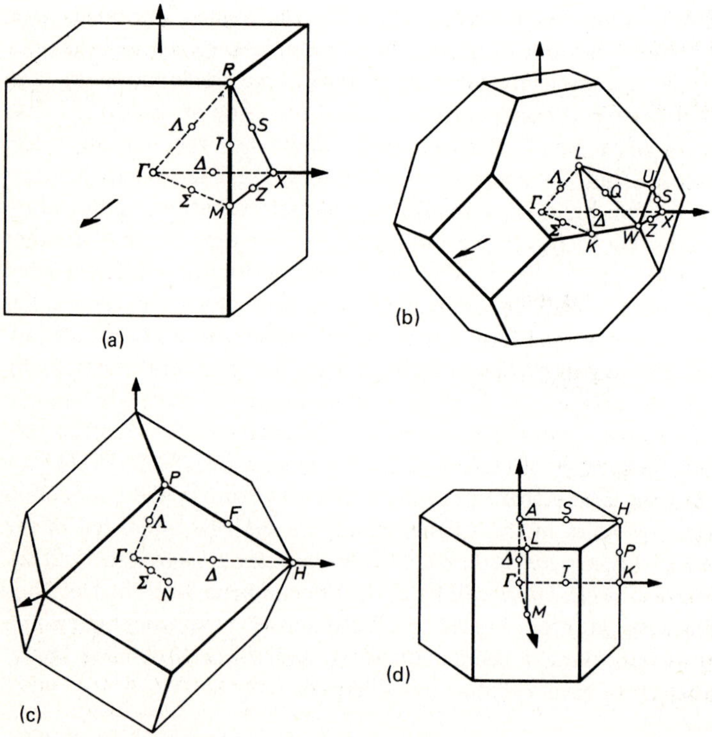

Fig. 2.22a-d. Brillouin zone for the (a) simple cubic, (b) face-centred cubic, (c) body-centred cubic, and (d) hexagonal lattice. The most important points and lines of symmetry are shown, together with their nomenclature.

tion every band in the band model would have to correspond to a level in the free atom.

This is often the case, and one talks of *s*-bands, *p*-bands, *d*-bands, etc. Fig. 2.23a shows schematically the origin of the 3*s*- and the 3*p*-bands of sodium. The atomic levels only split up weakly for large lattice constants. With decreasing atomic separation the bands become increasingly broad, and at the actual lattice constant of sodium both bands overlap. This is the case only when the atomic states in the crystal remain at least approximately unchanged. An opposing example is provided by diamond where the *s*- and *p*-states of the free C-atoms become sp^3-hybrid levels in the crystal, on account of the adjusted valences to the four nearest neighbours (see Sec. 8.2.3). In this case the *s*- and *p*-bands only remain separate at large lattice constants (Fig. 2.23b). There is then a region in which the bands overlap. Through the hybridization they then split up again into two separate bands with mixed *s*- and *p*-character.

We return to such questions in Section 2.2.14 when we use group theory to examine the symmetry behaviour of wave functions. These remarks are only intended here to relate individual bands of the band structures to be discussed to levels in the atoms. This is particularly useful in order to find out whether a given solid is a metal or an insulator (semiconductor). In the two examples in Fig. 2.23 it is clear that sodium must be a metal, and diamond, on the other hand, an insulator. Even if its *s*- and *p*-bands were not to overlap, sodium would be a metal, for the 3*s*-level in the free atom contains only one electron so that only half of the states in the *s*-band are occupied. In diamond,

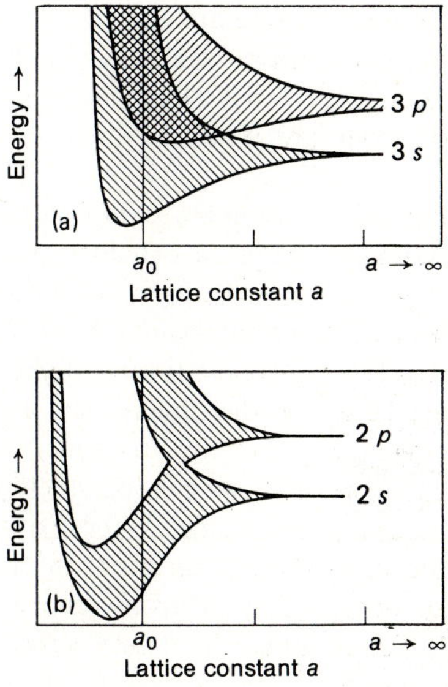

Fig. 2.23a and b. Formation of a band structure from discrete levels of isolated atoms by bringing the atoms together into a crystal. (a) In sodium, *s*- and *p*-levels lead to *s*- and *p*-bands which overlap in the crystal. (b) In diamond, as the atoms approach, there is a rearrangement of the *s*- and *p*-levels to hybrid sp^3-states. Two sp^3-bands are formed, separated by an energy gap (schematic).

the overlapping $2s/2p$-bands split into two new bands, each of which takes half of the states, that is, four per atom. The four valence electrons of the C-atoms in diamond just fill the lower lying band.

We now turn to the band structure of important metals. If we look first to the *monovalent alkali metals*, we find relatively simple valence band structures. The Fermi surfaces are approximately spherical; the free electron approximation is therefore justified. As we advance from Li through Na, K to heavier alkali metals, the bands of the d-electrons (d-bands) move higher and higher and influence the shape of the Fermi surfaces.

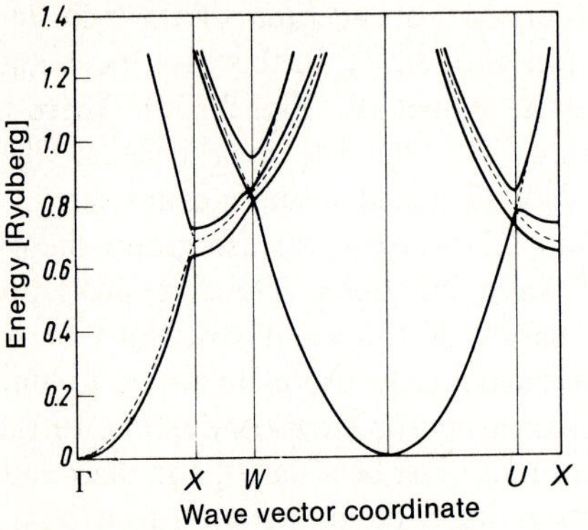

Fig. 2.24. The valence bands of aluminium (from *Harrison* [13]).

Even for *polyvalent metals* one often finds very simple relationships. As an instructive example we shall discuss the band structure of aluminium. Fig. 2.24 shows its shape along important lines of symmetry in the Brillouin zone (see Fig. 2.22b). The bands which we could expect if the valence electrons ($3s$- and $3p$-electrons of the free atom) were completely free are shown with a dashed line. There is no evidence here of any connection with levels in the free atom. This is characteristic of many metals. The inner electrons of the ion core occupy filled bands; the valence electrons are practically free. We shall be able to explain this in Section 2.2.16.

In spite of this, the Fermi surfaces of aluminium are quite complicated. This is because the approximately spherical Fermi surface in the extended zone scheme lies outside the first Brillouin zone. Fig. 2.25 shows this in a section through $\boldsymbol{k}$-space which along with the midpoint contains the points X, U, L, and K (Fig. 2.22b). The shape of the free electron Fermi sphere is only slightly distorted in the vicinity of Bragg reflections. However the sphere cuts the 2nd, 3rd and 4th Brillouin zone. If we reduce these zones to the first zone, we get the Fermi surfaces illustrated in Fig. 2.26. The first zone is completely filled with electrons. The second contains only electrons outside (!) the Fermi surface shown, the third zone only inside the cigar-shaped surfaces. In the fourth zone finally there are only small regions occupied by electrons (pockets of electrons). In this figure we have used the Fermi sphere for free electrons and have thus neglected the small distortions produced by Bragg reflections. We shall see below that to a large extent the shape of these Fermi surfaces

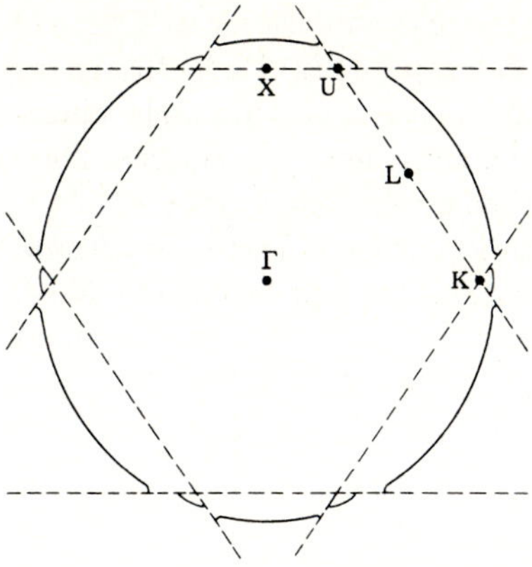

Fig. 2.25. The Fermi sphere in aluminium penetrates the second, third, and fourth Brillouin zones. Its form deviates only slightly from the free electron case in the regions of Bragg reflections (from *Harrison* [13]).

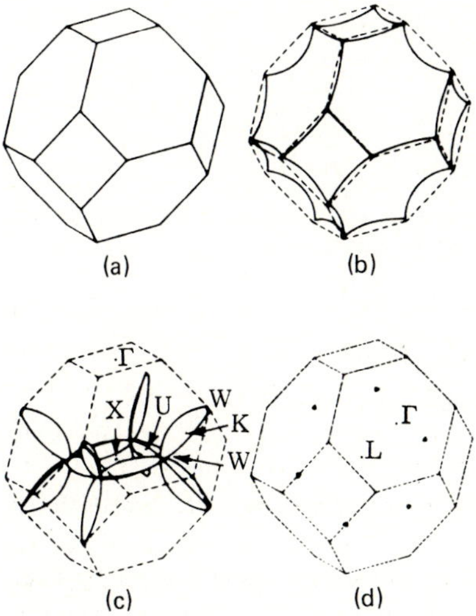

Fig. 2.26a-d

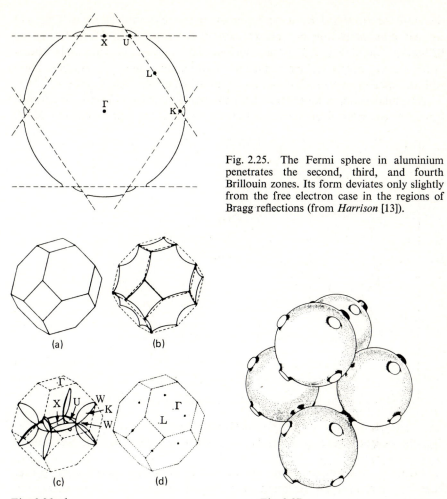

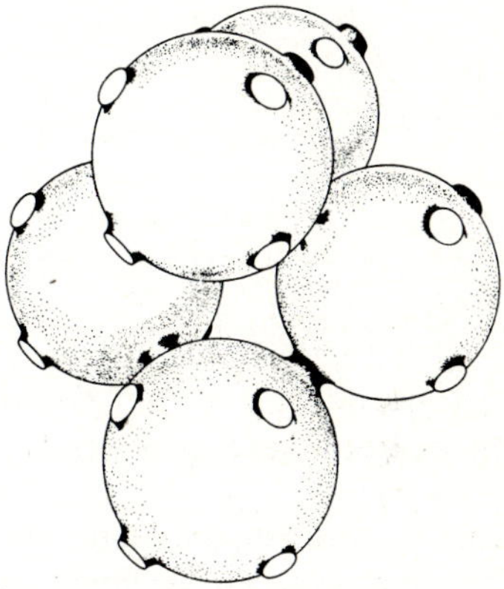

Fig. 2.27

Fig. 2.26a-d. Reduction of the Fermi sphere of aluminium to the first Brillouin zone (distortions of the sphere near the Bragg reflections have been neglected). (a) to (d): Reduction of the four Brillouin zones (see Fig. 2.25). In order to make the Fermi surfaces in (c) and (d) easier to show, the Brillouin zone has in each case been displaced by half a reciprocal lattice vector in the repeated zone scheme (from *Harrison* [13]).

Fig. 2.27. The Fermi surface of copper in the repeated zone scheme (from *Mackintosh* [113a]).

produced by reducing the free electron Fermi sphere can be detected experimentally. In doing so, a terminology has developed for Fermi surfaces which describes them according to their form as monsters, needles, cigars, lenses, discs, four-winged butterflies, etc. The monsters in the third zone in Fig. 2.26 touch the surface of the zone. In the repeated zone scheme $\boldsymbol{k}$-space will therefore be traversed by a net of interconnected Fermi surfaces. One can see this interconnection particularly clearly in copper (Fig. 2.27). Inside the Brillouin zone of Fig. 2.22b these surfaces are spheres but with slight bulges in the vicinity of the surface of the zone and, in consequence, connected there with the spheres of the neighbouring zones in the repeated zone scheme.

As a final example we consider now the *transition metals*. These are different from the examples given so far in that the *d*-bands are not completely filled.

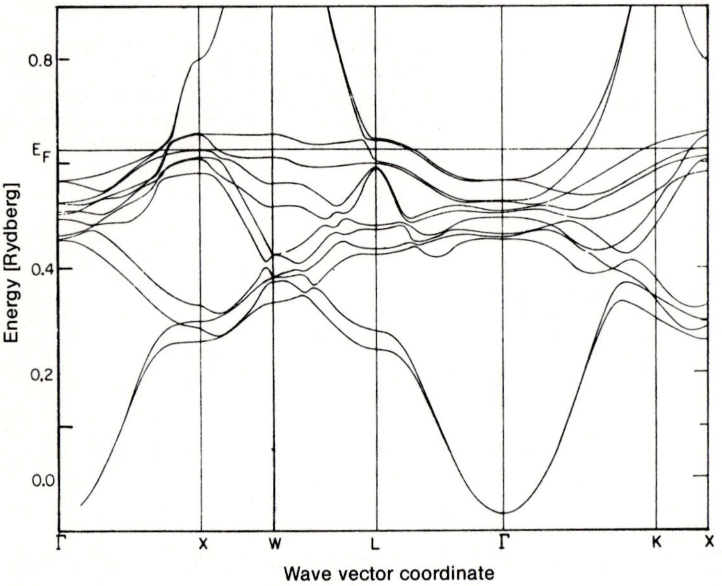

Fig. 2.28. The band structure of nickel [from E. I. Zornberg: Phys. Rev. **B1**, 244 (1970)].

Figure 2.28 shows the band structure of nickel along the most important lines of symmetry in the Brillouin zone. We can distinguish parabolae, starting from Γ and continuing above E_F. They represent the band of the 4*s*-electrons. Superimposed on this is a multitude of overlapping *d*-bands. The Fermi energy lies just below the maximum of the uppermost *d*-band. Fig. 2.29 presents the density of states in the energy region of the *d*-bands. Its complicated structure shows the overlapping of numerous narrow bands. Above the *d*-bands, only the density of states of the 4*s*-band remains.

To a good approximation we can assume that all transition metals have roughly the same band structure and with it the same density of states. The difference between Fe, Co, and Ni (and further Cu) lies only in the different extent to which the bands are filled. In Fe a considerable portion of the *d*-bands is empty, in Ni only a small portion, while in Cu the Fermi energy is situated in the 4*s*-band. We return to this in Section 3.4.5 (Fig. 3.16).

All the important methods used to experimentally determine the shape of the Fermi surface involve the motion of electrons in a magnetic field since such

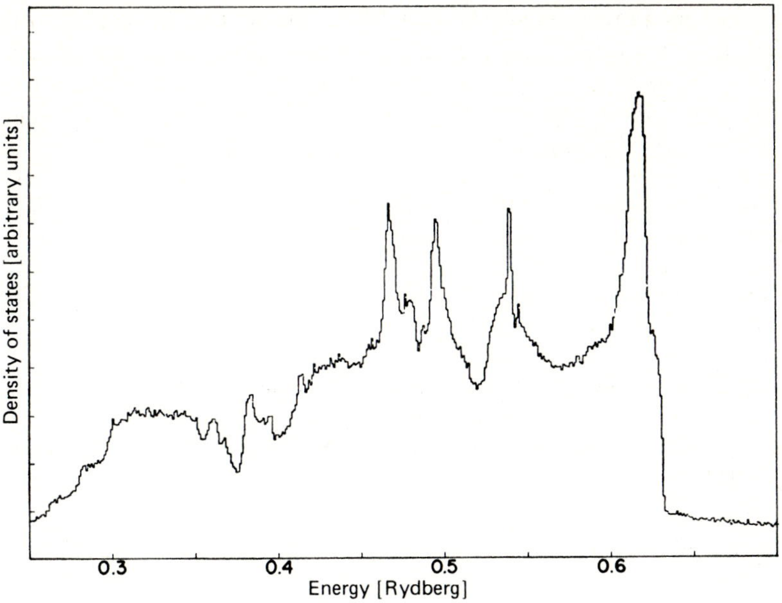

Fig. 2.29. Density of states of nickel in the energy range of the *d*-bands [from E. I. Zornberg: Phys. Rev. **B1**, 244 (1970)].

motion always takes place on a surface of constant energy. For other methods of determination the reader is referred to the literature cited at the end of this section. We are concerned here only with the most important method: the de Haas-van Alphen effect. We have already examined the main aspects of this effect as an example of free electron behaviour in Section 2.1.7. We therefore need only ascertain what changes there are in the results of Section 2.1.7 when the electrons no longer execute circular orbits but rather follow orbits of complicated shape in the plane perpendicular to the field.

Let us start with the motion of an electron round any type of orbit in $\boldsymbol{k}$-space. The orbit is the intersection of a plane perpendicular to the magnetic field and a surface of constant energy. The area $F_{\boldsymbol{k}}$ enclosed by the orbit is given by

$$F_{\boldsymbol{k}} = \int_0^E dE' \oint_{E'=\text{const}} \frac{dk}{\nabla_{\boldsymbol{k}} E'_\perp} . \tag{2.153}$$

The frequency of rotation is obtained from the equation

$$\hbar \dot{\boldsymbol{k}} = e\boldsymbol{v} \times \boldsymbol{B} \qquad \text{or hence} \quad \hbar \dot{k} = e v_\perp B , \tag{2.154}$$

where $\boldsymbol{k}$ points in the direction of the orbit and $\boldsymbol{v}_\perp$ normal to it and to the magnetic field. By separation of the variables k and t, and integration, we obtain the circulation time T_c

$$T_c = \frac{\hbar}{eB}\oint\frac{dk}{v_\perp} = \frac{\hbar}{eB}\oint\frac{dk}{\nabla_k E_\perp}. \tag{2.155}$$

By comparison of (2.153) and (2.155), the rotation frequency (cyclotron resonance frequency) is

$$\omega_c = \frac{2\pi}{T_c} = \frac{2\pi eB}{\hbar^2}\left(\frac{dF_k}{dE}\right)^{-1}. \tag{2.156}$$

This becomes the cyclotron resonance frequency defined for free electrons in Section 2.1.6 when we put $F_k = \pi k_E^2 = 2\pi mE/\hbar^2$.

'We now consider the quantization of these orbits. From the Bohr quantum condition follows

$$\left|\oint \boldsymbol{p}\cdot d\boldsymbol{q}\right| = 2\pi\hbar(\nu + \gamma), \tag{2.157}$$

where γ is a phase constant which for free electrons has the value 1/2. For the momentum we have to put here $\hbar\boldsymbol{k} + e\boldsymbol{A}$ and for the space vector the radius vector of the orbit in space in the plane perpendicular to the magnetic field. We then have for the first part of the integral

$$\oint \hbar\boldsymbol{k}\cdot d\boldsymbol{r}_\perp = e\oint(\boldsymbol{r}\times\boldsymbol{B})\cdot d\boldsymbol{r}_\perp = -e\boldsymbol{B}\cdot\oint\boldsymbol{r}_\perp\times d\boldsymbol{r} = -2eBF_r \tag{2.158}$$

and for the second part

$$\oint e\boldsymbol{A}\cdot d\boldsymbol{r}_\perp = e\int\nabla\times\boldsymbol{A}\cdot d\boldsymbol{f} = eBF_r, \tag{2.159}$$

together therefore

$$F_r = \frac{2\pi\hbar}{eB}(\nu + \gamma). \tag{2.160}$$

Here F_r is the area enclosed by the orbit in space. Through (2.154) F_r is related to F_k by the factor $(eB/\hbar)^2$, so that we have

$$F_k = \frac{2\pi eB}{\hbar}(\nu + \gamma). \tag{2.161}$$

With the aid of Fig. 2.8 we now remind ourselves of the explanation for the de Haas-van Alphen effect of free electrons. In that case the continuous distribution of states in $\boldsymbol{k}$-space was confined in a magnetic field on to concentric cylinders. The cross sections of the cylinders were the orbital surfaces $F_{\boldsymbol{k}} = \pi k_{\perp}^2$ allowed by the quantization. Oscillations in the magnetic susceptibility always arose when the surface of a cylinder left the Fermi sphere and the electrons in states on the surface fell back into states on the cylinder surface below.

We can now carry over exactly the same interpretation to the case where the Fermi surface has any shape. The quantized orbital surfaces $F_{\boldsymbol{k}}$ are not circular; accordingly the cross sections of the concentric tubes are no longer circular. However this changes nothing in the argument. As the magnetic field is increased every time a "tube" leaves the Fermi surface, there is an abrupt change of free energy and with it a change of magnetization. The period of the de Haas-van Alphen oscillations is therefore determined by the *extremal cross sections of the Fermi surface* normal to the magnetic field. When we consider Fig. 2.27 we find different types of extremal orbits depending on the orientation of the magnetic field. The most important types are given in Fig. 2.30. Several extremal orbits are possible for certain directions. The oscillations are then made up of a superposition of different frequencies.

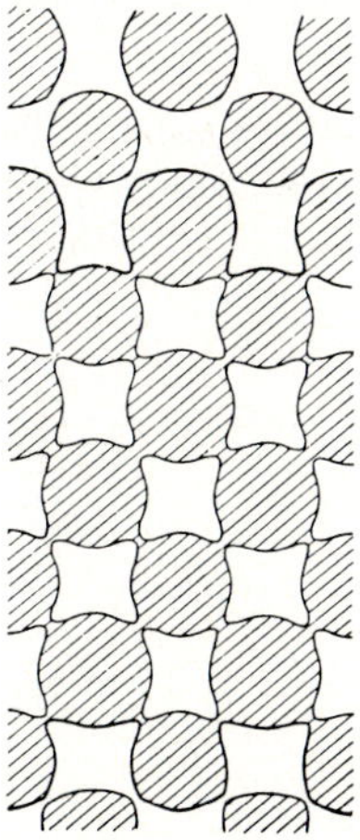

Fig. 2.30. Fermi surfaces of copper in the repeated zone scheme, in a plane in $\boldsymbol{k}$-space which is slightly inclined to the (001)-plane. With a magnetic field normal to this plane, the electrons move along the lines of intersection of the plane with the Fermi surface. One can distinguish between closed orbits which encircle occupied states (electron orbits) and those which encircle unoccupied states (hole orbits). The direction of circulation is opposite in the two cases. In addition to these types of orbits, the figure shows an open orbit. Extremal orbits, which are detected in the de Haas-van Alphen effect, are here above all the circular electron orbits round the Fermi sphere and the narrow connections of the spheres with one another ("belly" and "bottleneck" orbits) and the hole orbits which touch four spheres ("rosette" and "dog's-bone" orbits) (from *Mackintosh* [113a]).

All these considerations are restricted to *closed orbits*. However, it is also possible to have open orbits which in the repeated zone scheme traverse $\boldsymbol{k}$-space. An example is shown in Fig. 2.30. For further discussion of Fermi surfaces and experimental methods, see among others the contributions by *Mackintosh* [113a], *Schoenberg* [65], *Harrison* [13,58], and *Ziman* [23].

2.2.12 The Band Structure of Semiconductors and Insulators

Following the discussion of metals, we turn now to semiconductors and insulators. Fig. 2.31 shows the band structure of the semiconductor silicon and

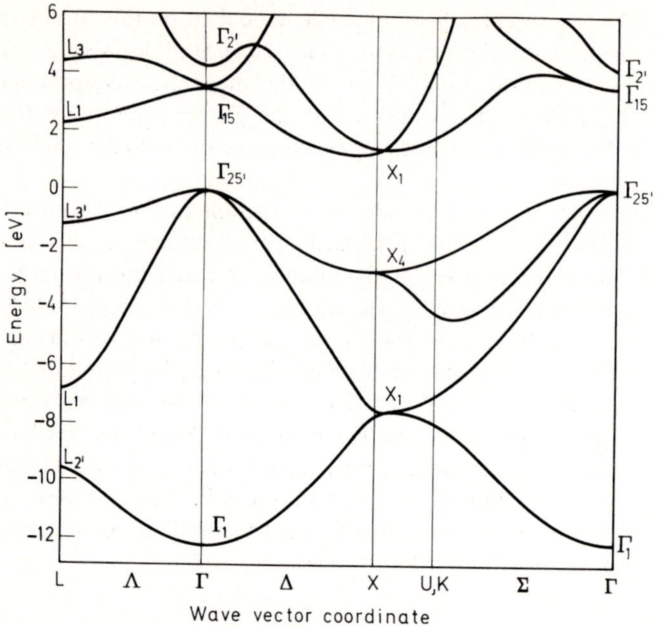

Fig. 2.31. Band structure of silicon along the most important axes of symmetry in the Brillouin zone. For an explanation of the symmetry symbols also shown see Section 2.2.14 [from J. R. Chelikowski, M. L. Cohen: Phys. Rev. B **14**, 556 (1976)]

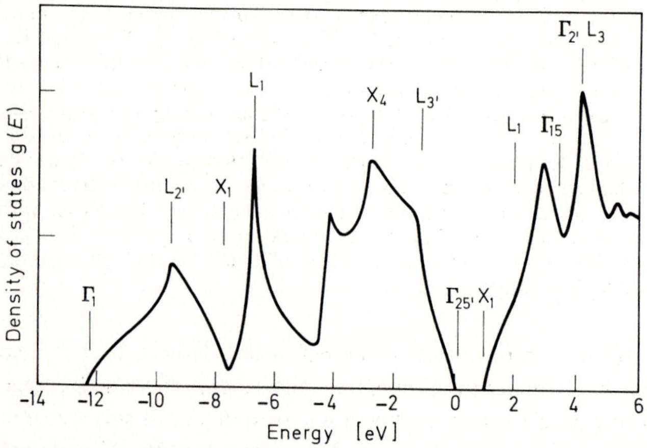

Fig. 2.32. Density of states in the valence and conduction bands of silicon. For an easier comparison with Fig. 2.31 the symbols of the high symmetry points of the band structure are inserted [from J. R. Chelikowski, D. J. Chadi, M. L. Cohen: Phys. Rev. B **8**, 2786 (1973)]

Fig. 2.32 its density of states. Along the most important axes of symmetry—the Brillouin zone is that in Fig. 2.22b—we find a number of overlapping subbands which fall into two groups separated by an *energy gap*. The lower group forms the subbands of the *valence band*, the upper those of the *conduction band*.

The width of the energy gap between the highest level of the valence band at Γ and the lowest of the conduction band along the Δ-axis is somewhat greater than 1 eV. At low temperatures the valence band is completely filled, the conduction band completely empty. Silicon then behaves like an insulator. If we compare the density of states in Fig. 2.32 with the band structure, we quickly see that certain regions in the individual subbands make particularly large contributions to the density of states. We shall postpone a discussion of these questions, in which the concept of the so-called critical points in the density of states becomes important, until Section 6.2.2 when we examine optical transitions between valence and conduction bands.

In semiconductors, as we have already emphasized, it is mainly the regions of energy near the top of the valence band and the bottom of the conduction band which are of interest, since they contain the levels which in thermal equilibrium are occupied by electrons and holes. In the vicinity of a band edge, i.e., an extremum of the function $E_n(\mathbf{k})$, one can expand this function and stop after the quadratic term

$$E(\mathbf{k}) = E(\mathbf{k}_0 + \boldsymbol{\kappa}) = E(\mathbf{k}_0) + \frac{1}{2}\sum_{\alpha\beta} \frac{\partial^2 E}{\partial k_\alpha \partial k_\beta} \kappa_\alpha \kappa_\beta \,, \tag{2.162}$$

where $\mathbf{k}_0$ gives the position of the band edge in the Brillouin zone. The factor associated with $\kappa_\alpha \kappa_\beta$ in the second term is the effective mass tensor of (2.121). If the surfaces of constant energy in the vicinity of a band edge are concentric ellipsoids and if it is permissible to terminate the expansion at the quadratic term, the charge carriers (electrons and holes) can be described as having a constant, possibly direction-dependent, effective mass. This approximation is often possible in semiconductor theory.

If in particular the effective mass is a scalar, i.e., independent of direction, the theoretical treatment becomes particularly simple. The electrons then behave like free electrons but with a different value of mass. The same is true for holes which are accorded a positive charge as well as a (scalar) positive effective mass. This very simple model is complicated by the following possibilities (see Fig. 2.33). In this figure, first, a band structure similar to Fig. 2.31 has been illustrated (the band structure of germanium). Individual regions have been lifted out. Let us begin with the point Γ, the midpoint of the Brillouin zone. If the lower edge of the conduction band lies at this point (Fig. 2.33a), we have in general an isotropic parabolic band, i.e., states which can be described by a directionally independent constant effective mass. Only with increasing energy does the $E(\mathbf{k})$ dependence deviate from a parabola, the effective mass then becoming energy dependent. This case is realized for instance in the semiconducting compound InSb.

The upper edge of the valence band in many semiconductors lies at $\mathbf{k} = 0$ (those of the group IV elements—diamond, Si, Ge, α-Sn—and the III–V compounds, for example). Here, however, two different subbands have a common extremum (Fig. 2.33d). Holes can then appear near the extremum in both

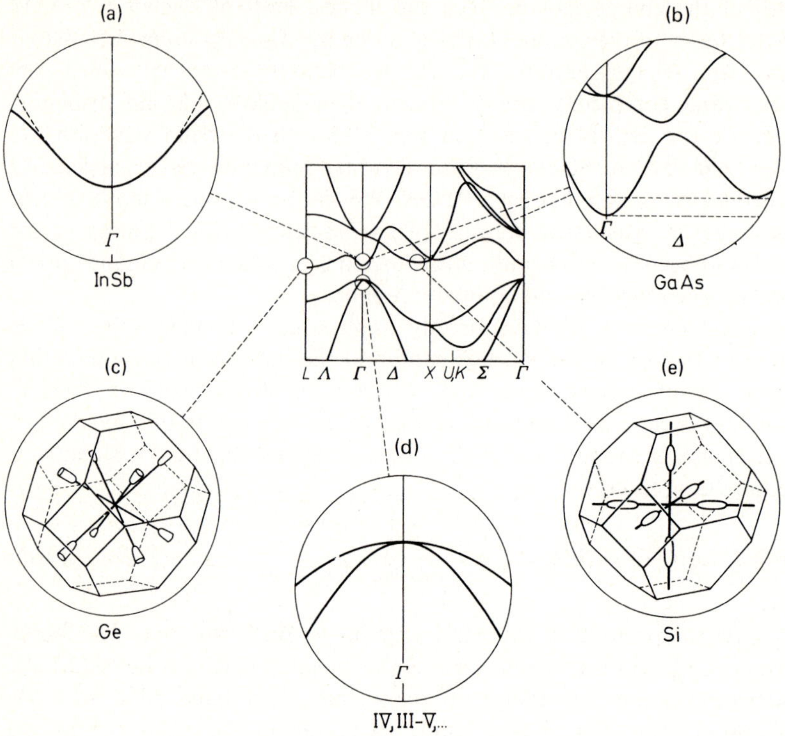

Fig. 2.33a-e. The most important details of the band structure of a semiconductor.

subbands. In this case the free charge carrier model has to take account of two different sorts of holes existing together.

If the extrema of a band lie outside $\boldsymbol{k} = 0$, then from symmetry considerations (see the following section) a number of equivalent extrema must be present. Surfaces of constant energy are often ellipsoidal in the vicinity of such extrema. This case, exemplified by Si and Ge, is shown in Figs. 2.33c, e.

Only Fig. 2.33b remains to be discussed. It shows the overlapping of two subbands with extrema which are close together in energy but at different points in the Brillouin zone. Here too we have to consider two sorts of carriers (electrons) with differing effective masses. Since the minima lie at different energies, the ratio between the concentrations of the two sorts of electrons will become temperature dependent. An example of this case is the band structure of GaAs.

In (2.147) we derived a Schrödinger equation for a wave packet constructed from Bloch functions, which no longer contains the periodic lattice potential. In place of the kinetic energy operator there appeared an operator $E_n(-\mathrm{i}\nabla)$, that is, an operator obtained from the band structure function by replacing $\boldsymbol{k}$ with $-\mathrm{i}\nabla$. An approximation of the form (2.162) however cannot immediately be used to construct the operator, for it operates on rapidly changing functions and it has not been proved a priori that the expansion of the operator can be terminated after the second term.

We consider the simpler case that $E_n(\boldsymbol{k})$ can be approximated by the expansion $E_n(\boldsymbol{k}) = E_{\min} + \hbar^2 k^2/2m^*$ in the vicinity of $\boldsymbol{k} = 0$. The electrons in these states then have a scalar effective mass. We form the wave packet only with Bloch functions from the region at the edge of the band in question. We assume that their lattice periodic part changes slowly

$$\psi_n(\boldsymbol{k}, \boldsymbol{r}) \approx u_n(0, \boldsymbol{r}) \exp(\mathrm{i}\boldsymbol{k}\cdot\boldsymbol{r}) = \psi_n(0, \boldsymbol{r}) \exp(\mathrm{i}\boldsymbol{k}, \boldsymbol{r}). \tag{2.163}$$

The wave packet then becomes

$$\psi = \sum_{k} c_n(\boldsymbol{k}, t) \exp(\mathrm{i}\boldsymbol{k}\cdot\boldsymbol{r})\psi_n(0, \boldsymbol{r}) \equiv F(\boldsymbol{r}, t)\psi_n(0, \boldsymbol{r}), \tag{2.164}$$

where the summation is limited to states with small $\boldsymbol{k}$. $F(\boldsymbol{r}, t)$ consists only of plane waves with large wavelength and is consequently a slowly varying function of $\boldsymbol{r}$. Making a rearrangement similar to (2.143), we then have

$$\begin{aligned} E_n(-\mathrm{i}\nabla)F(\boldsymbol{r}, t)\psi_n(0, \boldsymbol{r}) &= \sum_{m} E_{nm}F(\boldsymbol{r} + \boldsymbol{R}_m, t)\psi_n(0, \boldsymbol{r} + \boldsymbol{R}_m) \\ &= \psi_n(0, \boldsymbol{r}) \sum_{m} E_{nm} \exp(\boldsymbol{R}_m\cdot\nabla)F(\boldsymbol{r}, t) = \psi_n(0, \boldsymbol{r})E_n(-\mathrm{i}\nabla)F(\boldsymbol{r}, t) \end{aligned} \tag{2.165}$$

and (2.147) becomes

$$[E_n(-\mathrm{i}\nabla) - e\phi]F(\boldsymbol{r}, t) = -\frac{\hbar}{\mathrm{i}}\frac{\partial}{\partial t}F(\boldsymbol{r}, t). \tag{2.166}$$

Now we can expand $E_n(-\mathrm{i}\nabla)$ in (2.166), and it follows that

$$\left(-\frac{\hbar^2}{2m^*}\nabla^2 - e\phi\right)F(\boldsymbol{r}, t) = -\frac{\hbar}{\mathrm{i}}\frac{\partial}{\partial t}F(\boldsymbol{r}, t), \tag{2.167}$$

i.e., we find exactly an equation for a "free" electron of effective mass m^* and charge $-e$ in an electric field $-\nabla\phi$. For this approximation the concept of effective mass is therefore justified. Now, however, we can see its limitations. First the expression which leads to (2.147) neglects all field-induced band-to-band transitions. The functions (2.163) are only taken from a small part of a band. The wave packet is therefore only incompletely described. A limited number of expansion functions is sufficient only when the wave packet can be taken to stretch over an extended region (of the order of a few lattice constants). This approximation is only justified if there is little change in the electric field accelerating the wave packet over this region. Thus eq. (2.167) only properly describes *electrons in the immediate vicinity of the band edges in weak slowly varying fields*. One designates this approximation, which leads to (2.167), the *effective-mass approximation* and (2.167) the effective-mass equation.

Equation (2.167) can be extended to cover anisotropic effective masses [(2.162)] and added magnetic fields. The reasoning is then much more involved and the limits of validity more difficult to perceive. We shall only refer the reader to the literature, e.g., *Callaway* [9] and *McLean* [112.22].

In many later sections we shall take up and extend the results of this section. Semiconductors and their properties will serve time and again as examples when we discuss the interaction of electrons in a solid with other elementary excitations, particularly with phonons and photons.

2.2.13 Consequences of the Invariance of the Hamiltonian to Symmetry Operations of the Space Group

The function $E_n(\boldsymbol{k})$ possesses numerous symmetries within the Brillouin zone. In order to determine them we associate operators with the $\{\alpha|\boldsymbol{a}\}$ by the following definition:

$$S_{\{\alpha|\boldsymbol{a}\}}f(\boldsymbol{r}) = f(\{\alpha|\boldsymbol{a}\}^{-1}\boldsymbol{r})\,. \tag{2.168}$$

The operators $T_{\boldsymbol{R}_l}$ defined in Section 2.2.5 are connected with these operators by $T_{\boldsymbol{R}_l} = S_{\{E|-\boldsymbol{R}_l\}} = S^{-1}_{\{E|\boldsymbol{R}_l\}}$.

The group of the $S_{\{\alpha|\boldsymbol{a}\}}$ is isomorphic to the space group of the $\{\alpha|\boldsymbol{a}\}$

$$\begin{aligned} S_{\{\alpha|\boldsymbol{a}\}}S_{\{\beta|\boldsymbol{b}\}}f(\boldsymbol{r}) &= S_{\{\alpha|\boldsymbol{a}\}}f(\{\beta|\boldsymbol{b}\}^{-1}\boldsymbol{r}) = f(\{\beta|\boldsymbol{b}\}^{-1}\{\alpha|\boldsymbol{a}\}^{-1}\boldsymbol{r}) \\ &= f((\{\alpha|\boldsymbol{a}\}\{\beta|\boldsymbol{b}\})^{-1}\boldsymbol{r}) = S_{\{\alpha|\boldsymbol{a}\}\{\beta|\boldsymbol{b}\}}f(\boldsymbol{r})\,. \end{aligned} \tag{2.169}$$

From consideration of (2.71) it can further be seen that

$$S^{-1}_{\{\alpha|\boldsymbol{a}\}}S_{\{E|\boldsymbol{R}_l\}}S_{\{\alpha|\boldsymbol{a}\}} = S_{\{\alpha|\boldsymbol{a}\}^{-1}\{E|\boldsymbol{R}_l\}\{\alpha|\boldsymbol{a}\}} = S_{\{E|\alpha^{-1}\boldsymbol{R}_l\}} \tag{2.170}$$

or

$$S_{\{E|\boldsymbol{R}_l\}}S_{\{\alpha|\boldsymbol{a}\}} = S_{\{\alpha|\boldsymbol{a}\}}S_{\{E|\alpha^{-1}\boldsymbol{R}_l\}}\,. \tag{2.171}$$

Since the scalar product of two vectors is unchanged if both vectors are subjected to an orthogonal transformation ($\boldsymbol{k}\cdot\alpha^{-1}\boldsymbol{R}_l = \alpha\boldsymbol{k}\cdot\boldsymbol{R}_l$), we find that

$$\begin{aligned} S_{\{E|\boldsymbol{R}_l\}}S_{\{\alpha|\boldsymbol{a}\}}\psi_n(\boldsymbol{k},\boldsymbol{r}) &= S_{\{\alpha|\boldsymbol{a}\}}S_{\{E|\alpha^{-1}\boldsymbol{R}_l\}}\psi_n(\boldsymbol{k},\boldsymbol{r}) \\ &= S_{\{\alpha|\boldsymbol{a}\}}\exp(-\mathrm{i}\boldsymbol{k}\cdot\alpha^{-1}\boldsymbol{R}_l)\psi_n(\boldsymbol{k},\boldsymbol{r}) = S_{\{\alpha|\boldsymbol{a}\}}\exp(-\mathrm{i}\alpha\boldsymbol{k}\cdot\boldsymbol{R}_l)\psi_n(\boldsymbol{k},\boldsymbol{r}) \\ &= \exp(-\mathrm{i}\alpha\boldsymbol{k}\cdot\boldsymbol{R}_l)S_{\{\alpha|\boldsymbol{a}\}}\psi_n(\boldsymbol{k},\boldsymbol{r}) \end{aligned} \tag{2.172}$$

and

$$S_{\{E|\boldsymbol{R}_l\}}\psi_n(\alpha\boldsymbol{k}, \boldsymbol{r}) = \exp(-\mathrm{i}\alpha\boldsymbol{k}\cdot\boldsymbol{R}_l)\psi_n(\alpha\boldsymbol{k}, \boldsymbol{r}). \tag{2.173}$$

A comparison of (2.172) and (2.173) shows that the functions $\psi_n(\alpha\boldsymbol{k}, \boldsymbol{r})$ and $S_{\{\alpha|\boldsymbol{a}\}}\psi_n(\boldsymbol{k}, \boldsymbol{r})$ are eigenfunctions of $S_{\{E|\boldsymbol{R}_l\}}$ with the same eigenvalues. Hence

$$\psi_n(\alpha\boldsymbol{k}, \boldsymbol{r}) = \lambda^{\{\alpha|\boldsymbol{a}\}}S_{\{\alpha|\boldsymbol{a}\}}\psi_n(\boldsymbol{k}, \boldsymbol{r}), \quad |\lambda^{\{\alpha|\boldsymbol{a}\}}|^2 = 1. \tag{2.174}$$

Finally we have

$$\begin{aligned} E_n(\alpha\boldsymbol{k}) &= \langle\psi_n(\alpha\boldsymbol{k}, \boldsymbol{r})H\psi_n(\alpha\boldsymbol{k}, \boldsymbol{r})\rangle = \langle S_{\{\alpha|\boldsymbol{a}\}}\psi_n(\boldsymbol{k}, \boldsymbol{r})HS_{\{\alpha|\boldsymbol{a}\}}\psi_n(\boldsymbol{k}, \boldsymbol{r})\rangle \\ &= \langle S^{-1}_{\{\alpha|\boldsymbol{a}\}}S_{\{\alpha|\boldsymbol{a}\}}\psi_n(\boldsymbol{k}, \boldsymbol{r})H\psi_n(\boldsymbol{k}, \boldsymbol{r})\rangle = \langle\psi_n(\boldsymbol{k}, \boldsymbol{r})H\psi_n(\boldsymbol{k}, \boldsymbol{r})\rangle = E_n(\boldsymbol{k}) \end{aligned} \tag{2.175}$$

or in general

$$E_n(\boldsymbol{k}) = E_n(\alpha\boldsymbol{k}). \tag{2.176}$$

This important result states that the function $E_n(\boldsymbol{k})$ in the Brillouin zone possesses the full symmetry of the point group $\{\alpha|0\}$, even when the lattice is not invariant to some of the $\{\alpha|0\}$. This shows us the importance of the point group of a lattice, independent of the properties of the space group. This result also states that the Brillouin zone has the full symmetry of the point group.

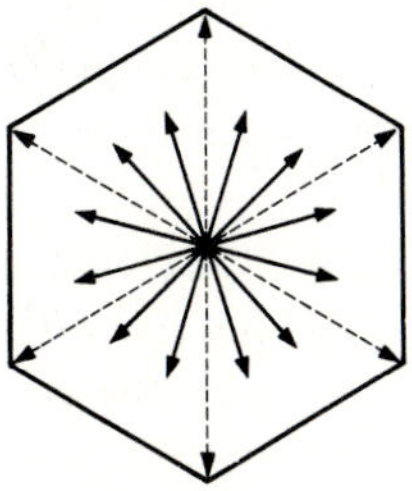

Fig. 2.34. Stars of two $\boldsymbol{k}$-vectors in the two-dimensional hexagonal point lattice.

It follows from (2.176) that all vectors $\boldsymbol{k}' = \alpha\boldsymbol{k}$ lead to the same energy. One designates the set of all $\boldsymbol{k}'$ as the *star of* $\boldsymbol{k}$. If all $\boldsymbol{k}' = \alpha\boldsymbol{k}$ are different vectors, one refers to $\boldsymbol{k}$ as a *general point* in the Brillouin zone. In this case the star of $\boldsymbol{k}$ contains as many prongs as the point group has elements. The points and lines of high symmetry in the Brillouin zone which are invariant to some of the $\{\alpha|0\}$ will be important in our later considerations. If the vector $\boldsymbol{k}$ is invariant to n of the g point group elements, its star has g/n prongs. Fig. 2.34 shows two stars for different $\boldsymbol{k}$-vectors in the Brillouin zone of the two-dimensional hexagonal point lattice.

We thus have the following symmetries for $E_n(\boldsymbol{k})$ (in the repeated zone scheme):

$$E_n(\boldsymbol{k}) = E_n(\boldsymbol{k} + \boldsymbol{K}_m) \tag{2.177}$$

$$E_n(\boldsymbol{k}) = E_n(-\boldsymbol{k}) \tag{2.178}$$

$$E_n(\boldsymbol{k}) = E_n(\alpha\boldsymbol{k}) . \tag{2.179}$$

Kramers' theorem (2.178) is contained in (2.179) if the point group contains the inversion I ($I\boldsymbol{k} = -\boldsymbol{k}$). Otherwise (2.178) represents an additional statement about the symmetries of the band structure function $E_n(\boldsymbol{k})$.

2.2.14 Irreducible Representations of Space Groups

Equations (2.177–179) contain an abundance of information on the symmetry properties of a band structure, particularly on the shape of surfaces of constant energy in the Brillouin zone. Along with qualitative statements, they also make possible a considerable simplification of the problem of quantitative determination of a band structure. They reduce the region in which the function $E_n(\boldsymbol{k})$ must be calculated to a portion of the Brillouin zone, the so called irreducible zone. The following questions are important in this context:

1) What symmetries have wave functions $|n\boldsymbol{k}\rangle$, and what degree of degeneracy have energy states $E_n(\boldsymbol{k})$ for given $\boldsymbol{k}$ and given lattice symmetry? The answer to this questions means a *classification of eigenvalues* similar to the classification of states in the free atom into *s*-, *p*-, *d*-states.

2) Suppose there is a degenerate energy level E_n at a given $\boldsymbol{k}$. Does the degeneracy remain when we go to an adjacent $\boldsymbol{k} + \boldsymbol{\kappa}$ or does the level split? This is related to the question whether in an atomic system of given symmetry, degeneracies are removed or not when a perturbation of lower symmetry is applied.

3) Suppose there are two states at energies $E_n(\boldsymbol{k})$ and $E_{n'}(\boldsymbol{k}')$. Are transitions possible between the two levels, or other interactions involving a matrix element $\langle n',\boldsymbol{k}'|L|n,\boldsymbol{k}\rangle$ (with given symmetry of the operator L), or do the two levels not combine for reasons of symmetry?

To answer such questions one needs some theorems from group theory, particularly those relating to irreducible representations of finite groups. We present now a short summary of those concepts which are most important for band theory. More detailed discussions are to be found in [43–49].

A set of matrices is called a *representation* of a group G if to each element A of G a matrix $D(A)$ corresponds so that if $AB = C$ it is also true that $D(A)D(B) = D(C)$. The correspondence need not be one to one; thus for $D(M)D(N) = D(P)$ it does not follow that $MN = P$.

We became familiar with such representations in Section 2.2.5. There a representation of the translation group was produced from a basis of f degenerate orthogonal eigenfunctions $\psi_{n\kappa}$. We saw that by transformation of the basis functions other equivalent representations were produced. Among these there was one representation which had only diagonal matrices.

One refers to a representation as *reducible* if it can be transformed into an equivalent representation in which all matrices take the form

$$\begin{Vmatrix} D^{(1)} & & & 0 \\ & D^{(2)} & & \\ & & D^{(3)} & \\ 0 & & & \cdots \end{Vmatrix}.$$

Apart from the "blocks" $D^{(i)}$ along the diagonal, all the matrix elements are zero. In this case the submatrices $D^{(i)}$ likewise are representations of the group. One says that D is reduced into a *direct sum* $D = D^{(1)} \oplus D^{(2)} \oplus D^{(3)} \oplus \ldots$ of representations of smaller dimension.

If the $D^{(i)}$ cannot be reduced further, they form *irreducible representations* of the group.

For a representation produced from a basis of degenerate eigenfunctions ψ_κ, this means that the equation

$$A\psi_\kappa = \sum_{\kappa'=1}^{f} D_{\kappa\kappa'}(A)\psi_{\kappa'}, \qquad \kappa = 1, \ldots, f \tag{2.180}$$

can by appropriate choice of new $\bar{\psi}_\kappa$ be put into the form

$$A\bar{\psi}_{\kappa_i} = \sum_{\kappa_i'=1}^{f_i} D^{(i)}_{\kappa_i\kappa_i'}\bar{\psi}_{\kappa_i'}, \qquad i = 1, \ldots, n, \quad \kappa_i = 1, \ldots, f_i, \quad \sum_i f_i = f. \tag{2.181}$$

The set of f degenerate basis functions $\bar{\psi}_\kappa$ is therefore divided into n subsets $\bar{\psi}_{\kappa_i}$ such that the operations A of the group transform a $\bar{\psi}_{\kappa_i}$ into a linear combination of $\bar{\psi}_{\kappa_i}$ only. As far as the operations of the group are concerned, only the $\bar{\psi}_{\kappa_i}$ are degenerate with each other.

Expressing things in this way, the transformation (2.85) and (2.88) of the representation of the translation group means the reduction of the representation λ to a direct sum of one-dimensional representations $\Lambda_{\kappa\kappa}$. That this transformation is possible follows from a theorem of group theory: an abelian group (i.e., a group whose elements commute: $AB = BA$) possesses only one-dimensional irreducible representations.

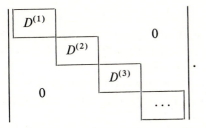

For the questions posed at the start of this section it is therefore important to know the irreducible representations of the symmetry group which belongs to a state $E_n(\boldsymbol{k})$ with a given $\boldsymbol{k}$. We need two further concepts for this.

We understand a *class* to be all the elements A of a group which are produced from one element A' by the formation of products $A = X^{-1}A'X$. X here runs through all elements of the group. One can show that each group can be uniquely resolved into classes.

We understand the *character* χ of a representation matrix to be its trace, i.e., the sum of all diagonal elements: $\chi(A) = \sum_i D_{ii}(A)$. All the matrices in a class have the same trace and therefore the same character.

The theory of group representations then states:

1) Every reducible representation can be transformed for all A into the direct sum $D(A) = n_1 D^{(1)}(A) \oplus n_2 D^{(2)}(A) \oplus n_3 D^{(3)}(A) \oplus \ldots$, where the $D^{(i)}$ are irreducible and the n_i are integers.

2) Every finite group of order g (with g elements) has a finite number of irreducible representations. This number equals the number of its classes.

3) The sum of the squares of the dimensions of the irreducible representations is equal to the order of the group

$$\sum_\alpha n_\alpha^2 = g \, . \tag{2.182}$$

4) A necessary and sufficient condition for a representation to be irreducible is

$$\sum_A |\chi(A)|^2 = g \, . \tag{2.183}$$

5) For the characters of the irreducible representations, the following "orthogonality relations" are valid:

$$\sum_A \chi_\alpha^*(A)\chi_\beta(A) = g\delta_{\alpha\beta} \, , \qquad \sum_\alpha \chi_\alpha^*(A)\chi_\alpha(A') = \frac{g}{h_A}\delta_{AA'} \, , \tag{2.184}$$

where the indices α, β indicate different representations of the same group and h_A is the number of elements in the class of A.

These theorems allow the determination of the number of irreducible representations of a (finite) group, the specification of all characters of all irreducible representations, and the determination of the symmetry properties of the basis functions which produce these representations. All the questions posed earlier in this section can then in principle be answered.

The group of the *translation operators* $T_{\boldsymbol{R}_l}$ is finite and of order N (for cyclic boundary conditions). Since the $T_{\boldsymbol{R}_l}$ can be interchanged, the group is abelian. Since $T_{\boldsymbol{R}_m}^{-1} T_{\boldsymbol{R}_l} T_{\boldsymbol{R}_m} = T_{\boldsymbol{R}_l}$ for any $T_{\boldsymbol{R}_m}$, every element forms a class of its own. There are N classes, therefore also N irreducible representations.

One can readily determine the form of these representations when one notes that, since $\boldsymbol{R}_l = l_1\boldsymbol{a}_1 + l_2\boldsymbol{a}_2 + l_3\boldsymbol{a}_3$, $T_{\boldsymbol{R}_l} = T^{l_1}_{\boldsymbol{a}_1}T^{l_2}_{\boldsymbol{a}_2}T^{l_3}_{\boldsymbol{a}_3}$. The translation group can therefore also be taken as a product of the groups of the "translations in the direction of the $\boldsymbol{a}_i$". Each such group has N_i elements $T^{l_i}_{\boldsymbol{a}_i}$ with $l_i = 0, 1, \dots N_i - 1$. The cyclic boundary conditions require $(T^{l_i}_{\boldsymbol{a}_i})^{N_i} = E$. Since the irreducible representations of the $T^{l_i}_{\boldsymbol{a}_i}$ are onedimensional and therefore numbers, for them $D(T^{l_i}_{\boldsymbol{a}_i})^{N_i} = 1$. This is fulfilled by the N_i different possibilities

$$D(T^{l_i}_{\boldsymbol{a}_i}) = [\exp(2\pi i n_i/N_i)]^{l_i}, \qquad (2.185)$$

with $n_i = 1, \dots N_i$ and

$$D(T_{\boldsymbol{R}_l}) = \exp\left[2\pi i\left(\frac{n_1 l_1}{N_1} + \frac{n_2 l_2}{N_2} + \frac{n_3 l_3}{N_3}\right)\right]. \qquad (2.186)$$

From (2.127) and (2.77) this is identical to

$$D(T_{\boldsymbol{R}_l}) = \exp(i\boldsymbol{k}\cdot\boldsymbol{R}_l), \qquad (2.187)$$

where $\boldsymbol{k}$ can take $N = N_1N_2N_3$ different values. The Bloch factors $\exp(i\boldsymbol{k}\cdot\boldsymbol{R}_l)$ are thus nothing other than the irreducible representations of the translation group.

We turn now to the irreducible representations of the space groups, limiting ourselves, however, to only the most important facts, which will be needed for general statements.

In connection with (2.176), we introduced the *star* of a vector $\boldsymbol{k}$ as the set of vectors $\boldsymbol{k}_i$ which is produced from one vector $\boldsymbol{k}$ by application of the elements of the point group $\{\alpha|0\}$. If the point group has g elements and the star n different $\boldsymbol{k}_i = \alpha\boldsymbol{k}$, the point group can be divided into n sets of g/n elements which convert $\boldsymbol{k}$ into a certain $\boldsymbol{k}_i$: $\beta_i\boldsymbol{k} = \boldsymbol{k}_i$. This is only true for points inside the Brillouin zone. On the surface of the Brillouin zone there are pairs of equivalent points which differ by a lattice vector $\boldsymbol{K}_m$. The condition for the β_i is therefore more exact: $\beta_i\boldsymbol{k} = \boldsymbol{k}_i + \boldsymbol{K}_m$. Among the sets $\{\beta_i|0\}$ is a set $\{\beta|0\}$ which, apart from a lattice vector $\boldsymbol{K}_m$, leaves $\boldsymbol{k}$ invariant. We define as the *group of the vector* $\boldsymbol{k}$ the operations $\{\beta|\boldsymbol{b}\}$ from $\{\alpha|\boldsymbol{a}\}$, for whose rotation component

$$\beta\boldsymbol{k} = \boldsymbol{k} + \boldsymbol{K}_m. \qquad (2.188)$$

We give as examples: a) For a general point of the Brillouin zone the group of $\boldsymbol{k}$ only contains the primitive translations. b) For $\boldsymbol{k} = 0$, $\boldsymbol{k}$ is invariant to all $\{\alpha|0\}$ and the group of $\boldsymbol{k}$ is the full space group.

If $\boldsymbol{k} \neq 0$ the group of the vector $\boldsymbol{k}$ is always a subgroup of the space group. Other elements of the space group, produced from the subgroup by applying a rotation β_i, are then allotted to the vector $\boldsymbol{k}_i$ of the stars.

For points inside the Brillouin zone the irreducible representations of the group of the vector $\boldsymbol{k}$ are given by

$$D(\{\beta|\boldsymbol{b}\}) = \exp(\mathrm{i}\boldsymbol{k}\cdot\boldsymbol{b})D(\beta)\,, \tag{2.189}$$

where $D(\beta)$ is an irreducible representation of the point group β, and $\boldsymbol{b}$ covers primitive and nonprimitive translations. The proof is straightforward when we remember that then $\boldsymbol{k}\cdot\boldsymbol{b}' = \beta^{-1}\boldsymbol{k}\cdot\boldsymbol{b}' = \boldsymbol{k}\cdot\beta\boldsymbol{b}'$.

$$\begin{aligned} D(\{\beta|\boldsymbol{b}\})D(\{\beta'|\boldsymbol{b}'\}) &= \exp(\mathrm{i}\boldsymbol{k}\cdot\boldsymbol{b})\exp(\mathrm{i}\boldsymbol{k}\cdot\boldsymbol{b}')D(\beta)D(\beta') \\ &= \exp(\mathrm{i}\boldsymbol{k}\cdot(\boldsymbol{b} + \beta\boldsymbol{b}'))D(\beta\beta') = D(\{\beta\beta'|\boldsymbol{b} + \beta\boldsymbol{b}'\}) = D(\{\beta|\boldsymbol{b}\}\{\beta'|\boldsymbol{b}'\})\,. \end{aligned} \tag{2.190}$$

Thus the matrices (2.189) satisfy the same multiplication rules as the space group itself. From the assumption of the irreducibility of the $D(\beta)$ it follows that the representation is irreducible.

For $\boldsymbol{k}$-vectors on the surface of the Brillouin zone ($\boldsymbol{K}_m \neq 0$), the proof can only be carried out when the space group is symmorphic, i.e., when it contains no nonprimitive translations. For then $\boldsymbol{k}\cdot\boldsymbol{R}_l = \beta^{-1}(\boldsymbol{k} + \boldsymbol{K}_m)\cdot\boldsymbol{R}_l = \boldsymbol{k}\cdot\beta\boldsymbol{R}_l + \boldsymbol{K}_m\cdot\beta\boldsymbol{R}_l = \boldsymbol{k}\cdot\beta\boldsymbol{R}_l +$ a multiple of 2π.

For nonsymmorphic groups there are complications which go beyond the bounds of the treatment presented here. In connection with this and other questions in this section see in particular *Koster* [101.5].

Equation (2.189) gives the irreducible representations for all $\{\beta|\boldsymbol{b}\}$ of the space group which belong to the group of vector $\boldsymbol{k}$. The $\{\alpha|\boldsymbol{a}\}$ which are not included in this belong to operations α of the point group which convert $\boldsymbol{k}$ into another $\boldsymbol{k}_i$ of the star. The irreducible representations belonging to these $\{\alpha|\boldsymbol{a}\}$ can be expressed by the $D(\{\beta|\boldsymbol{b}\})$. We do not want to enter into this here since we do not need it for later discussions. For us it is only important that all irreducible representations of the space group for a given vector $\boldsymbol{k}$ (with the exception of the special case noted above) are defined by the irreducbile representations of the translation group $\exp(\mathrm{i}\boldsymbol{k}\cdot\boldsymbol{R}_l)$ and the point group belonging to $\boldsymbol{k}$.

The general statements of the previous sections can now be extended by the following special statements about the connection of the irreducible representations of a space group with the band structure of a solid having the symmetries of the space group:

1) There is an irreducible representation associated with every state $E_n(\boldsymbol{k})$. The dimensions of the possible irreducible representations for a given $\boldsymbol{k}$ give the possible degeneracies for this $\boldsymbol{k}$.

2) Since the eigenfunctions belonging to a state $E_n(\boldsymbol{k})$ can serve as basis functions for the associated irreducible representation, the transformation properties, i.e., the symmetries of the $\psi_n(\boldsymbol{k}, \boldsymbol{r})$, follow from knowledge of the irreducible representations.

3) If one goes from a point $\boldsymbol{k}$ to an adjacent point $\boldsymbol{k} + \boldsymbol{\kappa}$ of lesser symmetry, an irreducible representation in $\boldsymbol{k}$ can become reducible in $\boldsymbol{k} + \boldsymbol{\kappa}$. A degenerate level will then split in going from $\boldsymbol{k}$ to $\boldsymbol{k} + \boldsymbol{\kappa}$.

In presentations of the band structure of a solid, one finds this information in the form of symbols (e.g., Γ_2, Δ_5, . . .). The letters here give the group of the vector $\boldsymbol{k}$ (see Figs. 2.22 and 2.31); the indices indicate the respective irreducible representation. These symbols thus provide much information on the symmetries and degeneracies of the wave functions of the eigenvalue concerned.

We want to illustrate this with an example: the symmetries of the two-dimensional hexagonal point lattice which we have used in earlier examples (Figs. 2.11, 2.14, 2.15, 2.17, and 2.18). We can now answer the following question: In Fig. 2.18 we compared the band structures for free and almost free electrons in the two-dimensional hexagonal point lattice and found that degeneracies of the bands in the free electron case are removed by the lattice potential. To what extent can we predict from symmetry considerations which degeneracies will be removed and how many individual components a degenerate level will split into? The space group of the two-dimensional hexagonal lattice consists of the primitive translations $\{E|n_1\boldsymbol{a}_1 + n_2\boldsymbol{a}_2\}$ where the two $\boldsymbol{a}_i$ are defined in Fig. 2.11. To these are added the twelve operations of the point group (designation C_{6v}) which leave a hexagon invariant. Nonprimitive translations do not arise. The space group is therefore symmorphic: $\{\alpha|\boldsymbol{R}_n\} = \{\alpha|0\}\{E|\boldsymbol{R}_n\}$. The operations α are (ordered in classes)

- E unit operation
- C_2 rotation by 180° about the midpoint of the hexagon
- C_3 two rotations by $\pm 120°$
- C_6 two rotations by $\pm 60°$
- σ three reflections at the lines through two opposite corners
- σ' three reflections at the lines through the midpoints of two opposite sides.

Altogether that makes twelve operations. Since the space group is symmorphic we need only pay attention to the irreducible representations of the point group $\{\alpha|0\}$. The translations give the Brillouin zone, the form of which we already know, and the $\boldsymbol{k}$-points it contains. The point group has six classes, hence also six irreducible representations. On account of $\sum_\alpha n_\alpha^2 = 12$ [(2.182)], with integer n_α there are two two-dimensional and four one-dimensional irreducible representations ($2^2 + 2^2 + 1^2 + 1^2 + 1^2 + 1^2 = 12$).

From equations (2.182)–(2.184) the characters of the irreducible representations may easily be determined. They are arranged in matrix-form in so-called *character tables*. For a compilation for all known space groups see e.g. [47] and the article by *Koster* in [101.5]. We do not need the character table for our example since we are only interested in the degeneracy of the bands and the band-connections along the most important symmetry axes. In this simple example all informations are already given:

The group of the $\boldsymbol{k}$-vector for the midpoint of the Brillouin zone Γ ($\boldsymbol{k} = 0$) is the full space group. Thus the associated point group has the six irreducible representations deduced above. We indicate the four one-dimensional irreducible representations by Γ_1, Γ_2, Γ_3, Γ_4 and the two two-dimensional ones by Γ_5 and Γ_6. The levels $E_n(\Gamma)$ are classified according to the six possible types of symmetry: levels Γ_1 to Γ_4 are non-degenerate, levels Γ_5 and Γ_6 are twofold symmetry-degenerate. Of course, the degeneracy may be higher in an actual case on account of other symmetry requirements (time reversal degeneracy, see next section) or by numerical coincidence (accidental degeneracy).

Let us now go to the T-axis (see Figs. 2.18 and 2.35 for the nomenclature). A $\boldsymbol{k}$-vector along T has as group the element E and one σ. That means two classes and two irreducible representations. There are therefore only non-degenerate bands along T. A $\boldsymbol{k}$-vector at the end of the T-axis (point K) has as group the symmetry operations E, $C_3(2)$, and $\sigma(3)$. Four of these operations take $\boldsymbol{k}$ to another corner, to a $\boldsymbol{k}' = \boldsymbol{k} + \boldsymbol{K}_m$. That gives three classes and as many irreducible representations (two one-dimensional and one two-dimensional). For all other points in the Brillouin zone there are only one-dimensional irreducible representations. Thus in our simple example, symmetry-degenerate levels occur only at Γ and at K. Bands which originate from such levels split up on the axes.

An analysis of the symmetry properties of the wave functions at Γ and K then shows the following in the free-electron case (Figs. 2.18 and 2.35): For free electrons, numerous bands are degenerate with each other at points of symmetry. At point Γ the lowest eigenvalue $E_1(\Gamma)$ is single, the next $[E_2(\Gamma)]$ six-fold

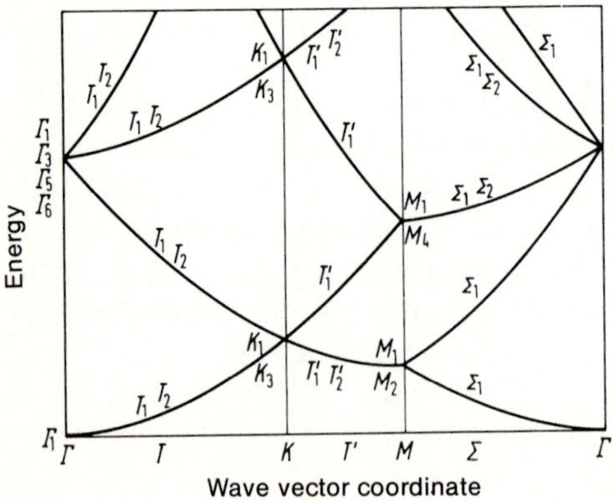

Fig. 2.35. Symmetry classification of the bands of free electrons in the two-dimensional hexagonal point lattice.

degenerate. $E_1(\Gamma)$ is of type Γ_1, $E_2(\Gamma)$ is a superposition of Γ_1, Γ_3, Γ_5, and Γ_6. At K the two lowest eigenvalues $E(K)$ are each triply degenerate. Through a perturbation, i.e., through a finite lattice potential, $E_2(\Gamma)$ splits into two single and two double bands, and $E(K)$ into one double and one single. All remaining degenerate levels at Γ and K then split up likewise in going away from these points, as is shown in Fig. 2.18. In this figure the assignments at the points of degeneracy are, however, arbitrary. Which bands are connected with each other, where accidental degeneracies perhaps remain, in which order the split levels at points of symmetry come to lie—all these questions are decided by the quantitative behaviour of the potential. They are answered only by numerical calculations.

2.2.15 Spin, Time Reversal

So far we have neglected spin in the Schrödinger equation for the electron in a periodic potential. Introducing it at first the number of levels $E_n(\boldsymbol{k})$ doubles because each $\boldsymbol{k}$ can then be occupied twice. These degeneracies can in part split again as a result of spin-orbit coupling.

To determine quantitatively the corrections to the analysis carried out so far, we supplement the Hamiltonian with the *spin-orbit coupling term*

$$\Sigma = \frac{\hbar^2}{4im^2c^2}\boldsymbol{\sigma}\cdot(\nabla V \times \nabla)\,. \tag{2.191}$$

Σ operates on spinors $\psi(\boldsymbol{r}, s)$ with the gradient in the parentheses operating on the space component and the spin operator $\boldsymbol{\sigma}$ on the spin component. If we write the $\psi(\boldsymbol{r}, s)$ as two-component spinors

$$\psi(\boldsymbol{r}, s) = \begin{vmatrix}\psi_1(\boldsymbol{r})\\ \psi_2(\boldsymbol{r})\end{vmatrix}, \tag{2.192}$$

the components of σ have the form

$$\sigma_x = \begin{vmatrix}0 & 1\\ 1 & 0\end{vmatrix}, \quad \sigma_y = \begin{vmatrix}0 & -i\\ i & 0\end{vmatrix}, \quad \sigma_z = \begin{vmatrix}1 & 0\\ 0 & -1\end{vmatrix}. \tag{2.193}$$

With $\varepsilon = \begin{vmatrix}1 & 0\\ 0 & 1\end{vmatrix}$, the complete Hamiltonian takes the form

$$H = \left[-\frac{\hbar^2}{2m}\nabla^2 + V(\boldsymbol{r})\right]\varepsilon + \frac{\hbar^2}{4im^2c^2}\boldsymbol{\sigma}\cdot(\nabla V \times \nabla)\,. \tag{2.194}$$

The elements of the space group have now to be extended likewise with operators which operate on the spin component. The operator $S_{\{\alpha|\boldsymbol{a}\}}$ does not commute with (2.191). It is therefore not sufficient to supplement the $S_{\{\alpha|\boldsymbol{a}\}}$ (like the spin-free terms of the Hamiltonian) with a factor ε. For this purpose we seek an operator D_α with the property that the products $D_\alpha S_{\{\alpha|a\}}$ commute with H. These products form the elements of the new space group, designated as the *double group*. The condition imposed on D_α is therefore

$$(D_\alpha S_{\{\alpha|a\}})H(\boldsymbol{r})(D_\alpha S_{\{\alpha|a\}})^{-1} = D_\alpha H(\{\alpha|\boldsymbol{a}\}\boldsymbol{r})D_\alpha^{-1} = H(\boldsymbol{r})\,. \tag{2.195}$$

This is satisfied by the matrix

$$D^{(1/2)} = \begin{pmatrix} \cos(\theta/2)\exp\left[-\frac{\mathrm{i}}{2}(\chi+\varphi)\right] & -\sin(\theta/2)\exp\left[-\frac{\mathrm{i}}{2}(\chi-\varphi)\right] \\ \sin(\theta/2)\exp\left[\frac{\mathrm{i}}{2}(\chi-\varphi)\right] & \cos(\theta/2)\exp\left[\frac{\mathrm{i}}{2}(\chi+\varphi)\right] \end{pmatrix}, \tag{2.196}$$

in which θ, χ, φ are the Euler angles of the rotation α. $D_\alpha^{(1/2)}$ is a special irreducible representation of the full rotation group. The matrices (2.196) have the property that a rotation by 2π about any axis transforms $D_\alpha^{(1/2)}$ into $-D_\alpha^{(1/2)}$. Then $D_\alpha^{(1/2)}S_{\{\alpha|a\}}$ and $-D_\alpha^{(1/2)}S_{\{\alpha|a\}}$ satisfy (2.195) and both can be taken as different elements of the extended group. The double group has therefore twice as many elements as the single space group. One can take the newly added elements as products of the old elements with the operation $\{\bar{E}|0\}$ (rotation by 2π). $\{\bar{E}|0\}$ changes the sign of $D_\alpha^{(1/2)}$. A rotation by 4π is needed for equivalence with the application of the unit element $\{E|0\}$.

With the number of elements, the dimensions of the irreducible representations of the single group double. As representations of the double group, however, these representations can be reducible and resolve into new irreducible representations of smaller dimension (*extra representations*). This corresponds to the fact that the energy eigenvalues, whose degeneracy is doubled by the introduction of the spin, may split by spin-orbit coupling, into separate eigenvalues of lower degeneracy.

We proved earlier that in the spin-free case $\psi(\boldsymbol{k}, \boldsymbol{r})$ is degenerate with $\psi(-\boldsymbol{k}, \boldsymbol{r})$ and $E(\boldsymbol{k})$ is degenerate with $E(-\boldsymbol{k})$. We now want to generalize this result. To this end we introduce the *time-reversal operator* K. It reverses the state of motion of a system. In the time-independent Schrödinger equation this means that K leaves the space operator invariant but reverses the sign of the momentum and the spin operators

$$K\boldsymbol{r}K^{-1} = \boldsymbol{r}\,, \qquad K\boldsymbol{p}K^{-1} = -\boldsymbol{p}\,, \qquad K\boldsymbol{\sigma}K^{-1} = -\boldsymbol{\sigma}\,. \tag{2.197}$$

This is satisfied by

$$K = -\mathrm{i}\sigma_y K_0\,, \tag{2.198}$$

where K_0 is defined by $K_0\psi = \psi^*$ and σ_y by (2.193).

Since $\boldsymbol{p} = (\hbar/\mathrm{i})\nabla$, K_0 causes $\boldsymbol{p}$ to change sign, while σ_y causes the spin to change sign ($\sigma_y \boldsymbol{\sigma} \sigma_y^{-1} = -\boldsymbol{\sigma}^*$). From (2.197) K leaves the Hamiltonian invariant $KH\psi = HK\psi$. $K\psi$ is therefore degenerate with ψ. For the spin-free case ($K = K_0$) this means that ψ is degenerate with ψ^*. With spin we have

$$K\begin{vmatrix}\psi\\0\end{vmatrix} = \begin{vmatrix}0\\\psi^*\end{vmatrix}, \quad K\begin{vmatrix}0\\\psi\end{vmatrix} = -\begin{vmatrix}\psi^*\\0\end{vmatrix}. \tag{2.199}$$

It follows from (2.199) that K reverses the spin and transforms the Bloch functions $\psi(\boldsymbol{k}, \boldsymbol{r})$ into $\psi^*(\boldsymbol{k}, \boldsymbol{r})$. The same argument as in (2.130) then shows that $\psi(\boldsymbol{k}, \boldsymbol{r})$ with "spin up" and $\psi(-\boldsymbol{k}, \boldsymbol{r})$ with "spin down" are degenerate. Eq. (2.130) [$E(\boldsymbol{k}) = E(-\boldsymbol{k})$] is therefore also valid here, with the extension that the two eigenvalues belong to states with opposite spins. This is the full content of Kramers' theorem mentioned at the end of Section 2.2.8.

"Time-reversal degeneracy" is not limited to this example. Eigenvalues which need not be degenerate by reasons of symmetry can be degenerate by time reversal.

2.2.16 Pseudopotentials

In Sections 2.2.4 and 2.2.6 we found our way to the band model through Bragg reflections. The continuous $E(\boldsymbol{k})$ spectrum of the free electron splits up into bands in the periodic potential of the lattice ions. Our approach in Section 2.2.6, however, was limited to weak potentials. In the Schrödinger equation $V(\boldsymbol{r})$ was only treated as a small perturbation in this approximation. The band structure followed from a first-order perturbation calculation as the solution of the secular determinant

$$\det\left|\left[\frac{\hbar^2}{2m}(\boldsymbol{k} + \boldsymbol{K}_n)^2 - E(\boldsymbol{k})\right]\delta_{mn} + V(\boldsymbol{K}_n - \boldsymbol{K}_m)\right| = 0\,. \tag{2.200}$$

$V(\boldsymbol{K}_i)$ is the ith Fourier component of the potential. Eq. (2.200) is a generalization of the analysis limited to two $\boldsymbol{K}_i$ in Section 2.2.6 [(2.101)].

In fact the lattice potential is not weak enough to be treated as a small perturbation. Accordingly the wave function $\psi_n(\boldsymbol{k}, \boldsymbol{r})$ does not have the character of a plane wave. From (2.92), ψ can always be described as a Bloch function, i.e., as a plane wave with lattice periodicity. The modulation factor $u_n(\boldsymbol{k}, \boldsymbol{r})$, however, will oscillate strongly like a wave function of the free atom near the lattice ions. On the other hand, in the region of weak potential between the ions, the Bloch functions will behave like plane waves.

It is not appropriate therefore to use an expression of the form (2.97), in which only plane waves are superimposed, since far too many plane waves would have to be taken to represent the Bloch functions. To remedy this deficiency, we start from the following line of thought. We divide the bands of a solid into two groups, the low-lying bands of core electrons and the valence and conduction bands. We assume that the former are relatively narrow and that their position is not much changed from the corresponding levels in the free atom. To a good approximation we can then take the states in these bands to be the core states of the free atom. The latter group however is the one which really interests us, and our goal is to calculate the eigenvalues of the Schrödinger equation for these bands $E_n(\boldsymbol{k})$.

The Bloch functions of the valence and conduction bands and the wave functions of the core states, as solutions of the same Schrödinger equation, must be orthogonal to each other. If we represent the core states by φ_j and their energy by $E_j(H\varphi_j = E_j\varphi_j)$, the condition for orthogonality

$$\langle\varphi_j|\psi\rangle = 0 \tag{2.201}$$

will be satisfied by the ansatz

$$\psi_n(\boldsymbol{k}, \boldsymbol{r}) = \chi_n(\boldsymbol{k}, \boldsymbol{r}) - \sum_j \langle\varphi_j|\chi_n\rangle\varphi_j\,, \tag{2.202}$$

for then

$$\langle\varphi_{j'}|\psi\rangle = \langle\varphi_{j'}|\chi\rangle - \sum_j \langle\varphi_j|\chi\rangle\langle\varphi_{j'}|\varphi_j\rangle = \langle\varphi_{j'}|\chi\rangle - \sum_j \langle\varphi_j|\chi\rangle\delta_{jj'} = 0\,. \tag{2.203}$$

If we choose plane waves for the $\chi_n(\boldsymbol{k}, \boldsymbol{r})$, (2.202) is called an *orthogonalized plane wave* (OPW). An expression of the form (2.97) with such OPWs can serve as the foundation for a quantitative determination of band structure. We return to this at the end of the section.

For the moment we leave open the form of the functions χ. By putting (2.202) into the Schrödinger equation, we have as the equation defining the $\chi_n(\boldsymbol{k}, \boldsymbol{r})$

$$H\chi_n - \sum_j \langle\varphi_j|\chi_n\rangle H\varphi_j = E_n(\boldsymbol{k})(\chi_n - \sum_j \langle\varphi_j|\chi_n\rangle\varphi_j) \tag{2.204}$$

or

$$H\chi_n + \sum_j [E_n(\boldsymbol{k}) - E_j]\varphi_j\langle\varphi_j|\chi_n\rangle = E_n(\boldsymbol{k})\chi_n\,. \tag{2.205}$$

If we introduce the integral operator $V_\mathrm{p} = \sum_j (E_n(\boldsymbol{k}) - E_j)\varphi_j\langle\varphi_j|$ we have the formally simple equation

$$(H + V_\mathrm{p})\chi_n = [H_0 + V(\boldsymbol{r}) + V_\mathrm{p}]\chi_n = E_n(\boldsymbol{k})\chi_n\,. \tag{2.206}$$

Compared with the original equation, the Bloch functions $\psi_n(\boldsymbol{k}, \boldsymbol{r})$ are now replaced by the *pseudo wave functions* $\chi_n(\boldsymbol{k}, \boldsymbol{r})$ and the potential $V(\boldsymbol{r})$ by the *pseudopotential* $V_{ps} = V(\boldsymbol{r}) + V_p$. The additional potential V_p will give a positive contribution since $E_n(\boldsymbol{k})$ is always larger than E_j. Since on the other hand $V(\boldsymbol{r})$ is negative, the contributions of the two terms partly compensate. The pseudopotential is therefore weaker than the actual potential. In any case it is nonlocal. V_p is an integral operator and in the $V_p\chi$ the wave function occurs in the integrand. We had previously a similar difficulty with the periodic potential in which nonlocal exchange terms appear. As in Section 1.3 we must therefore replace the nonlocal terms by a local approximation.

In (2.202) χ can be supplemented by any linear combination $\sum_i a_i\varphi_i$ without altering the left-hand side, for additive terms with core functions combined in any way always disappear on the right-hand side of (2.202). The pseudopotential and the associated wave function are therefore not uniquely determined. This helps greatly to simplify the solution of the pseudopotential wave equation. By a variational technique we can determine the coefficients a_i in the additive part of χ in such a way that either the pseudopotential is as small as possible or the pseudo wave functions are as slowly varying as possible. The two requirements are approximately equivalent. The wave functions optimized in this way can then be approximated by a few terms of a superposition of plane waves (2.97).

With the introduction of the pseudopotential we have achieved the following advance. The Hamiltonian $H = H_0 + V$ has been replaced by a new Hamiltonian $H_{ps} = H_0 + V_{ps}$ which abandons the lower lying bands as of no interest to us but provides the *same eigenvalues* $E_n(\boldsymbol{k})$ as H in the region of the valence and conduction bands. The associated wave functions χ are however slowly varying and accordingly can be better approximated to by plane waves.

If we put the expression (2.97) for χ we arrive at a secular equation (2.200) in which the $V(\boldsymbol{K}_n - \boldsymbol{K}_m)$ are now Fourier components of the pseudopotential,

$$V_{ps}(\boldsymbol{K}_n - \boldsymbol{K}_m) = \frac{1}{V_g}\int \exp(-\mathrm{i}(\boldsymbol{k} + \boldsymbol{K}_n)\cdot\boldsymbol{r})V_{ps}\exp(\mathrm{i}(\boldsymbol{k} + \boldsymbol{K}_m)\cdot\boldsymbol{r})d\tau\,. \qquad (2.207)$$

In this form V_{ps} is still an integral operator. If we approximate to it with a local function $V_{ps}(\boldsymbol{r})$, we find

$$V_{ps}(\boldsymbol{q}) = \frac{1}{V_g}\int V_{ps}(\boldsymbol{r})\exp(-\mathrm{i}\boldsymbol{q}\cdot\boldsymbol{r})d\tau \qquad (\boldsymbol{q} = \boldsymbol{K}_n - \boldsymbol{K}_m)\,. \qquad (2.208)$$

It is useful to resolve $V_{ps}(\boldsymbol{r})$ into contributions from the individual lattice ions. For several basis atoms in the Wigner-Seitz cell, which we distinguish by the translations $\boldsymbol{R}_\alpha$, let

$$V_{ps}(\boldsymbol{r}) = \sum_{\alpha n} v_\alpha(\boldsymbol{r} - (\boldsymbol{R}_n + \boldsymbol{R}_\alpha))\,. \qquad (2.209)$$

$V_{ps}(\boldsymbol{q})$ can then be written as

$$V_{ps}(\boldsymbol{q}) = \frac{1}{V_g} \sum_{\alpha n} \int v_\alpha(\boldsymbol{r} - \boldsymbol{R}_n - \boldsymbol{R}_\alpha) \exp(-i\boldsymbol{q} \cdot \boldsymbol{r}) d\tau$$
$$= \frac{N}{V_g} \sum_\alpha \exp(-i\boldsymbol{q} \cdot \boldsymbol{R}_\alpha) \int \exp(-i\boldsymbol{q} \cdot \boldsymbol{r}) v_\alpha(\boldsymbol{r}) d\tau = \sum_\alpha \exp(-i\boldsymbol{q} \cdot \boldsymbol{R}_n) F_\alpha . \tag{2.210}$$

The factor F_α is called the *form factor*. If all basis atoms are identical, F_α is independent of α. $V_{ps}(\boldsymbol{q})$ is then the product of a *structure factor*, involving only the lattice symmetry, and a form factor, involving only the potential.

When the pseudopotential is sufficiently smooth, only Fourier coefficients with small $\boldsymbol{q}$ are significant. The determinant (2.200) then has small dimension. Band structures may be calculated by this method if one knows the form factors F_α. There are three possible ways to determine them:

1) One calculates the pseudopotential directly from the true potential. This method is then equivalent to the superposition of OPWs referred to above, and there is no advantage in introducing a pseudopotential.
2) The form factors are determined by empirical matching to experimental data.
3) One uses *model potentials*, i.e., simple expressions for the potentials with one or more free parameters which are later fixed by matching.

We do not want to go into the details here and refer to the literature cited below. As an example, Fig. 2.36 shows a model potential V_{ps} in which the potential was replaced by a constant value within a given distance of the lattice ions and the distance chosen so that the pseudo wave function at these places agrees with the Bloch function.

In addition to the obvious simplification of the Schrödinger equation resulting from the introduction of pseudopotential, these results show that it is no surprise if valence electrons in metals behave approximately like free electrons. Of course the basis of the approximation used here must be satisfied, namely that all the electrons in the metal can be clearly divided into core electrons and valence electrons. If higher lying *d*-bands occur, these methods, in the form presented here, break down.

The book by *Harrison* [58] is the best source of further information on the concept of pseudopotentials and its application in many areas of solid-state physics. Anothes useful reference is *Heine* et al. in [101.24].

Both the *OPW method* and the *pseudopotential method*, with empirical matching of form factors to experiment, play a major role in the quantitative calculation of band structures. We cannot go into this set of problems here. For presentations in that area we would refer the reader to *Ziman* [101.26], *Callaway* [55], *Louks* [61], *Treusch* and *Rössler* in [111c.35], *Ziman* in [113a], the review articles in [101.24], and the conference reports [115].

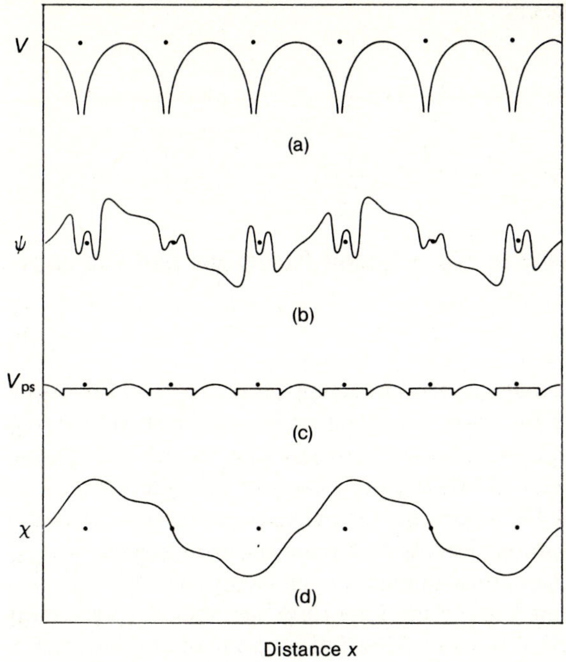

Fig. 2.36a-d. Comparison of (a) a given potential and (b) the associated wave function with (c) a model pseudopotential and (d) the corresponding pseudowave function (from *Harrison* [13]).

3. Elementary Excitations

3.1 The Interacting Electron Gas: Quasi-Electrons and Plasmons

3.1.1 Introduction

In the last chapter we considered the noninteracting electron gas and electrons in a periodic potential. In both cases the Hamiltonian could be split into a sum of one-particle operators. We now want to improve on this approximation by taking *Coulomb interaction* into account. Parts of this interaction were indeed already included in the Hamiltonian (1.32), but we lost important features of it in making the approximations leading to the one-electron model.

We shall consider the Coulomb interaction explicitly in the following section. In doing so, we shall learn some important methods for describing many-body systems. The difficulties in taking full account of the interaction will lead us to the concept of the *quasi-particle* which we already introduced in Section 1.1 and which we can now define more precisely. Sections 3.1.3 and 3.1.4 are then concerned with approximations, first of all the Hartree-Fock approximation of Section 1.3 and then an approximation in which the Coulomb interaction is divided into two components, one a short-range and the other a long-range interaction. There, along with quasi-particles, we shall become familiar with *collective excitations* of the electron gas, i.e., with the *plasmons*. The results of both approximations then allow in Section 3.1.5 a more general discussion of quasi-particles in the theory of the interacting electron gas.

In Sections 3.2, 3.3, and 3.4 we shall then encounter other quasi-particles and collective excitations which are important in solid-state theory.

For further reading on the content of the following sections we refer the reader most of all to the books by *Kittel* [16], *Pines* [27] and *Nakajima* et al. [107.12] and to the presentations in [36–42, 107.7]. However, many other texts and monographs in the reference list will also be found useful.

3.1.2 The Coulomb Interaction

We shall treat the interacting electron gas in the context of the jellium model. We are therefore concerned with the Hamiltonian (1.7). Starting from the noninteracting electron gas, we take the Coulomb interaction as a *perturbation*. From the general principles of quantum mechanical perturbation theory we have then to ask the following questions:

1) Is the perturbation sufficiently small for a perturbation calculation (of first or second order) to provide the necessary corrections?
2) If such a calculation is possible, how are the energies of the stationary states affected and what consequences do the corrections have for the dynamics of the electron gas?
3) Do the stationary states of the noninteracting electron gas remain stationary states when the perturbation (or a part of it) is considered? If not, to what extent can we continue to use features which describe the noninteracting electron gas as an approximation when account is taken of the Coulomb interaction?

We shall discuss these three questions below.

We begin by converting the operator (1.7)

$$H = -\sum_i \frac{\hbar^2}{2m}\nabla_i^2 + \frac{1}{8\pi\varepsilon_0}\sum_{ii'}\frac{e^2}{|\boldsymbol{r}_i - \boldsymbol{r}_{i'}|} + H_+ \tag{3.1}$$

into the occupation number representation. From (A.31) and (A.33) we have

$$\begin{aligned} H = &\sum_{\substack{\lambda\lambda' \\ \sigma_\lambda\sigma_{\lambda'}}} \langle\lambda'| -\frac{\hbar^2}{2m}\nabla^2|\lambda\rangle c^+_{\lambda'\sigma_{\lambda'}} c_{\lambda\sigma_\lambda} \\ &+ \frac{1}{8\pi\varepsilon_0}\sum_{\substack{\lambda\lambda' \\ \mu\mu' \\ \sigma_\lambda\sigma_{\lambda'}\sigma_\mu\sigma_{\mu'}}} \langle\lambda'\mu'|\frac{e^2}{|\boldsymbol{r}-\boldsymbol{r}'|}|\lambda\mu\rangle c^+_{\lambda'\sigma_{\lambda'}} c^+_{\mu'\sigma_{\mu'}} c_{\lambda\sigma_\lambda} c_{\mu\sigma_\mu} + H_+ \,. \end{aligned} \tag{3.2}$$

We take the eigenfunctions for the matrix elements to be the plane waves $|\lambda\rangle = (1/\sqrt{V_g})\exp(\mathrm{i}\boldsymbol{k}_\lambda\cdot\boldsymbol{r})$ of the unperturbed system multiplied by a spin function. Since spin does not appear explicitly in the Hamiltonian, we can immediately carry out the sum over the spins in (3.2). Since the spin functions are orthogonal, this leads to a factor $\delta_{\sigma_\lambda\sigma_{\lambda'}}$ in the first term and one $\delta_{\sigma_\lambda\sigma_{\lambda'}}\delta_{\sigma_\mu\sigma_{\mu'}}$ in the second term. The matrix elements can then be evaluated, and (3.2) becomes

$$H = \sum_{\lambda\sigma_\lambda} E_\lambda c^+_{\lambda\sigma_\lambda} c_{\lambda\sigma_\lambda} + \frac{1}{2}\sum_{\substack{\lambda\mu\nu \\ \sigma_\lambda\sigma_\mu}} V_\nu c^+_{\lambda-\nu,\sigma_\lambda} c^+_{\mu+\nu,\sigma_\mu} c_{\mu\sigma_\mu} c_{\lambda\sigma_\lambda} + H_+ = H_0 + H' + H_+ \tag{3.3}$$

with $\boldsymbol{k}_\lambda - \boldsymbol{k}_{\lambda'} = \boldsymbol{k}_\nu$ and

$$E_\lambda = \frac{\hbar^2 k_\lambda^2}{2m}, \qquad V_\nu = \frac{e^2}{V_g\varepsilon_0 k_\nu^2} = \frac{e^2}{V_g\varepsilon_0}\frac{1}{(\boldsymbol{k}_\lambda - \boldsymbol{k}_{\lambda'})^2}\,. \tag{3.4}$$

V_ν is the Fourier transform of the Coulomb potential as we can readily see from the relationship

$$\int \frac{e^{ik\cdot r - \beta r}}{r} d\tau = \frac{4\pi}{k^2 + \beta^2} \tag{3.5}$$

with $\beta = 0$.

We now inquire about the energy of the *ground state* of the interacting electron gas. We construct the eigenvector of the ground state "filled Fermi sphere" in accordance with (A.18) by creating electrons in all states $k < k_F$ of the vacuum state

$$|0\rangle = \prod_{\substack{k<k_F \\ \sigma}} c_{k\sigma}^{+}|\text{vac}\rangle . \tag{3.6}$$

The energy of the ground state then follows from a perturbation calculation

$$E = \langle 0|H_0|0\rangle + \langle 0|H'|0\rangle + \sum_i \frac{\langle 0|H'|i\rangle\langle i|H'|0\rangle}{E_i - E_0} + \cdots + \langle 0|H_+|0\rangle \tag{3.7}$$

if this series converges.

The individual contributions to (3.7) can be interpreted as processes in which electrons are created and destroyed. Since creation and annihilation operators always appear in pairs, every pair can be taken as an electron transition from one stationary state to another, with change of wave vector (momentum). In this description Coulomb interaction consists of elementary processes in which momentum is exchanged between electrons in the gas.

Interactions of this kind are often presented pictorially. In such *graphs*, electrons of a given momentum $\boldsymbol{k}_\lambda$ are represented by lines (Fig. 3.1a) and momentum exchange between electrons by dashed lines. Arrows on the electron lines show the direction of time. Thus, for example, the interaction described by the second term in (3.3) is represented by two electron lines $\boldsymbol{k}_\lambda$ and $\boldsymbol{k}_\mu$ before the interaction and by two lines $\boldsymbol{k}_\lambda - \boldsymbol{k}_\nu$ and $\boldsymbol{k}_\mu + \boldsymbol{k}_\nu$ after it (Fig. 3.1b). Fig. 3.1c describes two successive interaction processes with momentum exchange $\boldsymbol{k}_\nu$ and $\boldsymbol{k}_\sigma$.

We use these diagrams in Fig. 3.2 to represent contributions to the perturbation expansion (3.7). In this we note that the initial and final states are both the ground state. The electron lines running out to the right must therefore be identical with those running in from the left. Fig. 3.1a then becomes the closed line of Fig. 3.2a. The simple interaction in Fig. 3.1b can be closed either by sequentially connecting the upper and lower lines as pairs or by connecting the right upper with the left lower and vice versa. That gives two possible interaction processes. We shall now discuss them in turn.

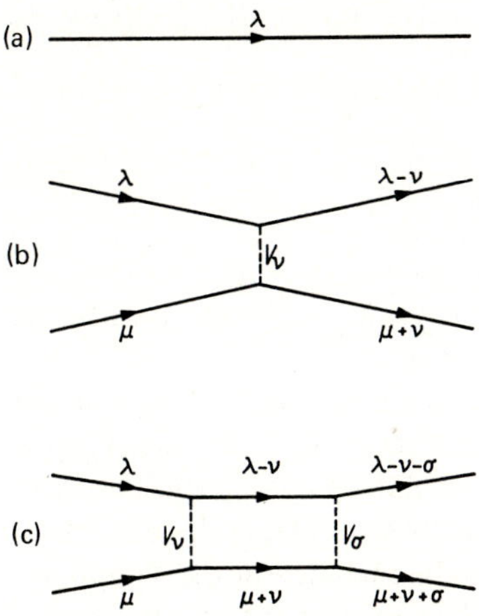

Fig. 3.1a-c. Graphs for the electron-electron interaction: (a) noninteracting electron with momentum $\boldsymbol{k}_\lambda$, (b) interaction between two electrons in which the momentum $\boldsymbol{k}_\nu$ is exchanged, (c) two successive interactions like (b).

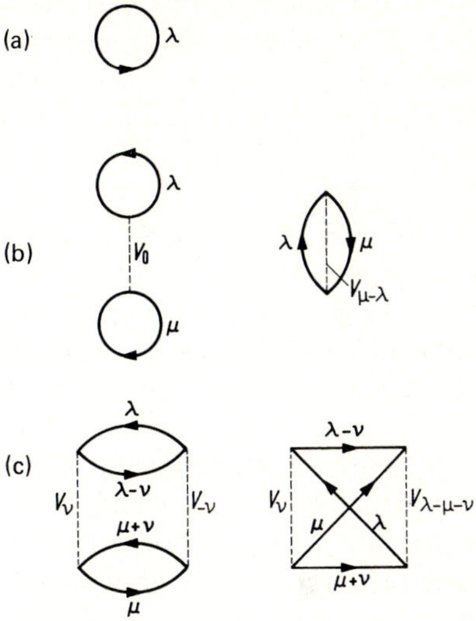

Fig. 3.2a–c. Contributions to the ground state energy
(a) $\langle 0|c_\lambda^+ c_\lambda|0\rangle$
(b) $\langle 0|(V_\nu/2)c_{\lambda-\nu}^+ c_{\mu+\nu}^+ c_\mu c_\nu|0\rangle$
(c) $\langle 0|(V_\sigma/2)c_{\lambda-\nu-\sigma}^+ c_{\mu+\nu+\sigma}^+ c_{\mu+\nu} c_{\lambda-\nu}|i\rangle$
$\langle i|(V_\nu/2)c_{\lambda-\nu}^+ c_{\mu+\nu}^+ c_\mu c_\lambda|0\rangle$

The *first term* in (3.7) is made up additively from contributions of the type shown in Fig. 3.2a. Since from (A.7) $c_\lambda^+ c_\lambda$ is the particle number operator n_λ, $\langle 0|H_0|0\rangle$ is the sum of the energies of all occupied states, i.e., the total energy of the *non-interacting electron gas*.

In accordance with Fig. 3.2b, two processes contribute to the *second term* in (3.7): $\boldsymbol{k}_\nu = 0$ and $\boldsymbol{k}_\nu = \boldsymbol{k}_\lambda - \boldsymbol{k}_\mu (\sigma_\lambda = \sigma_\mu)$.

$$
\begin{aligned}
E_1 &= \langle 0| \sum_{\substack{\lambda\mu \\ \sigma_\lambda\sigma_\mu}} \frac{V_0}{2} c_{\lambda\sigma_\lambda}^+ c_{\mu\sigma_\mu}^+ c_{\mu\sigma_\mu} c_{\lambda\sigma_\lambda}|0\rangle + \langle 0| \sum_{\substack{\lambda\mu \\ \sigma_\lambda}} \frac{V_{\mu-\lambda}}{2} c_{\mu\sigma_\lambda}^+ c_{\lambda\sigma_\lambda}^+ c_{\mu\sigma_\lambda} c_{\lambda\sigma_\lambda}|0\rangle \\
&= \sum_{\substack{\lambda\mu \\ \sigma_\lambda\sigma_\mu}} \frac{V_0}{2} n_\lambda n_\mu - \sum_{\substack{\lambda\mu \\ \sigma_\lambda}} \frac{V_{\mu-\lambda}}{2} n_\lambda n_\mu \,.
\end{aligned}
\tag{3.8}
$$

The first term on the right is equal to $V_0/2$ times a sum over all occupied states $\boldsymbol{k}_\lambda$ and $\boldsymbol{k}_\mu$. From (3.4) this contribution diverges since V_0 is infinite. A more exact analysis—such as we shall consider in the following section—shows that this term compensates the contribution from the positive background $\langle 0|H_+|0\rangle$. The inclusion of this term and the contribution from H_+ mean a change-over from the noninteracting electron gas to the *Hartree approximation* (1.18). We shall also show this in the next section.

The second term on the right in (3.8) will prove to be the exchange interaction of the *Hartree-Fock equation* (1.23). On account of the factor $n_\lambda n_\mu$, the summation is again only over states inside the Fermi sphere. If we replace the summation by an integration in $\boldsymbol{k}$-space we have

$$-\sum_{\substack{\lambda\mu \\ \sigma\lambda}} \frac{V_{\mu-\lambda}}{2} n_\lambda n_\mu = \sum_{\lambda\sigma\lambda} n_\lambda \left[-\frac{e^2}{8\pi^3 \varepsilon_0} \int_{k_F} \frac{d\tau_\mu}{(\boldsymbol{k}_\lambda - \boldsymbol{k}_\mu)^2} \right]. \tag{3.9}$$

In spite of the singularity in the integrand for $\boldsymbol{k}_\lambda - \boldsymbol{k}_\mu \to 0$, this contribution is finite. The result of the integration is given in (3.11).

The *third term* in (3.7) is made up of processes of the type shown in Fig. 3.2c. Two electrons ($\boldsymbol{k}_\lambda$ and $\boldsymbol{k}_\mu$) are raised into states $\boldsymbol{k}_\lambda - \boldsymbol{k}_\nu$ and $\boldsymbol{k}_\mu + \boldsymbol{k}_\nu$ outside the Fermi sphere and fall back again from this *virtual* intermediate state either into their own original places or into each other's places. In accordance with (3.7), the denominator $E_i - E_0$ has also to be included in the summation over all intermediate states. When we convert the summation to an integration we quickly see that for small momentum exchange ($\boldsymbol{k}_\nu$, $\boldsymbol{k}_\sigma \to 0$) the integrand becomes so large that the integral (logarithmically) diverges. The second-order contribution to the energy of the ground state (3.7) becomes infinite. The same is true for all contributions of higher order. The perturbation calculation does *not* converge.

This divergence is clearly the result of the long range of the Coulomb interaction. For an exponentially screened interaction, we see from (3.5) that the Fourier transform of the potential remains finite as $\boldsymbol{k} \to 0$. The divergence then disappears.

We have thus found a preliminary answer to the first two questions posed at the start of this section. The Coulomb interaction in the Hamiltonian (3.1) cannot be treated rigorously by perturbation techniques. The contributions of second and higher order diverge. There are two possible ways to circumvent this difficulty.

By also using the graphical analysis of Fig. 3.2 for interactions of higher order, partial sums of the (in toto diverging) perturbation series can be added up. This way leads to the more general field theoretical methods of many-body theory, in particular to the formalism of Green's functions. Such aids are of immense importance to solid-state theory. However within the confines of this book it is not appropriate for us to involve ourselves in such a highly complicated mathematical formalism. We shall be making use of diagrams of the type shown in Figs. 3.1 and 3.2 later when looking at interactions in solids. But in the mathematical treatment we shall restrict ourselves to the formalism of the occupation number representation.

A second way to remove the divergence in (3.7) is by the introduction of a screened Coulomb interaction. This is made possible by an artifice. We shall describe this method in detail in Section 3.1.4.

We close this section with a qualitative discussion of the third question posed at the beginning. The stationary states of the free electron gas are certainly not stationary states of the Hamiltonian (3.1). This means that an electron in state $\boldsymbol{k}$, σ of the unperturbed system can be *scattered* into another state by Coulomb interaction. It has a certain *lifetime* in a state $\boldsymbol{k}$, σ, this lifetime being determined by the scattering probability.

Let us look particularly at an electron outside the filled Fermi sphere. Through the interaction it can in a "collision process" (Fig. 3.1b) exchange energy and momentum with an electron in the Fermi sphere. Before the collision let the energy of the electron be $E(\boldsymbol{k})$ and that of its collision partner $E_1(\boldsymbol{k}_1)$. Afterwards let the energies be $E_2(\boldsymbol{k}_2)$ and $E_3(\boldsymbol{k}_3)$. Here E, E_2, $E_3 > E_F$, and $E_1 < E_F$. Since energy and momentum must be conserved in the collision, i.e., $E + E_1 = E_2 + E_3$ and $\boldsymbol{k} + \boldsymbol{k}_1 = \boldsymbol{k}_2 + \boldsymbol{k}_3$, the scattering probability is strongly dependent on energy. The Pauli principle restricts the region of $\boldsymbol{k}$-space in which $\boldsymbol{k}_2$ and $\boldsymbol{k}_3$ can lie. Pauli principle and energy conservation already indicate that for $E = E_F + \delta E$, E_1 is limited to a region δE below the Fermi surface, and E_2 and E_3 to a region δE above the surface. Momentum conservation places further limits on this region. For small δE the scattering probability, and with it the reciprocal lifetime, is proportional to $(\delta E)^2$. We therefore have the following important result: Regardless of the strength of the interaction, electrons in a state directly at the Fermi surface have an infinite lifetime. Far away from E_F, on the other hand, the lifetime is short and the energy of the state considerably broadened. The concept of approximately interaction-free electrons then becomes pointless.

Before we pursue these lines of thought in Section 3.1.5, we must look more closely at the different approximations involved in the description of the interacting electron gas. This we do in the next two sections.

3.1.3 The Hartree-Fock Approximation for the Electron Gas

The Hartree-Fock equation (1.23)—like the Schrödinger equation for the non-interacting electron gas—is solved by plane waves (2.4). We see this immediately if we put the wave function (2.4) into (1.23). With (1.7) (jellium model) and with the abbreviation $\boldsymbol{r}' - \boldsymbol{r} = \boldsymbol{r}''$, we have

$$\left\{-\frac{\hbar^2}{2m}\nabla^2 + \frac{e^2}{4\pi\varepsilon_0}\sum_{\substack{k\\(\neq j)}}\frac{1}{V_g}\int\frac{d\tau''}{r''} - \frac{e^2}{4\pi\varepsilon_0}\sum_{\substack{k(\neq j)\\ \parallel}}\frac{1}{V_g}\right.$$
$$\left.\times\int\frac{\exp\left(\mathrm{i}(\boldsymbol{k}_k - \boldsymbol{k}_j)\cdot\boldsymbol{r}''\right)}{r''}d\tau'' + H_+\right\}\psi = E\psi\,. \qquad (3.10)$$

The first integral on the left is the interaction of the jth electron with $N - 1$ others, which are uniformly distributed about the volume V_g. The term H_+ gives the interaction of the same electron with the positive background (N

uniformly distributed positive charges). The two terms compensate each other except for the insignificant contribution of the interaction with *one* charge distributed over V_g. Thus the *Hartree equation* [(3.10) without the third term on the left] leads back to the free electron gas. We have already mentioned this result in the previous section.

On the other hand, the third term on the left in (3.10) implies an interaction which we have already discussed qualitatively in Section 1.3. If we carry out the integration using (3.5), we find a summation over terms of the form $\text{const}/(\boldsymbol{k}_k - \boldsymbol{k}_j)^2 (k \neq j)$ which can easily be evaluated. To do so, we take V_g to be so large that the sum over discrete values of $\boldsymbol{k}$ becomes an integration over the Fermi sphere $[(1/V_g)\sum_{\boldsymbol{k}} \ldots = \int g(\boldsymbol{k}) \ldots d\tau_{\boldsymbol{k}}]$. From (3.10) we then have the energy eigenvalue of this electron as

$$\begin{aligned} E(\boldsymbol{k}) &= \frac{\hbar^2 k^2}{2m} - \frac{e^2}{8\pi^3 \varepsilon_0} \int_{k_F} \frac{d\tau_{\boldsymbol{k}'}}{(\boldsymbol{k}' - \boldsymbol{k})^2} \\ &= \frac{\hbar^2 k^2}{2m} - \frac{e^2 k_F}{8\pi^2 \varepsilon_0} \left(2 + \frac{k_F^2 - k^2}{k k_F} \ln \left| \frac{k + k_F}{k - k_F} \right| \right). \end{aligned} \tag{3.11}$$

We are still assuming here that the number of electrons with spin up and spin down is equal. This is so if we assume that all states in the Fermi sphere are occupied and all outside are empty.

The second term on the right in (3.11) is clearly a term of the sum (3.9), i.e., the contribution of one electron to the first-order perturbation energy of the ground state of the electron gas [(3.7)].

Since the wave functions of the Hartree-Fock approximation are plane waves, the relationship $\boldsymbol{p} = \hbar \boldsymbol{k}$ is valid. Eq. (3.11) then shows that the energy and momentum of a "Hartree-Fock electron" are not related by the classical equation $E = p^2/2m$.

To understand this better we consider the exchange charge density (1.29), which from (3.10) here takes the form

$$\rho_j^{\text{HF}}(\boldsymbol{r}, \boldsymbol{r}') = -\frac{e}{V_g} \sum_{k\|} \exp\left(\mathrm{i}(\boldsymbol{k}_j - \boldsymbol{k}_k) \cdot (\boldsymbol{r} - \boldsymbol{r}')\right). \tag{3.12}$$

We again replace summation by integration over the Fermi sphere, assume that $N/2$ electrons in the sphere have the same spin, and find (again with $\boldsymbol{r}' - \boldsymbol{r} = \boldsymbol{r}''$)

$$\rho_j^{\text{HF}}(\boldsymbol{r}'') = \frac{3}{2} \frac{eN}{V_g} \exp(\mathrm{i}\boldsymbol{k}_j \cdot \boldsymbol{r}'') \frac{1}{(k_F r'')^3} (k_F r'' \cos k_F r'' - \sin k_F r''). \tag{3.13}$$

This still depends on the $\boldsymbol{k}$-vector of the jth electron. With an approach like that of (1.31) we finally find that

$$\bar{\rho}^{\text{HF}}(r) = \frac{9}{2} \frac{eN}{V_g} \frac{(k_F r \cos k_F r - \sin k_F r)^2}{(k_F r)^6}. \tag{3.14}$$

Thus the charge density that a Hartree-Fock electron "sees" is given by

$$\rho - \bar{\rho}^{\mathrm{HF}} = \frac{eN}{V_{\mathrm{g}}}\left[1 - \frac{9}{2}\frac{(k_{\mathrm{F}}r\cos k_{\mathrm{F}}r - \sin k_{\mathrm{F}}r)^2}{(k_{\mathrm{F}}r)^6}\right]. \tag{3.15}$$

This function is illustrated in Fig. 3.3. The concentration of electrons with the same spin is reduced in the vicinity of the electron considered, while the electrons of opposite spin are uniformly distributed. One expresses this fact by saying that the electron is surrounded by an *exchange hole* which carries the charge $+e$. The Pauli principle brings in a *correlation* of electrons with the same spin.

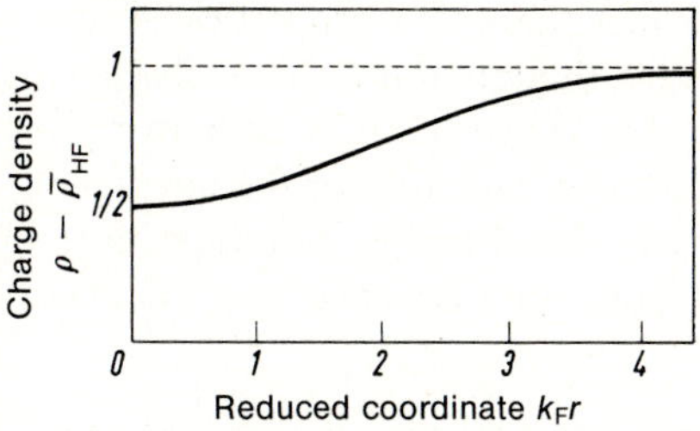

Fig. 3.3. Exchange hole in the vicinity of an electron in the Hartree-Fock-approximation [3.15].

The electron is accompanied by its exchange hole in its motion through the crystal. Since this implies continuous rearrangement of the surrounding electrons it is little wonder that the free electron energy-momentum relation no longer holds for a Hartree-Fock electron. The parentheses in (3.11) have a value between 4 (for $k = 0$) and 2 (for $k = k_{\mathrm{F}}$). The second term is therefore always negative. The relation $E = \hbar^2k^2/2m$ can be formally maintained if, as in the band model, we replace the electron mass m by an *effective mass* m^* which then from (3.11) is dependent on k and always has a value greater than m. The effective mass of the electron has thus been increased by having to carry the exchange hole around with it.

The mean energy of the electron gas is also altered by the exchange interaction. By integration of (3.11) over the Fermi sphere, we again find a contribution $(3/5)E_{\mathrm{F}}$ for the first term [see (2.36)]. The second term makes a contribution $E_{\mathrm{exch}} = -3e^2k_{\mathrm{F}}/8\pi^2\varepsilon_0 = (3/4)E^F_{\mathrm{exch}}$, where E^F_{exch} is the value of the exchange energy at the Fermi surface.

3.1.4 Screening, Plasmons

In Section 3.1.2 we saw that it is not possible to consider the full Coulomb interaction beyond the first-order contribution. The reason lies in the long range of this interaction. We now want to show that the Coulomb interaction can be replaced by a screened interaction. The long-range component which is thereby neglected then reemerges in the form of additional collective oscillations of the electron gas.

For a moment we shall take the electron gas to be a uniformly distributed density of space charge. If we now introduce an additional negative charge at a point $\boldsymbol{r}$, two things happen. By Coulomb repulsion, charge is driven away from the immediate vicinity of the negative point charge. This rearrangement means a positive charge-cloud around it relative to the average charge density of the electron gas This in turn means a *screening* of the charge of the electron.

The rearrangement, however, will only be the final state of a dynamical process. First the negative charge around the electron will be repelled. On account of the long-range of the Coulomb potential, the rearrangement will initially extend too far, the charge will flow back, etc. *Collective oscillations* appear which correspond to compression waves in the electron gas.

We want to examine now the extent to which these two properties of an electron gas, namely screening of the Coulomb interaction of individual charges and collective oscillations resulting from the long range of the Coulomb potential, can be described explicitly by the Schrödinger equation for the interacting electron gas. We proceed from the classical Hamiltonian. The transition to quantum mechanics will not be made until later. We then have

$$H = \sum_i \frac{p_i^2}{2m} + \frac{e^2}{8\pi\varepsilon_0} \sum_{\substack{i,j \\ i \neq j}} \frac{1}{|\boldsymbol{r}_i - \boldsymbol{r}_j|}. \tag{3.16}$$

The explicit introduction of screening would be made possible by adding to the second term a factor $\exp(-k_c|\boldsymbol{r}_i - \boldsymbol{r}_j|)$ with initially undefined screening constant. However, this would mean neglect of the long-range component of the Coulomb potential and we would have to examine its effect separately. A better way is the following: We write the second term of (3.16) as a Fourier series

$$\frac{e^2}{8\pi\varepsilon_0} \sum_{\substack{i,j \\ i \neq j}} \frac{1}{|\boldsymbol{r}_i - \boldsymbol{r}_j|} = \frac{e^2}{2V_g\varepsilon_0} \sum_{\substack{i,j \\ i \neq j}} \sum_{k}{}' \frac{(\exp i\boldsymbol{k}\cdot(\boldsymbol{r}_i - \boldsymbol{r}_j))}{k^2}, \tag{3.17}$$

where we have again omitted the term $k = 0$. For a screened Coulomb potential we would, from (3.5), replace k^2 here with $k^2 + k_c^2$. However, almost the same series is obtained if we keep (3.17) but limit the summation to the terms $k > k_c$. Fig. 3.4 shows the full Coulomb potential and the division of (3.17) into a short-range ($k > k_c$) and a long-range ($k < k_c$) component. One can see that this division is reasonable.

For what follows we can therefore start with the Hamiltonian

$$H = \sum_{i,j} \frac{p_i^2}{2m} + \frac{e^2}{2V_g\varepsilon_0} \sum_{i,j} \left(\sum_{k<k_c}{}' + \sum_{k>k_c} \right) \frac{\exp(i\boldsymbol{k}\cdot(\boldsymbol{r}_i - \boldsymbol{r}_j))}{k^2} - \frac{Ne^2}{2V_g\varepsilon_0} \sum_{k}{}' \frac{1}{k^2}, \tag{3.18}$$

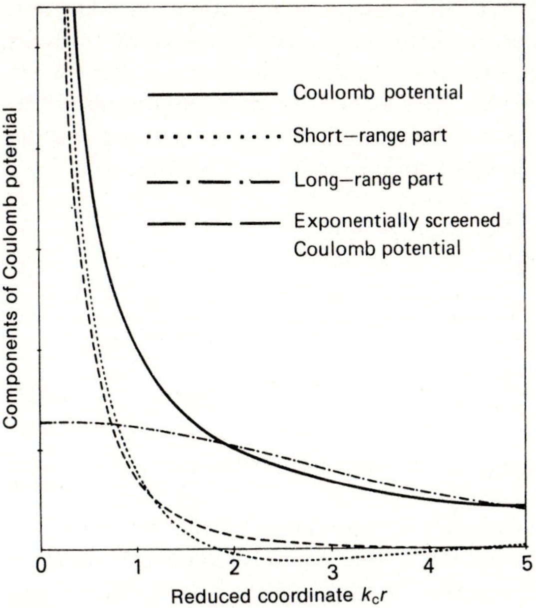

Fig. 3.4. The Coulomb potential and its division into a short-range component ($k > k_c$) and a long-range component ($k < k_c$). For comparison an exponentially screened potential [$\sim\exp(-k_c r)$] is also shown (from *Haug* [14]).

where we have included the term $i = j$ in the second summation and removed it again by means of the third term.

Along with the screening, we want to find out about the collective oscillations, which describe the motion of the electrons in the field produced by their own Coulomb potential. We describe this field by the vector potential $\boldsymbol{A}(\boldsymbol{r}_i)$ where $\boldsymbol{r}_i$ is the location of the ith electron. We express $\boldsymbol{A}(\boldsymbol{r}_i)$ as a Fourier series

$$\boldsymbol{A}(\boldsymbol{r}_i) = \frac{1}{\sqrt{V_g \varepsilon_0}} \sum_{\boldsymbol{k}}{}' \frac{\boldsymbol{k}}{k} Q_{\boldsymbol{k}} \exp(\mathrm{i}\boldsymbol{k}\cdot\boldsymbol{r}_i), \tag{3.19}$$

in which we have noted that $\boldsymbol{A}$ is an irrotational field. The factor in front of the summation appears for reasons of expediency. Since $\boldsymbol{A}$ should be real, we have for the Fourier coefficients that $(\boldsymbol{k}/k)Q_{\boldsymbol{k}}^* = (-\boldsymbol{k}/k)Q_{-\boldsymbol{k}}$; hence $Q_{\boldsymbol{k}}^* = -Q_{-\boldsymbol{k}}$. From (3.19) the electric field follows as

$$\boldsymbol{E} = -\dot{\boldsymbol{A}} = -\frac{1}{\sqrt{V_g \varepsilon_0}} \sum_{\boldsymbol{k}} \frac{\boldsymbol{k}}{k} \dot{Q}_{\boldsymbol{k}} \exp(\mathrm{i}\boldsymbol{k}\cdot\boldsymbol{r}_i) = -\frac{1}{\sqrt{V_g \varepsilon_0}} \sum_{\boldsymbol{k}} \frac{\boldsymbol{k}}{k} P_{\boldsymbol{k}}^* \exp(\mathrm{i}\boldsymbol{k}\cdot\boldsymbol{r}_i) \tag{3.20}$$

where we have also put $\dot{Q}_{\boldsymbol{k}} = P_{\boldsymbol{k}}^* (P_{\boldsymbol{k}}^* = -P_{-\boldsymbol{k}})$.

The $Q_{\boldsymbol{k}}$ and $P_{\boldsymbol{k}}$ (as canonical conjugates) can be taken as collective coordinates of the field which describe the Coulomb interaction of the electrons. We can also express the Hamiltonian (3.16) in terms of these fields by replacing the $\boldsymbol{p}_i$ by $\boldsymbol{p}_i + e\boldsymbol{A}(\boldsymbol{r}_i)$ and the interaction term by the energy $(\varepsilon_0/2)\int E^2 d\tau$. When we have written all the added terms as Fourier series, we can also combine both possibilities and leave the components $k > k_c$ as in (3.18) and describe only the $k < k_c$ components in terms of the fields. This leads to

$$\begin{aligned} H = \sum_i \frac{1}{2m}\left(\boldsymbol{p}_i + \frac{e}{\sqrt{V_g\varepsilon_0}}\sum_{k<k_c}\frac{\boldsymbol{k}}{k}Q_k \exp(\mathrm{i}\boldsymbol{k}\cdot\boldsymbol{r}_i)\right)^2 \\ + \frac{e^2}{2V_g\varepsilon_0}\sum_{ij}\sum_{k>k_c}\frac{\exp[\mathrm{i}\boldsymbol{k}\cdot(\boldsymbol{r}_i - \boldsymbol{r}_j)]}{k^2} - \frac{Ne^2}{2V_g\varepsilon_0}\sum_k{}'\frac{1}{k^2} \\ + \frac{1}{2V_g}\sum_{k,k'<k_c}{}'\frac{\boldsymbol{k}\cdot\boldsymbol{k}'}{kk'}P_k P_{k'}\int \exp[\mathrm{i}(\boldsymbol{k}+\boldsymbol{k}')\cdot\boldsymbol{r}_i]d\tau_i\,. \end{aligned} \tag{3.21}$$

Since $\int \exp[\mathrm{i}(\boldsymbol{k}+\boldsymbol{k}')\cdot\boldsymbol{r}]d\tau = V_g\delta_{\boldsymbol{k},-\boldsymbol{k}'}$ the last term here can be arranged further into $(1/2)\sum_{k<k_c}P_k^*P_k$.

We go over now from the classical Hamiltonian to the corresponding operator. We take the $\boldsymbol{p}_i$ and $\boldsymbol{r}_i$ and the P_k and Q_k to be canonical conjugate operators satisfying the commutation relation $[p_{i\nu}, r_{j\mu}] = -\mathrm{i}\hbar\delta_{ij}\delta_{\mu\nu}$ and $[P_k, Q_{k'}] = -\mathrm{i}\hbar\delta_{\boldsymbol{k},\boldsymbol{k}'}$. We then have in the first term of the brackets

$$\boldsymbol{p}_i \exp(\mathrm{i}\boldsymbol{k}\cdot\boldsymbol{r}_i) + \exp(\mathrm{i}\boldsymbol{k}\cdot\boldsymbol{r}_i)\boldsymbol{p}_i = 2\boldsymbol{p}_i\exp(\mathrm{i}\boldsymbol{k}\cdot\boldsymbol{r}_i) - \hbar\boldsymbol{k}\exp(\mathrm{i}\boldsymbol{k}\cdot\boldsymbol{r}_i) \tag{3.22}$$

and hence after a few rearrangements ($\omega_p^2 \equiv ne^2/m\varepsilon_0$, $n = N/V_g$)

$$\begin{aligned} H = \sum_i \frac{p_i^2}{2m} + \frac{e^2}{2V_g\varepsilon_0}\sum_{ij}\sum_{k>k_c}\frac{\exp(\mathrm{i}\boldsymbol{k}\cdot(\boldsymbol{r}_i - \boldsymbol{r}_j))}{k^2} - \frac{ne^2}{2\varepsilon_0}\sum_{k>k_c}\frac{1}{k^2} \\ + \frac{1}{2}\sum_{k<k_c}{}'\left(P_k^*P_k + \omega_p^2 Q_k^*Q_k - \frac{ne^2}{\varepsilon_0 k^2}\right) \\ + \frac{e}{\sqrt{V_g\varepsilon_0}m}\sum_{k<k_c}Q_k\frac{\boldsymbol{k}}{k}\cdot\sum_i\left(\boldsymbol{p}_i - \frac{\hbar\boldsymbol{k}}{2}\right)\exp(\mathrm{i}\boldsymbol{k}\cdot\boldsymbol{r}_i) \\ + \frac{e^2}{2V_g\varepsilon_0 m}\sum_{\substack{k,k'<k_F \\ -\boldsymbol{k}\neq\boldsymbol{k}'}}\frac{\boldsymbol{k}\cdot\boldsymbol{k}'}{kk'}Q_kQ_{k'}\sum_i\exp(\mathrm{i}(\boldsymbol{k}+\boldsymbol{k}')\cdot\boldsymbol{r}_i)\,. \end{aligned} \tag{3.23}$$

Before we interpret this Hamiltonian we must note the following. The introduction of the Q_k and P_k along with the $\boldsymbol{r}_i$ and $\boldsymbol{p}_i$ has raised the number of degrees of freedom of the system. There must therefore be just as many supplementary conditions connecting the two sets of coordinates. For this we use the equation $\nabla\cdot\boldsymbol{E} - \rho/\varepsilon_0 = 0$. If we expand the charge density ρ in a Fourier series, we have for every Fourier component

$$P_k - \mathrm{i}\sqrt{\frac{e^2}{V_g\varepsilon_0 k^2}}\sum_j \exp(\mathrm{i}\boldsymbol{k}\cdot\boldsymbol{r}_j) = 0\,. \tag{3.24}$$

There are as many equations here as additional pairs P_k, Q_k we have introduced. In the transition to quantum mechanics, (3.24) becomes an operator equation. We then require that the operator on the left side of (3.24) disappears when applied to a wave function.

Equation (3.23) reduces to three components. The first line is the Hamiltonian of an *electron gas with screened interaction*. The second line describes the collective oscillations of the electron gas (*plasma oscillations*). It has the form of a sum over Hamiltonians of individual harmonic oscillators of frequency ω_p. We interpret the energy quanta of these oscillators as the quanta of the oscillations of the electron gas, or *plasmons*. The third and fourth lines then give the interaction between screened electrons and plasmons. One sees this from the fact that the collective coordinates Q_k appear together with the electron coordinates $\boldsymbol{r}_i$ in the terms on these two lines. One usually omits the term on the fourth line with the justification that the summation over the $\boldsymbol{r}_i$ disappears since the electron locations are statistically distributed. This approximation is referred to as the *random phase approximation*.

With the introduction of the screened electrons, we have become familiar with a new type of quasi-electrons, and the plasmons are a further example of the collective excitations referred to in Section 1.1. We shall consider the characteristics of such collective excitations later. Here we want to uncover some aspects of the elementary excitations and arrive at further statements about the interaction terms in (3.23) by converting (3.23) into the occupation number representation.

We look initially at the first term. It becomes

$$H_{\mathrm{el},1} = \sum_i \frac{p_i^2}{2m} = -\frac{\hbar^2}{2m}\sum_i \nabla_i^2 = -\frac{\hbar^2}{2m}\sum_{\lambda\lambda'} \langle\lambda'|\nabla^2|\lambda\rangle c_{\lambda'}^+ c_\lambda \,. \tag{3.25}$$

When we insert plane waves, the matrix element becomes

$$\langle\lambda'|\nabla^2|\lambda\rangle = -k_\lambda^2 \delta_{\lambda'\lambda} \,. \tag{3.26}$$

Hence

$$H_{\mathrm{el},1} = \sum_{\boldsymbol{k}} \frac{\hbar^2 k^2}{2m} c_{\boldsymbol{k}}^+ c_{\boldsymbol{k}} \,. \tag{3.27}$$

Since in this section $\boldsymbol{k}$ defines the state including spin, the sum over $\boldsymbol{k}$ in (3.27) implies a sum over the spins as well. As long as spin is unimportant, as here and in what follows, we shall not specifically mention it.

Equation (3.27) is the Hamiltonian of the noninteracting electron gas. The electron-electron interaction is given by the second and third term. It becomes

$$H_{\mathrm{el},2} = \sum_{k>k_c} \frac{e^2}{2V_g\varepsilon_0 k^2} \sum_{\substack{ij\\ i\neq j}} \exp(\mathrm{i}\boldsymbol{k}\cdot(\boldsymbol{r}_i - \boldsymbol{r}_j))$$

$$= \sum_{k>k_c} \frac{e^2}{2V_g\varepsilon_0 k^2} \sum_{\lambda'\mu'\lambda\mu} \langle\lambda'\mu'|\exp(\mathrm{i}\boldsymbol{k}\cdot(\boldsymbol{r}_1 - \boldsymbol{r}_2))|\lambda\mu\rangle c^+_{\lambda'}c^+_{\mu'}c_\mu c_\lambda, \tag{3.28}$$

or since

$$\langle\cdots\rangle = \frac{1}{V_g^2}\int \exp(\mathrm{i}(\boldsymbol{k}_\lambda + \boldsymbol{k} - \boldsymbol{k}_{\lambda'})\cdot\boldsymbol{r}_1)\exp(\mathrm{i}(\boldsymbol{k}_\mu - \boldsymbol{k} - \boldsymbol{k}_{\mu'})\cdot\boldsymbol{r}_2)d\tau_1 d\tau_2$$

$$= \delta_{\boldsymbol{k}_{\lambda'},\boldsymbol{k}_\lambda+\boldsymbol{k}}\delta_{\boldsymbol{k}_{\mu'},\boldsymbol{k}_\mu-\boldsymbol{k}}, \tag{3.29}$$

$$H_{\mathrm{el},2} = \sum_{k>k_c} \frac{e^2}{2V_g\varepsilon_0 k^2} \sum_{\lambda\mu} c^+_{\boldsymbol{k}_\lambda+\boldsymbol{k}}c^+_{\boldsymbol{k}_\mu-\boldsymbol{k}}c_{\boldsymbol{k}_\mu}c_{\boldsymbol{k}_\lambda}. \tag{3.30}$$

In this form, $H_{\mathrm{el},2}$ represents a screened Coulomb interaction, which is described by individual interaction processes in which the momentum $\boldsymbol{k}$ is transferred from the μth electron to the λth electron (annihilation of two electrons with momentum $\boldsymbol{k}_\mu$, $\boldsymbol{k}_\lambda$ and creation of two electrons with momentum $\boldsymbol{k}_\lambda + \boldsymbol{k}$ and $\boldsymbol{k}_\mu - \boldsymbol{k}$).

We have already transformed the fourth term in (3.23) in the Appendix. In this representation it describes the energy (less the self-energy) of a gas of noninteracting bosons. These are the plasmons, i.e., the quanta of the collective oscillations.

In the fifth term, in accordance with (A.3), we replace Q_k with $\sqrt{\hbar/2\omega_p}(a_{\boldsymbol{k}} + a^+_{-\boldsymbol{k}})$. From (A.31) the factor $(1/2m)\sum_i (2\boldsymbol{k}\cdot\boldsymbol{p}_i - \hbar k^2)\exp(\mathrm{i}\boldsymbol{k}\cdot\boldsymbol{r}_i)$ becomes

$$\sum_{\lambda'\lambda} c^+_{\lambda'}c_\lambda\left(\frac{\boldsymbol{k}\cdot\boldsymbol{k}_\lambda}{m} + \frac{\hbar^2 k^2}{2m}\right)\delta_{\boldsymbol{k}_{\lambda'},\boldsymbol{k}_\lambda+\boldsymbol{k}}.$$

We then have

$$H_{\mathrm{el-pl}} = \sum_{k<k_c}\sqrt{\frac{e^2\hbar^3}{8V_g\varepsilon_0\omega_p m^2 k^2}} \sum_{\boldsymbol{k}_\lambda} (2\boldsymbol{k}\cdot\boldsymbol{k}_\lambda + k^2)(a_{\boldsymbol{k}}c^+_{\boldsymbol{k}_\lambda+\boldsymbol{k}}c_{\boldsymbol{k}_\lambda} + a^+_{-\boldsymbol{k}}c^+_{\boldsymbol{k}_\lambda+\boldsymbol{k}}c_{\boldsymbol{k}_\lambda}). \tag{3.31}$$

The electron-plasmon interaction is represented here by processes in which a momentum $\boldsymbol{k}$ is transferred to an electron with absorption of a plasmon $\boldsymbol{k}$ (first term) or emission of a plasmon $-\boldsymbol{k}$ (second term).

We had omitted the last term in (3.23) in the random phase approximation. The occupation number representation gives a new interpretation here. The term $Q_k Q_{k'}\sum_i \exp[\mathrm{i}(\boldsymbol{k} + \boldsymbol{k}')\cdot\boldsymbol{r}_i]$ gives, apart from a numerical factor,

$$(a_{\boldsymbol{k}} + a^+_{-\boldsymbol{k}})(a_{\boldsymbol{k}'} + a^+_{-\boldsymbol{k}'})c^+_{\boldsymbol{k}_{\lambda'}}c_{\boldsymbol{k}_\lambda}\langle\lambda'|\exp[\mathrm{i}(\boldsymbol{k} + \boldsymbol{k}')\cdot\boldsymbol{r}_i]|\lambda\rangle. \tag{3.32}$$

The matrix element again takes care of momentum conservation ($\boldsymbol{k} + \boldsymbol{k}' + \boldsymbol{k}_\lambda - \boldsymbol{k}_{\lambda'} = 0$). The relevant part of (3.32) becomes

$$(a_{\boldsymbol{k}}a_{\boldsymbol{k}'} + a^+_{-\boldsymbol{k}}a^+_{-\boldsymbol{k}'} + a^+_{-\boldsymbol{k}}a_{\boldsymbol{k}'} + a_{\boldsymbol{k}}a^+_{-\boldsymbol{k}'})c^+_{\boldsymbol{k}+\boldsymbol{k}'+\boldsymbol{k}_\lambda}c_{\boldsymbol{k}_\lambda}\,. \tag{3.33}$$

This term describes processes in which a momentum $\boldsymbol{k} + \boldsymbol{k}'$ is transferred to the λth electron with simultaneous absorption of two plasmons $\boldsymbol{k}, \boldsymbol{k}'$, or emission of two plasmons $-\boldsymbol{k}, -\boldsymbol{k}'$, or absorption and emission of plasmons. The neglect of this interaction thus means the neglect of all processes involving more than one plasmon.

In all these considerations we have characterized the division into individual and collective excitations of the electron gas by a parameter k_c. This parameter is still undefined. It can be determined by comparing the theoretically calculated energy of the electron gas with experimental values (cf. Sec. 8.2.6).

An estimate of k_c follows by comparing the dispersion relation for plasmons with that for pair excitations of the electron gas. From Section 2.1.3 (Fig. 2.3), the maximum energy of a pair excitation for a given κ is $E_{max} = (\hbar^2/2m) \times [(\kappa + k_F)^2 - k_F^2] = \hbar^2\kappa^2/2m + \hbar\kappa v_F$ (with $v_F = \hbar k_F/m$ the velocity of an electron on the Fermi sphere). On the other hand, the energy of a plasmon is $E_{pl} = \hbar\omega_p$, to which in the next higher order of approximation a term proportional to κ^2 is added. For $\kappa < \kappa_p$ [with κ_p from $E_{max}(\kappa_p) = E_{pl}(\kappa_p)$], $E_{pl} > E_{max}$ (Fig. 3.5). Plasmons can therefore neither be excited by individual pair excitations of the electron gas, nor can they decay into them. For $\kappa > \kappa_p$ ($E_{max} > E_{pl}$), we have the region in which pair excitations are possible. κ_p thus broadly defines the boundary between collective excitations and particle excitations of the electron gas. If one puts $\kappa_p = k_c$, one has for small κ the often-used estimate $k_c \approx \omega_p/v_F$. k_c then still depends through ω_p and v_F on the density of the electron gas.

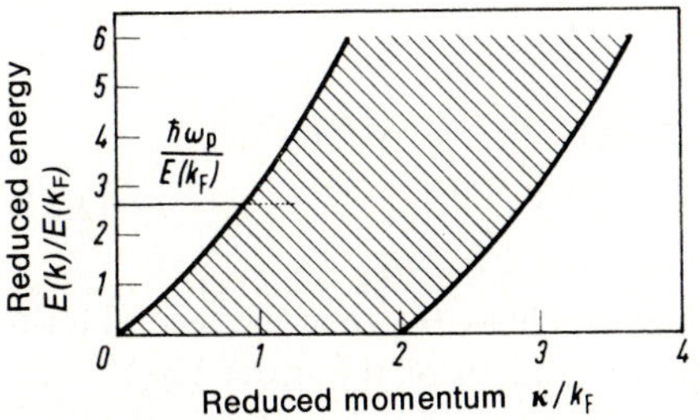

Fig. 3.5. Pair excitations (Fig. 2.3) and plasma branch.

The fact that it is impossible to excite plasmons by particle excitations of the gas itself in the stable region $k < k_c$ also means that when we discuss particle excitations, plasmon effects can mostly be ignored. Such effects do become important when sufficient energy is supplied from outside, for example by the transit of fast electrons through a solid. Fast electrons suffer characteristic energy losses in metals due to plasmon excitation.

3.1.5 Quasi-Electrons

The introduction of plasmons as additional collective oscillations of the electrons brings with it at the same time a screening of the Coulomb interaction. The fundamental difficulties with the treatment of the electron-electron interaction by perturbation theory are thereby removed. Instead, a new interaction, the *electron-plasmon interaction*, has entered into the picture.

This interaction can be largely eliminated by a further artifice. We shall present the approach in general terms since we shall want to make use of it at various places in this book. In addition it will allow us to grasp the concept of *quasi-particles* more precisely.

We consider a system of electrons (fermions) and bosons with mutual interaction. The bosons can be the plasmons of the previous section or other collective excitations, like the photons, the phonons, or the magnons which we shall discuss later. Electron-electron or boson-boson interactions will be neglected for simplicity. We write the Hamiltonian in the form

$$H = \sum_{\boldsymbol{k}} E(\boldsymbol{k}) c_{\boldsymbol{k}}^{+} c_{\boldsymbol{k}} + \sum_{\boldsymbol{q}} \hbar\omega_{\boldsymbol{q}} a_{\boldsymbol{q}}^{+} a_{\boldsymbol{q}} + \sum_{\boldsymbol{k}\boldsymbol{q}} M_{\boldsymbol{q}} (a_{-\boldsymbol{q}}^{+} + a_{\boldsymbol{q}}) c_{\boldsymbol{k}+\boldsymbol{q}}^{+} c_{\boldsymbol{k}} \equiv H_0 + H_1 , \tag{3.34}$$

in which the interaction matrix element may only depend on the momentum $\boldsymbol{q}$ transferred in the interaction. We subject (3.34) now to a canonical transformation: $H_s = e^{-s} H e^{s}$. The transformed Hamiltonian possesses the same eigenvalues as H. Expanding e^s we have

$$\begin{aligned} H_s = e^{-s} H e^{s} &= (1 - s + \cdots) H (1 + s + \cdots) \\ &= H + [H, s] + \frac{1}{2}[[H, s], s] + \cdots \\ &= H_0 + (H_1 + [H_0, s]) + \frac{1}{2}[(H_1 + [H_0, s]), s] + \frac{1}{2}[H_1, s] + \cdots . \end{aligned} \tag{3.35}$$

If we choose s to be of the same order of magnitude as H_1, then the second term on the right is of the order H_1, the third and fourth of the order $(H_1)^2$, and the terms omitted of higher order. If we now choose s such that $H_1 + [H_0, s] = 0$, the second and third terms on the right in (3.35) disappear. Apart from a contribution of higher order, the electron-boson interaction H_1 is eliminated.

For the operator s we choose the form

$$s = \sum_{\boldsymbol{k}\boldsymbol{q}} M_{\boldsymbol{q}} (\alpha a_{-\boldsymbol{q}}^{+} + \beta a_{\boldsymbol{q}}) c_{\boldsymbol{k}+\boldsymbol{q}}^{+} c_{\boldsymbol{k}} \tag{3.36}$$

with as yet undefined coefficients α and β. Putting this into $H_1 + [H_0, s] = 0$ yields

$$\alpha^{-1} = E(\boldsymbol{k}) - E(\boldsymbol{k}+\boldsymbol{q}) - \hbar\omega_{\boldsymbol{q}}, \qquad \beta^{-1} = E(\boldsymbol{k}) - E(\boldsymbol{k}+\boldsymbol{q}) + \hbar\omega_{\boldsymbol{q}}. \tag{3.37}$$

If we use (3.36) and (3.37) in (3.35), the new interaction term $\frac{1}{2}[H_1, s]$ becomes a sum of terms which contain operator products of the form

$$a^{(+)}_{\pm\boldsymbol{q}} a^{(+)}_{\pm\boldsymbol{q}'} c^{+}_{\boldsymbol{k}'+\boldsymbol{q}'} c_{\boldsymbol{k}'} c^{+}_{\boldsymbol{k}+\boldsymbol{q}} c_{\boldsymbol{k}}. \tag{3.38}$$

They describe processes in which two electrons from states $\boldsymbol{k}$ and $\boldsymbol{k}'$ transfer into states $\boldsymbol{k}+\boldsymbol{q}$ and $\boldsymbol{k}'+\boldsymbol{q}'$. From momentum conservation we can put $\boldsymbol{q}' = -\boldsymbol{q}$. The four combinations $a^{+}_{-\boldsymbol{q}} a^{+}_{\boldsymbol{q}}$, $a_{\boldsymbol{q}} a_{-\boldsymbol{q}}$, $a_{-\boldsymbol{q}} a^{+}_{-\boldsymbol{q}}$, $a^{+}_{\boldsymbol{q}} a_{\boldsymbol{q}}$ then remain as boson factors in (3.38). Of these latter, the first two combinations describe two-boson processes. The two other combinations can be written in the form $1 + n_{-\boldsymbol{q}}$ and $n_{\boldsymbol{q}}$. They appear in (3.38) together with the factors $c^{+}_{\boldsymbol{k}'-\boldsymbol{q}} c_{\boldsymbol{k}'} c^{+}_{\boldsymbol{k}+\boldsymbol{q}} c_{\boldsymbol{k}}$. Taking account of the commutation relations for the c-operators, we find among the combinations (3.38) one which does not contain any a-operators

$$c^{+}_{\boldsymbol{k}+\boldsymbol{q}} c^{+}_{\boldsymbol{k}'-\boldsymbol{q}} c_{\boldsymbol{k}'} c_{\boldsymbol{k}} \qquad \boldsymbol{k}' \neq \boldsymbol{k}, \boldsymbol{k}+\boldsymbol{q} \tag{3.39}$$

and for $\boldsymbol{k}' = \boldsymbol{k}+\boldsymbol{q}$ combinations which only contain particle number operators

$$n_{\boldsymbol{k}}(1 - n_{\boldsymbol{k}+\boldsymbol{q}}) \quad \text{and} \quad n_{\boldsymbol{k}}(1 - n_{\boldsymbol{k}+\boldsymbol{q}}) n_{\boldsymbol{q}}. \tag{3.40}$$

Terms of the form (3.39) mean an additional contribution to the electron-electron interaction which is not written down in (3.34). We shall see in Section 5.1 that they describe an electron-electron interaction by exchange of virtual bosons.

The terms (3.40) contain the particle number operator for the electrons; the second term also has that for the bosons. The first term can be added to the term $\sum_{\boldsymbol{k}} E_{\boldsymbol{k}} n_{\boldsymbol{k}}$ in H_0 and means—as we shall discuss in more detail further below—a *renormalization* of the electron mass. The second terms in (3.40) can correspondingly be interpreted as contributions to a renormalization of the boson frequency.

In order to determine this renormalization quantitatively, we do not want to follow the rather complicated route through the canonical transformation (3.35). It is simpler to start from (3.7) and calculate the second-order interaction energy. If we put H_1 from (3.34) into the equation

$$E_n^{(1)} = E_n^{(0)} + \langle n|H_1|n\rangle + \sum_{m(\neq n)} \frac{|\langle m|H_1|n\rangle|^2}{E_n^{(0)} - E_m^{(0)}} \tag{3.41}$$

and use state vectors of the form $|\ldots n_{\boldsymbol{k}} \ldots; \ldots n_{\boldsymbol{q}} \ldots\rangle$, we find

$$E^{(1)} = E^{(0)} + \sum_{\boldsymbol{kq}} |M_{\boldsymbol{q}}|^2 (1 - n_{\boldsymbol{k}+\boldsymbol{q}}) n_{\boldsymbol{k}} \times \left\{ \frac{n_{\boldsymbol{q}}}{E^{(0)}(\boldsymbol{k}) - E^{(0)}(\boldsymbol{k}+\boldsymbol{q}) + \hbar\omega_{\boldsymbol{q}}} + \frac{n_{-\boldsymbol{q}} + 1}{E^{(0)}(\boldsymbol{k}) - E^{(0)}(\boldsymbol{k}+\boldsymbol{q}) - \hbar\omega_{\boldsymbol{q}}} \right\}. \tag{3.42}$$

We modify this equation by replacing the occupation numbers with their mean values. We further take steady-state conditions and put $\bar{n}_{-\boldsymbol{q}} = \bar{n}_{\boldsymbol{q}}$. Finally we combine the sum terms in a different way. We then have

$$E^{(1)} = E^{(0)} + \sum_{\boldsymbol{kq}} |M_{\boldsymbol{q}}|^2 \bar{n}_{\boldsymbol{k}} \times \left\{ \frac{1 - \bar{n}_{\boldsymbol{k}+\boldsymbol{q}}}{E^{(0)}(\boldsymbol{k}) - E^{(0)}(\boldsymbol{k}+\boldsymbol{q}) - \hbar\omega_{\boldsymbol{q}}} + \frac{2[E^{(0)}(\boldsymbol{k}) - E^{(0)}(\boldsymbol{k}+\boldsymbol{q})]\bar{n}_{\boldsymbol{q}}}{[E^{(0)}(\boldsymbol{k}) - E^{(0)}(\boldsymbol{k}+\boldsymbol{q})]^2 - (\hbar\omega_{\boldsymbol{q}})^2} \right\}. \tag{3.43}$$

We find the single particle energies of the electrons and the bosons by differentiating this with respect to the $\bar{n}_{\boldsymbol{k}}$ and the $\bar{n}_{\boldsymbol{q}}$,

$$E^{(1)}(\boldsymbol{k}) = E^{(0)}(\boldsymbol{k}) + \sum_{\boldsymbol{q}} |M_{\boldsymbol{q}}|^2 \times \frac{[E^{(0)}(\boldsymbol{k}) - E^{(0)}(\boldsymbol{k}+\boldsymbol{q}) + \hbar\omega_{\boldsymbol{q}}](1 - 2\bar{n}_{\boldsymbol{k}+\boldsymbol{q}}) + 2[E^{(0)}(\boldsymbol{k}) - E^{(0)}(\boldsymbol{k}+\boldsymbol{q})]\bar{n}_{\boldsymbol{q}}}{[E^{(0)}(\boldsymbol{k}) - E^{(0)}(\boldsymbol{k}+\boldsymbol{q})]^2 - (\hbar\omega_{\boldsymbol{q}})^2} \tag{3.44}$$

$$\hbar\omega_{\boldsymbol{q}}^{(1)} = \hbar\omega_{\boldsymbol{q}}^{(0)} + 2|M_{\boldsymbol{q}}|^2 \sum_{\boldsymbol{k}} \bar{n}_{\boldsymbol{k}} \frac{E^{(0)}(\boldsymbol{k}) - E^{(0)}(\boldsymbol{k}+\boldsymbol{q})}{[E^{(0)}(\boldsymbol{k}) - E^{(0)}(\boldsymbol{k}+\boldsymbol{q})]^2 - (\hbar\omega_{\boldsymbol{q}})^2}. \tag{3.45}$$

The change in the one-electron energy due to the interaction is made up of three terms. The first term would also contribute for $\bar{n}_{\boldsymbol{q}} = 0$. We see from (3.41) and (3.44) that it is concerned with processes in which a boson is emitted and then absorbed again through a virtual intermediate state. The term therefore represents the virtual boson cloud which the electron carries with it. The other terms extends this result to the case where $\bar{n}_{\boldsymbol{q}} \neq 0$. Eq. (3.44) can be written formally in the form

$$E(k) = \frac{\hbar^2 k^2}{2m} + \Delta E(k) \equiv \frac{\hbar^2 k^2}{2m^*(k)}. \tag{3.46}$$

The electron with its boson cloud thus behaves like a noninteracting particle with a k-dependent *effective mass* m^*.

A closer discussion of this mass renormalization shows further that $E(\boldsymbol{k})$ is altered particularly in the vicinity of the Fermi surface. For $k = k_F$, $E(\boldsymbol{k})$ does indeed remain unchanged. However, below the Fermi surface $E(\boldsymbol{k})$ is displaced to higher values, and above it to lower values; dE/dk is therefore smaller. Or, put another way, the electron velocity is reduced in the region of the Fermi energy.

Equation (3.45) means a renormalization of the boson frequency. It is concerned here with the virtual excitation of individual electrons out of the Fermi sphere, or differently expressed, with the creation of virtual electron-hole pairs.

All these results can now been applied to the screened electron-plasmon system. The electron-plasmon interaction (3.31) can be eliminated by a suitable canonical transformation. This leads to renormalization of the electron mass. The plasmon frequency becomes $\boldsymbol{k}$-dependent. Its value ω_p is only the limiting value for $\boldsymbol{k} = 0$. We shall leave the reader to discover the transformation operator and carry out the calculation as an exercise.

In this chapter we have defined various *quasi-electrons* which differ from free electrons in a way which depends on the approximation used to include parts of the electron-electron interaction: the *Hartree-Fock electron*, the *screened electron*, and the *screened and renormalized electron.* In each case a part of the interaction to which the "bare particle" is subjected is incorporated into the mass of the "dressed particle". The residual interaction was so much reduced by this that the quasi-electron—at least close to the Fermi surface—could be defined as a stable quasi-particle.

The quasi-particle concept is not defined uniquely in the literature. It comes originally from the *Landau theory of quantum fluids.* If one describes a system of fermions with a strong fermion-fermion interaction it is often not possible or not helpful to start from the free fermion gas as the zeroth approximation. In spite of this it is almost always possible to find a set of single-particle states which can form the basis of a treatment of the interaction by perturbation methods. The particles associated with these states are then the quasi-particles. A fermion system is called normal if the quasi-particles become real fermions when the interaction is removed adiabatically. This is so in all cases considered in this section. We refer to the literature (e.g. [36–47]) for a presentation of this theory which is of basic importance for a quantitative understanding of the quasi-particle concept.

In what follows we shall interpret every electron as a quasi-particle or quasi-electron if parts of an interaction are incorporated into its characteristics. Thus the *Bloch electron* of the band model also comes within this definition. We shall come to know a further quasi-electron, the *polaron*, in Section 4.1.3.

Plasmons belong to the *collective excitations*, which have no corresponding real particles when the interactions are removed. They are bosons. We shall meet further collective excitations in the following chapters.

We combine quasi-particles and collective excitations under the title *elementary excitations* by which we shall often refer to them in this book. *Excitons*

(Sec. 3.2) and *Cooper pairs* (Sec. 5.2) also belong to the elementary excitations. They cannot however be uniquely associated with the two subgroups mentioned, a point which we shall return to later.

Before discussing the most important of the other elementary excitations, namely the phonons, magnons, and excitons, we want to expand our discussion of the interacting electron gas by considering the dielectric constant of the gas.

3.1.6 The Dielectric Constant of the Electron Gas

The screened electron-electron interaction problem can also be treated from an entirely different point of view, that of the dielectric behaviour of the electron gas under an external perturbation.

We start with a homogeneous noninteracting electron gas of concentration $n = N/V_g$. If we apply an external perturbation $V_a(\boldsymbol{r}, t)$, fluctuations in the electron gas are induced ($n = n_0 + \delta n$) which are connected with an internal potential through the Poisson equation

$$\nabla^2 V_i(\boldsymbol{r}, t) = -\frac{e^2}{\varepsilon_0}\delta n(\boldsymbol{r}, t). \tag{3.47}$$

The total potential acting on the electron is then $V(\boldsymbol{r}, t) = V_a(\boldsymbol{r}, t) + V_i(\boldsymbol{r}, t)$. The part V_i describes the screening effect of the electron gas on an electron at the point $\boldsymbol{r}$. The ratio between V_a and V can be defined as the dielectric constant of the electron gas. It will emerge that this ratio is frequency- and wave number-dependent. Accordingly, in this approach the electron-electron interaction is described by a *frequency- and wave number-dependent dielectric constant*.

We describe the equilibrium state of the electron gas (index 0) by the statistical operator ρ_0 of (2.26). From (2.29) we have

$$H_0|\boldsymbol{k}\rangle = -\frac{\hbar^2}{2m}\nabla^2|\boldsymbol{k}\rangle = E(\boldsymbol{k})|\boldsymbol{k}\rangle, \qquad \rho_0|\boldsymbol{k}\rangle = f_0(\boldsymbol{k})|\boldsymbol{k}\rangle, \tag{3.48}$$

where f_0 is the Fermi distribution.

In the presence of the time-dependent external perturbation $H = H_0 + V(\boldsymbol{r}, t)$, $\rho = \rho_0 + \delta\rho$, and since $i\hbar\dot{\rho} = [H, \rho]$ we have for ρ

$$i\hbar\delta\dot{\rho}|\boldsymbol{k}\rangle = [H, \rho]|\boldsymbol{k}\rangle = ([H_0, \delta\rho] + [V, \rho_0])|\boldsymbol{k}\rangle, \tag{3.49}$$

where we have neglected terms of the order $V\delta\rho|\boldsymbol{k}\rangle$ on the right-hand side.

The next step is to form matrix elements between two states $|\boldsymbol{k}'\rangle$ and $|\boldsymbol{k}\rangle$. Eq. (3.49) becomes

$$i\hbar\langle\boldsymbol{k}'|\delta\dot{\rho}|\boldsymbol{k}\rangle = [E(\boldsymbol{k}') - E(\boldsymbol{k})]\langle\boldsymbol{k}'|\delta\rho|\boldsymbol{k}\rangle - [f_0(\boldsymbol{k}') - f_0(\boldsymbol{k})]\langle\boldsymbol{k}'|V|\boldsymbol{k}\rangle. \tag{3.50}$$

The matrix element $\langle \boldsymbol{k}'|V|\boldsymbol{k}\rangle$ is the $\boldsymbol{q}$th Fourier component of the potential

$$V_{\boldsymbol{q}}(t) = \frac{1}{V_g}\int \exp(-i\boldsymbol{q}\cdot\boldsymbol{r})V(\boldsymbol{r}, t)d\tau \tag{3.51}$$

with $\boldsymbol{q} = \boldsymbol{k}' - \boldsymbol{k}$.

It is convenient at this point to give the external perturbation a definite time dependence. We choose

$$V_a(\boldsymbol{r}, t) = V_a(\boldsymbol{q}, \omega) \exp[i(\boldsymbol{q}\cdot\boldsymbol{r} - \omega t)] \exp(\alpha t), \tag{3.52}$$

i.e., a Fourier component of a general time-dependent perturbation. The factor $\exp(\alpha t)$ takes care of the fact that the system adiabatically follows the perturbation, which grows exponentially from the time it is applied at $t = -\infty$. In the final result we can then let α approach the limit zero.

We can take the same $\boldsymbol{r}$- and t-dependence for the screening potential V_i and for V itself. Since $\delta\rho$ will then also have this time dependence, it follows from (3.50) that

$$[E(\boldsymbol{k} + \boldsymbol{q}) - E(\boldsymbol{k}) + i\hbar(i\omega - \alpha)]\langle \boldsymbol{k} + \boldsymbol{q}|\delta\rho|\boldsymbol{k}\rangle = [f_0(\boldsymbol{k} + \boldsymbol{q}) - f_0(\boldsymbol{k})]V_{\boldsymbol{q}}(t) \tag{3.53}$$

and $V_{\boldsymbol{q}}(t)$ becomes $[V_a(\boldsymbol{q}, \omega) + V_i(\boldsymbol{q}, \omega)] \exp(-i\omega t + \alpha t)$.

So far we have not yet made use of the Poisson equation (3.47). The $\boldsymbol{q}$th component of (3.47) becomes

$$-q^2 V_{i\boldsymbol{q}}(t) = -\frac{e^2}{\varepsilon_0}\delta n_{\boldsymbol{q}}. \tag{3.54}$$

We shall have to link $\delta n_{\boldsymbol{q}}$ with the matrix elements of $\delta\rho$. To do this we use the fact that from (2.24) the particle concentration $\delta n(\boldsymbol{r}, t)$ is given by

$$\begin{aligned}
\delta n(\boldsymbol{r}_0, t) &= \mathrm{Trace}\,\{\delta(\boldsymbol{r} - \boldsymbol{r}_0)\delta\rho\} = \sum_{\boldsymbol{k}\boldsymbol{k}'} \langle \boldsymbol{k}|\delta(\boldsymbol{r} - \boldsymbol{r}_0)|\boldsymbol{k}'\rangle\langle \boldsymbol{k}'|\delta\rho|\boldsymbol{k}\rangle \\
&= \frac{1}{V_g}\sum_{\boldsymbol{k}\boldsymbol{k}'} \exp[i(\boldsymbol{k}' - \boldsymbol{k})\cdot\boldsymbol{r}_0]\langle \boldsymbol{k}'|\delta\rho|\boldsymbol{k}\rangle \\
&= \frac{1}{V_g}\sum_{\boldsymbol{q}} \exp(i\boldsymbol{q}\cdot\boldsymbol{r}_0)\sum_{\boldsymbol{k}} \langle \boldsymbol{k} + \boldsymbol{q}|\delta\rho|\boldsymbol{k}\rangle = \sum_{\boldsymbol{q}} \exp(i\boldsymbol{q}\cdot\boldsymbol{r}_0)\delta n_{\boldsymbol{q}}.
\end{aligned} \tag{3.55}$$

The last line in (3.55) gives the connection we are looking for. Eqs. (3.53), (3.54) and (3.55) together then lead to

$$\begin{aligned}
V_i(\boldsymbol{q}, \omega) &= V(\boldsymbol{q}, \omega) - V_a(\boldsymbol{q}, \omega) \\
&= \frac{e^2}{V_g\varepsilon_0 q^2}\sum_{\boldsymbol{k}} \frac{f_0(\boldsymbol{k} + \boldsymbol{q}) - f_0(\boldsymbol{k})}{E(\boldsymbol{k} + \boldsymbol{q}) - E(\boldsymbol{k}) - \hbar\omega - i\hbar\alpha} V(\boldsymbol{q}, \omega).
\end{aligned} \tag{3.56}$$

Equation (3.56) links the Fourier components of the external potential $V_a(\boldsymbol{q}, \omega)$ with those of the local potential $V(\boldsymbol{q}, \omega)$. We can then define the dielectric constant by

$$V(\boldsymbol{q}, \omega) = \frac{V_a(\boldsymbol{q}, \omega)}{\varepsilon(\boldsymbol{q}, \omega)} \tag{3.57}$$

with

$$\varepsilon(\boldsymbol{q}, \omega) = 1 - \lim_{\alpha \to 0} \frac{e^2}{V_g \varepsilon_0 q^2} \sum_{\boldsymbol{k}} \frac{f_0(\boldsymbol{k} + \boldsymbol{q}) - f_0(\boldsymbol{k})}{E(\boldsymbol{k} + \boldsymbol{q}) - E(\boldsymbol{k}) - \hbar\omega - i\hbar\alpha}. \tag{3.58}$$

Eq. (3.58) is called the *Lindhard equation* for the dielectric constant of the electron gas.

This equation provides the most important results derived by other means in Section 3.1.4.

We first resolve the dielectric constant into its real and imaginary parts. Using the relationship

$$\lim_{\alpha \to 0} \frac{1}{z - i\alpha} = P\left(\frac{1}{z}\right) + i\pi\delta(z) \qquad \left[P\left(\frac{1}{z}\right) = \text{principal part of } \frac{1}{z}\right], \tag{3.59}$$

we find

$$\varepsilon_1(\boldsymbol{q}, \omega) = 1 - \frac{e^2}{V_g \varepsilon_0 q^2} \sum_{\boldsymbol{k}} P\left[\frac{f_0(\boldsymbol{k} + \boldsymbol{q}) - f_0(\boldsymbol{k})}{E(\boldsymbol{k} + \boldsymbol{q}) - E(\boldsymbol{k}) - \hbar\omega}\right], \tag{3.60}$$

$$\varepsilon_2(\boldsymbol{q}, \omega) = \frac{\pi e^2}{V_g \varepsilon_0 q^2} \sum_{\boldsymbol{k}} [f_0(\boldsymbol{k} + \boldsymbol{q}) - f_0(\boldsymbol{k})]\delta(E(\boldsymbol{k} + \boldsymbol{q}) - E(\boldsymbol{k}) - \hbar\omega). \tag{3.61}$$

The imaginary part of the dielectric constant is connected with the absorption constant of the electron gas. We see from the δ-function in (3.61) that absorption always occurs when the energy of the incident wave $\hbar\omega$ is equal to the difference in energy between two states $\boldsymbol{k}$ and $\boldsymbol{k} + \boldsymbol{q}$. Pair excitations of the kind described in Figs. 2.3 and 3.5 are therefore involved in the absorption. The conservation of energy indicated by the δ-function limits the pair excitations here to the hatched region in Figs. 2.3 and 3.5. Above this region $\hbar\omega > (\hbar^2/2m) \times (k_F + q)q$. This region of vanishing absorption is characterized by the fact that $\hbar\omega$ is larger than every energy difference $E(\boldsymbol{k} + \boldsymbol{q}) - E(\boldsymbol{k})$ appearing in the denominator of (3.60). Eq. (3.60) can easily be rearranged for this region. We divide the summation over $\boldsymbol{k}$ into two parts and introduce new summation indices $\boldsymbol{k} + \boldsymbol{q}$ and $-\boldsymbol{k}$ into the first and second part, respectively. Then we bring the two parts together again and neglect all energy differences between states

$\boldsymbol{k} + \boldsymbol{q}$ and $\boldsymbol{k}$. To a first approximation we then find that

$$\varepsilon_1(\omega) = 1 - \frac{e^2 n}{\varepsilon_0 m \omega^2} = 1 - \frac{\omega_p^2}{\omega^2}. \tag{3.62}$$

$\varepsilon_1(\omega)$ goes to zero at the plasma resonance frequency ω_p. According to the definition of the dielectric constant in (3.57), this means that even an infinitesimally small external perturbation will produce strong internal fields. Thus the electron gas oscillates collectively.

We have thereby obtained plasma oscillations and the pair excitations of Fig. 3.5 out of (3.58). Finally we want to show the screening behaviour of the electron gas. We limit ourselves in this to the static case, i.e., $\omega = 0$.

We begin with an approximation for small $\boldsymbol{q}$. We expand the energy and the distribution function as

$$E(\boldsymbol{k} + \boldsymbol{q}) = E(\boldsymbol{k}) + \boldsymbol{q} \cdot \nabla_{\boldsymbol{k}} E + \cdots, \quad f_0(\boldsymbol{k} + \boldsymbol{q}) = f_0(\boldsymbol{k}) + \frac{\partial f_0}{\partial E} \boldsymbol{q} \cdot \nabla_{\boldsymbol{k}} E + \cdots \tag{3.63}$$

and obtain

$$\begin{aligned} \varepsilon_1(\boldsymbol{q}, 0) &= 1 + \frac{e^2}{\varepsilon_0 q^2} \frac{1}{V_g} \sum_{\boldsymbol{k}} \left(-\frac{\partial f_0}{\partial E}\right) = 1 + \frac{e^2}{\varepsilon_0 q^2} \int d\tau_{\boldsymbol{k}} g(\boldsymbol{k}) \delta(E - E_F) \\ &= 1 + \frac{e^2}{\varepsilon_0 q^2} g(E_F). \end{aligned} \tag{3.64}$$

If from (2.7) and (2.9) we replace $g(E_F)$ in (3.64) by E_F and n, we have finally

$$\varepsilon_1(\boldsymbol{q}, 0) = 1 + \frac{3e^2 n}{2\varepsilon_0 q^2 E_F} \equiv 1 + \frac{\lambda^2}{q^2}. \tag{3.65}$$

For a discussion of this equation we assume that $V_a(\boldsymbol{r})$ is the potential of an electron at $\boldsymbol{r} = 0$: $V_a(\boldsymbol{r}) = -e^2/4\pi\varepsilon_0 r\ [V_a(\boldsymbol{q}) = -e^2/\varepsilon_0 V_g q^2]$. Then from (3.5), (3.57) and (3.65) we have

$$V(\boldsymbol{q}) = -\frac{e^2}{q^2 + \lambda^2} \frac{1}{V_g \varepsilon_0} \quad \text{hence} \quad V(r) = -\frac{e^2}{4\pi\varepsilon_0 r} e^{-\lambda r}. \tag{3.66}$$

The charge is therefore exponentially screened. The screening constant is λ. This result is limited to small $\boldsymbol{q}$. For any $\boldsymbol{q}$ (3.64) is replaced by the expression

$$\varepsilon_1(\boldsymbol{q}, 0) = 1 + \frac{e^2}{\varepsilon_0 q^2} g(E_F) \left(\frac{1}{2} + \frac{1 - \eta^2}{4\eta^2} \ln \left| \frac{1 + \eta}{1 - \eta} \right| \right), \quad \eta = \frac{q}{2k_F}. \tag{3.67}$$

We do not want to discuss this further here, nor do we want to go into the further approximations for $\varepsilon(\boldsymbol{q}, \omega)$. We refer the reader particularly to the monographs by *Kittel* [16], *Pines* [27], and *Ziman* [23] for a discussion of these questions and a deeper understanding of the electron-electron interaction.

3.2 Electron-Hole Interaction in Semiconductors and Insulators: Excitons

3.2.1 Introduction

In Section 2.1.3 we came across pair excitations of the electron gas. The transition of an electron from the filled Fermi sphere into a state $k > k_F$ was described as the creation of an *electron-hole pair.*

In the free electron gas, occupied states in the Fermi sphere and unoccupied states outside it are directly adjacent. Pair excitations of infinitesimally small energy are therefore possible. However in insulators (semiconductors), from the results in Section 2.2 we know that between the ground state (filled valence band, empty conduction band) and the first excited state (one electron-hole pair) there is an excitation energy E_G, the energy gap.

The electron in the conduction band and the hole in the valence band are quasi-particles with opposite charge. In the band model approximation there is no interaction between them. In this chapter we want to go beyond this approximation and examine the pair excitation "electron-hole pair with mutual interaction". It is called an *exciton.* We shall derive it first from the basic equations of Chapter 1 and discuss its most important properties. The exciton will become important when we discuss the electron interaction with other elementary excitations, particularly in the optical phenomena in Chapter 6.

The following references are relevant: the monographs of *Knox* [102.5], the articles by *Elliott* and *Haken* in [118]; further the appropriate sections in *Anderson* [24], *Haug* [14], and *Maradudin* [117]. Many of the general textbooks given in the reference list contain sections on excitons.

3.2.2 The Ground State of the Insulator in Bloch and Wannier Representation

The theory of the band model is concerned with the one-electron Schrödinger equation (1.32). This differs from the Hartree-Fock equation (1.23) in that the mutual Coulomb- and exchange-interaction of the electron gas is averaged. In this way the electrons can be decoupled. They move in a common mean potential. The energy of a Bloch state given by the function $E_n(\boldsymbol{k})$ is independent of the distribution of electrons over the other states. In this approximation the electrons are thus regarded as noninteracting quasi-particles which occupy a given energy spectrum in accordance with Fermi statistics.

The pair excitation "electron-hole pair" then has the energy given by the difference between the Bloch state of the electron in the conduction band and the Bloch state of the hole in the valence band. To improve on this approximation, we note that in the Hartree-Fock approximation *before* the averaging which leads to (1.32), there is a difference between the energy of an excited electron for interaction with all the electrons in the ground state ($N + 1$ electron problem) and for interaction with the $N - 1$ electrons remaining in the valence band (N electron problem). This difference is precisely the electron-hole interaction we are looking for.

The difference is not too significant in the case of the *free electron gas*. There the energy spectrum of pair excitations is continuous, starting from zero energy. The creation of an electron-hole pair in an *insulator* leads, however, when we consider the interaction, to new "bound" states below the energy E_G. These are the exciton states.

In the quantitative treatment of the exciton problem we restrict ourselves for simplicity to a Bravais lattice with bivalent atoms. The $2N$ electrons of the outermost shell of the lattice atoms completely fill a valence band. At an energy E_G above the highest level lies the lowest level of the conduction band. We shall take an interest first of all in the ground state of the system.

The one-electron wave functions have not yet been specified in the Slater-determinants (1.19). They follow as solutions of the Hartree-Fock equation. We shall not go too far wrong if we replace these unknown (and difficult to calculate) functions by the solutions of the averaged equation (1.32), i.e., if we use Bloch functions as one-electron functions in the Slater determinant. The $\varphi_i(\boldsymbol{q}_k)$ are products of each of these Bloch functions with a spin function. For the ground state we need the N Bloch functions (given by the N $\boldsymbol{k}$-vectors of the Brillouin zone) of the valence band (band index m): $|m\boldsymbol{k}\rangle$. The energy of the ground state is given by (1.20). We can sum the spins immediately. As a result of the orthogonality of the spin functions, we have in the first term in (1.20) as well as a residual sum over all $\boldsymbol{k}$-vectors a factor 2 which accounts for the two spin directions, and in the second term as well as the two $\boldsymbol{k}$-summations a factor 4 which accounts for the four possible spin combinations, and in the final term again a factor 2 since only electrons of identical spin contribute to the exchange energy. Altogether we have for the energy of the ground state, in *Bloch representation*,

$$E_0 = 2\sum_{\boldsymbol{k}}\Bigg[\langle m\boldsymbol{k}| -\frac{\hbar^2}{2m}\nabla^2 + V(\boldsymbol{r})|m\boldsymbol{k}\rangle + \sum_{\boldsymbol{k}'(\neq \boldsymbol{k})}\langle m\boldsymbol{k}, m\boldsymbol{k}'|g|m\boldsymbol{k}, m\boldsymbol{k}'\rangle$$
$$-\frac{1}{2}\sum_{\boldsymbol{k}'(\neq \boldsymbol{k})}\langle m\boldsymbol{k}, m\boldsymbol{k}'|g|m\boldsymbol{k}', m\boldsymbol{k}\rangle\Bigg] \tag{3.68}$$

with $g(\boldsymbol{r}, \boldsymbol{r}') = e^2/4\pi\varepsilon_0|\boldsymbol{r} - \boldsymbol{r}'|$.

It is convenient to rearrange this expression a little. The Bloch functions are solutions of (1.32): $[-(\hbar^2/2m)\nabla^2 + U(\boldsymbol{r})]|m\boldsymbol{k}\rangle = E_m(\boldsymbol{k})|m\boldsymbol{k}\rangle$. The potential

$U(\mathbf{r})$ which appears here is the sum of the lattice potential $V(\mathbf{r})$ and the mean interaction $W(\mathbf{r})$ $[U(\mathbf{r}) = V(\mathbf{r}) + W(\mathbf{r})]$. Hence the first term on the right in (3.68) can be put into the form

$$\langle m\mathbf{k}|E_m(\mathbf{k}) - W(\mathbf{r})|m\mathbf{k}\rangle . \tag{3.69}$$

In addition to the total energy, we consider the one-electron energy which follows from the Hartree-Fock equation when Bloch functions are used. The one-electron energy is

$$W_m(\mathbf{k}) = \langle m\mathbf{k}|E_m(\mathbf{k}) - W(\mathbf{r})|m\mathbf{k}\rangle + \sum_{\boldsymbol{\kappa}} 2\langle m\mathbf{k}, m\boldsymbol{\kappa}|g|m\mathbf{k}, m\boldsymbol{\kappa}\rangle - \sum_{\boldsymbol{\kappa}} \langle m\mathbf{k}, m\boldsymbol{\kappa}|g|m\boldsymbol{\kappa}, m\mathbf{k}\rangle . \tag{3.70}$$

The sum over the spin has already been taken in (3.70). The energy of the ground state (3.68) is accordingly the sum over all one-particle energies (3.70), where the interaction terms are to be only half counted.

In contrast to the $E_m(\mathbf{k})$, the one-particle energies $W_m(\mathbf{k})$ in the Bloch representation depend on the occupation of other states.

Along with the description in terms of Bloch functions, another form of description is often suitable: the *Wannier representation*. It uses Wannier functions (2.104) instead of Bloch functions in the Slater determinants

$$a_m(\mathbf{R}_n, \mathbf{r}) = \frac{1}{\sqrt{N}} \sum_{\mathbf{k}} \exp(-i\mathbf{k}\cdot\mathbf{R}_n)\psi_m(\mathbf{k}, \mathbf{r}) = \frac{1}{\sqrt{N}} \sum_{\mathbf{k}} \exp[i\mathbf{k}\cdot(\mathbf{r} - \mathbf{R}_n)]u_m(\mathbf{k}, \mathbf{r}) . \tag{3.71}$$

If we form a Slater determinant with the Wannier functions, the Bloch functions $B_{jl} = \psi_m(\mathbf{k}_l, \mathbf{r}_j)$ are replaced by Wannier functions $W_{jl} = a_m(\mathbf{R}_l, \mathbf{r}_j)$. If we now multiply the Slater determinant in the Bloch representation with the unitary transformation matrix $U[U_{li} = N^{-1/2} \exp(-i\mathbf{k}_l\cdot\mathbf{R}_i)]$, the elements B_{jl} are replaced by the elements $\sum_l B_{jl}U_{li}$. From (3.71), however, these are just the elements W_{ji} of the Slater determinant in the Wannier representation. Thus one representation can go over into the other via a unitary transformation, and the energy of the ground state is the same in both approximations.

3.2.3 Excited States, the Exciton Representation

We now consider an electron from a state $m, \mathbf{k}, s$ in the filled valence band being raised into a state $n, \mathbf{k}', s'$ in the otherwise empty conduction band. Since in the ground state the total quasi-momentum and total spin are zero, the excited state has the quasi-momentum $\mathbf{K} = \mathbf{k}' - \mathbf{k}$ and spin $\hbar(s' - s)$.

The ground state is described by a Slater determinant with Bloch functions $\psi_m(\boldsymbol{k}_i, \boldsymbol{r}_j, s)$. To describe the excited state we replace the Bloch functions in the $(\boldsymbol{k}, s)$ column of the determinant by conduction band functions $\psi_n(\boldsymbol{k}'_i, \boldsymbol{r}_j, s')$.

The energy of such an excitation is readily obtained when we recognize that an excited electron sees a different potential to a valence electron. The energy of the ground state is given by (3.68). The removal of a valence electron from state m, $\boldsymbol{k}$, s makes a contribution $-W_m(\boldsymbol{k})$, where $W_m(\boldsymbol{k})$ is given by (3.70) and in this expression is to be summed over *all* valence states m, $\boldsymbol{k}$. The addition of an electron in a conduction band state n, $\boldsymbol{k}'$, s' makes three contributions to the energy, which we shall examine separately. First the one-electron energy $W_n(\boldsymbol{k}')$, which comes from (3.70) by replacing m, $\boldsymbol{k}$ by n, $\boldsymbol{k}'$ and summing again over all $\boldsymbol{k}$ in the valence band. This contribution contains the interaction with a filled valence band. We therefore subtract the interaction of the conduction band electron with the electron pair m, $\boldsymbol{k}$, $+s$. That gives a contribution $-2\langle n\boldsymbol{k}', m\boldsymbol{k}|g|n\boldsymbol{k}', m\boldsymbol{k}\rangle + \langle n\boldsymbol{k}', m\boldsymbol{k}|gm\boldsymbol{k}, m\boldsymbol{k}'\rangle$. The interaction of the electrons n, $\boldsymbol{k}'$, s' and m, $\boldsymbol{k}$, $-s$ then remains. Here we have to distinguish between the two possible spin directions. From the analogous helium problem we know that in the first excited state $[1s(1)2s(2)]$ a triplet state results for parallel spin and a singlet state for antiparallel spin. On account of the requirement of antisymmetry imposed on the wave function by an exchange of the two electrons, the spin components of the wave function are chosen symmetric or antisymmetric: $\alpha(1)\alpha(2)$, $\beta(1)\beta(2)$, $(1/\sqrt{2})[\alpha(1)\beta(2) \pm \beta(1)\alpha(2)]$. Correspondingly we have also to choose here suitable linear combinations of Slater determinants in order to obtain states of defined multiplicity.

When we do this, we find as interaction energy the Coulomb interaction of the pair plus (in the singlet state) or minus (in the triplet state) the exchange energy. This contribution and the second contribution partly cancel each other, and altogether the excitation energy is

$$\Delta W = W_n(\boldsymbol{k}') - W_m(\boldsymbol{k}) - \langle n\boldsymbol{k}', m\boldsymbol{k}|g|n\boldsymbol{k}', m\boldsymbol{k}\rangle + 2\delta_s\langle n\boldsymbol{k}', m\boldsymbol{k}|g|m\boldsymbol{k}, m\boldsymbol{k}'\rangle \tag{3.72}$$

where δ_s is equal to 1 for the singlet state and equal to 0 for the triplet state.

Equation (3.72) explicitly contains only the $\boldsymbol{k}$-vectors of the nonsaturated valence electron and of the conduction electron. It suits us at this point in the description to go over to the picture of the electron-hole pair as a pair excitation. Following the considerations of Section 2.1.3, we put for the $\boldsymbol{k}$-vector and the spin of the hole the values of the missing electron $\boldsymbol{k}_h = \boldsymbol{k}$, $s_h = s$. We have to regard that in this designation the resultant $\boldsymbol{k}$-vector and spin of the valence band after removal of the electron equals $-\boldsymbol{k}$, $-s$. For the $\boldsymbol{k}$-vector and spin of the exciton then follows $\boldsymbol{K} = \boldsymbol{k}' - \boldsymbol{k} = \boldsymbol{k}_e - \boldsymbol{k}_h$, $\sigma = \hbar(s' - s) = \hbar(s_e - s_h)$. In this designation we give the matrix elements between the spin symmetrized determinants considered above with the Hamiltonian (1.13), which contain as diagonal elements the expressions (3.72)

$$\begin{aligned}\langle n\boldsymbol{k}_\mathrm{e}, m\boldsymbol{k}_\mathrm{h}|H|n\boldsymbol{k}_\mathrm{e}', m\boldsymbol{k}_\mathrm{h}'\rangle = {} & \delta_{\boldsymbol{k}_\mathrm{h}\boldsymbol{k}_\mathrm{h}'}\delta_{\boldsymbol{k}_\mathrm{e}\boldsymbol{k}_\mathrm{e}'}[E_0 + W_n(\boldsymbol{k}_\mathrm{e}) + W_m(\boldsymbol{k}_\mathrm{h})] \\ & + \delta_{\boldsymbol{k}_\mathrm{e}-\boldsymbol{k}_\mathrm{h},\boldsymbol{k}_\mathrm{e}'-\boldsymbol{k}_\mathrm{h}'}(2\delta_\mathrm{s}\langle n\boldsymbol{k}_\mathrm{e}, m\boldsymbol{k}_\mathrm{h}'|g|m\boldsymbol{k}_\mathrm{h}, n\boldsymbol{k}_\mathrm{e}'\rangle \\ & - \langle n\boldsymbol{k}_\mathrm{e}, m\boldsymbol{k}_\mathrm{h}'|g|n\boldsymbol{k}_\mathrm{e}', m\boldsymbol{k}_\mathrm{h}\rangle)\,. \end{aligned} \tag{3.73}$$

Eq. (3.73) contains nondiagonal elements. Eq. (3.72) does not therefore give the eigenvalues of the problem. For these, (3.73) must be diagonalized. In other words, individual Slater determinants are not sufficient to describe an excited state. Rather is such a state described by a superposition of different determinants.

Because of the translational symmetry of the problem, one can only superimpose determinants which describe states of the same $\boldsymbol{K} = \boldsymbol{k}_\mathrm{e} - \boldsymbol{k}_\mathrm{h}$. This is also expressed in the δ-symbol for the nondiagonal elements in (3.73), for if one replaces all the $\boldsymbol{r}$ in the Slater determinant of an excited state by $\boldsymbol{r} + \boldsymbol{R}_n$, one obtains factors of the form $\exp[\mathrm{i}(\boldsymbol{k}_1 + \boldsymbol{k}_2 + \cdots + \boldsymbol{k}_N)\cdot\boldsymbol{R}_n]$. Since $\sum_i \boldsymbol{k}_i = 0$ for a full band, these factors are equal to $\exp[\mathrm{i}(\boldsymbol{k}_\mathrm{e} - \boldsymbol{k}_\mathrm{h})\cdot\boldsymbol{R}_n] = \exp(\mathrm{i}\boldsymbol{K}\cdot\boldsymbol{R}_n)$. Since the solutions to the many-electron problem must also have a definite quasi-momentum, only those Slater determinants for which $\boldsymbol{K}$ has a fixed value can be superimposed.

It is not helpful to diagonalize (3.73) in the Bloch representation. It is better to consider first the corresponding expression in the Wannier representation. We start from Slater determinants with Wannier functions (3.71) for the ground state and then replace the Wannier functions of the $(\boldsymbol{R}_i, s)$th column by conduction band functions $a_n(\boldsymbol{R}_i', \boldsymbol{r}_j, s')$. The difference to the Bloch representation is then clear. The Bloch functions are characterized by the $\boldsymbol{k}$-vector, the Wannier function by the lattice position $\boldsymbol{R}_i$. The exciton in Bloch representation is characterized by its quasi-momentum $\boldsymbol{K}$. Correspondingly, in the Wannier representation the characteristic parameter is $\boldsymbol{\beta} = \boldsymbol{R}_i' - \boldsymbol{R}_i$, the separation between electron and hole. Both parameters are important in describing excitons. They are combined in the *exciton representation.*

The wave functions of this representation are given by

$$\begin{aligned}\Phi_{mn}(\boldsymbol{K}, \boldsymbol{\beta}) &= \frac{1}{\sqrt{N}}\sum_{\boldsymbol{k}} \exp(-\mathrm{i}\boldsymbol{k}\cdot\boldsymbol{\beta})\Phi_{mn}(\boldsymbol{k} - \boldsymbol{K}, \boldsymbol{k}) \\ &= \frac{1}{\sqrt{N}}\sum_{\boldsymbol{R}} \exp(\mathrm{i}\boldsymbol{K}\cdot\boldsymbol{R})\Phi_{mn}(\boldsymbol{R}, \boldsymbol{R} + \boldsymbol{\beta}) \end{aligned} \tag{3.74}$$

where the Φ_{mn} on the right-hand side in (3.74) are Slater determinants in the Bloch or Wannier representation. Through (3.74), the matrix elements can be converted from the Bloch to the exciton representation. One obtains

$$\begin{aligned}\langle mn\boldsymbol{K}\boldsymbol{\beta}|H|mn\boldsymbol{K}\boldsymbol{\beta}'\rangle &= E_0\delta_{\boldsymbol{\beta}\boldsymbol{\beta}'} + \frac{1}{N}\sum_{\boldsymbol{k}} \exp[\mathrm{i}\boldsymbol{k}\cdot(\boldsymbol{\beta}-\boldsymbol{\beta}')][W_n(\boldsymbol{k}) - W_m(\boldsymbol{k}-\boldsymbol{K})] \\ &\quad + 2\delta_s\langle n\boldsymbol{\beta}, m\boldsymbol{O}|g|m\boldsymbol{O}, n\boldsymbol{\beta}'\rangle - \langle n\boldsymbol{\beta}, m\boldsymbol{O}|g|n\boldsymbol{\beta}', m\boldsymbol{O}\rangle \\ &\quad + \sum_{\boldsymbol{R}(\neq 0)} \exp(\mathrm{i}\boldsymbol{K}\cdot\boldsymbol{R})(2\delta_s\langle n\boldsymbol{\beta}, m\boldsymbol{R}|g|m\boldsymbol{O}, n\,\boldsymbol{R}+\boldsymbol{\beta}'\rangle \\ &\quad - \langle n\boldsymbol{\beta}, m\boldsymbol{R}|g|n\,\boldsymbol{R}+\boldsymbol{\beta}', m\boldsymbol{O}\rangle)\,. \end{aligned} \tag{3.75}$$

Here $\boldsymbol{O}$ means the (arbitrarily chosen) origin of coordinates. The diagonalization of (3.75) is achieved by superposition of wave functions (3.74)

$$\Psi_{mn\nu\boldsymbol{K}} = \sum_{\boldsymbol{\beta}} U_{mn\nu\boldsymbol{K}}(\boldsymbol{\beta})\Phi_{mn}(\boldsymbol{K}, \boldsymbol{\beta})\,. \tag{3.76}$$

The eigenvalue problem is then

$$\sum_{\boldsymbol{\beta}'} \langle mn\boldsymbol{K}\boldsymbol{\beta}|H|mn\boldsymbol{K}\boldsymbol{\beta}'\rangle U_{mn\nu\boldsymbol{K}}(\boldsymbol{\beta}') = EU_{mn\nu\boldsymbol{K}}(\boldsymbol{\beta}) \tag{3.77}$$

and the eigenvalues are determined by the zeros of the determinant

$$\det|\langle mn\boldsymbol{K}\boldsymbol{\beta}|H|mn\boldsymbol{K}\boldsymbol{\beta}'\rangle - E\delta_{\boldsymbol{\beta}\boldsymbol{\beta}'}| = 0\,. \tag{3.78}$$

In treating this problem further, it suits us to consider the limiting cases of weak and strong electron-hole interaction. This is done in the next two sections. The generality of the results is so far limited only by the assumption of a Bravais lattice and the restriction to *one* (filled in the ground state) valence band and *one* conduction band.

3.2.4 Wannier Excitons

Of the interaction terms in (3.75), the Coulomb interaction between electron and hole $\langle n\boldsymbol{\beta}, m\boldsymbol{O}|g|n\boldsymbol{\beta}, m\boldsymbol{O}\rangle = e^2/4\pi\varepsilon_0\beta$ is the most important. At least this term has the longest range. In the case where electron and hole are separated by several lattice constants, it is reasonable in a first approximation to neglect all other interaction terms or to consider them later by perturbation methods. Excitons described by this approximation are called *Wannier excitons.*

For Wannier excitons we write (3.77) in the form

$$\begin{aligned}\sum_{\boldsymbol{\beta}'}\left\{E_0\delta_{\boldsymbol{\beta}\boldsymbol{\beta}'} + \frac{1}{N}\sum_{\boldsymbol{k}} \exp[\mathrm{i}\boldsymbol{k}\cdot(\boldsymbol{\beta}-\boldsymbol{\beta}')][W_n(\boldsymbol{k}) - W_m(\boldsymbol{k}-\boldsymbol{K})] - \frac{e^2}{4\pi\varepsilon_0\beta}\delta_{\boldsymbol{\beta}\boldsymbol{\beta}'}\right\} \\ \times\, U_{\nu\boldsymbol{K}}(\boldsymbol{\beta}') = EU_{\nu\boldsymbol{K}}(\boldsymbol{\beta})\,. \end{aligned} \tag{3.79}$$

To convert the second term, we use the same artifice as in Section 2.2.9 [(2.142, 143)]. With

$$U(\boldsymbol{\beta}') = \sum_{\boldsymbol{\kappa}} \exp(\mathrm{i}\boldsymbol{\kappa}\cdot\boldsymbol{\beta}')G(\boldsymbol{\kappa}) \quad \text{and} \quad W_n(\boldsymbol{k}) = \sum_m \exp(\mathrm{i}\boldsymbol{k}\cdot\boldsymbol{R}_m)W_{nm} \tag{3.80}$$

we have

$$\frac{1}{N}\sum_{\beta'}\sum_{k}\exp[ik\cdot(\beta-\beta')]W_n(k)U(\beta')$$
$$=\frac{1}{N}\sum_{\beta',k,\kappa,m}W_{nm}\exp[ik\cdot R_m+ik\cdot\beta+i(\kappa-k)\cdot\beta']G(\kappa)$$
$$=\sum_{\kappa,m}W_{nm}\exp(i\kappa\cdot R_m+i\kappa\cdot\beta)G(\kappa)$$
$$=\sum_{\kappa,m}W_{nm}\exp(R_m\cdot\nabla_\beta)\exp(i\kappa\cdot\beta)G(\kappa)=W_n(-i\nabla_\beta)U(\beta), \tag{3.81}$$

and we obtain

$$\left[W_n(-i\nabla_\beta)-W_m(-i\nabla_\beta-K)-\frac{e^2}{4\pi\varepsilon_0\beta}\right]U_{\nu K}(\beta)=(E-E_0)U_{\nu K}(\beta) \tag{3.82}$$

or, finally, with the help of the transformation $F_{\nu K}(\beta)=\exp(-iK\cdot\beta/2)U_{\nu K}(\beta)$

$$\left[W_n\left(-i\nabla_\beta+\frac{K}{2}\right)-W_m\left(-i\nabla_\beta-\frac{K}{2}\right)-\frac{e^2}{4\pi\varepsilon_0\beta}\right]F_{\nu K}(\beta)=(E-E_0)F_{\nu K}(\beta)\,. \tag{3.83}$$

Equation (3.83) corresponds to (2.147). We use the effective mass approximation to examine it. For the $W_{n,m}(k)$ in the vicinity of the lower edge of the conduction band and upper edge of the valence band we put

$$W_n(k)=E_n+\frac{\hbar^2k^2}{2m_n}\quad\text{and}\quad W_m(k)=E_m-\frac{\hbar^2k^2}{2m_p}\,. \tag{3.84}$$

Using the reduced mass $\mu^{-1}=m_n^{-1}+m_p^{-1}$ and the energy gap $E_G=E_n-E_m$, we find

$$\left[-\frac{\hbar^2}{2\mu}\nabla^2-\frac{e^2}{4\pi\varepsilon_0\beta}-\frac{\hbar^2}{2i}\left(\frac{1}{m_p}-\frac{1}{m_n}\right)K\cdot\nabla\right]F=\left(E-E_0-E_G-\frac{\hbar^2K^2}{8\mu}\right)F\,. \tag{3.85}$$

The third term here can be eliminated by the transformation

$$F=\exp\left(\frac{i}{2}\frac{m_n-m_p}{m_n+m_p}K\cdot\beta\right)F'$$

and the final result is

$$\left(-\frac{\hbar^2}{2\mu}\nabla^2-\frac{e^2}{4\pi\varepsilon_0\beta}\right)F'=\left[E-E_0-E_G-\frac{\hbar^2K^2}{2(m_n+m_p)}\right]F'\,. \tag{3.86}$$

This equation is formally identical to the Schrödinger equation for the free hydrogen atom. Consequently it has the eigenvalues

$$E = E_0 + E_n(\boldsymbol{K}),\ E_n(\boldsymbol{K}) = E_G - \frac{\mu e^4}{32\pi^2\varepsilon_0^2\hbar^2 n^2} + \frac{\hbar^2 K^2}{2(m_n + m_p)}. \tag{3.87}$$

$E_n(\boldsymbol{K})$, the exciton energy, consists of the energy difference E_G between the two bands less the binding energy (level spectrum corresponding to the hydrogen problem), and the kinetic energy of the centre of gravity of the exciton.

We can improve the approximation used here phenomenologically through the interpretation of (3.86). Eq. (3.86) describes two particles of opposite charge, with mutual Coulomb interaction and with effective masses m_n and m_p. The effect of the electron-hole interaction on the other charged particles in the vicinity is not considered. For large separations between electron and hole the crystal can be taken as a homogeneous medium with static dielectric constant $\varepsilon(0)$ in which the electron-hole pair moves. The interaction is thereby reduced by a factor $1/\varepsilon(0)$. Polarization of the lattice ions is the prime cause of this. With reducing separation between electron and hole, the rotation frequency of each particle round the other increases. Since the polarization of the lattice means an excitation of optical phonons, this polarization will disappear when the rotation frequency becomes higher than the frequency of the longitudinal optical phonons (cf. Sec. 3.3.8). However there still remains then the polarization of the valence electrons, whose screening effect on the Coulomb field is described by the high-frequency dielectric constant $\varepsilon(\infty)$. If electron and hole come still closer together, the electrons too will eventually be unable to follow the polarization effect of the pair. However this only starts to happen at such small separations (a few atomic distances) that it is also then no longer possible to neglect exchange-interaction.

In the context of the Wannier approximation it is therefore expedient to modify the Coulomb interaction in (3.86) by introduction of an *effective dielectric constant* ε^*. It has a value $\varepsilon(0)$ for large $\boldsymbol{\beta}$, and $\varepsilon(\infty)$ for small $\boldsymbol{\beta}$. In the transition zone (less than 50 lattice constants for typical crystals in which Wannier excitons are observed), the $\boldsymbol{\beta}$-dependence of ε^* is relatively complicated. The theoretical approach to this is made possible by first taking account of the polarization effect of the electron on its surroundings by introducing new quasi-particles, the polarons, and then constructing the exciton from polarons. We shall not examine the polaron concept until Chapter 4.

The introduction of an effective dielectric constant alters (3.86) to

$$\left(-\frac{\hbar^2}{2\mu}\nabla^2 - \frac{e^2}{4\pi\varepsilon_0\varepsilon^*\beta}\right)F' = (E - E_0 - E_G - E_{\mathrm{kin}})F'. \tag{3.88}$$

The binding energy in (3.87) is thereby reduced by a factor ε^{*-2}.

Summarizing, we can interpret the result in this way: In the excitation of electrons out of the valence band and into the conduction band, the one-electron approximation of the band model does not take account of the Coulomb interaction between the excited electron and the hole left in the valence band. It leads—if we restrict ourselves to transitions which maintain the electron $\boldsymbol{k}$-vector, i.e., to excitons with $\boldsymbol{K} = 0$—to an excitation spectrum similar to that of hydrogen, with limiting energy E_G. A further energy contribution must be added for indirect transitions ($\boldsymbol{K} \neq 0$). For small $\boldsymbol{K}$ it can be described in accordance with (3.87) as the kinetic energy of the motion of the centre of gravity of the exciton. This is not valid in all cases where indirect exciton transitions are important. Indirect transitions can become important when the extrema of conduction and valence bands do not lie at the same $\boldsymbol{k}$-vector. However then the effective mass expression (3.84) is not valid and (3.87) has to be corrected on two counts. The last term takes a much more complicated form. The last term but one (Rydberg term) is modified at least by a $\boldsymbol{K}$-dependent effective mass.

3.2.5 Frenkel Excitons

The Wannier exciton, as the limiting case of weak electron-hole interaction, leads to the continuum model in which electron and hole move in a homogeneous dielectric. The opposite case is an atomic model in which electron and hole are localized at the same lattice position (*Frenkel exciton*). It is then more appropriate to talk in terms of excited states of individual atoms and to leave the band model concept out of the picture. The most important property that this limiting case has in common with the Wannier exciton is the possibility of successive transfer of excitation energy from one lattice position to another, i.e., the movement of an exciton through the crystal. We limit our discussion to this problem.

In the Slater determinants of the excited states we use Wannier functions or atomic functions. If we consider a transition at lattice position $\boldsymbol{O}$ from the ground state 0 into the excited state 1, the matrix elements (3.75) are

$$\begin{aligned}\langle 01, \boldsymbol{KO}|H|01, \boldsymbol{KO}\rangle = E_0 &+ \frac{1}{N}\sum_{\boldsymbol{k}}[W_1(\boldsymbol{k}) - W_0(\boldsymbol{k}-\boldsymbol{K})] \\ &+ 2\delta_s\langle 1\boldsymbol{O}, 0\boldsymbol{O}|g|0\boldsymbol{O}, 1\boldsymbol{O}\rangle - \langle 1\boldsymbol{O}, 0\boldsymbol{O}|g|1\boldsymbol{O}, 0\boldsymbol{O}\rangle \\ &+ \sum_{\boldsymbol{R}(\neq 0)} \exp(\mathrm{i}\boldsymbol{K}\cdot\boldsymbol{R})(2\delta_s\langle 1\boldsymbol{O}, 0\boldsymbol{R}|g|0\boldsymbol{O}, 1\boldsymbol{R}\rangle \\ &- \langle 1\boldsymbol{O}, 0\boldsymbol{R}|g|1\boldsymbol{R}, 0\boldsymbol{O}\rangle). \qquad (3.89)\end{aligned}$$

The second term in the lattice summation [last two lines of (3.89)] is a Coulomb term which describes the interaction between the charge-clouds $a_1^*(\boldsymbol{r})a_1(\boldsymbol{r}-\boldsymbol{R})$ and $a_0^*(\boldsymbol{r}-\boldsymbol{R})a_0(\boldsymbol{r})$. The effective charge of these charge-clouds is given by the overlap of the wave functions involved and falls exponentially as the wave functions with increasing $\boldsymbol{R}$. The first term in the lattice summation only appears for singlet states. It refers to a Coulomb interaction between charge-clouds $a_1^*(\boldsymbol{r})a_0(\boldsymbol{r})$ and $a_0^*(\boldsymbol{r}-\boldsymbol{R})a_1(\boldsymbol{r}-\boldsymbol{R})$. The effective charge of this cloud does not alter with distance $\boldsymbol{R}$. The interaction here is given by multipole terms and does not decay exponentially with $\boldsymbol{R}$. Clearly this term means a displacement of excitation energy from lattice position $\boldsymbol{O}$ to lattice position $\boldsymbol{R}$, which is just the exciton movement mentioned above.

The requirements needed to obtain Frenkel excitons are satisfied most readily in crystals with large lattice constant and small dielectric constant, e.g., molecular crystals, but also in ionic crystals and solid rare gases.

3.2.6 Excitons as Elementary Excitations

In our treatment of excitons we have not so far gone into the question of how far excitons can be taken as elementary excitations and, in particular, whether these pair excitations are fermions or bosons. We shall examine these questions using the occupation number representation.

We define creation and annihilation operators c. In the conduction band (index n) $c_{n\boldsymbol{k}}^+$ and $c_{n\boldsymbol{k}}$ will create or destroy an electron with the $\boldsymbol{k}$-vector $\boldsymbol{k}$. In the valence band $c_{m\boldsymbol{k}}$ and $c_{m\boldsymbol{k}}^+$ will likewise create or destroy a hole. These operators are appropriate if we begin with the Bloch representation. If we prefer the Wannier representation, the corresponding operators $c_{n\boldsymbol{R}}^+$, $c_{n\boldsymbol{R}}$, $c_{m\boldsymbol{R}}^+$, $c_{m\boldsymbol{R}}$ describe creation and annihilation of particles at lattice position $\boldsymbol{R}$.

We obtain the commutation relations for these operators by application of products of two operators on the wave functions of the ground state or of an excited state. For the $c_{n\boldsymbol{R}}^{(+)}$ we then find that all $c_{n'\boldsymbol{R}'}^+$ commute with all $c_{n''\boldsymbol{R}''}^+$ just as all $c_{n'\boldsymbol{R}'}$ commute with all $c_{n''\boldsymbol{R}''}$. The $c_{n'\boldsymbol{R}'}^+$ commute with the $c_{n''\boldsymbol{R}''}$ if either $n' \neq n''$ or $\boldsymbol{R}' \neq \boldsymbol{R}''$. On the other hand, the c^+ and c with identical indices anticommute

$$[c_{n'\boldsymbol{R}}^+, c_{n'\boldsymbol{R}}]_+ = 1\,. \tag{3.90}$$

This follows, for example, from the fact that $c_{n'\boldsymbol{R}}^+ c_{n'\boldsymbol{R}}|0\rangle$ is zero if n' is the conduction band; on the other hand it is equal to $|0\rangle$ if $n' = m$. The opposite is true for $c_{n'\boldsymbol{R}} c_{n'\boldsymbol{R}}^+$. Hence, together we have $(c_{n'\boldsymbol{R}}^+ c_{n'\boldsymbol{R}} + c_{n'\boldsymbol{R}} c_{n'\boldsymbol{R}}^+)|0\rangle = |0\rangle$ or—since this equation can be proved for each wave function—the operator equation (3.90).

The $c_{n\boldsymbol{R}}$ are operators in the Wannier representation. Analogous to (3.74) we therefore convert to the exciton representation and introduce operators

$$\begin{aligned} b_{\beta\boldsymbol{K}}^+ &= \frac{1}{\sqrt{N}}\sum_{\boldsymbol{k}} \exp(-\mathrm{i}\boldsymbol{\beta}\cdot\boldsymbol{k})c_{n\boldsymbol{k}}^+ c_{m,\boldsymbol{k}-\boldsymbol{K}} = \frac{1}{\sqrt{N}}\sum_{\boldsymbol{R}} \exp(\mathrm{i}\boldsymbol{K}\cdot\boldsymbol{R})c_{n,\boldsymbol{R}+\boldsymbol{\beta}}^+ c_{m\boldsymbol{R}} \\ b_{\beta\boldsymbol{K}} &= \frac{1}{\sqrt{N}}\sum_{\boldsymbol{k}} \exp(\mathrm{i}\boldsymbol{\beta}\cdot\boldsymbol{k})c_{m,\boldsymbol{k}-\boldsymbol{K}}^+ c_{n\boldsymbol{k}} = \frac{1}{\sqrt{N}}\sum_{\boldsymbol{R}} \exp(-\mathrm{i}\boldsymbol{K}\cdot\boldsymbol{R})c_{m\boldsymbol{R}}^+ c_{n,\boldsymbol{R}+\boldsymbol{\beta}}\,. \end{aligned} \tag{3.91}$$

These operators create or annihilate excitons with quasi-momentum $\boldsymbol{K}$ and electron-hole separation $\boldsymbol{\beta}$. The transformation means the transition from individual excitations with defined $\boldsymbol{k}$ or $\boldsymbol{R}$ to collective excitations in which only $\boldsymbol{K}$ and $\boldsymbol{\beta}$ are given. Corresponding to (3.76), we then construct general *exciton operators* from linear combinations of these operators

$$b_{\nu\boldsymbol{K}}^{(+)} = \sum_{\boldsymbol{\beta}} U_{\nu\boldsymbol{K}}(\boldsymbol{\beta}) b_{\boldsymbol{\beta K}}^{(+)} . \tag{3.92}$$

By considering the commutation relations of these $b_{\nu\boldsymbol{K}}^{(+)}$, one can show that the excitons have boson character. The proof is simpler if carried out for the limiting case of Frenkel excitons. We shall restrict ourselves to this. As in the previous section we put $\boldsymbol{\beta} = 0$. We have then just one term left in (3.92). If we further put $m = 0$ and $n = 1$ then, since the operators for $\boldsymbol{K} \neq \boldsymbol{K}'$ clearly commute,

$$\begin{aligned}[b_{\boldsymbol{K}}, b_{\boldsymbol{K}'}^{+}] &= \delta_{\boldsymbol{KK}'}[b_{\boldsymbol{K}}, b_{\boldsymbol{K}}^{+}] \\ &= \frac{1}{N}\sum_{\boldsymbol{R}} [c_{0\boldsymbol{R}}^{+}c_{0\boldsymbol{R}}(1 - c_{1\boldsymbol{R}}^{+}c_{1\boldsymbol{R}}) - c_{1\boldsymbol{R}}^{+}c_{1\boldsymbol{R}}(1 - c_{0\boldsymbol{R}}^{+}c_{0\boldsymbol{R}})]\delta_{\boldsymbol{KK}'} .\end{aligned} \tag{3.93}$$

The product operators appearing in the last line are the particle number operators $N_{1\boldsymbol{R}}$ and $N_{0\boldsymbol{R}}$. They have the eigenvalues 1 or 0 depending on whether an electron at $\boldsymbol{R}$ is in an excited state or the ground state. The sum of the operators $N_{1\boldsymbol{R}} + N_{0\boldsymbol{R}}$ is correspondingly equal to the unit operator. (In our model there is an electron in either excited or ground state of the atom at every lattice position). On the other hand, the product of the two operators is clearly zero. Eq. (3.93) then becomes

$$[b_{\boldsymbol{K}}, b_{\boldsymbol{K}}^{+}] = \delta_{\boldsymbol{KK}'}\frac{1}{N}\sum_{\boldsymbol{R}}(1 - 2N_{1\boldsymbol{R}}) = \delta_{\boldsymbol{KK}'}\left(1 - \sum_{\boldsymbol{R}}\frac{2N_{1\boldsymbol{R}}}{N}\right) . \tag{3.94}$$

The last term is of the order of "the number of excitons divided by the number of unexcited lattice atoms". When one is concerned with the lowest excitations of the insulator, this term can be neglected as small by comparison with 1. In this approximation excitons are therefore bosons.

High exciton excitations has recently become an area of particular interest (Bose-Einstein condensation of excitons, exciton molecules, electron-hole drops). For further reading on this we refer the reader to the review article by *Haken* and *Nikitine* [111d.73].

3.3 Ion-Ion Interaction: Phonons

3.3.1 Introduction

Having so far taken an interest only in electron motion, we want to turn now to the motion of the ion lattice itself.

Lattice dynamics is important in many areas of solid-state physics. As a result of their thermal motion, the lattice ions vibrate about their equilibrium positions. The forces which oppose this motion are those of the chemical bond. All the elastic properties, compressibility, the propagation of acoustic waves are related to it. These features are mostly described in the context of a continuum theory by disregarding the atomic structure of the lattice.

The continuum approximation is a limiting case of the microscopic theory which looks at the dynamics of the lattice ions themselves. To discuss this theory we shall first of all set up equations of motion for the lattice ions within the framework of classical mechanics and derive from these the energy and frequency of the "normal modes of oscillations" of the lattice. In describing the dispersion relations of these normal modes, we shall again meet up with the reciprocal lattice, the Brillouin zone concept, and other mathematical aids already introduced in the previous chapter. We shall be able to draw many parallels with earlier results which will allow us to limit the discussion necessary in this chapter. Section 3.3.2 is dedicated to a classical description of lattice vibrations.

If we supply thermal energy to a lattice ion, it will rapidly be distributed throughout the entire lattice by the mutual interaction between the ions. Local excitations will therefore lead to *collective vibrations* of the whole ion system. It is therefore appropriate to use collective coordinates (normal coordinates) for the mathematical description. The lattice vibrations can be readily quantized in this new representation. The associated quanta are elementary excitations called *phonons*. Phonons are bosons and thus call for the application of a different sort of statistics to electrons. Section 3.3.3 is devoted to the phonon concept.

We come to grips with our first application of this concept in Section 3.3.4 when we consider the energy content of lattice vibrations and specific heat. Section 3.3.5 then presents an overview of the calculation of phonon dispersion relations.

The phonon dispersion spectrum allows us to determine the associated density of states in a way similar to that in which the density of one-electron states followed from the band model. The close similarity means that the treatment of this question in Section 3.3.6 can be kept short.

In the last two sections of this chapter we examine the limiting case of lattice vibrations of very long wavelength. If the wavelength is large compared with the atomic separation, the microscopic structure of the solid can be neglected. This gives us, therefore, a link to the classical continuum theory.

In the approximation used throughout, the potential energy of a lattice ion is expanded in powers of the instantaneous deviation from its equilibrium position and only the first nonvanishing (harmonic) teım is taken. This is the *harmonic approximation.* With it the Hamiltonian can be resolved into a sum of independent terms with the form of Hamiltonians of harmonic oscillators. This is the basis of the quantization and with it the description of lattice vibrations as a noninteracting phonon gas. The inclusion of higher anharmonic terms in the expansion means an *interaction between the phonons.* This forms the content of a later chapter (Chap. 7).

The topic of lattice vibrations and the harmonic approximation is treated in many publications. Special attention is given to lattice dynamics in the books by *Maradudin*, *Montroll*, and *Weiss* [102.3]; *Wallis* [120]; *Stevenson* [121]; *Enns* and *Haering* [133.II]; *Bak* [66]; and *Born* and *Huang* [67]. An excellent overview is also given by *Cochran* and *Cowley* [106, XXV/20], by *Leibfried* in [113a] and *Parrott* in [117]. The book series [101] contains numerous special articles. Phonon dispersion relations are collected in a book by *Bilz* and *Kress* [107.10].

3.3.2 The Classical Equations of Motion

We shall now consider ion-ion interaction and in doing so we shall ignore electron motion. The model is therefore that of (1.8): the lattice ions vibrate about their equilibrium positions. Between them forces are present which correlate the individual motions. The electron system is replaced by a spatially uniform negative space charge (inverse jellium model). Again we consider the ion lattice within a volume V_g with cyclic boundary conditions. Let the number of Wigner-Seitz cells in this volume be N. Let the equilibrium positions of the ions be $\boldsymbol{R}_{n\alpha} = \boldsymbol{R}_n + \boldsymbol{R}_\alpha$. Here $\boldsymbol{R}_n$ is a suitable reference point inside the Wigner-Seitz cell and the $\boldsymbol{R}_\alpha$ are the vectors from this point to the αth basis atom. The index α will run from 1 to r for a basis made up of r ions. Let the instantaneous displacement of the $n\alpha$th ion from its equilibrium position be given by the time-dependent vector $\boldsymbol{s}_{n\alpha}(t)$.

In this section we consider the classical problem. The classical Hamiltonian is made up of the kinetic energy of all the ions and of their interaction energy. The *kinetic energy* is

$$T = \sum_{n\alpha i} \frac{M_\alpha}{2} \dot{s}_{n\alpha i}^2 \qquad n = 1, \ldots, N, \quad \alpha = 1, \ldots, r, \quad i = 1, 2, 3. \tag{3.95}$$

M_α here is the mass of the αth basis atom. The index i distinguishes the three cartesian coordinates of the vector $\boldsymbol{s}_{n\alpha}$.

We expand the *potential energy* in increasing powers of the displacement $s_{n\alpha i}$. The first (constant) term in this expansion is the potential energy of the ion lattice in equilibrium. We shall omit this component along with the negative background since it does not contribute to the dynamics of lattice vibrations which is our sole interest here.

The second term in the expansion is linear in the $s_{n\alpha i}$. Since we are expanding about an equilibrium position, this term must disappear. The third term is quadratic in the displacement and has the form

$$\frac{1}{2}\sum_{\substack{n\alpha i\\ n'\alpha' i'}} \frac{\partial^2 V}{\partial R_{n\alpha i}\partial R_{n'\alpha' i'}} s_{n\alpha i}s_{n'\alpha' i'} = \frac{1}{2}\sum_{\substack{n\alpha i\\ n'\alpha' i'}} \Phi_{n\alpha i}^{n'\alpha' i'} s_{n\alpha i}s_{n'\alpha' i'}\,. \tag{3.96}$$

The matrix $\Phi_{n\alpha i}^{n'\alpha' i'}$ appearing here has $3rN$ rows and columns. We terminate the expansion at this first nonvanishing term (*harmonic approximation*). We shall not look at corrections to this approximation until Chapter 7. We recognize the meaning of the $\Phi_{n\alpha i}^{n'\alpha' i'}$ from the equations of motion

$$M_\alpha \ddot{s}_{n\alpha i} = -\frac{\partial V}{\partial s_{n\alpha i}} = -\sum_{n'\alpha' i'} \Phi_{n\alpha i}^{n'\alpha' i'} s_{n'\alpha' i'}\,. \tag{3.97}$$

$\Phi_{n\alpha i}^{n'\alpha' i'}$ is hereafter the force in the i-direction on the αth ion in the nth elementary cell when the α'th ion in the n'th cell is displaced by unit distance in the i'-direction.

The $\Phi_{n\alpha i}^{n'\alpha' i'}$ are called atomic force constants. They are linked by a large number of symmetry relations. First we recognize from (3.96) that the force constants are symmetric

$$\Phi_{n\alpha i}^{n'\alpha' i'} = \Phi_{n'\alpha' i'}^{n\alpha i}\,. \tag{3.98}$$

Furthermore they are clearly real. We obtain further relations when we use the fact that the potential energy must be invariant to an (infinitesimal) translation or rotation of the crystal. Let the translation be given by $s_{n\alpha i} = \delta s_i$ for all n, α, i and the rotation by $s_{n\alpha i} = \sum_k \delta\omega_{ik} R_{n\alpha k} (\delta\omega_{ik} = -\delta\omega_{ki})$. These operations are not allowed to result in any forces on the lattice ions, so the right-hand side in (3.97) must disappear. This leads to

$$\sum_{i'} \delta s_{i'} \sum_{n'\alpha'} \Phi_{n\alpha i}^{n'\alpha' i'} = 0 \quad \text{and} \quad \sum_{i'k'} \delta\omega_{i'k'} \sum_{n'\alpha'} \Phi_{n\alpha i}^{n'\alpha' i'} R_{n'\alpha' k'} = 0 \tag{3.99}$$

for translation and rotation, respectively, and hence to the symmetry relations

$$\sum_{n\alpha} \Phi_{n\alpha i}^{n'\alpha' i'} = 0 \tag{3.100}$$

and

$$\sum_{n\alpha} \Phi_{n'\alpha' i'}^{n\alpha i} R_{n\alpha k} = \sum_{n\alpha} \Phi_{n'\alpha' i'}^{n\alpha k} R_{n\alpha i}\,. \tag{3.101}$$

Along with these general relations, many others can be derived by use of lattice symmetry. We want to go into these later, but for the moment we shall leave lattice symmetry out of our treatment. We shall then be in a better position later to distinguish between general statements and those which spring from lattice symmetry.

We are looking for solutions of the equations of motion which are periodic in time. To do this we put

$$s_{n\alpha i}(t) = \frac{1}{\sqrt{M_\alpha}} u_{n\alpha i} \exp(-i\omega t) \quad \text{with time-independent } u_{n\alpha i}. \tag{3.102}$$

Here we have drawn out a factor $M_\alpha^{-1/2}$. Using this expression, the equations of motion become

$$\omega^2 u_{n\alpha i} = \sum_{n'\alpha' i'} D_{n\alpha i}^{n'\alpha' i'} u_{n'\alpha' i'} \qquad \left(D \equiv \frac{\Phi}{\sqrt{M_\alpha M_{\alpha'}}}\right). \tag{3.103}$$

This is an eigenvalue equation for the real symmetric matrix $D_{n\alpha i}^{n'\alpha' i'}$ with $3rN$ real eigenvalues ω_j^2. The ω_j can only be either real or purely imaginary. The latter possibility can be eliminated since (3.102) would then lead to $s_{n\alpha i}$ which increase or decrease continuously with time.

The eigenvectors $u_{n\alpha i}$ of (3.103) are correspondingly characterized by the index j: $u_{n\alpha i}^{(j)}$, i.e., for each ω_j there are $3rN$ $u_{n\alpha i}^{(j)}$. They are called *normal modes*.

We now take account of the translation symmetry of the lattice. This requires that the $\Phi_{n\alpha i}^{n'\alpha' i'}$ (or $D_{n\alpha i}^{n'\alpha' i'}$) can depend not on the cell indices n' and n separately, but only on their difference $n' - n$: $\Phi_{n\alpha i}^{n'\alpha' i'} = \Phi_{\alpha i}^{\alpha' i'}(n' - n)$.

With this condition and the expression

$$u_{n\alpha i} = c_{\alpha i} \exp(i\boldsymbol{q} \cdot \boldsymbol{R}_n) \tag{3.104}$$

(3.103) becomes

$$\omega^2 c_{\alpha i} = \sum_{\alpha' i'} \left\{ \sum_{n'} \frac{1}{\sqrt{M_\alpha M_{\alpha'}}} \Phi_{\alpha i}^{\alpha' i'}(n' - n) \exp[i\boldsymbol{q} \cdot (\boldsymbol{R}_n - \boldsymbol{R}_{n'})] \right\} c_{\alpha' i'} \tag{3.105}$$

or since the summation over n' can be changed into a summation over $n' - n$

$$\omega^2 c_{\alpha i} = \sum_{\alpha' i'} D_{\alpha i}^{\alpha' i'}(\boldsymbol{q}) c_{\alpha' i'}, \tag{3.106}$$

with $D_{\alpha i}^{\alpha' i'}(\boldsymbol{q})$ defined by (3.105).

Lattice periodicity has thus reduced the system of $3rN$ equations (3.103) to a system of $3r$ equations. This system then has only $3r$ eigenvalues, that is, $3r$ ω_j. However these are functions of the vector $\boldsymbol{q}$

$$\omega = \omega_j(\boldsymbol{q}) \qquad j = 1, \ldots, 3r. \tag{3.107}$$

For each ω_j (3.106) has a solution $c_{\alpha i} = e_{\alpha i}^{(j)}(\boldsymbol{q})$. These solutions can be combined to vectors. They are defined except for a common factor which can be chosen such that the $\boldsymbol{e}_{\alpha}^{(j)}(\boldsymbol{q})$ are normalized (and orthogonal to each other).

For the displacements $\boldsymbol{s}_{n\alpha}(t)$ we then have, as special solutions of the equations of motion (3.97),

$$s_{n\alpha}^{(j)}(\boldsymbol{q}, t) = \frac{1}{\sqrt{M_\alpha}} \boldsymbol{e}_{\alpha}^{(j)}(\boldsymbol{q}) \exp\left(\mathrm{i}[\boldsymbol{q}\cdot\boldsymbol{R}_n - \omega_j(\boldsymbol{q})t]\right), \tag{3.108}$$

from which we can construct the general solutions.

Before looking more closely at (3.108) we want to examine the dispersion relations (3.107). $\hbar\omega_j$ is an energy, $\boldsymbol{q}$ is a vector in reciprocal space. The function $\omega_j(\boldsymbol{q})$ therefore has the same significance for lattice vibrations as the function $E_n(\boldsymbol{k})$ has for the motion of the electrons in the periodic lattice. We can adopt all the important results concerning qualitative properties from Section 2.2.

1) The function $\omega_j(\boldsymbol{q})$ is periodic in $\boldsymbol{q}$-space. So we need only consider one Brillouin zone, the form of which is given by the point group of the crystal.
2) The set of $\boldsymbol{q}$-values is finite as a result of the cyclic boundary conditions imposed on the crystal. If V_g contains N elementary cells, then there are N values of $\boldsymbol{q}$ in the Brillouin zone. Since j can take $3r$ values, there are $3rN$ different $\omega_j(\boldsymbol{q})$, that is, as many as the crystal has internal degrees of freedom.
3) $\omega_j(\boldsymbol{q})$ is an analytic function of $\boldsymbol{q}$ in the Brillouin zone in the same sense as $E_n(\boldsymbol{k})$ is analytic. However, whereas the index n in $E_n(\boldsymbol{k})$ can take any number of integer values, j has only $3r$ different values; $\omega_j(\boldsymbol{q})$ has $3r$ *branches*.
4) In the Brillouin zone $\omega_j(\boldsymbol{q})$ has the same symmetries as the band structure $E_n(\boldsymbol{k})$. In addition to the symmetries resulting from the space group of the crystal, we have by time reversal symmetry $\omega_j(\boldsymbol{q}) = \omega_j(-\boldsymbol{q})$.

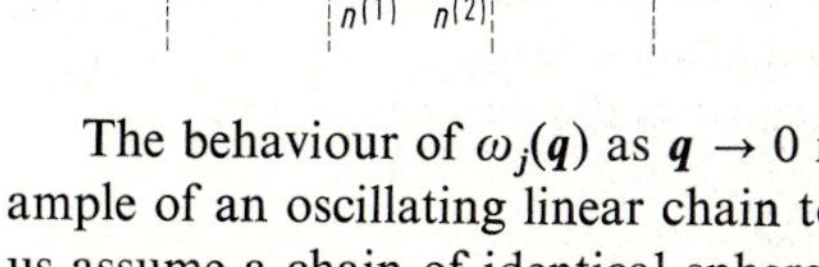

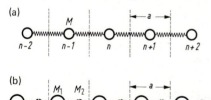

Fig. 3.6a and b. The linear chain (a) without and (b) with a basis.

The behaviour of $\omega_j(\boldsymbol{q})$ as $\boldsymbol{q} \to 0$ is important. We shall take the simple example of an oscillating linear chain to look at the most important aspects. Let us assume a chain of identical spheres linked together by springs of force constant f (Fig. 3.6a). Let a further be the separation between the spheres in equilibrium and let s_n be the displacement of the nth sphere from its equilibrium position. Eq. (3.97) then becomes

$$M\ddot{s}_n = -f(s_n - s_{n+1}) + f(s_{n-1} - s_n). \tag{3.109}$$

In accordance with (3.102) and (3.104) we put

$$s_n = \frac{1}{\sqrt{M}} c \exp[\mathrm{i}(qan - \omega t)]. \tag{3.110}$$

We then find

$$\omega^2 M = f[2 - \exp(-\mathrm{i}qa) - \exp(\mathrm{i}qa)] \tag{3.111}$$

and

$$\omega = 2\sqrt{\frac{f}{M}}\left|\sin\frac{qa}{2}\right| \tag{3.112}$$

so that ω is a periodic function of q. The first period (Brillouin zone) lies between $-\pi/a$ and $+\pi/a$ (Fig. 3.7).

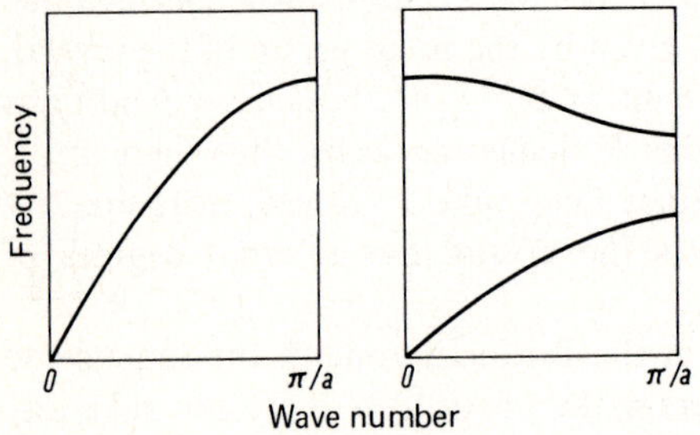

Fig. 3.7. Dispersion relations $\omega(q)$ for the linear chain without and with a basis (left and right, respectively).

If there are two atoms in the elementary cell, instead of (3.109) we have the equations (see Fig. 3.6b for the notation)

$$\begin{aligned} M_1 \ddot{s}_n^{(1)} &= -f(2s_n^{(1)} - s_n^{(2)} - s_{n-1}^{(2)}), \\ M_2 \ddot{s}_n^{(2)} &= -f(2s_n^{(2)} - s_{n+1}^{(1)} - s_n^{(1)}). \end{aligned} \tag{3.113}$$

Using the following expressions

$$\begin{aligned} s_n^{(1)} &= \frac{1}{\sqrt{M_1}} c_1 \exp\left(\mathrm{i}\left[q\left(n - \frac{1}{4}\right)a - \omega t\right]\right) \\ s_n^{(2)} &= \frac{1}{\sqrt{M_2}} c_2 \exp\left(\mathrm{i}\left[q\left(n + \frac{1}{4}\right)a - \omega t\right]\right) = \exp\left(\mathrm{i}\frac{qa}{2}\right)\frac{c_2}{c_1} s_n^{(1)} \sqrt{\frac{M_1}{M_2}} \end{aligned} \tag{3.114}$$

we find

$$\begin{aligned} -\omega^2\sqrt{M_1}\, c_1 &= -\frac{2f}{\sqrt{M_1}} c_1 + \frac{2f}{\sqrt{M_2}} c_2 \cos\frac{qa}{2}, \\ -\omega^2\sqrt{M_2}\, c_2 &= -\frac{2f}{\sqrt{M_2}} c_2 + \frac{2f}{\sqrt{M_1}} c_1 \cos\frac{qa}{2} \end{aligned} \tag{3.115}$$

and as the solutions of the determinant

$$\begin{vmatrix} \dfrac{2f}{\sqrt{M_1}} - \omega^2\sqrt{M_1} & -\dfrac{2f}{\sqrt{M_2}}\cos\dfrac{qa}{2} \\ -\dfrac{2f}{\sqrt{M_1}}\cos\dfrac{qa}{2} & \dfrac{2f}{\sqrt{M_2}} - \omega^2\sqrt{M_2} \end{vmatrix} = 0\,, \tag{3.116}$$

we finally arrive at

$$\omega_\pm^2 = f\left(\frac{1}{M_1} + \frac{1}{M_2}\right) \pm f\sqrt{\left(\frac{1}{M_1} + \frac{1}{M_2}\right)^2 - \frac{4}{M_1M_2}\sin^2\frac{qa}{2}}\,. \tag{3.117}$$

ω has therefore two branches, $\omega_+(q)$ and $\omega_-(q)$, which take the values $[2f(1/M_1 + 1/M_2)]^{1/2}$ and 0, respectively, at $q = 0$, and the values $(2f/M_1)^{1/2}$ and $(2f/M_2)^{1/2}$, respectively, at $q = \pm\pi/a$ (Fig. 3.7).

For the two limits $q = 0$ and $\pm\pi/a$, the ratio of the amplitudes c_2/c_1 follows from (3.115) as

$$\begin{aligned} c_2/c_1 &= +(M_2/M_1)^{1/2} \quad \text{for} \quad q = 0 \quad \text{and} \quad \omega = \omega_- \\ &= -(M_1/M_2)^{1/2} \quad \text{for} \quad q = 0 \quad \text{and} \quad \omega = \omega_+ \end{aligned} \tag{3.118}$$

$$\begin{aligned} &= \infty \quad \text{for} \quad q = \pm\pi/a \quad \text{and} \quad \omega = \omega_- \\ &= 0 \quad \text{for} \quad q = \pm\pi/a \quad \text{and} \quad \omega = \omega_+\,. \end{aligned} \tag{3.119}$$

These limiting cases correspond to typical modes of oscillation. Since $|q| = 2\pi/\lambda$, $q = 0$ means oscillations of infinite wavelength. All elementary cells move in the same sense. For $\omega = \omega_-$ the displacements of the two basis atoms in the cell are in the same direction, for $\omega = \omega_+$ they are in opposite directions.

The first case is the limiting case of an acoustic wave. The branch disappearing at $q = 0$ is correspondingly called the *acoustic branch*. The second type of vibration is readily excited optically in ionic crystals. The associated branch is therefore called the *optical branch*.

For $q = \pm\pi/a$ the basis atoms of one sort (M_1 or M_2) are at the nodes of the vibrations of wavelength $2a$. If each elementary cell has r basis atoms, then in addition to the acoustic branch there are $r - 1$ optical branches.

We have said nothing in (3.109) about whether these equations are meant to describe *transverse* vibrations (displacements perpendicular to the chain) or *longitudinal* vibrations (displacements along the chain). Eq. (3.109) applies to both situations as long as the vibration amplitudes are small. The meaning of the force constant f is different though for the two cases. For small amplitudes, each three-dimensional vibration of the chain can be resolved into three independent components, one longitudinal and two transverse. The two transverse vibrations take place in two perpendicular planes whose line of intersection is the equilibrium position of the chain.

Thus we see that in the general case the function $\omega_j(q)$ for the oscillating chain has three acoustic and $3(r - 1)$ optical branches.

If we now go from this example to a crystal with r atoms per elementary cell, we find that qualitatively the picture is the same. $\omega_j(\boldsymbol{q})$ is made up of three acoustic branches (degenerate at $\boldsymbol{q} = 0$) and $3(r - 1)$ optical branches. These branches are now functions of a vector $\boldsymbol{q}$. The degeneracy of the pair of transverse branches in the one-dimensional case only remains at points or along lines of high symmetry in the Brillouin zone. At a general point $\boldsymbol{q}$ all $3r$ branches are separated.

The expression "optical branch" cannot always be taken literally. Optical vibrations can in some cases not be excited optically, e.g., if the two basis atoms have the same effective charge. Again, away from $\boldsymbol{q} = 0$ the vibrations of an optical branch are not always out of phase and those of an acoustic branch not always in phase. Complicated mixed modes of the two limiting cases (3.118) can arise here. Likewise, the lattice vibrations are only strictly longitudinal or transverse at points or along lines of high symmetry.

3.3.3 Normal Coordinates, Phonons

According to (3.95) and (3.96), the Hamiltonian function for lattice vibrations has the form

$$H = \sum_{n\alpha i} \frac{M_\alpha}{2} \dot{s}_{n\alpha i}^2 + \frac{1}{2} \sum_{\substack{n\alpha i \\ n'\alpha' i'}} \Phi_{n\alpha i}^{n'\alpha' i'} s_{n\alpha i} s_{n'\alpha' i'} . \tag{3.120}$$

The $s_{n\alpha i}(t)$ are linear combinations of the particular solutions (3.108)

$$s_{n\alpha i}(t) = \frac{1}{\sqrt{NM_\alpha}} \sum_{j\boldsymbol{q}} Q_j(\boldsymbol{q}, t) e_{\alpha i}^{(j)}(\boldsymbol{q}) \exp(\mathrm{i}\boldsymbol{q} \cdot \boldsymbol{R}_n) \tag{3.121}$$

where the time-dependent exponential factor in (3.108) has been included in the $Q_i(\boldsymbol{q}, t)$ and a factor $1/\sqrt{N}$ has been separated out.

By putting (3.121) into (3.120), the Hamiltonian can be expressed in terms of the *normal coordinates* Q_j. For the conversion we use

$$\sum_n \exp[\mathrm{i}(\boldsymbol{q} - \boldsymbol{q}') \cdot \boldsymbol{R}_n] = N\delta_{\boldsymbol{q}' - \boldsymbol{q}, \boldsymbol{K}_m} \tag{3.122}$$

where $\delta_{\boldsymbol{q}, \boldsymbol{K}_m}$ is equal to unity, when $\boldsymbol{q}$ equals zero or a reciprocal lattice vector, and otherwise disappears.

Since the $s_{n\alpha i}(t)$ must be real, we have

$$e_{\alpha i}^{*(j)}(\boldsymbol{q}) Q_j^*(\boldsymbol{q}, t) = e_{\alpha i}^{(j)}(-\boldsymbol{q}) Q_j(-\boldsymbol{q}, t) . \tag{3.123}$$

We satisfy this by the requirements

$$e_{\alpha i}^{*(j)}(\boldsymbol{q}) = e_{\alpha i}^{(j)}(-\boldsymbol{q}) \quad \text{and} \quad Q_j^*(\boldsymbol{q}, t) = Q_j(-\boldsymbol{q}, t). \tag{3.124}$$

Here we have complex $e_{\alpha i}$. Following (3.106), the $\boldsymbol{e}_\alpha$ can be chosen mutually orthogonal. We must then have further for the $e_{\alpha i}$

$$\sum_{\alpha i} e_{\alpha i}^{*(j)}(\boldsymbol{q}) e_{\alpha i}^{(j')}(\boldsymbol{q}) = \delta_{jj'}. \tag{3.125}$$

After much calculation, (3.122–125) yield

$$H = \frac{1}{2} \sum_{j\boldsymbol{q}} [\dot{Q}_j^*(\boldsymbol{q}, t)\dot{Q}_j(\boldsymbol{q}, t) + \omega_j^2 Q_j^*(\boldsymbol{q}, t) Q_j(\boldsymbol{q}, t)]. \tag{3.126}$$

By introducing normal coordinates, the Hamiltonian resolves into a sum of $3rN$ individual terms. The coupled individual oscillations of the ions are formally replaced by decoupled collective oscillations. The normal coordinates used here are complex. One can also choose real normal coordinates instead.

One finds the conjugate momentum P to the Q^* from the Lagrange function $L = T - V$ according to

$$P_j(\boldsymbol{q}, t) = \frac{\partial L}{\partial \dot{Q}_j^*(\boldsymbol{q}, t)} = \dot{Q}_j(\boldsymbol{q}, t). \tag{3.127}$$

We then find that

$$H = \frac{1}{2} \sum_{j\boldsymbol{q}} [P_j^*(\boldsymbol{q}, t) P_j(\boldsymbol{q}, t) + \omega_j^2(\boldsymbol{q}) Q_j^*(\boldsymbol{q}, t) Q_j(\boldsymbol{q}, t)]. \tag{3.128}$$

The Hamilton equations yield ($\dot{P} = -\partial H/\partial Q^*$)

$$\dot{P}_j(\boldsymbol{q}, t) = \ddot{Q}_j(\boldsymbol{q}, t) = -\omega_j^2(\boldsymbol{q}) Q_j(\boldsymbol{q}, t). \tag{3.129}$$

The equation of motion in normal coordinates is then

$$\ddot{Q}_j(\boldsymbol{q}, t) + \omega_j^2(\boldsymbol{q}) Q_j(\boldsymbol{q}, t) = 0 \tag{3.130}$$

which is formally identical with the equation of motion of a harmonic oscillator of frequency $\omega_j(\boldsymbol{q})$.

The transition to a quantum mechanical description is now easy to execute. We have only to interpret the P and Q as operators which have been subjected to the commutation relations

$$[Q_j(\boldsymbol{q}), P_{j'}(\boldsymbol{q}')] = \mathrm{i}\hbar \delta_{\boldsymbol{q}\boldsymbol{q}'} \delta_{jj'}. \tag{3.131}$$

The Hamiltonian operator (3.128) together with the commutation relations (3.131) corresponds exactly to (A.1) and (A.2) of the Appendix. We can therefore interpret the quantized collective oscillations as elementary excitations. They are called *phonons*. The introduction of creation and annihilation operators leads to a Hamiltonian of the form

$$H = \sum_{j\boldsymbol{q}} \hbar\omega_j(\boldsymbol{q})\left[a_j^+(\boldsymbol{q})a_j(\boldsymbol{q}) + \frac{1}{2}\right]. \tag{3.132}$$

Each of the states defined by a pair $(\boldsymbol{q}, j)$ is occupied by $n_j(\boldsymbol{q})$ phonons of energy $\hbar\omega_j(\boldsymbol{q})$, where the $n_j(\boldsymbol{q})$ are the eigenvalues of the operator $a_j^+(\boldsymbol{q})a_j(\boldsymbol{q})$ appearing in (3.132). The contribution which such a state (a normal mode) makes to the total energy is $n_j(\boldsymbol{q})\hbar\omega_j(\boldsymbol{q})$ and the total energy (including the zero point energy) is

$$E = \sum_{j\boldsymbol{q}} \hbar\omega_j(\boldsymbol{q})\left[n_j(\boldsymbol{q}) + \frac{1}{2}\right]. \tag{3.133}$$

In the harmonic approximation used here, the phonon gas described by (3.132) consists of noninteracting phonons. It is therefore appropriate to make a comparison with the noninteracting electron gas of Section 2.1. The basic difference between the two cases lies in the fact that electrons are *fermions* and phonons are *bosons*. Each state in the lattice vibration spectrum can therefore be occupied by any number of (indistinguishable) phonons. In addition, the number of phonons depends on the energy content of the lattice vibrations, i.e., on the temperature. At $T = 0$, no phonons are excited and the lattice has its zero point energy only.

Consequently the statistics are different. We are not interested, as in Section 2.1.4, in the distribution of N indistinguishable fermions among given energy states at a given temperature, but in the number of excited bosons in the oscillator states as a function of temperature. The probability P_n that a particular oscillator has the excitation energy $E_n = \hbar\omega(n + \frac{1}{2})$ is proportional to $\exp(-E_n/k_B T)$. Since $\sum_n P_n = 1$, this probability is given explicitly by

$$P_n = \frac{\exp(-E_n/k_B T)}{\sum\limits_n \exp(-E_n/k_B T)} = \frac{\exp(-n\hbar\omega/k_B T)}{\sum\limits_n \exp(-n\hbar\omega/k_B T)}. \tag{3.134}$$

Since $\sum_n x^n = (1 - x)^{-1}$ (all summations run from $n = 0$ to $n = \infty$), it follows that

$$P_n = \exp(-n\hbar\omega/k_B T)\{1 - \exp[-\hbar\omega/k_B T]\}. \tag{3.135}$$

We then find for the mean energy of the oscillator

$$\bar{E} = \sum_n E_n P_n = E_0 + \sum_n n\hbar\omega P_n. \tag{3.136}$$

Since $\sum_n nx^n = x/(1-x)^2$, we finally have

$$\bar{E} = \frac{\hbar\omega}{\exp\left(\dfrac{\hbar\omega}{k_B T}\right) - 1} + \frac{\hbar\omega}{2}. \tag{3.137}$$

The mean occupation probability of an oscillator, i.e., the mean number of phonons in state j, $\boldsymbol{q}$ is then given by

$$\bar{n}_j(\boldsymbol{q}) = \frac{1}{\exp\left[\hbar\omega_j(\boldsymbol{q})/k_B T\right] - 1} \qquad (\textit{Bose distribution}). \tag{3.138}$$

Following Section 3.3.2, the spectrum of the lattice vibrations resolves into $3r$ branches (index j), each of which can be represented as a function in $\boldsymbol{q}$-space. Since there is a state associated with each (quasi-discrete) point $\boldsymbol{q}$ of each branch, we have to distinguish between phonons of the different branches. Depending of the behaviour of a branch at $\boldsymbol{q} = 0$ and the polarization of the normal mode, we distinguish between *acoustic* and *optical*, *longitudinal* and *transverse phonons*. Since they show different properties when interacting with other quasi-particles and collective excitations, we shall note the different types, as far as necessary: TA-, TO-, LA-, LO-phonons.

We close this section with two fundamental remarks. The transition to normal coordinates, and the decoupling of the Hamiltonian into independent normal modes thereby achieved, was made possible by the fact that the Hamiltonian (3.120) is a positive definite quadratic form. Such a form can be diagonalized. We could therefore have already made the transition to quantum mechanics and introduced phonons by following on from (3.103), and before the explicit consideration of lattice periodicity. The appearance of elementary excitations at this point is not therefore linked to the properties of the lattice. The division of all ω_j into branches which can be presented in one Brillouin zone of $\boldsymbol{q}$-space, on the other hand, is a consequence of the periodicity of the lattice.

If one had not terminated the expansion (3.96) at the second term, it would not have been possible to diagonalize. The consideration of higher, anharmonic terms therefore brings *phonon-phonon interaction* into the picture (Chap. 7).

3.3.4 The Energy Content of the Lattice Vibrations, Specific Heat

The total energy of the lattice vibrations at a temperature T is given from (3.133) and (3.138) by

$$E = \sum_{j\boldsymbol{q}} \left[\frac{1}{\exp(\hbar\omega_j/k_B T) - 1} + \frac{1}{2}\right] \hbar\omega_j(\boldsymbol{q}). \tag{3.139}$$

The summation over all $\boldsymbol{q}$ and over all branches j can be readily evaluated in two limiting cases.

1) *High temperature.* When $k_B T$ is large compared with $\hbar\omega_j$, the exponential function in the denominator can be expanded and it follows that

$$E = \sum_{jq} k_B T\left[1 + \frac{1}{12}\left(\frac{\hbar\omega_j(\boldsymbol{q})}{k_B T}\right)^2 + \cdots\right] \approx 3rNk_B T. \tag{3.140}$$

To a first approximation each of the $3rN$ oscillators makes a contribution $k_B T$ to the total energy (Dulong-Petit law). This is the classical result. Quantum effects do not appear.

2) *Low temperature.* In this case we cannot put $k_B T$ small in comparison with $\hbar\omega_j$ since all values of frequency from zero up to $\hbar\omega_j \gg k_B T$ appear. On the other hand, (3.139) allows us to conclude that frequencies for which $\hbar\omega_j \gg k_B T$ make no contribution. We can therefore limit the summation to the three acoustic branches. Here too only the lowest frequencies will be important. For these the dispersion relation $\omega_j = \omega_j(\boldsymbol{q})$ can be approximated by the linear expression $\omega_j(\boldsymbol{q}) = s_j(\theta, \varphi)q$. We then have

$$E = \sum_{jq} \frac{\hbar s_j q}{\exp(\hbar s_j q/k_B T) - 1} + \text{zero point energy } E_0 . \tag{3.141}$$

For sufficiently large V_g we can replace the summation over discrete $\boldsymbol{q}$-points by an integration in $\boldsymbol{q}$-space: $\sum_{\boldsymbol{q}} = [V_g/(2\pi)^3] \int d\tau_{\boldsymbol{q}}$. Eq. (3.141) then becomes

$$\begin{aligned} E - E_0 &= \frac{V_g}{(2\pi)^3} \sum_j \int \frac{\hbar s_j q}{\exp(\hbar s_j q/k_B T) - 1} d\tau_q \\ &= \frac{V_g}{(2\pi)^3} \frac{(k_B T)^4}{\hbar^3} \sum_j \int \frac{d\Omega}{s_j^3} \int_0^\infty \frac{x^3 dx}{e^x - 1} . \end{aligned} \tag{3.142}$$

The upper integration limit can be made infinity since large values of x make no contribution to the integrand. If we now average s_j^{-3} over all directions and branches we find

$$E = \frac{\pi^2}{10} \frac{V_g}{\hbar^3} \frac{(k_B T)^4}{\bar{s}^3} + \text{zero point energy} \tag{3.143}$$

in which the value $\pi^4/15$ has been inserted for the integral over all x.

These approximations are insufficient for the intermediate temperature range. We note first of all that a function of $\omega_j(\boldsymbol{q})$ is involved in the sum over $\boldsymbol{q}$ in (3.139). By replacing the summation with an integration one can immediately rearrange it as an integration over ω_j. To do this we introduce, as in Section

2.2.10, a *density of states* $g(\omega)$. Thus

$$g(\omega)d\omega = \frac{V_g}{(2\pi)^3} \int_{\omega=\text{const}} \frac{df_q}{|\nabla_q \omega_j(\boldsymbol{q})|} d\omega. \tag{3.144}$$

In contrast to Section 2.2.10, we have not normalized $g(\omega)$ here to V_g. With (3.144), we have for any function $F(\omega)$

$$\sum_q F(\omega) = \frac{V_g}{(2\pi)^3} \int F(\omega) d\tau_q = \int_0^\infty F(\omega) g(\omega) d\omega. \tag{3.145}$$

For the approximation $\omega_j(\boldsymbol{q}) = \bar{s}_j q$ (with s_j already averaged over all angles) we have for example

$$g_j(\omega_j)d\omega_j = \frac{V_g}{2\pi^2} \frac{\omega_j^2 d\omega_j}{\bar{s}_j^3} \tag{3.146}$$

and

$$\begin{aligned} E - E_0 &= \sum_j \int_0^\infty \frac{\hbar\omega_j}{\exp(\hbar\omega_j/k_B T) - 1} g_j(\omega_j) d\omega_j \\ &= \sum_j \frac{V_g}{2\pi^2} \frac{(k_B T)^4}{\hbar^3} \frac{1}{\bar{s}_j^3} \int_0^\infty \frac{x^3}{e^x - 1} dx. \end{aligned} \tag{3.147}$$

But this is exactly the result we have just obtained above for the low temperature limit. In this formulation we can at least correct it in one respect. The integration of the density of states (3.144) over all ω has to give exactly N $\boldsymbol{q}$-values for a branch [(3.145) with $F = 1$].

In the linear approximation we must therefore terminate the spectrum $\omega_j(\boldsymbol{q})$ at a frequency ω_D (*Debye frequency*) in order that this condition be met. This leads to the equation

$$\frac{V_g}{2\pi^2} \frac{1}{\bar{s}_j^3} \frac{\omega_{jD}^3}{3} = N, \tag{3.148}$$

or with $q_{jD} = \omega_{jD}/\bar{s}_j$ independent of j

$$q_{jD} = \left(\frac{6\pi^2 N}{V_g}\right)^{1/3} = (6\pi^2 n)^{1/3}. \tag{3.149}$$

Now N/V_g is the reciprocal volume of a Wigner-Seitz cell and is therefore equal to the volume of a Brillouin zone divided by $(2\pi)^3$. If we insert this into (3.149) it follows immediately that q_D is the radius of a sphere with volume equivalent to that of the Brillouin zone. The Debye approximation used here therefore consists of three approximations to the spectrum $\omega_j(\boldsymbol{q})$: 1) neglect of optical branches, 2) linear approximation of acoustic branches, and 3) replacement of the Brillouin zone by a sphere of equivalent volume and assumption of directional-independence of the linear approximation in this sphere.

In our approximation the Debye correction means replacing the upper limit ∞ in the integral (3.142) by $\hbar\omega_\mathrm{D}/k_\mathrm{B}T$, and hence the multiplication of (3.143) with a temperatuıe-dependent factor

$$\frac{15}{\pi^4}\int_0^{\hbar\omega_\mathrm{D}/k_\mathrm{B}T}\frac{x^3}{e^x-1}dx = f\left(\frac{\hbar\omega_\mathrm{D}}{k_\mathrm{B}T}\right) = f\left(\frac{\theta_\mathrm{D}}{T}\right). \tag{3.150}$$

Here we have also introduced the *Debye temperature* θ_D through $k_\mathrm{B}\theta_\mathrm{D} = \hbar\omega_\mathrm{D}$.

We want to touch here on a model introduced by Einstein, in preparation for its application later. In this it is assumed that only one oscillation frequency occurs: $\omega_j(\boldsymbol{q}) = \omega_\mathrm{E}$. Then for the density of states

$$g(\omega_j)d\omega_j = N\delta(\omega_j - \omega_\mathrm{E})d\omega_j, \tag{3.151}$$

and from (3.147) follows that

$$E - E_0 = \frac{N\hbar\omega_\mathrm{E}}{\exp(\hbar\omega_\mathrm{E}/k_\mathrm{B}T) - 1}. \tag{3.152}$$

Coarse as this approximation appears to be, it is nevertheless important for extending the Debye approximation. From Fig. 3.7 we see that, for the linear chain with a basis, the Debye approximation gives a very good description for the acoustic branch. For the optical branch, however, it is clear that an assumption of constant frequency for all optical phonons will be more accurate than a linear approximation of the Debye type.

In making a comparison between theory and experiment, the change of total energy with temperature, i.e., the *specific heat*, will prove to be more important than the total energy itself.

We do not have to distinguish between c_p and c_v in the harmonic approximation since this approximation does not include thermal expansion of the lattice. We obtain the specific heat from the Debye approximation by differentiating (3.147), (3.150) with respect to the temperature.

$$c_\mathrm{D}(T) = 3Nk_\mathrm{B}f_\mathrm{D}\left(\frac{\theta_\mathrm{D}}{T}\right) \quad \text{with} \quad f_\mathrm{D}(x) = \frac{3}{x^3}\int_0^x \frac{y^4 e^y dy}{(e^y-1)^2}. \tag{3.153}$$

For large θ_D/T (low temperatures), $f_\mathrm{D}(x)$ is approximately given by $4\pi^4/5x^3$. In this approximation the specific heat is then proportional to T^3 (Debye's T^3-law).

If we further introduce the Einstein temperature θ_E corresponding to the Debye temperature and a factor 3 for the three optical branches, we then find in the Einstein approximation that

$$c_E(T) = 3Nk_B f_E\left(\frac{\theta_E}{T}\right) \qquad \text{with} \quad f_E(x) = \frac{x^2 e^x}{(e^x - 1)^2}. \tag{3.154}$$

The temperature dependence of both approximations is given in Fig. 3.8.

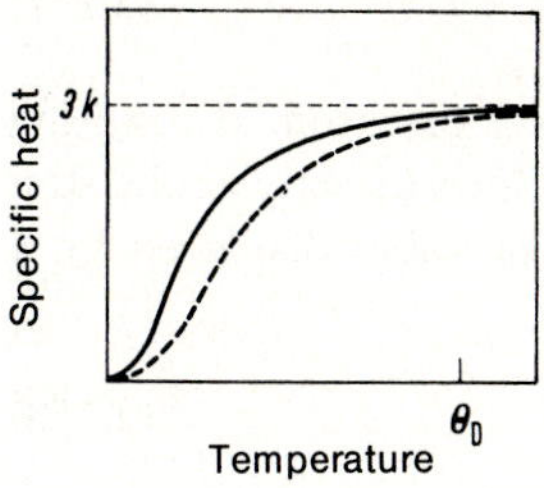

Fig. 3.8. Specific heat according to the Debye (upper curve) and Einstein (lower curve) approximations.

3.3.5 Calculation of Phonon Dispersion Relations

As already pointed out in Section 3.3.2, the function $\omega_j(\boldsymbol{q})$ can be represented in a Brillouin zone of $\boldsymbol{q}$-space in a way corresponding to that in which a band structure $E_n(\boldsymbol{k})$ is represented in a Brillouin zone of $\boldsymbol{k}$-space. In particular the same symmetry requirements apply.

Before we consider the calculation of such dispersion curves, let us draw attention to a few important points by taking the simple case of a two-dimensional square lattice.

The function $\omega_j(\boldsymbol{q})$ follows from (3.106) when the $3r \times 3r$ matrix $D_{\alpha i}^{\alpha' i'}(\boldsymbol{q})$ is known. The latter in turn is found from the force constants according to the equation

$$D_{\alpha i}^{\alpha' i'}(\boldsymbol{q}) = \sum_{n'} \frac{1}{\sqrt{M_\alpha M_{\alpha'}}} \Phi_{n\alpha i}^{n'\alpha' i'} \exp\left[\mathrm{i}\boldsymbol{q}\cdot(\boldsymbol{R}_n - \boldsymbol{R}_{n'})\right]. \tag{3.155}$$

In our model we assume elastic (spring) forces between neighbouring atoms, as in the one-dimensional case in Section 3.3.2. In accordance with the assumption that the forces are strongest between nearest neighbours and diminish with increasing distance to the next-nearest neighbour, third-nearest neighbour, etc., we limit ourselves to consideration of the nearest and next-nearest (Fig. 3.9). It would not be realistic to restrict our consideration to nearest neighbours only, if only because a square lattice (just like a cubic lattice in space) would not be stable against shear forces if elastic forces only exist between directly adjacent neighbours.

We consider now a particular atom $n = 0$. If the n'th atom is moved from its equilibrium position by a vector distance $\boldsymbol{s}_{n'}$, the ith component of force acting on the atom is given by

$$F_{n'\to 0} = a_{n'} e_{n'} (e_{n'} \cdot s_{n'}) . \tag{3.156}$$

Here $a_{n'}$ is the force constant of the particular spring and $e_{n'}$ is the unit vector which gives the direction of $R_{n'}$. If we compare this expression with (3.97), in which we can omit the index α for the consideration of Bravais lattices, we find for the force constants

$$\Phi_{0i}^{n'i'} = -a_{n'} e_{n'i} e_{n'i'} \quad (n' \neq 0) . \tag{3.157}$$

The force constant $\Phi_{0i}^{0i'}$, which describes the force on the atom 0 resulting from the displacement of this atom by a vector s_0, follows directly as the sum over all the forces which would arise if all other atoms were displaced by a vector $-s_0$

$$\Phi_{0i}^{0i'} = -\sum_{n'(\neq 0)} \Phi_{0i}^{n'i'} = \sum_{n'(\neq 0)} a_{n'} e_{n'i} e_{n'i'} . \tag{3.158}$$

From (3.157) and (3.158) we can readily give all the force constants for our particular example. Let the force constants to the four nearest neighbours be f_1 and to the four next-nearest neighbours be f_2, and let the lattice constant be a. We then have

$$\begin{aligned}
&\Phi_{01}^{11} = \Phi_{01}^{21} = \Phi_{02}^{32} = \Phi_{02}^{42} = -f_1 , \\
&\Phi_{0i}^{5i'} = \Phi_{0i}^{6i'} = -\frac{f_2}{2} \qquad (i,i' = 1, 2) , \\
&\Phi_{0i}^{7i} = \Phi_{0i}^{8i} = -\frac{f_2}{2} \qquad (i = 1, 2) , \\
&\Phi_{0i}^{7j} = \Phi_{0i}^{8j} = +\frac{f_2}{2} \qquad (i \neq j = 1, 2) \\
&\Phi_{01}^{01} = \Phi_{02}^{02} = 2(f_1 + f_2) \\
&\text{all other } \Phi_{0i}^{n'i'} = 0 .
\end{aligned} \tag{3.159}$$

The numbering of the individual neighbours is shown in Fig. 3.9.

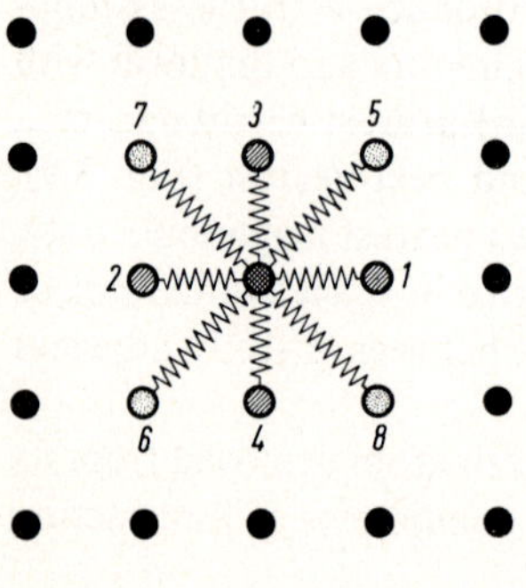

Fig. 3.9. Elastic forces to the nearest and next-nearest neighbours in a two-dimensional square lattice.

The $D_i^{i'}(\boldsymbol{q})$ can be calculated through (3.159) and inserted into (3.106). Eq. (3.106) is a system of equations which can only be solved if the determinant

$$|D_i^{i'}(\boldsymbol{q}) - \omega^2 \delta_{ii'}| = 0 \qquad (3.160)$$

vanishes. This leads in our example to

$$\begin{vmatrix} [f_1(1 - \cos q_x a) + f_2(1 - \cos q_x a \cos q_y a)] - \frac{M}{2}\omega^2 & \sin q_x a \sin q_y a \\ \sin q_x a \sin q_y a & [f_1(1 - \cos q_y a) + f_2(1 - \cos q_x a \cos q_y a)] - \frac{M}{2}\omega^2 \end{vmatrix} = 0 . \qquad (3.161)$$

q_x and q_y are the components of $\boldsymbol{q}$ in the (two-dimensional) Brillouin zone. For the two-dimensional square lattice this is itself a square of side length π/a.

The determinantal condition (3.161) leads to two solutions of $\omega_j(\boldsymbol{q})$ corresponding to the two branches possible in the two-dimensional Bravais lattice. We shall give them for a few points and lines of symmetry. The symmetry symbols are those of Fig. 2.22a (Brillouin zone of the cubic lattice in $\boldsymbol{k}$-space) which are contained within the square $k_z = 0$. The midpoint is therefore Γ; the Δ-axis leads to the midpoint (X) of a side of the square, and the Σ-axis to a corner of the square (M). The Z-axes are the sides of the square X-M-X-. . . . In the two-dimensional case, the point groups associated with the individual points and lines are of course others.

From (3.161) we have

$$\begin{aligned}
\Delta\text{-axis } (\Gamma \to X)\colon \quad \omega_1 &= \left[\frac{2}{M}(f_1 + f_2)(1 - \cos q_x a)\right]^{1/2} \\
\omega_2 &= \left[\frac{2}{M} f_2(1 - \cos q_x a)\right]^{1/2} \\
\Sigma\text{-axis } (\Gamma \to M)\colon \quad \omega_1 &= \left[\frac{2}{M}[f_1(1 - \cos q_x a) + f_2(1 - \cos 2q_x a)]\right]^{1/2} \\
\omega_2 &= \left[\frac{2}{M} f_1(1 - \cos q_x a)\right]^{1/2} \qquad (3.162) \\
Z\text{-axis } (X \to M)\colon \quad \omega_1 &= \left[\frac{2}{M}[2f_1 + f_2(1 - \cos q_y a)]\right]^{1/2} \\
\omega_2 &= \left[\frac{2}{M}[f_1 + f_2 - (f_1 - f_2) \cos q_y a]\right]^{1/2} .
\end{aligned}$$

The dispersion curves are shown in Fig. 3.10 for an arbitrary choice of the ratio f_2/f_1 ($f_2/f_1 = 2$). Fig. 3.10 further shows the symbols of the irreducible

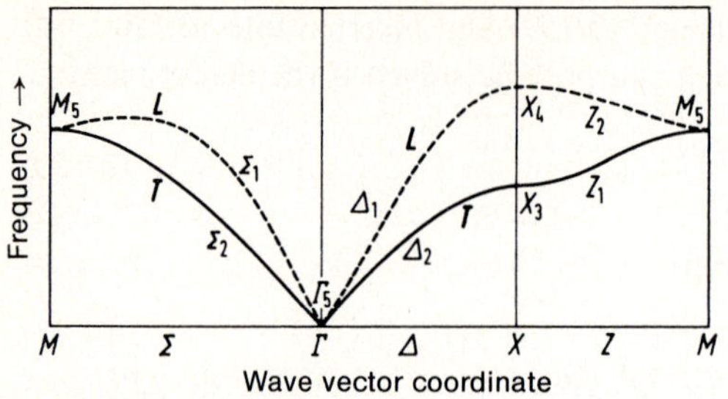

Fig. 3.10. Dispersion curves for the square net, considering spring forces between nearest and next-nearest neighbours ($f_1/f_2 = 2$). Ordinate: frequency in arbitrary units, abscissa: wave vector along symmetry lines of the Brillouin zone.

representations which are associated with the ω_i at points and lines of high symmetry. We shall not involve ourselves in the group theory side here. It is analogous to the example treated in Section 2.2.14. On the other hand, one further remark is important.

If we insert the solutions $\omega_j(\boldsymbol{q})$ into (3.106), we can determine the $c_i^{(j)}$. Then for example we find on the Δ-axis for ω_1: $c_x^{(1)} = 1$, $c_y^{(1)} = 0$. The ω_1-branch thus describes longitudinal lattice vibrations. Correspondingly we find transverse vibrations for ω_2 ($c_x^{(2)} = 0$, $c_y^{(2)} = 1$). On the Σ-axis, too, the upper branch is associated with longitudinal vibrations and the lower branch with transverse. On the other hand, the vibrations with a $\boldsymbol{q}$-vector on the Z-axes cannot be classified in this way. The division into longitudinal and transverse vibrations is therefore possible only for particular $\boldsymbol{q}$-values.

Calculations have been carried out for many solids in an analogous way to the model treated above. The method is however limited in its applicability by the following facts:

1) A model in which the chemical bond between atoms is simulated by the forces of a spring can only be successful if the bonding forces are central. This is the case in ionic crystals where electrostatic forces between the lattice ions are responsible for the chemical bond. For a lattice based on covalent bonds the bonding forces are directional and therefore dependent on the bond angle. Such forces can only be poorly represented by simple springs. In spite of this, quite good results are obtained here.

2) The replacement of lattice ions by fixed spheres connected by springs leaves the polarizability of the ions and their compressibility out of the picture.

3) Bonding forces do not act only between nearest neighbours. In principle they must also be considered to involve even ions which are very far apart. However, this introduces more and more parameters into the results and these have to be determined by matching with experimental observations. It is then pointless to refine the model with more free parameters than can be determined by experimental measurements.

Within the limits of the spring representation, the simplest model is that of *rigid ions*. Here, however, the number of parameters needed is very large if one cannot restrict oneself to nearest neighbours only. For germanium, for example, calculations show that contributions up to the sixth-nearest neighbour have to be included in order to obtain satisfactory results.

The theory is substantially improved when one also takes account of the polarizability of lattice ions. This can be modelled by describing the valence electrons of the ions by a massless, negatively charged shell which is bound to the positive ions by isotropic elastic forces (*shell model*). By considering in Ge nearest neighbours only, satisfactory agreement with experiment was obtained using the shell model with only five free parameters. However here too the number of parameters rises rapidly when one attempts to explain details of the vibrational spectrum.

A further important improvement to the model is brought about when one takes account of the compressibility of lattice ions by assuming a compressible shell ("*breathing*" *shell model*). In many instances this refinement would lead to an almost quantitative agreement with experiment.

In fitting values to the parameters, we can first of all make use of the elastic constants and the dielectric constant (Secs. 3.3.7 and 3.3.8). These will not usually prove to be sufficient to determine all the free parameters. One therefore looks to experimentally determined vibration spectra and matches parameter values to them. Dispersion curves can be measured by inelastic neutron scattering, the method being restricted to solids with not too large a scattering cross section.

Figure 3.11 shows the branches of the function $\omega_j(\boldsymbol{q})$ for diamond, along the most important axes within and on the surface of the Brillouin zone. The most significant features in Fig. 3.11 can be derived from symmetry considerations alone, as we learned in Section 2.2.14.

We do not want here to go into all the finer points which can arise in calculating dispersion curves. One such point would be the consideration of angle-dependent forces in covalent crystals.

In metals, screening of the ion-ion interaction by the electron gas of the

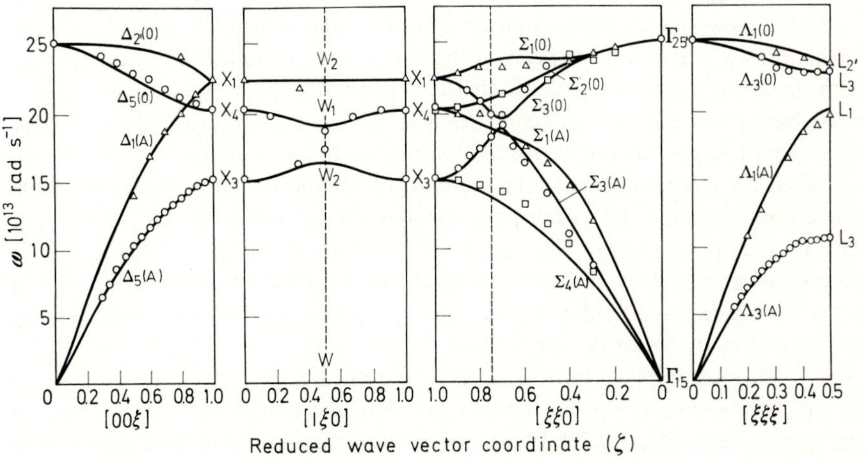

Fig. 3.11. Dispersion curves for diamond. Ordinate: frequency ν, abscissa: wave vector along high symmetry axes [from J. L. Warren et al.: Phys. Rev. **158**, 805 (1967)]

valence electrons is important. This therefore represents one place where electron-ion interaction has to be incorporated into the theory.

In principle the force constants can also be derived from the Schrödinger equation for the entire problem. For metals, the concept of pseudopotentials helps here to develop an approximation procedure for the calculation of dispersion curves which does not require the introduction of nonphysical spring forces. We can only point here to the literature covering this, in particular to *Harrison* [58].

3.3.6 The Density of States

Knowledge of the dispersion curves for the entire Brillouin zone, i.e., the function $\omega_j(\boldsymbol{q})$, allows us to calculate the density of states in accordance with (3.144).

For the one-dimensional case of the linear chain with two atoms in its basis, Fig. 3.12 shows the density of states for two different mass ratios. For equal masses (left), the density of states corresponds to the case of the simple linear chain. There is only one acoustic branch. The upper branch appearing in the figure is only the continuation of the lower acoustic branch, for, if one takes the linear chain with one atom per elementary cell to be a chain of double the lattice constant with two identical atoms per elementary cell, the associated Brillouin zone is only half as large. The acoustic branch in the usual representation has then to be reduced to the first Brillouin zone in the new representation, and forms an upper branch connected to the lower branch at the surface of the (one-dimensional) Brillouin zone. If one then assumes infinitesimally different masses for the two atoms in the basis, the branches separate and the upper branch becomes a "true optical branch". From this short discussion we can already see that the words "optical" and "acoustic", just like "transverse" and "longitudinal" are only meaningful in limiting cases.

In the two- and the three-dimensional cases, the density of states of different branches can overlap. The spectrum becomes more complicated in comparison with the one-dimensional case, where the density of states consists of the separate contributions of the two branches. As with the band structure $E_n(\boldsymbol{k})$, one needs the function $\omega_j(\boldsymbol{q})$ for the entire Brillouin zone for a calculation of the density of states. We shall not go into the methods of calculation here. Such spectra are often approximated by combining the density of states of the Debye and Einstein approximations. Fig. 3.12a shows that an acoustic branch can be roughly reproduced by a Debye spectrum [approximation $\omega_{\mathrm{ac}}(q) \propto q$]. Optical branches are often very flat. If one totally neglects the frequency dependence, the optical branch can be approximated by a δ-function with an Einstein frequency. If one combines both, one can represent the density of states spectrum to the point where it can be used to calculate the temperature dependence of the specific heat. Fig. 3.12b presents such an approximation for tungsten and for lithium. In combining Debye and Einstein terms, the Debye temperature must

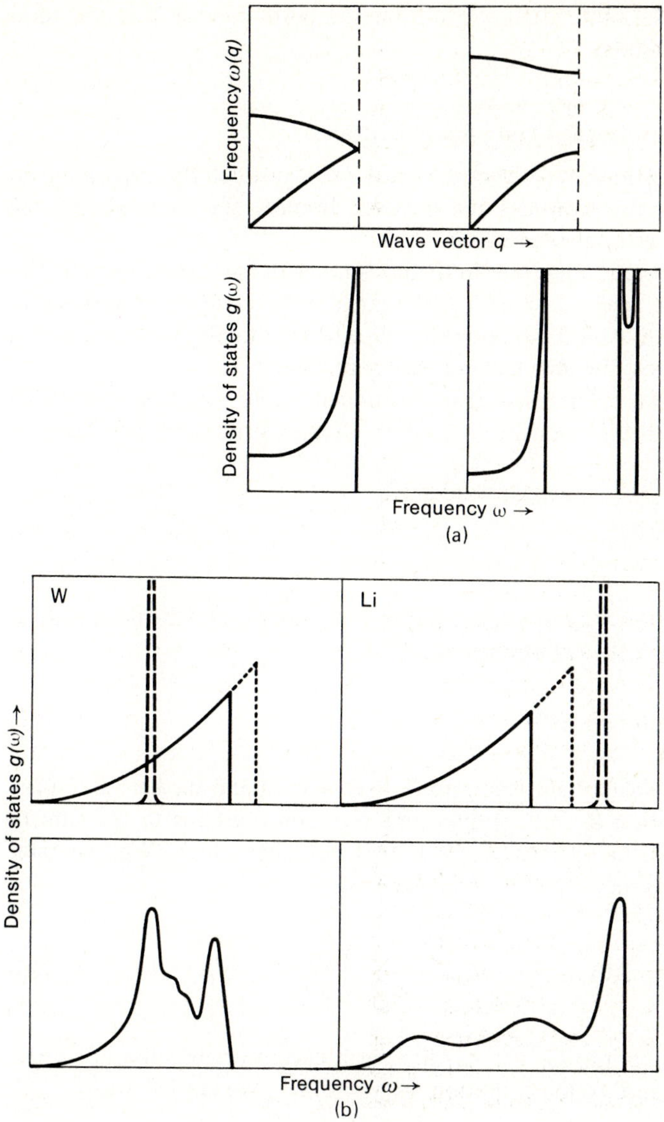

Fig. 3.12. (a) Dispersion curves $\omega(q)$ and density of states $g(\omega)$ for the linear chain with two atoms in the basis, for different mass relationships ($M_1 = M_2, M_1 \neq M_2$). (b) Approximation to the density of states of tungsten and lithium using Debye and Einstein terms (from *Leibfried* [106.VII/1]).

be lowered so that the summation over all states in both branches gives the total number of normal modes.

3.3.7 The Long Wavelength Limit: Acoustic Branch

The limiting case of long wavelengths (small $\boldsymbol{q}$) is particularly interesting on several counts. We first consider the acoustic branch. We shall discuss the optical branch in the following section.

In the acoustic branch all atoms in the basis move in the same direction. For long wavelength the vibration amplitudes change only slowly from elementary cell to elementary cell. Atomic structure has then no part to play and a transition to a continuum is possible. We execute this as follows.

First we can think of the basis atoms as united at the centre of mass (total mass M). We have therefore only to consider a Bravais lattice and the equations of motion become

$$M\ddot{s}_{ni} = -\sum_{n'i'} \Phi_{ni}^{n'i'} s_{n'i'} \,. \tag{3.163}$$

We now define a (slowly changing) *displacement field* $\boldsymbol{s}(\boldsymbol{r}, t)$ which is identical with the discrete $\boldsymbol{s}_n(t)$ at the lattice points

$$\boldsymbol{s}(\boldsymbol{r} = \boldsymbol{R}_n, t) = \boldsymbol{s}_n(t) \,. \tag{3.164}$$

We insert this displacement field into (3.163). We expand the field $\boldsymbol{s}$ at location $\boldsymbol{R}_{n'}$ about location $\boldsymbol{R}_n = 0$, and assume that contributions to the sum in (3.163) only come from regions where $\boldsymbol{s}$ does not alter much. We can then terminate the expansion at the first nonvanishing term

$$s_{n'i'} = s_{i'}(\boldsymbol{R}_{n'}) = s_{i'}(0) + \sum_j \frac{\partial s_{i'}}{\partial r_j} R_{n'j} + \frac{1}{2} \sum_{kl} \frac{\partial^2 s_{i'}}{\partial r_k \partial r_l} R_{n'k} R_{n'l} \,. \tag{3.165}$$

Putting this expansion into (3.163), the first term disappears because of (3.100) and the second because of the condition $\Phi_{0i}^{n'i'} = \Phi_{0i}^{-n'i'}$. We are left with

$$M\ddot{s}_{ni} = -\frac{1}{2} \sum_{\substack{n'i'\\kl}} \Phi_{ni}^{n'i'} R_{n'k} R_{n'l} \frac{\partial^2 s_{i'}}{\partial r_k \partial r_l} \,. \tag{3.166}$$

If we multiply this equation by the density $\rho = M/V_{\mathrm{WSC}}$ (V_{WSC} = volume of a Wigner-Seitz cell) and introduce the abbreviation

$$C_{ii'kl} = \sum_{n'} \Phi_{ni}^{n'i'} R_{n'k} R_{n'l} \tag{3.167}$$

we finally find

$$\rho \ddot{s}_i = \sum_{i'kl} C_{ii'kl} \frac{\partial^2 s_{i'}}{\partial r_k \partial r_l} . \tag{3.168}$$

The $C_{ii'kl}$ possess a range of symmetries. In particular

$$C_{iklm} = C_{lmik} . \tag{3.169}$$

We do not want to derive this relation for the general case. If the forces between all lattice components are central, it follows from a general symmetry relation which we shall only state here. For central forces the potential can be written as

$$V = \sum_{ss'} v(|\boldsymbol{R}_{s'} - \boldsymbol{R}_s|) \tag{3.170}$$

and the force constants become

$$\Phi_{ni}^{n'i'} = g(|\boldsymbol{R}_{n'} - \boldsymbol{R}_n|) R_{n'i} R_{n'i'} . \tag{3.171}$$

If we put this in (3.167), we see that in this special case all the indices of the $C_{ii'kl}$ are interchangeable.

Hence we can also write (3.168) as

$$\rho \ddot{s}_i = \sum_k \sum_{mn} C_{ikmn} \frac{\partial}{\partial r_k} \frac{1}{2} \left(\frac{\partial s_m}{\partial r_n} + \frac{\partial s_n}{\partial r_m} \right) . \tag{3.172}$$

This however is precisely the equation of motion of an elastic continuum.

$(1/2)[(\partial s_m/\partial r_n) + (\partial s_n/\partial r_m)]$ is the strain tensor ε_{mn}, related by Hooke's law to the stress tensor

$$\sigma_{ik} = \sum_{mn} C_{ikmn} \varepsilon_{mn} . \tag{3.173}$$

We can then identify the *elastic tensor* C_{ikmn} of (3.173) with the corresponding tensor of (3.172) and obtain

$$\rho \ddot{s}_i = \sum_k \frac{\partial}{\partial r_k} \sigma_{ik} . \tag{3.174}$$

The connection with the mechanics of continua is thereby found.

We can use these results to match unknown parameters in the calculation of force constants (e.g., for the shell model), using the components of the elastic tensor which can be measured.

Stress and strain tensors, and hence also the elastic tensor, are symmetric. One usually combines the groups of indices 11, 22, 33, 23, 13, 12 into new indices 1, 2, 3, 4, 5, 6 and writes (3.173) then in the form

$$\sigma_\alpha = \sum_\beta C_{\alpha\beta}\varepsilon_\beta \,. \tag{3.175}$$

The (symmetric) 6 × 6-matrix of the $C_{\alpha\beta}$ has 21 independent elements (elastic constants). For reasons of symmetry many of these constants are zero or equal to each other.

A cubic crystal for example is described by three elastic constants $C_{11} = C_{22} = C_{33}$, $C_{12} = C_{23} = C_{31}$, $C_{44} = C_{55} = C_{66}$, while all other C_{ik} are zero.

An isotropic solid is described by two elastic constants. For these it is customary to use the three cubic elastic constants just mentioned, with the condition $C_{11} = C_{12} + 2C_{44}$.

More complicated crystals can have up to 21 different elastic constants. With central forces, i.e., where (3.171) applies, this number is reduced to 15 by further relations (*Cauchy relations*).

In the lattice dynamics of the previous section we looked particularly at solutions of the equations of motion which have wavelike character. We therefore seek here too solutions of (3.172) with the form

$$\boldsymbol{s} = \boldsymbol{e} \exp\left[\mathrm{i}(\boldsymbol{q}\cdot\boldsymbol{r} - \omega t)\right] . \tag{3.176}$$

For isotropic media this leads to the equation

$$\rho\omega^2 \boldsymbol{e} = (C_{12} + C_{44})\boldsymbol{q}(\boldsymbol{q}\cdot\boldsymbol{e}) + C_{44}q^2\boldsymbol{e} \,. \tag{3.177}$$

It has one solution which corresponds to longitudinal waves ($\boldsymbol{q} \parallel \boldsymbol{e}$) and two which correspond to transverse waves ($\boldsymbol{q} \perp \boldsymbol{e}$). The dispersion relations for the two cases are

$$\rho\omega_\mathrm{L}^2 = (C_{12} + 2C_{44})q^2 = C_{11}q^2 \quad \text{and} \quad \rho\omega_\mathrm{T}^2 = C_{44}q^2 \,. \tag{3.178}$$

In both cases ω is proportional to q. The waves propagate with the *longitudinal* or *transverse velocity of sound*, respectively,

$$c_\mathrm{L} = \sqrt{C_{11}/\rho} \,, \qquad c_\mathrm{T} = \sqrt{C_{44}/\rho} \,. \tag{3.179}$$

We note that the isotropic linear relation between ω and q in this approximation just corresponds to the Debye approximation. The constant $\bar{s}_j^{-3}$ used there becomes equal to c_L^{-3} or c_T^{-3} (as appropriate) for each of the three acoustic branches. Averaging over all three branches gives

$$\bar{s}^{-3} = c_\mathrm{L}^{-3} + 2c_\mathrm{T}^{-3} \,. \tag{3.180}$$

This is the expression to be inserted into the density of states (3.146).

We shall also discuss briefly the corresponding relations for cubic crystals. Here we have

$$\rho \ddot{s}_x = C_{11}\frac{\partial^2 s_x}{\partial x^2} + (C_{12} + C_{44})\left(\frac{\partial^2 s_y}{\partial x \partial y} + \frac{\partial^2 s_z}{\partial x \partial z}\right) + C_{44}\left(\frac{\partial^2 s_x}{\partial y^2} + \frac{\partial^2 s_x}{\partial z^2}\right). \tag{3.181}$$

The two further equations for the y- and z-components of $\boldsymbol{s}$ follow by cyclic interchange of x, y, z.

Here, too, longitudinal and transverse solutions only appear in particular directions, while in a general direction $\boldsymbol{e}$ is neither perpendicular nor parallel to $\boldsymbol{q}$.

3.3.8 The Long Wavelength Limit: Optical Branch

In the long wavelength limit of the optical branch, the basis atoms in the individual Wigner-Seitz cells vibrate relative to one another, while the motion in adjacent cells is practically identical. If we look in particular at a solid with two oppositely charged ions in its basis, the two ion sublattices vibrate rigidly opposite to each other.

We shall examine the instantaneous deviation $\boldsymbol{s}_\pm$ of an ion. Let its effective charge be $\pm e^*$. The displacement can then also be described formally by the introduction of a dipole with moment $\pm e^* \boldsymbol{s}_\pm$ (addition of a charge $\pm e^*$ at position $\boldsymbol{s}_\pm$ and a charge $\mp e^*$ at the original position to compensate the charge on the ion). Thus the polarization of a cell is given by $e^*(\boldsymbol{s}_+ - \boldsymbol{s}_-) \equiv e^*\boldsymbol{s}$. The charge displacement can however also lead to internal fields which induce further dipole moments in the lattice ions. This brings a further contribution to the polarization of the cell, of magnitude αE_{eff} ($\alpha = \alpha_+ + \alpha_-$). $\alpha_\pm$ is the polarizability of the ions and E_{eff} the effective field at the ion location. This local field is related to the macroscopic field for simple cubic lattices by the well-known relation

$$E_{\text{eff}} = E + \frac{1}{3\varepsilon_0}P. \tag{3.182}$$

To derive this equation one takes a sphere of sufficiently large radius around the ion observed, and considers the region outside this sphere to be a homogeneous dielectric. Within the sphere the contributions of the other lattice ions to the effective field are summed. These contributions cancel for cubic lattices. Only the contribution given in (3.182) then remains.

Taking (3.182) to be valid, the total polarization for N Wigner-Seitz cells in the volume V_g is

$$\boldsymbol{P} = \frac{N}{V_g}(e^*\boldsymbol{s} + \alpha \boldsymbol{E}_{\text{eff}}) = \frac{N}{V_g}\frac{e^*\boldsymbol{s} + \alpha \boldsymbol{E}}{1 - N\alpha/3V_g\varepsilon_0}. \tag{3.183}$$

The equations of motion for the displacements are given by

$$
\begin{aligned}
M_+\ddot{\boldsymbol{s}}_+ &= -k(\boldsymbol{s}_+ - \boldsymbol{s}_-) + e^*\boldsymbol{E}_{\text{eff}} \\
M_-\ddot{\boldsymbol{s}}_- &= +k(\boldsymbol{s}_+ - \boldsymbol{s}_-) - e^*\boldsymbol{E}_{\text{eff}}
\end{aligned}
\tag{3.184}
$$

where k is the proportionality constant of the restoring force. With the reduced mass $\overline{M} = M_+M_-/(M_+ + M_-)$ (3.184) leads to

$$
\overline{M}\ddot{\boldsymbol{s}} = -k\boldsymbol{s} + e^*\boldsymbol{E}_{\text{eff}}\,. \tag{3.185}
$$

If we express the effective field here in terms of the macroscopic field, we obtain two equations linking $\boldsymbol{s}$, $\boldsymbol{E}$, and $\boldsymbol{P}$. It is convenient to replace the vector $\boldsymbol{s}$ by the vector $\boldsymbol{w} = \sqrt{N\overline{M}/V_g}\boldsymbol{s}$. Eqs. (3.183) and (3.184) then become

$$
\begin{aligned}
\ddot{\boldsymbol{w}} &= b_{11}\boldsymbol{w} + b_{12}\boldsymbol{E} \\
\boldsymbol{P} &= b_{21}\boldsymbol{w} + b_{22}\boldsymbol{E}
\end{aligned}
\tag{3.186}
$$

with symmetric coefficient matrix ($b_{12} = b_{21}$).

We can relate the coefficients to parameters which can be measured experimentally. In the static case, $\ddot{\boldsymbol{w}} = 0$ and hence

$$
\boldsymbol{P} = \left(b_{22} - \frac{b_{12}b_{21}}{b_{11}}\right)\boldsymbol{E} = [\varepsilon(0) - 1]\varepsilon_0\boldsymbol{E}\,. \tag{3.187}
$$

$\varepsilon(0)$ here is the *static dielectric constant.*

For externally applied electric fields of very high frequency the ions are no longer able to follow the rapidly changing forces. We then have $\boldsymbol{w} = 0$ and, with $\varepsilon(\infty)$ the dielectric constant for this limiting case,

$$
\boldsymbol{P} = b_{22}\boldsymbol{E} = [\varepsilon(\infty) - 1]\varepsilon_0\boldsymbol{E}\,. \tag{3.188}
$$

We now look for solutions of (3.186) of the type $\exp\,[\mathrm{i}(\boldsymbol{q}\cdot\boldsymbol{r} - \omega t)]$. External fields are not involved. We divide $\boldsymbol{w}$ into an irrotational and a divergence-free component ($\boldsymbol{w} = \boldsymbol{w}_{\text{L}} + \boldsymbol{w}_{\text{T}}$, $\nabla \times \boldsymbol{w}_{\text{L}} = 0$, $\nabla\cdot\boldsymbol{w}_{\text{T}} = 0$). The two components correspond exactly in our plane wave approximation to longitudinal and transverse waves, respectively. We further take account of the condition $\nabla\cdot\boldsymbol{D} = \nabla\cdot(\varepsilon_0\boldsymbol{E} + \boldsymbol{P}) = 0$. From this we then have

$$
\nabla\cdot(\varepsilon_0\boldsymbol{E} + b_{21}\boldsymbol{w} + b_{22}\boldsymbol{E}) = 0 \tag{3.189}
$$

with solution

$$
\boldsymbol{E} = -\frac{b_{21}}{\varepsilon_0 + b_{22}}\boldsymbol{w}_{\text{L}}\,. \tag{3.190}
$$

Putting this into the first (3.186) we find

$$\ddot{w}_T + \ddot{w}_L = b_{11}(w_T + w_L) - \frac{b_{12}b_{21}}{\varepsilon_0 + b_{22}} w_L . \tag{3.191}$$

If we separate the equation into irrotational and divergence-free components we obtain equations

$$\ddot{w}_T = b_{11} w_T , \qquad \ddot{w}_L = \left(b_{11} - \frac{b_{12}b_{21}}{\varepsilon_0 + b_{22}}\right) w_L = b_{11} \frac{\varepsilon(0)}{\varepsilon(\infty)} w_L . \tag{3.192}$$

If we call the frequencies of the transverse and longitudinal waves ω_T and ω_L, respectively (these are just the limiting frequencies of the corresponding optical branches as $\boldsymbol{q}$ goes to zero), we finally have

$$\begin{aligned} b_{11} &= -\omega_T^2 = -\omega_L^2 \frac{\varepsilon(\infty)}{\varepsilon(0)} , \qquad b_{22} = [\varepsilon(\infty) - 1]\varepsilon_0 , \\ b_{12} &= b_{21} = \omega_T\{[\varepsilon(0) - \varepsilon(\infty)]\varepsilon_0\}^{1/2} . \end{aligned} \tag{3.193}$$

Therefore

$$\omega_L^2 = \frac{\varepsilon(0)}{\varepsilon(\infty)} \omega_T^2 . \tag{3.194}$$

This last relationship is known as the *Lyddane-Sachs-Teller* relation. For ionic crystals with two atoms in the basis of a Wigner-Seitz cell, it relates the limiting frequencies of the two types of optical vibrations. In the form given here the relation is of course restricted to cubic crystals, in accordance with the remarks made following (3.182).

The classical treatment of optical vibrations given here is important for the interaction between phonons and photons discussed in Section 4.2.

3.4 Spin-Spin Interaction: Magnons

3.4.1 Introduction

We have mostly neglected the spin of the electrons and lattice ions in the elementary excitations examined so far. Apart from a short discussion in Section 2.2.15 of the effect of spin-orbit coupling on the band structure of a solid, spin has only entered our considerations through the Pauli principle. The Pauli principle is responsible for the exchange interaction (Sec. 1.3) which we took account of in the one-electron Schrödinger equation (1.32). On the other hand, we have not yet paid any attention to the spin of the lattice ions. The exchange interaction can give rise to collective excitations of the ion spin system, known as spin waves. The associated quanta are called *magnons*.

Collective excitations are low-lying excited states above a ground state. The ground state of the spin system is therefore important. If all spins are arranged in the same direction, the solid is a *ferromagnet*. In *ferrimagnets* and *antiferromagnets*, spins are only aligned in different sublattices of the solid. In the next section we shall look at spin waves in ferromagnets and in this simple example become familiar with the basis of the magnon concept. We can then readily extend these results to ferri- and antiferromagnetism. This we do in Section 3.4.3.

In examining elementary excitations in magnetic solids, we are only touching on a fraction of the important magnetic phenomena. We therefore extend our considerations in Section 3.4.4 with a short description of the molecular field approximation, which is important in explaining the properties of a ferromagnet near the Curie temperature.

It is not only the spin of the lattice ions which is significant where magnetism is concerned. We saw earlier that the division into lattice ions and valence electrons is an idealization which does not always hold. This is particularly so when *d*-electrons enter into the picture. In Section 3.4.5 we therefore discuss aspects of ordered magnetism involving valence and conduction electrons.

Since we want to put the emphasis of the presentation on elementary excitations, we shall only cover a part of the wide field of magnetism in this chapter. For more comprehensive presentations we refer the reader to the general textbooks and monographs [78–83]. For more extensive treatments of spin waves in particular, we recommend the article by *Keffer* [106, XVIII/2], the contribution by *Elliott* in [117], and the relevant chapters of the book by *Harrison* [13] and *Kittel* [16]. There are numerous contributions on magnetism in transition metals in the conference proceedings [102.37] (see also *Biondi* in [113a]). Finally we also point to the review articles by *Kittel* [101.22] and *Nagamiya* [101.20].

3.4.2 Spin Waves in Ferromagnets: Magnons

In our description of excited states we have so far always referred to a ground state in which the net spin of the valence electrons is zero. The reason for this was that each of the energy states in the one-electron approximation can be occupied by two electrons with opposite spins. The ground state was regarded as that state in which all the energy levels were filled up to a limiting energy E_F, and all levels above that were empty. Such a state has neither net spin nor net momentum.

This argument need no longer apply when the mutual interaction of the electrons is taken into account. This can be easily deduced from the Hartree-Fock approximation. From the result in Section 3.1.3, a Hartree-Fock electron has a mean kinetic energy proportional to k_F^2 and a mean exchange energy proportional to $-k_F$ (if the exchange integral itself is positive). The energy of the spin-compensated ground state is then $N(ak_F^2 - bk_F)$. If we align all the spins parallel, the electrons occupy states in a sphere of double the volume in $\boldsymbol{k}$-space. The energy of such a ferromagnetic state becomes $N(a2^{2/3}k_F^2 - b2^{1/3}k_F)$. This energy lies below the energy of the spin-compensated state if k_F is less than $0.44b/a$. The ferromagnetic state is therefore favoured in an electron gas of low density (small Fermi sphere radius).

Coulomb interaction has been ignored in this description. It prevents spin alignment when the density is low. In spite of this, the example shows that exchange interaction can be responsible for the spin correlation observed in *ferromagnets* (uniformly aligned spins), and in *antiferromagnets* and *ferrimagnets* (different alignment of the correlated spin system in different sublattices).

We take one further step in this consideration of the ferromagnetic Hartree-Fock electron gas. The energy of the ground state is given by (1.20), where all wave functions have identical spin. Since the spin functions are orthogonal, this means that they just drop out of (1.20) and the $\boldsymbol{q}_i$ can be replaced by $\boldsymbol{r}_i$. We now examine an excited state in which the spin of *one* electron is reversed. The energy of this state is obtained from (1.20) by associating with $N - 1$ electrons the spin function $\alpha(j)$, and with the ith electron the spin function $\beta(i)$. All the exchange integrals connecting this one electron with all other electrons then disappear. The difference in energy between the excited and ground states is

$$E_i - E_0 = \frac{e^2}{8\pi\varepsilon_0} \sum_{j(\neq i)} \int \frac{\varphi_j^*(\boldsymbol{r}_1)\varphi_j(\boldsymbol{r}_2)\varphi_i^*(\boldsymbol{r}_2)\varphi_i(\boldsymbol{r}_1)}{|\boldsymbol{r}_1 - \boldsymbol{r}_2|}\, d\tau_1 d\tau_2 \equiv \frac{1}{2} \sum_{j(\neq i)} J_{ij}\,. \tag{3.195}$$

The excited state $|i\rangle$ considered is degenerate with all states $|n\rangle$ in which another electron likewise has opposite spin. The solution of the Schrödinger equation for an excited state with this energy has therefore to be constructed as a linear combination of all $|n\rangle$: $\Phi = \sum_n a_n|n\rangle$.

We have here a behaviour analogous to that of lattice vibrations where kinetic energy fed to one lattice ion spreads itself by Coulomb interaction over all ions. The excitation which results can be described by wavelike states, and correspondingly the problem being considered here has wavelike solutions [$a_n \sim \exp(i\boldsymbol{k}\cdot\boldsymbol{r}_n)$]. The energy needed to reverse a spin is distributed throughout the entire spin system (*spin waves*, Fig. 3.13). Spin waves can be quantized like lattice waves. Thus we have the *magnons* as new collective excitations.

We do not however want to use the Hartree-Fock equations of the free electron gas to study this new type of elementary excitation, but rather a more general approach. The spins, whose correlation in ferromagnetism and related phenomena leads to a spontaneous magnetic moment, are mostly localized on the lattice ions. Furthermore several electrons contribute to the total ion spin. The ferromagnetic state is then the result of exchange interaction between the total spins of the different ions.

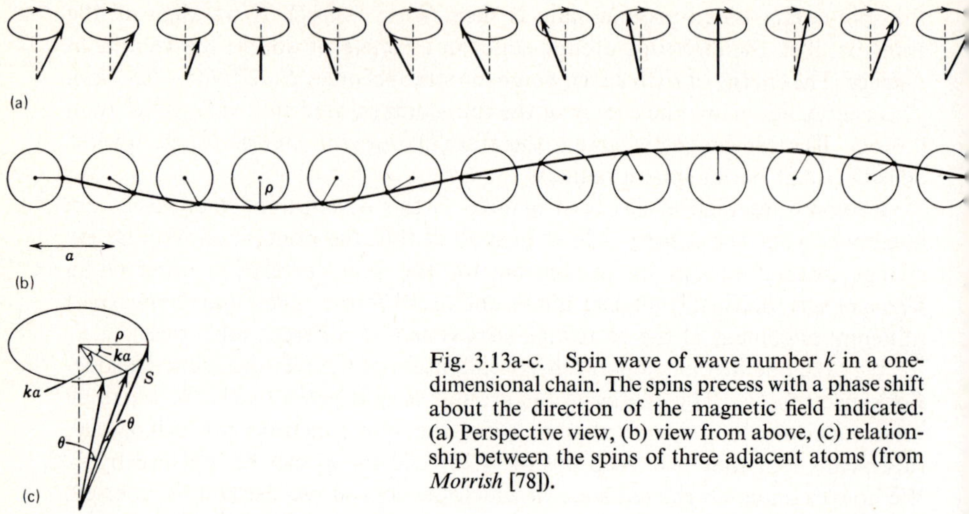

Fig. 3.13a-c. Spin wave of wave number k in a one-dimensional chain. The spins precess with a phase shift about the direction of the magnetic field indicated. (a) Perspective view, (b) view from above, (c) relationship between the spins of three adjacent atoms (from *Morrish* [78]).

The exchange Hamiltonian can be formally replaced by an operator introduced by Heisenberg

$$H = -\sum_{ij}' J_{ij} \boldsymbol{S}_i \cdot \boldsymbol{S}_j . \tag{3.196}$$

Here J_{ij} is an exchange integral. The $\boldsymbol{S}_i$ are the vector spin operators of the ith lattice ion. The sum is over all pairs of lattice ions.

For the case $s = 1/2$ the spin operators in (3.196) are given by the Pauli matrices

$$S_x = \frac{1}{2}\begin{vmatrix} 0 & 1 \\ 1 & 0 \end{vmatrix}, \qquad S_y = \frac{1}{2}\begin{vmatrix} 0 & -\mathrm{i} \\ \mathrm{i} & 0 \end{vmatrix}, \qquad S_z = \frac{1}{2}\begin{vmatrix} 1 & 0 \\ 0 & -1 \end{vmatrix} \tag{3.197}$$

with the commutation relations

$$[S_\lambda, S_\mu] = \mathrm{i}S_\nu \qquad \lambda,\mu,\nu = x, y, z \text{ and cycl}. \tag{3.198}$$

The Pauli matrices operate on spinors $\begin{vmatrix} \alpha \\ \beta \end{vmatrix}\!\Big\rangle$. For the spin functions α, β follows

$$S_z\alpha = \frac{1}{2}\alpha, \qquad S_z\beta = -\frac{1}{2}\beta, \qquad S^2\alpha = \frac{3}{4}\alpha, \qquad S^2\beta = \frac{3}{4}\beta. \tag{3.199}$$

α thus describes the state with $s_z = +1/2$ and β the state with $s_z = -1/2$. The eigenvalue of the operator S^2 is $s(s+1) = 3/4$.

For a total spin $s = n/2$, a spin operator S with $[S_\lambda, S_\mu] = iS_\nu$ can correspondingly be introduced. The matrices associated with its components are then $n + 1$-dimensional and there are $n + 1$ spin functions. The eigenvalues of S_z are $-s, -s + 1, \ldots, s - 1, s$ and of S^2: $s(s + 1)$.

In place of S_x and S_y, one often introduces the operators $S_+ = S_x + iS_y$ and $S_- = S_x - iS_y$. For the case $s = 1/2$ one then has

$$S_+\alpha = 0\,, \qquad S_+\beta = \alpha\,, \qquad S_-\alpha = \beta\,, \qquad S_-\beta = 0\,. \tag{3.200}$$

S_+ thus converts a minus-spin into a plus-spin and vice versa. For $s > 1/2$, the S_+ raise the total spin by unity and the S_- lower it by unity.

With these spin operators we now calculate the expectation value of the operator (3.196) in the case $s = 1/2$ for a pair of indices i, j. We find

$$E_{\uparrow\uparrow} = -J_{ij}\langle\alpha_i\alpha_j|\boldsymbol{S}_i\cdot\boldsymbol{S}_j|\alpha_i\alpha_j\rangle, \qquad E_{\uparrow\downarrow} = -J_{ij}\langle\alpha_i\beta_j|\boldsymbol{S}_i\cdot\boldsymbol{S}_j|\alpha_i\beta_j\rangle\,, \tag{3.201}$$

or using (3.197) and taking account of the orthonormalization of the spin functions α and β,

$$E_{\uparrow\uparrow} = -\frac{1}{4}J_{ij}\,, \qquad E_{\uparrow\downarrow} = +\frac{1}{4}J_{ij}\,. \tag{3.202}$$

The difference in energy between these two possibilities is therefore $J_{ij}/2$ and the difference between a state in which all spins are aligned and one in which the ith spin is reversed is $\sum_{j(\neq i)} J_{ij}/2$. This agrees with (3.195).

The exchange interaction is therefore formally reproduced by the operator (3.196) just as though it were explicitly a spin-spin interaction.

Since the exchange interaction between nearest neighbours is dominant, one usually restricts consideration to these terms and therefore includes in the summation over j only those terms formed by an $\boldsymbol{R}_j = \boldsymbol{R}_i + \boldsymbol{R}_\delta$ where $\boldsymbol{R}_\delta$ is a vector to the nearest neighbour ($\delta = 1, 2, \ldots \nu$) of the ith ion. If further one immediately assumes $J_{i,i+\delta} = J$ for all δ (thereby limiting oneself to simple lattices) the effective interaction becomes

$$H = -J\sum_{i,\delta} \boldsymbol{S}_i\cdot\boldsymbol{S}_{i+\delta}\,. \tag{3.203}$$

We now assume that the ion spins in the ground state are so aligned that their z-components have the maximum values s. We describe the wave function for the ground state as a product of spin functions $|s\rangle_n$ which represent the spin of the nth ion in state s: $\Phi_0 = \prod_n |s\rangle_n$. We introduce spin-raising and spin-lowering operators S_+ and S_- so that the Hamiltonian (3.203) takes the form

$$H = -J\sum_{ij}{}' \left[S_{iz}S_{jz} + \frac{1}{2}(S_{i+}S_{j-} + S_{i-}S_{j+})\right] \qquad (j = i + \delta)\,. \tag{3.204}$$

Application of this operator to the ground state yields

$$E_0 = -s^2 J \sum_{i,i+\delta} 1 = -Js^2 \nu N, \tag{3.205}$$

since the application of a spin-increase operator to a function with maximum spin must lead to zero. ν is again the number of nearest neighbours of the ion.

We consider now the state $\Phi_m = S_{m-} \prod_n |s\rangle_n$ in which the mth spin is diminished by one. We then have

$$H\Phi_m = -J \sum_{ij}{}' \left[S_{iz} S_{jz} S_{m-} + \frac{1}{2}(S_{i+} S_{j-} S_{m-} + S_{i-} S_{j+} S_{m-}) \right] \Phi_0 . \tag{3.206}$$

We can rearrange the product of spin operators on the right of (3.206) using the commutation relations which follow from (3.198): $[S_+, S_-] = 2S_z$, $[S_-, S_z] = S_-$, $[S_z, S_+] = S_+$. We then obtain

$$H\Phi_m = E_0 \Phi_m + 2Js \sum_\delta (\Phi_m - \Phi_{m+\delta}) . \tag{3.207}$$

Φ_m is not therefore an eigenstate of H. Such a state must rather be made up of all degenerate $\Phi_m = S_{m-}\Phi_0$: $\Phi = \sum_m a_m \Phi_m$. In view of the translation invariance of the lattice the a_m have the form $\exp(\mathrm{i}\boldsymbol{k}\cdot\boldsymbol{R}_m)$. It then follows that

$$H\Phi = \sum_m \exp(\mathrm{i}\boldsymbol{k}\cdot\boldsymbol{R}_m) H\Phi_m = [E_0 + 2J\nu s(1 - \gamma_k)]\Phi , \tag{3.208}$$

where

$$\gamma_{\boldsymbol{k}} = \frac{1}{\nu} \sum_\delta \exp(\mathrm{i}\boldsymbol{k}\cdot\boldsymbol{R}_\delta) . \tag{3.209}$$

The energy of the excited state is therefore

$$E_{\boldsymbol{k}} = E_0 + 2J\nu s(1 - \gamma_{\boldsymbol{k}}) , \tag{3.210}$$

where $\boldsymbol{k}$ (for cyclic boundary conditions) is limited to the N values inside a Brillouin zone in $\boldsymbol{k}$-space.

For small $\boldsymbol{k}$ this becomes

$$E_{\boldsymbol{k}} = E_0 + Js \sum_\delta (\boldsymbol{k}\cdot\boldsymbol{R}_\delta)^2 . \tag{3.211}$$

Eqs. (3.210) and (3.211) are dispersion relations for spin waves.

To quantize the spin waves, we start out from the following line of thought: All spins are aligned in the ground state. Their z-components have the maximum value $s_z = s$. We can describe an excited state by specifying by how many units the s_z differ from the maximum value. If we call this number n and add the index of the ion in question, then each state is described by specifying the $n_1, n_2, \ldots n_N$ ($n_i = 0, 1, 2, \ldots, 2s$) and we can describe this state in an occupation number representation by a state vector $|n_1, n_2, \ldots, n_N\rangle$ for bosons (see Appendix). Correspondingly we can introduce creation and annihilation operators in accordance with (A.15). $a_j^+ a_j$ is then an operator whose eigenstates describe the departures of the spin of the jth ion from the maximum value.

The a^+ and a can readily be obtained from the S_+ and S_- introduced earlier. For the $S_\pm$ we find from the commutation relations

$$
\begin{aligned}
S_{j+}|n_j\rangle &= \sqrt{2s+1-n_j}\sqrt{n_j}|n_j-1\rangle \\
S_{j-}|n_j\rangle &= \sqrt{2s-n_j}\sqrt{n_j+1}|n_j+1\rangle \\
S_{jz}|n_j\rangle &= (s-n_j)|n_j\rangle ,
\end{aligned}
\tag{3.212}
$$

where only the state of the jth ion has been given in the wave functions.

A comparison of (3.212) and (A.15) then shows the link between the a^+, a and the S_+, S_-, and S_z

$$
S_+ = \sqrt{2s - a^+ a}\, a , \qquad S_- = a^+ \sqrt{2s - a^+ a} , \qquad S_z = s - a^+ a . \tag{3.213}
$$

The expansion of the roots has to be inserted into the operators on the right side of the first two equations here.

We can use these relations to rewrite the Hamiltonian (3.204) in terms of creation and annihilation operators (*Holstein-Primakoff transformation*).

It is however convenient to go a step further. The a_j^+ and a_j change the spin of the jth ion. We saw already that through the exchange interaction such a change in spin is propagated through the whole spin system. We have therefore to include a transformation to creation and annihilation operators of the spin wave quanta. This corresponds to the transition from atomic to normal coordinates which we carried out for lattice vibrations before the quantization.

The corresponding transformation here is

$$
a_j^* = \frac{1}{\sqrt{N}} \sum_{\boldsymbol{k}} \exp(\mathrm{i}\boldsymbol{k}\cdot\boldsymbol{R}_j) b_{\boldsymbol{k}}^+ , \qquad a_j = \frac{1}{\sqrt{N}} \sum_{\boldsymbol{k}} \exp(-\mathrm{i}\boldsymbol{k}\cdot\boldsymbol{R}_j) b_{\boldsymbol{k}} . \tag{3.214}
$$

The new operators then satisfy the same commutation relations

$$
[b_{\boldsymbol{k}}, b_{\boldsymbol{k}'}^+] = \delta_{\boldsymbol{k}\boldsymbol{k}'} , \qquad [b_{\boldsymbol{k}}^+, b_{\boldsymbol{k}'}^+] = [b_{\boldsymbol{k}}, b_{\boldsymbol{k}'}] = 0 \tag{3.215}
$$

and we also have

$$
\sum_j a_j^+ a_j = \sum_{\boldsymbol{k}} b_{\boldsymbol{k}}^+ b_{\boldsymbol{k}} . \tag{3.216}
$$

Transforming the Hamiltonian to the $b_{\boldsymbol{k}}^+$, $b_{\boldsymbol{k}}$ presents the problem that in (3.213) $a_j^+ a_j$, i.e., sums over products of the $b_{\boldsymbol{k}}^+$ and $b_{\boldsymbol{k}}$, occur under the square root. If one restricts oneself to small departures from the ground state—and this is the only region where the concept of elementary excitations is reasonable—and hence to small n_j, one can terminate the expansion of the root early. S_+ is then a series with operators of the form $b_{\boldsymbol{k}}$, $b_{\boldsymbol{k}}^+ b_{\boldsymbol{k}'} b_{\boldsymbol{k}''}$, etc., S_- a series with operators $b_{\boldsymbol{k}}^+$, $b_{\boldsymbol{k}}^+ b_{\boldsymbol{k}'}^+ b_{\boldsymbol{k}''}$, etc. Additionally exponential functions appear containing summations over $\boldsymbol{k}$, $\boldsymbol{k}'$, $\boldsymbol{k}''$ in the exponent.

In the sum over the i, j in (3.204), relations between these $\boldsymbol{k}$, $\boldsymbol{k}'$, $\boldsymbol{k}''$ then follow so that finally H can be written as a series, whose terms up to the fourth order in the $b_{\boldsymbol{k}}$ have the following form for a lattice with a centre of inversion ($\gamma_{\boldsymbol{k}} = \gamma_{-\boldsymbol{k}}$):

$$H = E_0 + \sum_{\boldsymbol{k}} 2Jvs(1 - \gamma_{\boldsymbol{k}}) b_{\boldsymbol{k}}^+ b_{\boldsymbol{k}}$$

$$+ \frac{vJ}{2N} \sum_{\boldsymbol{k}\boldsymbol{k}'\boldsymbol{\kappa}} (\gamma_{\boldsymbol{k}-\boldsymbol{\kappa}} + \gamma_{\boldsymbol{k}'} - 2\gamma_{\boldsymbol{k}-\boldsymbol{\kappa}-\boldsymbol{k}'}) b_{\boldsymbol{k}-\boldsymbol{\kappa}}^+ b_{\boldsymbol{k}'+\boldsymbol{\kappa}}^+ b_{\boldsymbol{k}'} b_{\boldsymbol{k}} + \cdots, \tag{3.217}$$

where $\gamma_{\boldsymbol{k}}$ is again given by (3.209).

The first term is the energy of the ground state, the second is the energy contained in the magnons. From (3.217) the energy of a magnon is the expression already given in (3.210)

$$\hbar\omega_{\boldsymbol{k}} = 2Jvs(1 - \gamma_{\boldsymbol{k}}) \,. \tag{3.218}$$

$b_{\boldsymbol{k}}^+ b_{\boldsymbol{k}}$ is the magnon particle number operator.

The other terms of (3.217) describe the magnon-magnon interaction. The third term in particular contains processes in which two magnons $\boldsymbol{k}$, $\boldsymbol{k}'$ are annihilated and two magnons $\boldsymbol{k} - \boldsymbol{\kappa}$, $\boldsymbol{k}' + \boldsymbol{\kappa}$ created, with conservation of total momentum, or put another way, processes in which momentum $\boldsymbol{\kappa}$ is transferred from one magnon to another.

This term also contains processes in which $\boldsymbol{\kappa} = 0$ or $\boldsymbol{k}' = \boldsymbol{k} - \boldsymbol{\kappa}$. Such terms contribute to the magnon energy (3.218) and can be interpreted as renormalization of the magnon energy through exchange interaction.

As a simple application of the results, we want to estimate the energy contained in the magnons and hence their contribution to the specific heat. The only difference compared with the case treated in Section 3.3.4 is the different form of the dispersion relations for phonons and magnons. While phonon energy increases linearly for small q, magnon energy increases as the square of k. Restricting ourselves to the isotropic case $\hbar\omega_k \sim k^2$, the magnon energy follows correspondingly as

$$E = \sum_{\boldsymbol{k}} \frac{\hbar\omega_{\boldsymbol{k}}}{\exp(\hbar\omega_{\boldsymbol{k}}/k_{\mathrm{B}}T) - 1} \propto \int d\tau_{\boldsymbol{k}} \frac{k^2}{\exp(\alpha k^2/T) - 1}$$

$$\propto \int_0^{k_{\max}} \frac{k^4 dk}{\exp(\alpha k^2/T) - 1} \,. \tag{3.219}$$

Here we have again replaced the summation in $\boldsymbol{k}$-space by an integral.

Since the estimate is only valid for low temperatures, when few magnons are excited, we can replace the upper limit k_{max} of the magnon dispersion spectrum in the integral with infinity. Rearrangement in terms of a dimensionless integration variable then leads to a factor $T^{5/2}$, and hence the specific heat is proportional to $T^{3/2}$, in agreement with experiment.

The temperature dependence of the magnetization can be calculated in a similar way. The departure of the magnetization from saturation $\Delta M = M(T) - M(0)$ is proportional to the average number of magnons $\sum_k \bar{n}_k$, i.e., from (3.219) it is proportional to an integral with k^2 rather then k^4 in the numerator of the integrand. This also leads here to a $T^{3/2}$ law. Corrections to both $T^{3/2}$-laws for specific heat and magnetization become necessary on several counts at higher temperatures. The range of validity is limited most of all by the magnon-magnon interaction, and by the replacement of (3.211) by an isotropic k^2-law. Since the concept of elementary excitations is only useful as long as mutual interactions between the excitations can be neglected, we shall not be interested here in better approximations. We shall turn our attention later to temperatures near the Curie temperature of a ferromagnet.

Our model so far restricts the results to solids which in the ground state have a spin system bound to the ions of a Bravais lattice. We therefore have to extend the theory in two directions.

1) Non-Bravais lattices. This includes the case of antiferro- and ferrimagnetism.
2) Ferromagnetic metals. In this case the spin of the nonlocalized valence electrons plays a major role.

3.4.3 Spin Waves in Lattices with a Basis, Ferri-, and Antiferromagnetism

For Bravais lattices, the dispersion relation (3.218) gives a $\boldsymbol{k}$-dependence of the magnon energy which, similar to an acoustic branch of the phonon spectrum, starts with zero energy at $\boldsymbol{k} = 0$ and increases up to the surface of the Brillouin zone. For *lattices with a basis* we must expect there to be other branches in the magnon spectrum, branches which correspond to the optical phonons. It will not be possible in such lattices to restrict the Heisenberg operator to exchange interactions between nearest neighbours. The different basis atoms form sublattices and, in addition to the interaction within a given sublattice, that between different sublattices is important. However, an extension of our model is also needed for other reasons. In most cases the ions in the individual sublattices will be different. The sublattices will then possess different total spin and often also different directions for the (individually parallel aligned) spin systems. The ground state will admittedly then show a magnetic moment. However, this will be the vector sum of the spins of the sublattices. For two sublattices with opposite spins it will therefore be the difference in the spins. Such a *ferrimagnet* reveals differences compared with a ferromagnet.

Before going into these questions we treat a simpler case which can already indicate what is important. Thus far we have assumed that the spins of nearest neighbours in a Bravais lattice of a ferromagnet are all aligned parallel in the ground state by exchange interaction. For this it is important that the exchange integral be positive. The case of negative exchange integral, on the other hand, is also possible. The preferred condition is then for nearest neighbours to have antiparallel spin. In the ground state—or so we assume at least initially—we then find two sublattices of identical atoms but with opposite spin directions. This is the case of an *antiferromagnet* with mutually compensating spontaneous magnetic moments of the two sublattices.

We can treat this model using the operator (3.203). Let the exchange integral between the lattice ions (assumed to be the same type) be negative. We denote its absolute value by J and write

$$H = +J \sum_{i,\delta} \boldsymbol{S}_i \cdot \boldsymbol{S}_{j+\delta} \,. \tag{3.220}$$

We run into a difficulty in constructing the wave function for the ground state. In the ferromagnetic case we could realize the ground state in only one way, namely by alignment of all spins in a preferred direction, which we chose as the z-axis. A negligibly small magnetic field could serve to set up the ground state, and this can be taken account of by an additive term in the Hamiltonian. In the same way we can define a preferred direction here too and call it the z-axis. However, we still have to decide which ions in the uniform (apart from spin setting) lattice we want to combine into the sublattice with spin $+$, and which into the sublattice with spin $-$. The two possibilities are degenerate, and in order to accentuate one, and thereby stabilize the state, we have to introduce a small finite magnetic field (*anisotropy field*), which is positive at the ions of the one sublattice and negative at the ions of the other. Fields such as these, which are small compared with the other internal fields (see further below), are also observed experimentally. One can take account of them in the Hamiltonian by an additive term of the sort

$$g\mu_B B_A \left(\sum_b S_{bz} - \sum_a S_{az} \right) . \tag{3.221}$$

Here we have designated the anisotropy field B_A and given the two sublattices the indices a and b. We shall not need the anisotropy field at first in our considerations, but we shall return to it later.

For each sublattice we must introduce creation and annihilation operators by means of a Holstein-Primakoff transformation. The transition to magnon operators is then possible, analogous to (3.214). Insertion into (3.220) and expansion of the roots finally yields instead of (3.217), with terms up to second order in the magnon operators,

$$H = -2NJvs^2 + 2Jvs\left[\sum_k (b_{ak}^+b_{ak} + b_{bk}^+b_{bk}) + \sum_k \gamma_k(b_{ak}^+b_{bk}^+ + b_{ak}b_{bk})\right]. \tag{3.222}$$

Here N is the number of ions in a sublattice. The first term is the energy of the undisturbed state. The second describes spin waves in the respective sublattices. The third corresponds to an interaction between the sublattices, in which a pair of magnons is respectively created or annihilated with zero effective spin change. This interaction can be removed by introducing magnon operators which describe combined spin waves in the two sublattices.

For this, one introduces the following operators:

$$\begin{aligned} c_{1k} &= u_k b_{ak} - v_k b_{bk}^+, \qquad & c_{1k}^+ &= u_k b_{ak}^+ - v_k b_{bk} \\ c_{2k}^+ &= u_k b_{bk} - v_k b_{ak}^+, \qquad & c_{2k} &= u_k b_{bk}^+ - v_k b_{ak} \end{aligned} \tag{3.223}$$

with real u_k, v_k, $u_k^2 - v_k^2 = 1$ and $[c_{1,2k}, c_{1,2k}^+] = 1$, $[c_{1k}, c_{2k}] = 0$.

The u_k and v_k are defined such that the coefficients of the mixed terms $c_{1k}c_{2k}$ and $c_{1k}^+c_{2k}^+$ become zero. We are then left with

$$H = -2NvJs(s+1) + \sum_k \hbar\omega_k(c_{1k}^+c_{1k} + c_{2k}^+c_{2k} + 1), \tag{3.224}$$

where

$$\hbar\omega_k = +2Jvs\sqrt{1-\gamma_k^2}. \tag{3.225}$$

Along with the energy of the ground state, the Hamiltonian contains magnon particle number operators. For each $\boldsymbol{k}$ there are two different magnons, distinguished by the indices 1, 2. Clearly the ground state of the system (no magnons) is

$$E_0 = -2NvJs(s+1) + 2Jvs\sum_k \sqrt{1-\gamma_k^2}. \tag{3.226}$$

If γ_k were zero, E_0 would be given by $E_0 = -2NJvs^2$, the energy of the strictly antiparallel-ordered lattice. Since $\gamma_k \neq 0$, the second term is less than $2NJvs$. The *ground state* is not therefore strictly aligned. Each sublattice has a small amount of disorder in its spin alignment.

Equation (3.225) is the dispersion relation for the antiferromagnetic magnons. The branches for the different magnons (indices 1 and 2) split up in an external magnetic field. For small $\boldsymbol{k}$, $\sqrt{1-\gamma_k^2} \propto k$ for simple lattices. In contrast to ferromagnetism the magnon energy increases here linearly with k.

Here we should point out a correction using the stabilizing anisotropy field. For the case in which an external magnetic field B and [in (3.225)] an anisotropy field B_A are included, the dispersion relations (3.218) and (3.225) are

$$\hbar\omega_{\boldsymbol{k}} = 2Jsv(1 - \gamma_{\boldsymbol{k}}) + 2\mu_B B \qquad \text{for ferromagnets} \tag{3.227a}$$

$$\hbar\omega_{\boldsymbol{k}} = 2Jsv\sqrt{\left(1 + \frac{\mu_B B_A}{Jsv}\right)^2 - \gamma_{\boldsymbol{k}}^2} \pm 2\mu_B B \quad \text{for antiferromagnets}\,. \tag{3.227b}$$

In both cases we can make B as small as we like. Furthermore, experimental results show B_A to be of the order of 1000 gauss, and therefore negligible compared with Jsv/μ_B (order of 10^6 gauss). For $k = 0$, however, $\gamma_{\boldsymbol{k}} = 1$ and (3.227b) becomes

$$\hbar\omega_0 = 2Jsv\sqrt{\frac{\mu_B B_A}{Jsv}\left(1 + \frac{\mu_B B_A}{Jsv}\right)} \approx \sqrt{4Jsv\mu_B B_A}\,. \tag{3.228}$$

Since B_A appears here multiplied by Jsv, $\hbar\omega_0$ can deviate markedly from zero. There is then an energy gap between ground state and the lowest excited state.

Specific heat and magnetization can now be calculated in a similar way to ferromagnets.

The theory of *ferrimagnetic magnons* can be correspondingly formulated. We restrict ourselves to the statement of the dispersion relations for the simplest

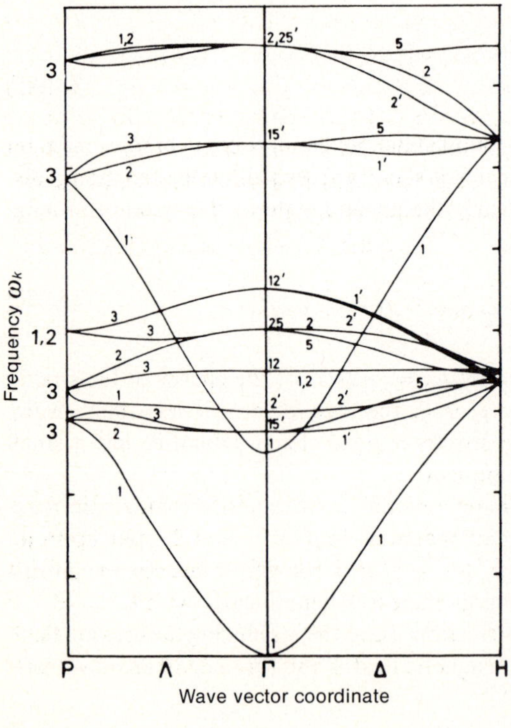

Fig. 3.14. Magnon dispersion curves for yttrium-iron-garnet ($Y_3Fe_5O_{12}$). The numbers indicate the symmetries of the individual branches [from W. Brinkman, R. J. Elliott: J. Appl. Phys. 37, 1458 (1966)].

case, that in the antiferromagnet considered above the spins of the sublattices have different values $s_a \neq s_b$. Again neglecting the anisotropy field we find

$$\hbar\omega_k = Jv[\sqrt{(s_a - s_b)^2 + 4s_a s_b(1 - \gamma_k^2)} \pm (s_a - s_b)]\,. \tag{3.229}$$

Equation (3.225) follows from this when we put $s_a = s_b$. The dispersion spectrum has two branches which, for $\boldsymbol{k} = 0$, have the values $\hbar\omega_0 = 0$ and $2Jv(s_a - s_b)$.

Ferrimagnets have in general a complicated lattice structure. In addition to the branches of the spectrum found here, we shall also find "optical" branches. Fig. 3.14 is an example of a complicated magnon dispersion spectrum. Consideration of the exchange between nearest neighbours in different sublattices and within a sublattice leads for YIG (yttrium iron garnet) to the fourteen branch spectrum shown. Group theory classifications have again to be used for such spectra. Space group symmetry is restricted here by the fact that similar ions with dissimilar spin alignment in the ground state are now regarded as different (*magnetic space groups*). To this are added symmetry operations in "spin-space" which leave the relative spin distribution of the lattice ions invariant. We cannot go into this aspect of group theory here.

3.4.4 Ferromagnetism Near the Curie Temperature

The concept of magnons as collective excitations without mutual interaction is of course only applicable to ferromagnetic problems when magnetization departs only a little from saturation magnetization. This is not, however, the only interesting region. The vicinity of the Curie temperature, above which spontaneous magnetiaztion vanishes, warrants particular attention. In this section, as an extension to spin wave theory, we want to show that the behaviour of a ferromagnet in this temperature range can likewise be explained by exchange interaction. The approximation used to do this is called the *molecular field approximation.*

The Hamiltonian for the exchange interaction (3.196), extended by an external magnetic field $\boldsymbol{B}$, is given by

$$H = -\sum_{ij}{}' J_{ij}\boldsymbol{S}_i\cdot\boldsymbol{S}_j - g\mu_B\boldsymbol{B}\cdot\sum_{i=1}^{N}\boldsymbol{S}_i\,. \tag{3.230}$$

The problem in solving a Schrödinger equation with this Hamiltonian is the nonlinearity of the first term. In the spin wave case, the difficulty could be circumvented by the Holstein-Primakoff transformation, with subsequent expansion of the root operator and inclusion of only the first term. In the present case we would at least have to take further terms in the expansion. A simpler approximation for (3.230) is a linearization of the operator such that one of the two spin operators is replaced by its mean value

$$H = -\sum_{i=1}^{N}\left(g\mu_B \boldsymbol{B} + \sum_{\substack{j=1 \\ (\neq i)}}^{N} J_{ij}\langle \boldsymbol{S}_j\rangle\right)\cdot \boldsymbol{S}_i\,. \tag{3.231}$$

Along with the external magnetic field there thus appears an *internal field* $\boldsymbol{B}_M = (1/g\mu_B)\sum_j J_{ij}\langle \boldsymbol{S}_j\rangle$.

Such an internal field was introduced early by Weiss to explain ferromagnetism (*Weiss field*). In isotropic media $\boldsymbol{B}_M$ will not depend on the index of the exchange integral. The mean value $\langle \boldsymbol{S}_j\rangle$ will further have the same direction as the magnetization $\boldsymbol{M}$: $\boldsymbol{M} = g\mu_B\langle \boldsymbol{S}_j\rangle N$, so that assuming exchange interaction with nearest neighbours only, it follows that

$$\boldsymbol{B}_M = \lambda \boldsymbol{M} \qquad \text{with} \quad \lambda = \nu J/g^2\mu_B^2 N\,. \tag{3.232}$$

The *Weiss constant* λ is therefore directly proportional to the exchange integral J.

Equation (3.232) is sufficient to calculate the Curie temperature and thereby to link λ with values which can be measured experimentally. To do this we must first return to the theory of magnetization of a paramagnetic material. We assume that the ions of a solid possess a magnetic moment $\boldsymbol{\mu}$. The thermal motion of the ions about their equilibrium positions results in a statistical distribution of all moments. If we apply an external field $\boldsymbol{B}$, the magnetization $\boldsymbol{M}$ depends on the ratio of the magnetic energy $\boldsymbol{\mu}\cdot\boldsymbol{B} = g\mu_B M_j B (M_j = j, j-1, \ldots, -j+1, -j)$ and the thermal energy $k_B T$

$$M = N\frac{\sum_{-j}^{+j} g\mu_B M_j \exp(g\mu_B M_j B/k_B T)}{\sum_{-j}^{+j} \exp(g\mu_B M_j B/k_B T)}\,. \tag{3.233}$$

The series can be summed to give

$$M = Ng\mu_B j B_j(y) \tag{3.234}$$

with $y = g\mu_B jB/k_B T$ and with the Brillouin function $B_j(y)$ given by

$$B_j(y) = \frac{2j+1}{2j}\coth\left(\frac{2j+1}{2j}y\right) - \frac{1}{2j}\coth\frac{y}{2j}\,. \tag{3.235}$$

For small y (weak magnetic fields), $B_j(y)$ can be replaced by the first term of a series: $B_j = (y/3)\cdot(j+1)/j$, and hence

$$M = \chi B \qquad \text{with} \quad \chi = \frac{Ng^2\mu_B^2 j(j+1)}{3k_B T} = \frac{C}{T}\,. \tag{3.236}$$

Eq. (3.236) is called the *Curie law* and C the *Curie constant*. The number $p_j = g\sqrt{j(j+1)}$ is designated the *effective number of magnetons*.

We can now apply (3.234) to the ferromagnetic case if we describe the interaction between the magnetic moments by an internal field. We have then only to add this internal field $\lambda \boldsymbol{M}$ to $\boldsymbol{B}$.

We obtain the spontaneous magnetic moment by putting the external field to zero

$$M = Ng\mu_B s B_s\left(\frac{g\mu_B s\lambda M}{k_B T}\right). \tag{3.237}$$

For $T = 0$, since $\coth y = 1$ for $y = \infty$, we have from this $M = Ng\mu_B s$ as saturation magnetization. M decreases with increasing temperature and finally approaches zero. In this limit we can use the following expansion for the Brillouin function:

$$B_s(y) = \frac{s+1}{s}\frac{y}{3} - \frac{(2s+1)^4 - 1}{(2s)^4}\frac{y^3}{45} \quad \text{for small } y. \tag{3.238}$$

Inserting this into (3.237), we find for the saturation magnetization a law of the form

$$M \sim \sqrt{T_c - T}, \qquad T_c = N\frac{g^2\mu_B^2\lambda}{3k_B}s(s+1) = \frac{s(s+1)vJ}{3k_B}. \tag{3.239}$$

Hence the magnetization disappears at the *Curie temperature* T_c. Above this temperature the solid is paramagnetic. Fig. 3.15 presents the temperature dependence of the spontaneous magnetization from (3.237), together with some experimental results for comparison.

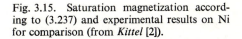

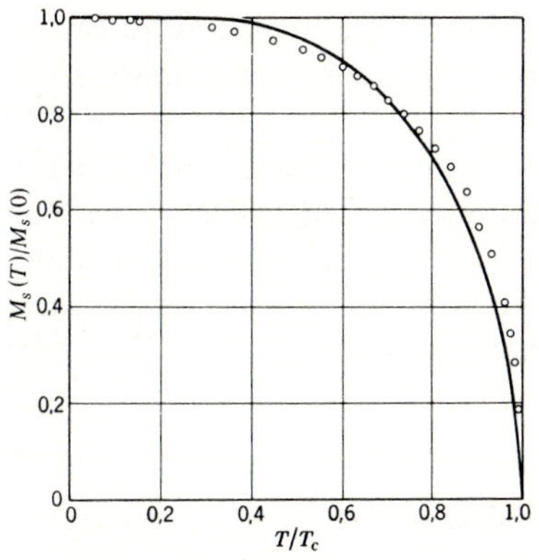

Fig. 3.15. Saturation magnetization according to (3.237) and experimental results on Ni for comparison (from *Kittel* [2]).

In the paramagnetic phase we can again describe the temperature dependence of magnetization by a Curie law of the form (3.236) if we include the internal field along with the external. We then have

$$M = \frac{C}{T}(B + \lambda M) \quad \text{or} \quad M = \frac{C}{T - C\lambda} B. \tag{3.240}$$

From (3.236) and (3.239) the Curie constant to be inserted here is equal to the Curie temperature divided by the Weiss constant λ. Hence

$$M = \chi B = \frac{C}{T - T_c} B. \tag{3.241}$$

This is the *Curie-Weiss law*. Above T_c, magnetic susceptibility increases as $(T - T_c)^{-1}$.

We can consider antiferromagnets and ferrimagnets in a similar way. Instead of an internal Weiss field, we have then to take different internal fields for the individual sublattices. We shall not present these treatments since they add nothing new to the general theory. Just as spin-wave theory is less important than molecular field theory near the Curie point, the reverse is true at low temperatures where the latter is too coarse. The low temperature approximation derived from (3.237) yields a saturation magnetization temperature dependence of the form $M(T)/M(0) = 1 - (1/s) \exp[-3T_c/(s + 1)T]$, which contradicts the experimentally proven $T^{3/2}$ law of spin-wave theory.

Thus we have to distinguish two areas of ordered magnetism, which have to be tackled by different methods. For small departures from the ground state, the method of elementary excitations is more important than all other approximations. At higher temperatures it is more advantageous to use semiclassical methods, which however can also be referred to the general exchange-interaction concept. This is not supposed to mean that the concept of elementary excitations is fundamentally inapplicable at high temperature. Many aspects of the behaviour of a ferromagnet near the Curie temperature can be clearly understood with magnons. One can even define *paramagnons* in the paramagnetic temperature region well above T_c.

A method which is in many respects better than the molecular field approximation chooses for the first term of the Hamiltonian (3.230), instead of a sum over products of spin operators $S_i \cdot S_j$, a sum over products of their z-components $S_{iz}S_{jz}$. This so-called *Ising model* plays an important role in the statistical theory of phase transitions, like the transition appearing here at T_c. This goes beyond the scope of this chapter.

3.4.5 Ordered Magnetism of Valence and Conduction Electrons, the Collective Electron Model

The model which we have considered so far is concerned with direct exchange interaction between localized spins of nearest neighbours. This assumes on the one hand that the electrons (of a lattice ion) contributing to the magnetic moment are sufficiently tightly bound for the ions to be taken as isolated, but on the other hand that the nearest neighbours are close enough for a noticeable exchange interaction to arise.

An exchange between magnetic ions often occurs over large distances in an insulator, the mechanism being that a paramagnetic ion in between communicates the interaction. If, for example, two metallic ions with unfilled *d*-shells are linked by an oxygen atom (e.g., MnO), each *d*-electron will interact with one of the two *p*-electrons of the outermost (spin-saturated) electron pair of the oxygen atom. Since the two spin directions of the two *p*-electrons are linked by the Pauli principle, this implies an effective interaction between the two *d*-electrons (*super exchange*).

A further possibility is *indirect exchange* in which localized spins of lattice ions interact with the conduction electrons of a metal. The information on the spin of a given ion is passed on by an electron to another ion. This indirect ion-ion interaction belongs to the so-called *Ruderman-Kittel interactions*. It plays the major role in the rare earths (Gd to Tm). These elements are distinguished by the fact that their ordered magnetism shows a multiplicity of different organizations. Excited states can also be described here by spin waves. For details we refer the reader to an article by *Cooper* [101.21].

The most important group of ferromagnetic metals is the transition metals (see here the conference volume [112.37]). Here the electrons whose spin gives rise to the ferromagnetism are *not* localized (*itinerant electrons*). The elements Fe, Co, Ni have per lattice atom 8, 9, 10 electrons in the uppermost bands. These are a superposition of 3*d*- and 4*s*-bands. We have already presented the density of states of Ni in Fig. 2.29. There are indications that one can assume approximately the same band structure, and hence the same density of states, for Cu and the transition metals. The elements are then distinguished only by the position of the Fermi energy relative to the band edges. Fig. 3.16 shows the density of states again schematically. For Cu, E_F lies above the filled *d*-bands.

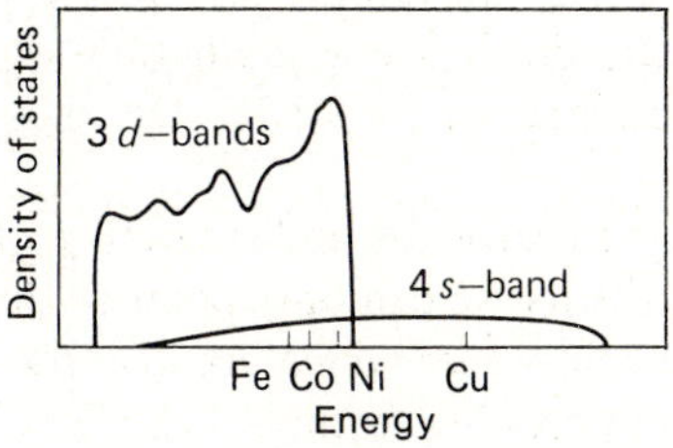

Fig. 3.16. Schematic representation of the density of states of the transition metals with the assumption that all these elements have approximately the same band structure (rigid band model). The *d*-bands are filled to higher energies in Fe, Co, Ni according to the number of valence electrons. In Cu, the Fermi energy lies above the *d*-bands, in the 4*s*-band.

For the transition metals, the *d*- and *s*-bands are only partly filled, Thus *d*-electrons, too, occupy states near the Fermi energy.

At the beginning of the chapter we saw that an electron gas in its ground state can become ferromagnetic if the reduction in energy through the exchange interaction of the aligned spins exceeds the increase in kinetic energy. We cannot use such a simple model in the present case since the valence electron gas almost fills the *d*-bands. The following simple model however can explain the main features of the ferromagnetism of transition metals (*Stoner's collective electron model*). We assume on the one hand that the Bloch states of the band model are maintained. However, we add the exchange energy to the energy of a Bloch state, describing it by an internal field. All states having one spin direction are then displaced in energy relative to the states having the other spin direction. If we further assume that the magnitude of this displacement is independent of $\boldsymbol{k}$, and depends only on the *s*-, *p*-, *d*-character of the states, then this means a rigid displacement of the states in a band with one spin direction relative to the states with the opposite spin direction. If the Fermi energy lies within such a band, the displacement causes a preponderance of electrons with one spin direction, and hence a spontaneous magnetic moment in the ground state.

Applied to the case of nickel this means: As a result of the overlapping of the *s*-band and *d*-bands, of the ten valence electrons of a Ni atom on average 9.46 electrons are in the *d*-bands and 0.54 in the *s*-band. The exchange interaction practically involves only the *d*-electrons. The exchange-interaction displacement of the two *d*-subbands proves to be so large that the one subband is fully filled and the other contains 0.54 holes per atom. The saturation magnetization is then $M = 0.54N\mu_B$. If the magnetization were to be caused by p localized electrons, it would be given by $M = pN\mu_B$. Thus the combination of band model and exchange interaction leads to noninteger *effective magneton numbers p*. This model is further supported by the experimental result presented in Fig. 3.17. We assume —as already mentioned above—that the density of states in Fig. 3.16 applies for all transition metals (and their alloys!), and hence that only the position of the Fermi energy determines the effective magneton number. Thus p depends only on the number of electrons in the *d*-band, which in turn is determined by the constituents of an alloy. In particular the ferromagnetic properties must disappear as soon as the *d*-bands become completely filled. These predictions are remarkably well confirmed by the results presented in Fig. 3.17. On these questions see also an article by *Friedel* in [65].

Such a model is, of course, only a first step towards an understanding of ferromagnetism in metals. For extension of the theory we can only point to the literature. In particular we shall not discuss the spin waves which are also possible in this case. We shall just mention here an interesting feature of the ferromagnetism of conduction electrons.

We consider an electron gas and put $E = \hbar^2k^2/2m$ for the one-electron energy. On top of this energy we superimpose an exchange energy which displaces the states of one spin direction by a constant amount V relative to the

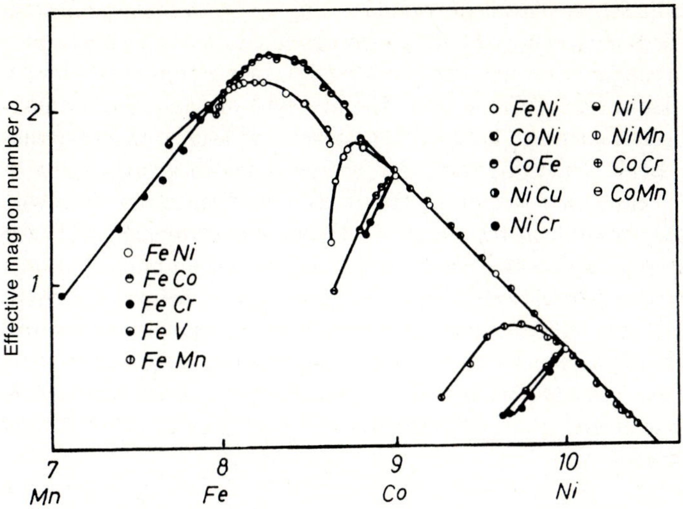

Fig. 3.17. Effective magneton number p for binary alloys of the transition metals as a function of the mean number of valence electrons (Slater-Pauling curve) [from J. Crangle, G. C. Hallam: Proc. Roy. Soc. **A272**, 119 (1963)].

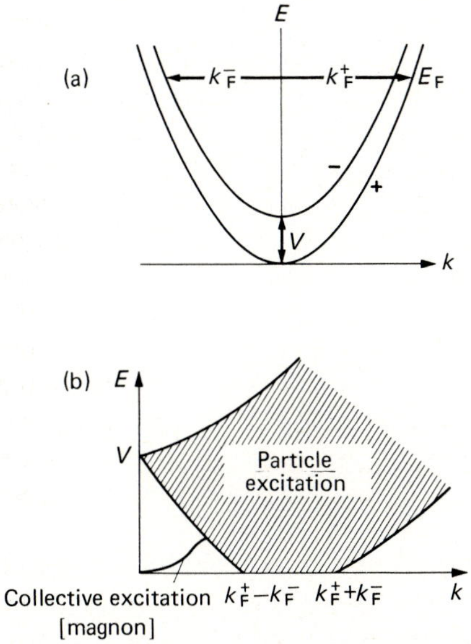

Fig. 3.18. (a) Shift of the free electron states having one spin direction relative to the states having the other spin direction by the exchange energy V. (b) Besides the excitation of an electron out of the Fermi sphere with conservation of spin, the spin can reverse in the transition. This is only possible if the relation between energy and momentum of the excitation is in the hatched area of the figure. Spin waves can disintegrate into these particle excitations if their energy lies above the threshold energy for the excitations. For a similar case see Fig. 3.5 (from *Elliott* [117]).

states of the other direction. This is illustrated in Fig. 3.18a. In the ground state let the states be filled up to energy E_F. It is then possible to set up excited states by excitation of a single electron, with conservation of electron spin or with simultaneous reversal of the electron spin. The first type of one-particle excitation has already been treated in Chapter 2. The other type leads to the spectrum presented in Fig. 3.18b. There the transition energy is shown as a function of the change of the electron $\boldsymbol{k}$-vector in the transition. Transitions involving change of $\boldsymbol{k}$ are only possible at the energy V. Transitions involving no change of energy are only possible on the Fermi surface between the limits $k_F^+ \pm k_F^-$. In addition to these one-particle excitations, spin waves, which below a threshold energy do not fall in the region of pair excitations, are possible in the electron gas. Above this threshold energy they can disintegrate into these one-particle excitations. We have already met this simultaneous existence of one-particle excitations and collective excitations in the treatment of plasmons (Fig. 3.5) at the end of Section 3.1.4.

4. Electron-Phonon Interaction: Transport Phenomena

4.1 The Interaction Processes

4.1.1 Introduction

In the band model description, *electrons* in a solid are quasi-particles which occupy one-electron states. They are described by Bloch functions $|n, \boldsymbol{k}, \sigma\rangle$ where n is the band index, $\boldsymbol{k}$ is the wave vector of the electron, and σ is the spin. In this chapter we shall be almost entirely concerned with electrons of one band, the conduction band. Accordingly we need only give the band index explicitly in a few cases. Since for transitions within the conduction band electron spin direction is in most cases maintained, we shall on many occasions describe the electron by its wave vector only.

Phonons are collective excitations of the lattice. The vibrational state of the lattice is characterized by the number of phonons in the individual oscillator states defined by the wave vector $\boldsymbol{q}$ and the branch j of the dispersion spectrum $\omega_j(\boldsymbol{q})$.

The basic *electron-phonon interaction* process is the annihilation (absorption) or creation (emission) of a phonon $(\boldsymbol{q}, j)$ with simultaneous change of the electron state from $|\boldsymbol{k}, \sigma\rangle$ to $|\boldsymbol{k} \pm \boldsymbol{q}, \sigma\rangle$. These two processes are illustrated in the top row of Fig. 4.1.

The two graphs for phonon emission and phonon absorption can at the same time describe two further processes if one transforms them slightly (second row in Fig. 4.1). If one thinks of a time axis running from left to right in these graphs and regards electrons running backwards in time as holes running forward in time, then the graphs respectively describe the recombination of an electron-hole pair with emission of a phonon, and the creation of an electron-hole pair with absorption of a phonon.

These four basic processes can be described quantum mechanically by a first-order perturbation calculation. From this, conservation laws follow for the total energy and for the sum of the wave vectors of the elementary excitations involved in the process.

The contributions from the perturbation calculations of higher order can be described as multiple-stage processes, which are made up of a number of successive basic processes. In contrast to the initial and final states, the intermediate states are not "real" (stationary) states of the system. They can be regarded as traversed in such a short time that—in view of the energy-time uncertainty relation—the energy conservation law need not apply. It only applies between

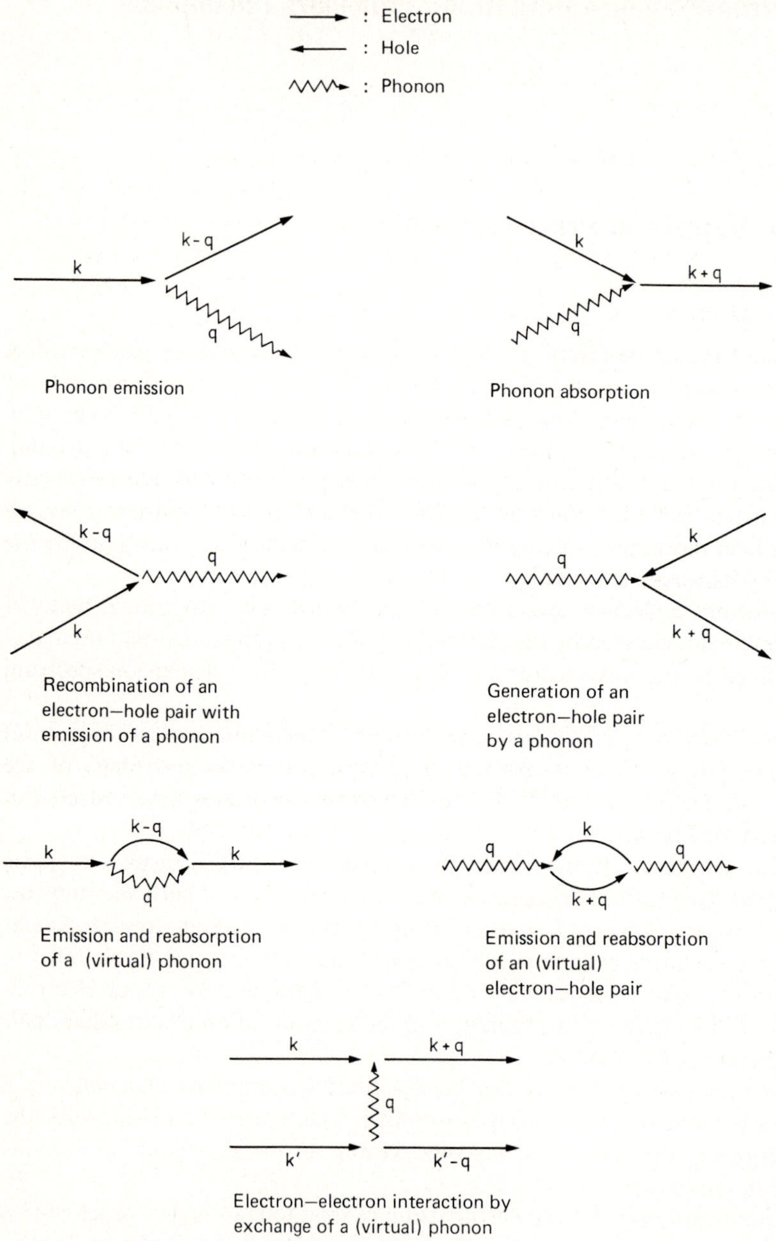

Fig. 4.1. Graphs for the electron-phonon interaction (Umklapp processes in which there is an additional change of the electron $\boldsymbol{k}$-vector by a primitive translation in the reciprocal lattice are not considered here, see Fig. 4.2.).

the initial and final states. One designates such states "*virtual states*". Of importance here are three further second-order processes illustrated by graphs in Fig. 4.1. The virtual emission and reabsorption of phonons by electrons implies a renormalization of the electron mass by the electron-phonon interaction (see Sec. 3.1.5).

Correspondingly, the creation of virtual electron-hole pairs by phonons means a renormalization of the phonon frequency. The exchange of a virtual phonon between two electrons represents an effective electron-electron interaction which must be taken into account along with the Coulomb interaction.

In the following two sections we shall look more closely at these interaction processes. In Section 4.1.2 we shall become familiar with the normal phonon emission and absorption processes by taking the interaction of Bloch electrons with acoustic phonons as an example, and we shall calculate the probability of an electron transition from one Bloch state to another.

Taking as an example the relatively strong coupling between electrons and optical phonons in polar solids, we shall consider the renormalization of the electron energy and mass, and thereby introduce the concept of the *polaron* as a new quasi-particle.

For the time being we shall ignore electron-electron interaction by exchange of virtual phonons. This is the underlying mechanism in superconductivity. We shall therefore consider it separately in Chapter 5.

4.1.2 Interaction of Electrons with Acoustic Phonons

In dividing up the Hamiltonian for the many-body problem in Section 1.2, along with the electron and ion terms, we expressed the electron-ion interaction in (1.4) as a sum of the contributions of the individual lattice ions

$$H_{\text{el-ion}} = \sum_{l,i} V_{\text{el-ion}}(\boldsymbol{r}_l - \boldsymbol{R}_i)\,. \tag{4.1}$$

Here $\boldsymbol{r}_l$ is the position of an electron and $\boldsymbol{R}_i$ that of an ion. In (1.6) we divided this operator up into the interactions of electrons with ions in their equilibrium positions, and the addition to this due to lattice vibrations

$$H_{\text{el-ion}} = H^0_{\text{el-ion}} + H_{\text{el-ph}}\,. \tag{4.2}$$

The first part describes the electron interaction with the periodic potential included in the one-electron band-model approximation. The second part is the *electron-phonon interaction* which couples the electron system with the lattice vibrations.

For an explicit formulation of this part we first of all replace $\boldsymbol{R}_i(t)$ with the more precise $\boldsymbol{R}_{n\alpha} + \boldsymbol{s}_{n\alpha}(t) = \boldsymbol{R}_n + \boldsymbol{R}_\alpha + \boldsymbol{s}_{n\alpha}(t)$ (see Sec. 3.3.2), that is, we divide up the ion position vector into the equilibrium position of the αth ion in the nth

Wigner-Seitz cell, and the instantaneous deviation of this ion from its equilibrium position.

In ansatz (4.1) we use an approximation which we shall subsequently be able to compare with other possible choices. We assume that the interaction potential depends only on the electron-ion separation, hence, that in its vibration about its equilibrium position the ion is subjected to a rigid displacement (*Nordheim's rigid ion model*). However, many of the statements derived in the following do not depend on this approximation.

Taking $\boldsymbol{s}_{n\alpha}$ to be small, we expand the potential (identified by an index α since it can be different for the different basis atoms) in powers of the deviations and break off after the linear term

$$V_\alpha(\boldsymbol{r}_l - \boldsymbol{R}_{n\alpha} - \boldsymbol{s}_{n\alpha}) = V_\alpha(\boldsymbol{r}_l - \boldsymbol{R}_{n\alpha}) - \boldsymbol{s}_{n\alpha} \cdot \nabla V_\alpha(\boldsymbol{r}_l - \boldsymbol{R}_{n\alpha}) . \tag{4.3}$$

The second term on the right (summed over all α, n, l) is the Hamiltonian for the electron-phonon interaction. From (3.121) we can rewrite the deviation $\boldsymbol{s}_{n\alpha}$ in normal coordinates, and from (A.3) express these in the occupation number representation. We then obtain

$$H_{\text{el-ph}} = -\sum_{\alpha n l} \frac{1}{\sqrt{NM_\alpha}} \sum_{jq} Q_{qj} \exp(\mathrm{i}\boldsymbol{q} \cdot \boldsymbol{R}_n) \boldsymbol{e}_\alpha^{(j)}(\boldsymbol{q}) \cdot \nabla V_\alpha(\boldsymbol{r}_l - \boldsymbol{R}_{n\alpha}) \tag{4.4}$$

with

$$Q_{qj} = \left(\frac{\hbar}{2\omega_{qj}}\right)^{1/2} (a_{-qj}^+ + a_{qj}) .$$

N is the number of Wigner-Seitz cells in V_g. The phonon component, written in the occupation number representation, shows that in this approximation two interactions are possible—one in which a phonon with wave number $\boldsymbol{q}$ in branch j is annihilated, and a second in which a phonon $-\boldsymbol{q}, j$ is created. The pseudo-momentum $-\boldsymbol{q}$ must be delivered by the electron system. One can therefore guess that both processes are accompanied by an electron transition from state $\boldsymbol{k}$ into state $\boldsymbol{k} + \boldsymbol{q}$.

To prove this assertion, we rearrange the electron part of (4.4) in terms of creation and annihilation operators for fermions. From (A.31) we have for a Hamiltonian consisting of the sum of one-electron operators

$$H = \sum_l h(\boldsymbol{r}_l) = \sum_{kk'\sigma} \langle \boldsymbol{k}'\sigma | h | \boldsymbol{k}\sigma \rangle c_{k'\sigma}^+ c_{k\sigma} . \tag{4.5}$$

We need only carry out one spin summation here since spin is conserved in the $\boldsymbol{k}$ to $\boldsymbol{k}'$ transitions considered. From (4.4), $h(\boldsymbol{r}_l)$—apart from $\boldsymbol{r}_l$-independent factors—is the operator $\nabla V_\alpha(\boldsymbol{r}_l - \boldsymbol{R}_n)$. The matrix element on the right in (4.5) is formed from Bloch functions. In future we can omit the index l in $\boldsymbol{r}_l$. If we express V_α as a Fourier series, we find the matrix element to be

$$\langle \boldsymbol{k}'\sigma|\nabla V_\alpha|\boldsymbol{k}\sigma\rangle = \sum_{\boldsymbol{\kappa}} \exp(-\mathrm{i}\boldsymbol{\kappa}\cdot\boldsymbol{R}_n)V_{\alpha\boldsymbol{\kappa}}\mathrm{i}\boldsymbol{\kappa}\langle \boldsymbol{k}'\sigma|\exp(\mathrm{i}\boldsymbol{\kappa}\cdot\boldsymbol{r})|\boldsymbol{k}\sigma\rangle\,. \tag{4.6}$$

The sum over n in (4.4) now only contains two exponential factors. Since

$$\sum_n \exp[\mathrm{i}(\boldsymbol{q}-\boldsymbol{\kappa})\cdot\boldsymbol{R}_n] = N\sum_{\boldsymbol{K}_m}\delta_{\boldsymbol{\kappa},\boldsymbol{q}+\boldsymbol{K}_m}\,,$$

only the term $\boldsymbol{\kappa} = \boldsymbol{q} + \boldsymbol{K}_m$ remains in the sum over $\boldsymbol{\kappa}$. If we insert Bloch functions for $|\boldsymbol{k}\rangle$ in the integral on the right in (4.6), and take note of the fact that apart from the factor $\exp[\mathrm{i}(\boldsymbol{k}+\boldsymbol{q}+\boldsymbol{K}_m-\boldsymbol{k}')\cdot\boldsymbol{r}]$ the integrand only contains the lattice periodic product $u_n^*(\boldsymbol{k}',\boldsymbol{r})u_n(\boldsymbol{k},\boldsymbol{r})$, it follows that the integral only differs from zero when $\boldsymbol{k}' = \boldsymbol{k}+\boldsymbol{q}+\boldsymbol{K}_m$. We are left with

$$H_{\mathrm{el,ph}} = -\sum_{\sigma\boldsymbol{k}\alpha\boldsymbol{K}_m j\boldsymbol{q}}\sqrt{\frac{N}{M_\alpha}}\,V_{\alpha,\boldsymbol{q}+\boldsymbol{K}_m}\mathrm{i}(\boldsymbol{q}+\boldsymbol{K}_m)\cdot\boldsymbol{e}_\alpha^{(j)}(\boldsymbol{q})\sqrt{\frac{\hbar}{2\omega_{\boldsymbol{q}j}}}$$
$$\times\int u_n^*(\boldsymbol{q}+\boldsymbol{K}_m+\boldsymbol{k},\boldsymbol{r})u_n(\boldsymbol{k},\boldsymbol{r})d\tau(a^+_{-\boldsymbol{q}j}+a_{\boldsymbol{q}j})c^+_{\boldsymbol{k}+\boldsymbol{q}+\boldsymbol{K}_m,\sigma}c_{\boldsymbol{k},\sigma}\,. \tag{4.7}$$

Here we have also put $n = n'$ since we can assume that, in making the transition from $\boldsymbol{k}$ to $\boldsymbol{k}'$, the electron remains in the same band.

We see first of all that with the emission of a phonon $\boldsymbol{q}$ or absorption of a phonon $-\boldsymbol{q}$, an electron makes the transition from state $\boldsymbol{k}$ into state $\boldsymbol{k}+\boldsymbol{q}+\boldsymbol{K}_m$. The statement made above is therefore not quite accurate. This has to do with the following situation: $\boldsymbol{k}$ and $\boldsymbol{k}'$ are reduced $\boldsymbol{k}$-vectors of the two Bloch functions $|\boldsymbol{k}\rangle$ and $|\boldsymbol{k}'\rangle$, and hence lie in the (first) Brillouin zone just like the $\boldsymbol{q}$-vector of the phonon. If one adds $\boldsymbol{q}$ to $\boldsymbol{k}$ vectorially, the resultant vector can lie outside the Brillouin zone, in a neighbouring zone in the repeated zone scheme. The point in the first Brillouin zone which is equivalent to the end-point of the vector $\boldsymbol{k}+\boldsymbol{q}$ is $\boldsymbol{k}'$, so that a corresponding $\boldsymbol{K}_m$ has to be added to $\boldsymbol{k}+\boldsymbol{q}$ (Fig. 4.2). This construction also shows that $\boldsymbol{K}_m$ is fixed for each $\boldsymbol{k},\boldsymbol{q}$-pair. The sum over $\boldsymbol{K}_m$ in (4.7) is therefore reduced to one term. If this $\boldsymbol{K}_m$ is zero, i.e., $\boldsymbol{k}$, $\boldsymbol{q}$ and $\boldsymbol{k}+\boldsymbol{q}$ all lie in the first Brillouin zone, the transition $\boldsymbol{k}\to\boldsymbol{k}'$ is called a *normal process* (*N-process*). If $\boldsymbol{K}_m \neq 0$, the transition is called an *Umklapp process* (*U-process*). For the latter, one must take note of the fact that the choice of the location of the Brillouin zone in $\boldsymbol{k}$-space is not unique. Depending on how $\boldsymbol{k}$-space is divided into Brillouin zones, a process can be either a normal or an Umklapp process. In spite of this the distinction will be important later (Secs. 4.2.2 and 7.4).

For our further discussions we want to make a few simplifying assumptions. First we restrict ourselves to *Bravais* lattices. The index α can then be omitted (only one atom in each Wigner-Seitz cell) and j only counts the different *acoustic*

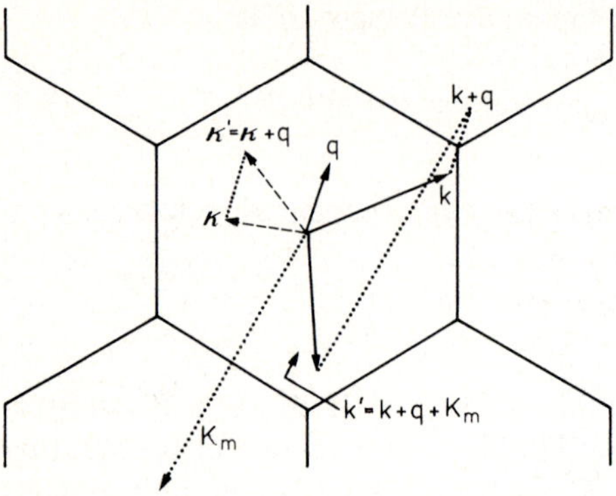

Fig. 4.2. Normal process ($\boldsymbol{\kappa}' = \boldsymbol{\kappa} + \boldsymbol{q}$) and Umklapp process ($\boldsymbol{k}' = \boldsymbol{k} + \boldsymbol{q} + \boldsymbol{K}_m$) in the Brillouin zone of the two-dimensional hexagonal lattice.

branches. Optical branches do not exist. We further restrict ourselves to *normal processes*, i.e., only to the possibilities illustrated in Fig. 4.1. $\boldsymbol{K}_m$ is then zero in (4.7). Finally we assume that phonons are either transverse or longitudinal. $\boldsymbol{e}^{(j)}$ is therefore either parallel or perpendicular to $\boldsymbol{q}$. With these restrictions we find that

$$H_{\text{el,ph}} = -\sum_{\sigma\boldsymbol{k}\boldsymbol{q}j} \mathrm{i}\sqrt{\frac{N\hbar}{2M\omega_{\boldsymbol{q}j}}}\, V_{\boldsymbol{q}}\boldsymbol{e}^{(j)}\cdot\boldsymbol{q}$$
$$\times \int u_n^*(\boldsymbol{k}+\boldsymbol{q},\boldsymbol{r})u_n(\boldsymbol{k},\boldsymbol{r})d\tau(a^+_{-\boldsymbol{q}j} + a_{\boldsymbol{q}j})c^+_{\boldsymbol{k}+\boldsymbol{q},\sigma}c_{\boldsymbol{k}\sigma}\,. \tag{4.8}$$

Since $\boldsymbol{e}^{(j)}\cdot\boldsymbol{q} = 0$ for $\boldsymbol{e}^{(j)} \perp \boldsymbol{q}$, we find from (4.8) that the first important result which this approximation leads to is the conclusion that only the *longitudinal acoustic phonons* are coupled to the electrons. We can therefore omit the summation over j and write $H_{\text{el-ph}}$ in the simpler form

$$H_{\text{el-ph}} = \sum_{\sigma\boldsymbol{k}\boldsymbol{q}} M_{\boldsymbol{k}\boldsymbol{q}}(a^+_{-\boldsymbol{q}} + a_{\boldsymbol{q}})c^+_{\boldsymbol{k}+\boldsymbol{q},\sigma}c_{\boldsymbol{k},\sigma}\,. \tag{4.9}$$

With the help of the operator (4.9) we can calculate the probability of an electron transition from state $|\boldsymbol{k}\rangle$ into state $|\boldsymbol{k}+\boldsymbol{q}\rangle$. The spin is unchanged in this transition and can therefore be ignored.

From the Dirac perturbation theory the transition probability ("golden rule") is

$$W(i \to f) = \frac{2\pi}{\hbar}|\langle f|H_{\text{el-ph}}|i\rangle|^2\delta(E_f - E_i)\,. \tag{4.10}$$

We characterize initial state $|i\rangle$ and final state $|f\rangle$ by the occupation numbers $n_{\boldsymbol{k}}$ and $n_{\boldsymbol{k}+\boldsymbol{q}}$ of the electron state involved in the transition, and those $n_{\boldsymbol{q}}$ and

n_{-q} of the phonon states

$$|n_{k+q}, n_k; n_q, n_{-q}\rangle \,. \tag{4.11}$$

We consider first of all a transition involving *absorption* of a phonon. We therefore apply the operator $c_{k+q}^+ c_k a_q$ to (4.11). From (A.15) and (A.23) we then obtain

$$\langle n_{k+q} + 1, n_k - 1; n_q - 1 | c_{k+q}^+ c_k a_q | n_{k+q}, n_k; n_q \rangle = \sqrt{(1 - n_{k+q}) n_k n_q}\,,$$
$$E_f - E_i = E(k + q) - E(k) - \hbar\omega_q\,, \tag{4.12}$$

for the case $n_k = 1$, $n_{k+q} = 0$. The matrix element vanishes in other cases.
Correspondingly we find for transitions involving emission of phonons

$$\langle n_{k+q} + 1, n_k - 1; n_{-q} + 1 | c_{k+q}^+ c_k a_{-q}^+ | n_{k+q}, n_k; n_{-q} \rangle$$
$$= \sqrt{(1 - n_{k+q}) n_k (n_{-q} + 1)} \tag{4.13}$$
$$E_f - E_i = E(k + q) - E(k) + \hbar\omega_q$$

again with the condition $n_k = 1$, $n_{k+q} = 0$.
Together we have therefore

$$W(k \to k + q) = \frac{2\pi}{\hbar} |M_{kq}|^2 (1 - n_{k+q}) n_k \{ n_q \delta(E(k + q) - E(k) - \hbar\omega_q)$$
$$+ (n_{-q} + 1) \delta(E(k + q) - E(k) + \hbar\omega_q) \}\,. \tag{4.14}$$

We have included the factor $(1 - n_{k+q}) n_k$ in (4.12) to (4.14) although it is unity since $n_k = 1$, $n_{k+q} = 0$. It becomes important when we no longer want to calculate the transition probability from *one* occupied state into *one* empty state, but want to consider a large number of states with a particular occupation probability. All n_k, n_{k+q}, n_q, n_{-q} in (4.14) have then to be replaced by their statistical mean values. If the electron and phonon systems are in equilibrium before the transition, these mean values are, respectively, the Fermi and Bose distributions. We shall return to this later when calculating transition rates in the Boltzmann equation.

The ansatz (4.1) for the interaction potential is not the only one possible. Eq. (4.1) describes rigid ions vibrating about their equilibrium positions. The electron shells of the ions will certainly be deformed by the vibration. We can obtain an expression which incorporates the deformation from the continuum approximation for lattice vibrations. In accordance with (3.164), we replace the discrete displacements $s_{n\alpha}(t)$ in the interaction potential $V(r_l - R_{n\alpha} - s_{n\alpha})$ by a displacement field $s(r, t)$. The change in potential then follows again by expanding the potential in terms of the small displacements

$$\delta V = -\boldsymbol{s}\cdot\nabla V(\boldsymbol{r}_l)\,. \tag{4.15}$$

This approximation, which as in Section 3.3.7 is only meaningful in the acoustic branch long wavelength limit, is due to Bloch. It yields the same general results as the rigid ion model. Only the factor $M_{\boldsymbol{kq}}$ in (4.14) is different.

A further ansatz, the Bardeen self-consistent potential, involves the assumption that in metals the core electrons of the ions are rigidly displaced by the vibrations, while the gas of conduction electrons rearranges itself according to the instantaneous positions of the lattice ions. As far as the interaction with a particular electron is concerned, this means nothing other than a screening of the potential of the rigid ions by the electron gas. This screening can be taken into account by dividing each Fourier component of the electron-ion interaction potential by a wave-number dependent dielectric constant (3.58). We do not want to involve ourselves in the somewhat complicated calculation of the transition probability (4.14) based on this approximation, and refer the reader to the literature for further details of the three approximations due to Nordheim, Bloch and Bardeen (e.g., *Ziman* [35]).

On the other hand, another possibility should be mentioned which is of great importance in semiconductors: the *deformation potential.* In the long wavelength acoustic branch limit, we consider once again the continuum approximation by a displacement field $\boldsymbol{s}(\boldsymbol{r}, t)$. Longitudinal acoustic waves are then compression waves in the continuum. A relative volume change $\Delta(\boldsymbol{r}, t) = \nabla\cdot\boldsymbol{s}$ is associated with a compression wave. A change in volume means a change in lattice constant, and hence in the band model parameters which depend on the lattice constant.

Within the framework of the effective mass approximation (see Sec. 2.2.9), one can divide the energy E of an electron or hole in a Bloch state of a band into the energy at the band edge (the lower edge of the conduction band E_C in the case of electrons), and the energy difference $E - E_C$. E_C can be interpreted as potential energy, and $E - E_C$ as kinetic energy. A periodic change of the lattice constant due to a compression wave will cause a periodic change in E_C. The potential energy of the electron therefore depends on position, and the perturbation $E_C(\boldsymbol{r}, t) - E_C^0 = \delta E_C$ is the perturbation potential which the electron-ion interaction induces. With $\boldsymbol{s} = \boldsymbol{s}_0 \exp[\mathrm{i}(\boldsymbol{q}\cdot\boldsymbol{r} - \omega t)]$ we then have

$$\delta E_C = \frac{\partial E_C}{\partial V}\delta V = V\frac{\partial E_C}{\partial V}\Delta = E_{1n}\nabla\cdot\boldsymbol{s} = \mathrm{i}E_{1n}\boldsymbol{s}\cdot\boldsymbol{q} \tag{4.16}$$

where we have introduced the deformation potential constant by the definition $E_{1n} = V(\partial E_C/\partial V)$. Since $\boldsymbol{s}$ is parallel to $\boldsymbol{q}$, we find here again an interaction operator similar to (4.9), in which the strength of the electron-phonon coupling is now determined by the deformation potential constant.

4.1.3 Electron-Phonon Interaction in Polar Solids, Polarons

The phonon spectrum of a Bravais lattice has only three acoustic branches. As a result the discussion of electron-phonon interaction in the previous section was restricted to longitudinal acoustic phonons. When we consider a lattice with a basis we have to distinguish two cases. If the basis atoms are identical (diamond structure, for example), the extension of the results of the previous section is not difficult, but it does not add anything basic to the previous discussion except the possibility of interaction with transverse phonons. If, on the other hand, the basis atoms have unequal charge (as in ionic crystals, for example), then associated with the optical vibrations there is a polarization which results in a strong coupling between electrons and LO-phonons.

We can see the most important features when we examine the motion of an electron in a polar lattice. The electron will polarize its surroundings and in its motion will carry the polarization cloud around with it (Fig. 4.3). Electron and polarization cloud together constitute a *quasi-particle*. The polarization of the surroundings means a distortion of the lattice, hence an excitation of optical phonons. The quasi-particle can be described as an electron surrounded by a cloud of (virtual) optical phonons. It is called a *polaron*. One of its most important properties is an increased inert mass.

The model describing a polaron will depend on whether the distortion of the lattice is limited to the immediate vicinity of the electron (*small polaron*) or whether it extends over several lattice constants (*large polaron*). The large

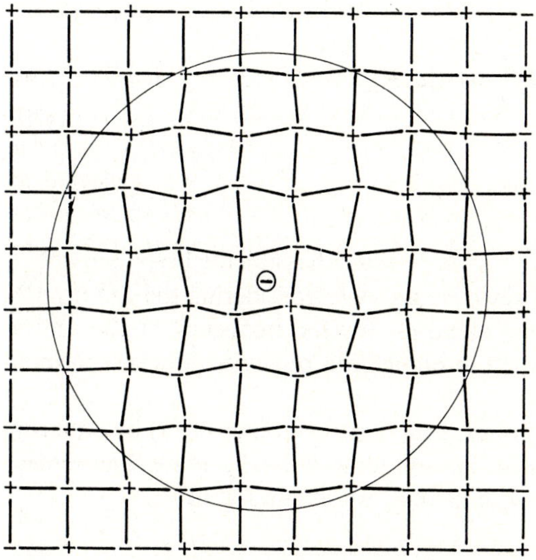

Fig. 4.3. An electron in an ionic crystal polarizes its surroundings by Coulomb interaction. Electron and lattice polarization (deformation) together constitute a quasi-particle (polaron).

polaron case is simpler to treat since one can then make use of the continuum approximation of Section 3.3.8.

We consider first the polarization in the continuum approximation. The second equation (3.186) gives the relation between $\boldsymbol{P}$, the displacement vector $\boldsymbol{w}$ (or $\boldsymbol{s}$), and the electric field $\boldsymbol{E}$. Since we are only considering longitudinal vibrations here, $\boldsymbol{E}$ is linked with $\boldsymbol{w}$ through (3.190). If one expresses the coefficients b_{ik} in terms of $\varepsilon(0)$ and $\varepsilon(\infty)$ and the limiting frequency of the longitudinal optical branch ω_{L}, it then follows from (3.186) that

$$\boldsymbol{P} = -\varepsilon_0 \boldsymbol{E} = \frac{\varepsilon_0 b_{21}}{\varepsilon_0 + b_{22}} \boldsymbol{w} = \left\{ \frac{N\overline{M}\omega_{\mathrm{L}}^2 \varepsilon_0}{V_g} \left[\frac{1}{\varepsilon(\infty)} - \frac{1}{\varepsilon(0)} \right] \right\}^{1/2} \boldsymbol{s} . \tag{4.17}$$

The interaction energy of an electron with the polarized medium is

$$H_{\text{el-ph}} = -\frac{e}{4\pi\varepsilon_0} \int \frac{\boldsymbol{P}(\boldsymbol{r}) \cdot (\boldsymbol{r} - \boldsymbol{r}_{\text{el}})}{|\boldsymbol{r} - \boldsymbol{r}_{\text{el}}|^3} d\tau . \tag{4.18}$$

We can use this expression directly as the electron-phonon interaction operator by inserting (4.17) and using the quantized form of $\boldsymbol{s}(\boldsymbol{r}, t)$.

To do this we must first of all connect $\boldsymbol{s}(\boldsymbol{r}, t)$, the difference in the displacements of the basis atoms, with the displacements $\boldsymbol{s}_{n\alpha}$. If we take a binary lattice with two oppositely charged ions in the Wigner-Seitz cell, $\boldsymbol{s}$ is equal to $\boldsymbol{s}_+ - \boldsymbol{s}_-$. For longitudinal vibrations the $\boldsymbol{e}_\alpha(\boldsymbol{q}) = \boldsymbol{e}_\alpha^*(-\boldsymbol{q})$ are parallel to $\boldsymbol{q}$. From (3.121) we then find for $\boldsymbol{s}_{n+} - \boldsymbol{s}_{n-}$

$$\begin{aligned} \boldsymbol{s}_n &= \boldsymbol{s}_{n+} - \boldsymbol{s}_{n-} = \frac{1}{\sqrt{N\overline{M}}} \sum_{\boldsymbol{q}} Q_{\boldsymbol{q}} \frac{\boldsymbol{q}}{q} \exp(\mathrm{i}\boldsymbol{q} \cdot \boldsymbol{R}_n) , \\ Q_{\boldsymbol{q}} &= \left(\frac{\hbar}{2\omega_{\mathrm{L}}} \right)^{1/2} (a_{-\boldsymbol{q}}^+ + a_{\boldsymbol{q}}) . \end{aligned} \tag{4.19}$$

$\overline{M}$ is the reduced mass $\overline{M}^{-1} = M_+^{-1} + M_-^{-1}$ which has already been used above. The summation over j is omitted since we are only considering the LO-branch. We have also inserted the limiting value ω_{L} for the frequency of the optical vibrations. The $\boldsymbol{q}$-dependence of the LO-branch can be neglected where the continuum approximation is valid.

Before making the transition from $\boldsymbol{s}_n$ to $\boldsymbol{s}(\boldsymbol{r})$, we rearrange (4.19) by replacing the summation variable $\boldsymbol{q}$ by $-\boldsymbol{q}$ in the sum term with $a_{-\boldsymbol{q}}^+$. If we then replace $\boldsymbol{R}_n$ by the vector $\boldsymbol{r}$ in accordance with (3.164), we obtain for $\boldsymbol{s}(\boldsymbol{r})$

$$\boldsymbol{s}(\boldsymbol{r}) = \sqrt{\frac{\hbar}{2N\overline{M}\omega_{\mathrm{L}}}} \sum_{\boldsymbol{q}} \frac{\boldsymbol{q}}{q} [a_{\boldsymbol{q}}^+ \exp(-\mathrm{i}\boldsymbol{q} \cdot \boldsymbol{r}) + a_{\boldsymbol{q}} \exp(\mathrm{i}\boldsymbol{q} \cdot \boldsymbol{r})] . \tag{4.20}$$

Only the two exponential factors are now $\boldsymbol{r}$-dependent and the integral in (4.18) is easy to carry out. Since

$$\int \frac{\exp(\pm \mathrm{i}\boldsymbol{q}\cdot\boldsymbol{r})(\boldsymbol{r}-\boldsymbol{r}_{\mathrm{el}})}{|\boldsymbol{r}-\boldsymbol{r}_{\mathrm{el}}|^3}\,d\tau = \mp 4\pi\mathrm{i}\frac{\boldsymbol{q}}{q^2}\exp(\pm\mathrm{i}\boldsymbol{q}\cdot\boldsymbol{r}_{\mathrm{el}}), \tag{4.21}$$

we then have for the electron-phonon interaction by inserting (4.20) in (4.17) and (4.17) in (4.18)

$$H_{\mathrm{el-ph}} = \mathrm{i}\left\{\frac{e^2\hbar\omega_{\mathrm{L}}}{2\varepsilon_0 V_{\mathrm{g}}}\left[\frac{1}{\varepsilon(\infty)}-\frac{1}{\varepsilon(0)}\right]\right\}^{1/2} \times \sum_{\boldsymbol{q}}\frac{1}{q}\left[a_{\boldsymbol{q}}\exp(\mathrm{i}\boldsymbol{q}\cdot\boldsymbol{r}_{\mathrm{el}}) - a_{\boldsymbol{q}}^{+}\exp(-\mathrm{i}\boldsymbol{q}\cdot\boldsymbol{r}_{\mathrm{el}})\right]. \tag{4.22}$$

The eigenstates of the electron system are altered by the electron-phonon interaction. We recall the general results of the Schrödinger perturbation theory. These tell us that the eigenfunctions and eigenvalues of a system described by the Hamiltonian H_0 will be changed by the introduction of a perturbation H' according to

$$\psi_n^{(1)} = \psi_n^{(0)} + \sum_{m(\neq n)}\frac{\langle m|H'|n\rangle}{E_n^{(0)}-E_m^{(0)}}\psi_m^{(0)} + \cdots, \tag{4.23}$$

$$E_n^{(1)} = E_n^{(0)} + \langle n|H'|n\rangle + \sum_{m(\neq n)}\frac{|\langle m|H'|n\rangle|^2}{E_n^{(0)}-E_m^{(0)}} + \cdots. \tag{4.24}$$

We apply these equations to a free electron which interacts with a polarizable medium. The zero-order wave functions are the plane waves $|\boldsymbol{k}\rangle$, extended by the vacuum state $|0\rangle$ of the phonon system. The zero-order energy is $E^{(0)}(\boldsymbol{k}) = \hbar^2k^2/2m$. The states over which the additional terms are summed are states in which an optical phonon of energy $\hbar\omega_{\mathrm{L}}$ and wave number $\boldsymbol{q}$ is emitted. Eqs. (4.23) and (4.24) then become

$$|\boldsymbol{k};0\rangle^{(1)} = |\boldsymbol{k};0\rangle^{(0)} + \sum_{\boldsymbol{q}}\frac{\langle \boldsymbol{k}-\boldsymbol{q},1_{\boldsymbol{q}}|H_{\mathrm{el-ph}}|\boldsymbol{k},0_{\boldsymbol{q}}\rangle}{E^{(0)}(\boldsymbol{k})-E^{(0)}(\boldsymbol{k}-\boldsymbol{q})-\hbar\omega_{\mathrm{L}}}|\boldsymbol{k}-\boldsymbol{q};1_{\boldsymbol{q}}\rangle^{(0)} \tag{4.25}$$

$$E^{(1)}(\boldsymbol{k}) = E^{(0)}(\boldsymbol{k}) + \sum_{\boldsymbol{q}}\frac{|\langle \boldsymbol{k}-\boldsymbol{q},1_{\boldsymbol{q}}|H_{\mathrm{el-ph}}|\boldsymbol{k},0_{\boldsymbol{q}}\rangle|^2}{E^{(0)}(\boldsymbol{k})-E^{(0)}(\boldsymbol{k}-\boldsymbol{q})-\hbar\omega_{\mathrm{L}}}. \tag{4.26}$$

Using (4.22), the matrix element appearing in both equations is

$$\langle \boldsymbol{k}-\boldsymbol{q},1_{\boldsymbol{q}}|H_{\mathrm{el-ph}}|\boldsymbol{k},0_{\boldsymbol{q}}\rangle = -\frac{\mathrm{i}}{q}\left\{\frac{e^2\hbar\omega_{\mathrm{L}}}{2\varepsilon_0 V_{\mathrm{g}}}\left[\frac{1}{\varepsilon(\infty)}-\frac{1}{\varepsilon(0)}\right]\right\}^{1/2} \equiv -\frac{C}{q}. \tag{4.27}$$

We determine first the energy (4.26).

If we replace the sum over $\boldsymbol{q}$ by an integral over the Brillouin zone in $\boldsymbol{q}$-space, we have

$$E^{(1)}(\boldsymbol{k}) = \frac{\hbar^2 k^2}{2m} + \frac{|C|^2 V_g}{(2\pi)^3} \int d\tau_{\boldsymbol{q}} \frac{1}{q^2} \left[\frac{\hbar^2}{2m} (2\boldsymbol{k}\cdot\boldsymbol{q} - q^2) - \hbar\omega_L \right]^{-1}. \tag{4.28}$$

We restrict ourselves in the following to the energy range $\hbar^2 k^2/2m < \hbar\omega_L$. The energy of the state "electron $\boldsymbol{k}$" is then insufficient to create the state "electron $\boldsymbol{k} - \boldsymbol{q}$ plus phonon $\boldsymbol{q}$". The electron can, however, emit and reabsorb virtual phonons for such short time intervals that, in view of the uncertainty principle, energy need not be conserved. While the zero-order approximation describes a free electron, the electron of (4.28) is surrounded by a cloud of virtual phonons. It is therefore the quasi-particle *polaron*.

The integrand in (4.28) can be expanded for small $\boldsymbol{k}$. We find

$$E^{(1)}(\boldsymbol{k}) = \frac{\hbar^2 k^2}{2m} - \alpha\left(\hbar\omega_L + \frac{\hbar^2 k^2}{12m} + \cdots\right) = -\alpha\hbar\omega_L + \frac{\hbar^2 k^2}{2m^{**}} + \cdots, \tag{4.29}$$

where

$$\alpha = \frac{e^2}{8\pi\varepsilon_0 \hbar\omega_L}\left(\frac{2m\omega_L}{\hbar}\right)^{1/2}\left[\frac{1}{\varepsilon(\infty)} - \frac{1}{\varepsilon(0)}\right] \tag{4.30}$$

and

$$m^{**} = \frac{m}{1 - \alpha/6}. \tag{4.31}$$

Compared with the corresponding values for a free electron, the energy of a polaron is reduced by $\alpha\hbar\omega_L$ and its mass by a factor m^{**}/m.

α represents the coupling parameter for the electron-phonon interaction (4.22). If α is small compared with unity (weak coupling), the perturbation calculation above is justified. If α is large compared with unity (strong coupling), other methods must be used. To obtain a feeling for what weak and strong coupling means here, we shall calculate the average number $\bar{N}$ of virtual phonons in a polaron. With the phonon particle number operator $\sum_{\boldsymbol{q}} a_{\boldsymbol{q}}^+ a_{\boldsymbol{q}}$ we find that

$$\bar{N} = \langle \boldsymbol{k}; 0|^{(1)} \sum_{\boldsymbol{q}} a_{\boldsymbol{q}}^+ a_{\boldsymbol{q}} |\boldsymbol{k}; 0\rangle^{(1)} = \sum_{\boldsymbol{q}} \frac{|\langle \boldsymbol{k} - \boldsymbol{q}, 1_{\boldsymbol{q}} | H_{\text{el-ph}} | \boldsymbol{k}, 0\rangle|^2}{[E^{(0)}(\boldsymbol{k}) - E^{(0)}(\boldsymbol{k} - \boldsymbol{q}) - \hbar\omega_L]^2}. \tag{4.32}$$

The sum can be transformed into an integral as in (4.28) and the matrix element expressed by (4.27). For small $\boldsymbol{k}$ we then have

$$\bar{N} = \alpha/2 . \tag{4.33}$$

α is therefore proportional to the average number of virtual phonons excited.

In semiconductors with partially polar bonds, the value of the coupling constant is typically less than unity. For the most important III–V compounds one finds values between 0.015 (InSb) and 0.080 (InP), for II–VI compounds 0.39 (CdTe) to 0.65 (CdS). In calculating these values, m has been replaced by the small effective electron mass m^* appropriate to these solids. For alkali halides, using the free electron mass m in (4.30), the α-values lie between 2.4 for LiI and 6.6 for RbBr.

The results in (4.29–33) are only valid for one limiting case, the *large polaron*. The lattice can only be regarded as a continuum when the lattice deformation extends over many lattice constants. The polar solids, for which the large polaron concept is of interest, are semiconductors and insulators in which the effective mass approximation applies. The use of plane waves for the matrix elements is therefore justified if we replace m by m^*. This is the second assumption which, along with the continuum approximation, limits the validity of the approximations used here.

We shall return to polaron theory for strong coupling and the theory of the "small" polaron in Section 8.3.5.

4.2 The Boltzmann Equation

4.2.1 Introduction

Through electron-phonon interaction, energy and momentum are exchanged between the electron system and the ion lattice. This interaction is of particular importance when one of the two systems is not in equilibrium. If one accelerates the electrons of a solid by an external electric field, the energy absorbed from the field will be passed on to the lattice by excitation of lattice vibrations, that is, by emission of phonons. Only in this way a steady state with current flow can be set up, in which acceleration of electrons by the field and slowing down by phonon emission are balanced. When the external field is switched off, the interaction processes ensure that equilibrium of the electron system is reestablished.

The electron-phonon interaction is not the only process which can dissipate the excess energy in the electron system. Scattering from lattice imperfections, grain boundaries, and surfaces can also play a part. Since in this chapter we are concerned with the undistorted, infinite solid, we can restrict our attention to the electron-phonon interaction.

Following what we have said above, the most important part of solid-state physics where electron-phonon interaction must be considered is the behaviour of a solid under the influence of external fields, i.e., *transport properties of solids*. Along with an electric field, the term "external field" embraces an additional magnetic field and a temperature gradient. Electrons driven by these forces carry charge and energy with them. Our goal is therefore the calculation of the flow of electric current and energy current caused by the fields.

These currents can be readily determined if the number of electrons with given momentum at a given location is known as a function of time. This *distribution function* follows from a differential equation, the so-called *Boltzmann equation*, which we shall set up in Section 4.2.2. The Boltzmann equation for the electrons can easily be solved if it is possible to describe the electron-lattice interaction through a relaxation time—the time constant of exponential decay of a disturbance of the electron system. If this is not possible, we have to adopt a variational technique for the solution of the Boltzmann equation. Both methods will be presented in Sections 4.2.3 and 4.2.4.

In addition to a disturbance of the electron system, we must also consider the disturbance of the phonon system induced by external forces (temperature gradients) and by interaction with the electrons. We must therefore add a Boltzmann equation for the phonons to that for the electrons. We shall include this in the treatment in Section 4.2.2, making the approximation however that the phonon system is in equilibrium when we calculate the electron distribution function. We shall return to the Boltzmann equation for the phonons in Chapter 7.

4.2.2 Boltzmann Equations for the Electron and Phonon Systems

The motion of electrons in a solid under the influence of external fields can be described by giving their position and momentum ($\boldsymbol{k}$-vector) as a function of time.

This statement needs to be qualified. To represent an electron, a wave packet has to be formed out of one-electron states. Such a packet is, however, extended in $\boldsymbol{r}$- and $\boldsymbol{k}$-space. Its mean diameter in $\boldsymbol{r}$-space $\Delta\boldsymbol{r}$ is connected with its extent in $\boldsymbol{k}$-space $\Delta\boldsymbol{k}$ by the uncertainty relation $\Delta\boldsymbol{r}\Delta\boldsymbol{k} = 1$. If the wave packet is localized in $\boldsymbol{k}$-space such that its overall dimensions are small compared with the mean radius of the Brillouin zone (about one reciprocal lattice constant), then its dimensions in real space will be large compared with a lattice constant. Thus the external fields (or other parameters which influence the electron, such as temperature gradients or inhomogeneities) must remain practically unchanged over the extent of the wave packet. One cannot describe the movement of an electron in the rapidly changing fields of the lattice ions in this way. The wave packet has therefore to be constructed with Bloch functions, which already incorporate the electron interaction with the periodic lattice potential. One must remember this restriction when speaking of an electron at location $\boldsymbol{r}$ and with $\boldsymbol{k}$-vector $\boldsymbol{k}$ (in a band n).

To describe the complete electron system, we introduce the distribution function $f_n(\boldsymbol{r}, \boldsymbol{k}, t)$, which gives the occupation probability of a "state" characterized by band index n, $\boldsymbol{k}$-vector $\boldsymbol{k}$, and space vector $\boldsymbol{r}$. More precisely, the product of distribution function, density of states, and volume element $d\tau_{\boldsymbol{r}} d\tau_{\boldsymbol{k}}$ of phase space gives the number of electrons in the volume elements ($\boldsymbol{r}$, $d\tau_{\boldsymbol{r}}$) of $\boldsymbol{r}$-space and ($\boldsymbol{k}$, $d\tau_{\boldsymbol{k}}$) of $\boldsymbol{k}$-space at time t (referred to the volume V_g). For a homogeneous solid in equilibrium, $f_n(\boldsymbol{r}, \boldsymbol{k}, t)$ is the Fermi distribution (2.30). As far as its $\boldsymbol{k}$-dependence is concerned, we can also identify the distribution function with the mean occupation number $\bar{n}_{\boldsymbol{k}}$ used in Section 4.1.3.

In future we shall omit the index n on the distribution function since we shall only be concerned with electrons in one band.

To calculate the distribution function $f(\boldsymbol{r}, \boldsymbol{k}, t)$ in given external fields, we examine its behaviour with time (Fig. 4.4). We consider a group of electrons in the volume element $d\tau_{\boldsymbol{r}} d\tau_{\boldsymbol{k}}$ in phase space. In the course of time this group will move through phase space. For the short time interval considered, let the form of the group (i.e., the associated volume element) remain essentially unchanged. Then the total differential quotient df/dt (i.e., the change in f in the coordinate system which moves with the group) would be zero if electrons were not scattered from ($\boldsymbol{k}$, $d\tau_{\boldsymbol{k}}$) to another ($\boldsymbol{k}'$, $d\tau_{\boldsymbol{k}'}$) and vice versa, by electron-phonon interaction. Such a process involving the emission or absorption of a phonon is often called a "lattice collision". Let the change in distribution function due to collisions be $\partial f/\partial t|_{\text{coll}}$. Then $df/dt = \partial f/\partial t|_{\text{coll}}$, or, if we replace the total differential quotient by the local one plus terms following from the implicit time dependence of f over $\boldsymbol{r}(t)$ and $\boldsymbol{k}(t)$,

$$\frac{df}{dt} = \frac{\partial f}{\partial t} + \dot{\boldsymbol{k}} \cdot \nabla_{\boldsymbol{k}} f + \dot{\boldsymbol{r}} \cdot \nabla_{\boldsymbol{r}} f = \left.\frac{\partial f}{\partial t}\right|_{\text{coll}}. \tag{4.34}$$

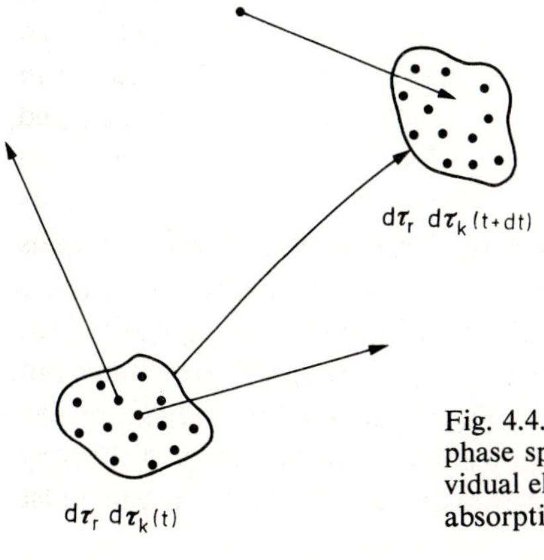

Fig. 4.4. A group of electrons moves rigidly through phase space under the influence of external fields. Individual electrons are scattered into or out of the group by absorption or emission of phonons.

In the steady state the local differential quotient is zero and we are left with

$$\dot{\boldsymbol{k}} \cdot \nabla_{\boldsymbol{k}} f + \dot{\boldsymbol{r}} \cdot \nabla_{\boldsymbol{r}} f = \left.\frac{\partial f}{\partial t}\right|_{\text{coll}} . \tag{4.35}$$

This is the usual form of the *Boltzmann equation* of the electron system and allows the distribution function to be calculated for given external fields and known electron-phonon interaction.

The interaction determines the form of the collision term on the right of (4.35). On the left side we replace $\dot{\boldsymbol{k}}$ by (2.137)

$$\dot{\boldsymbol{k}} = -\frac{e}{\hbar}(\boldsymbol{E} + \boldsymbol{v} \times \boldsymbol{B}) . \tag{4.36}$$

The external electric field and magnetic field thus appear in the picture. We replace $\dot{\boldsymbol{r}}$ according to (2.120) by the group velocity of the wave packet

$$\dot{\boldsymbol{r}} = \frac{1}{\hbar} \nabla_{\boldsymbol{k}} E(\boldsymbol{k}) . \tag{4.37}$$

We also insert (4.37) for the velocity appearing in (4.36). Finally we note that the distribution function f (since it contains the Fermi distribution as equilibrium component) is temperature dependent. The factor $\nabla_{\boldsymbol{r}} f$ therefore includes possible temperature gradients.

We can set up a corresponding Boltzmann equation for the phonon system. From the mean occupation numbers $\bar{n}_{\boldsymbol{q}}$ we define a distribution function g for the phonons. Its equilibrium value g_0 is the Bose distribution (3.138). Over a temperature gradient, g can become space dependent: $g = g(\boldsymbol{r}, \boldsymbol{q}, t)$. Analogously we find

$$\dot{\boldsymbol{r}} \cdot \nabla_{\boldsymbol{r}} g = \left.\frac{\partial g}{\partial t}\right|_{\text{coll}} \tag{4.38}$$

since a gradient of g in space is the only driving force for the phonons. Similar to (4.37), we can insert for the velocity $\dot{\boldsymbol{r}}$ the gradient of $\omega(\boldsymbol{q})$ with respect to $\boldsymbol{q}$.

We turn now to the collision terms in the two Boltzmann equations (4.35) and (4.38). They describe the changes in occupation of the electron states and the changes of the number of phonons in a normal mode of the lattice, respectively. Both terms are sums (or differences) of transition probabilities of the form considered in (4.14). We write the electron transition probability in general in the form

$$\begin{aligned} W(\boldsymbol{k} \to \boldsymbol{k} + \boldsymbol{q}) &= W^0_{\mathrm{a}\boldsymbol{q}}[1 - f(\boldsymbol{k} + \boldsymbol{q})]f(\boldsymbol{k})g(\boldsymbol{q})\delta(E(\boldsymbol{k} + \boldsymbol{q}) - E(\boldsymbol{k}) - \hbar\omega_{\boldsymbol{q}}) \\ &= W_{\mathrm{a}\boldsymbol{q}}(\boldsymbol{k} \to \boldsymbol{k} + \boldsymbol{q}) \quad \text{for absorption of a phonon } \boldsymbol{q} \end{aligned} \tag{4.39}$$

$$\begin{aligned} W(\boldsymbol{k} \to \boldsymbol{k} + \boldsymbol{q}) &= W^0_{\mathrm{e}-\boldsymbol{q}}[1 - f(\boldsymbol{k} + \boldsymbol{q})]f(\boldsymbol{k})[g(-\boldsymbol{q}) + 1] \\ &\quad \times \delta(E(\boldsymbol{k} + \boldsymbol{q}) - E(\boldsymbol{k}) + \hbar\omega_{\boldsymbol{q}}) \\ &= W_{\mathrm{e}-\boldsymbol{q}}(\boldsymbol{k} \to \boldsymbol{k} + \boldsymbol{q}) \quad \text{for emission of a phonon } -\boldsymbol{q}\,. \end{aligned} \tag{4.40}$$

We then find for the collision term in (4.35)

$$\begin{aligned} \left.\frac{\partial f}{\partial t}\right|_{\text{coll}} = \sum_{\boldsymbol{q}} [&W_{\mathrm{e}\boldsymbol{q}}(\boldsymbol{k} + \boldsymbol{q} \to \boldsymbol{k}) + W_{\mathrm{a}-\boldsymbol{q}}(\boldsymbol{k} + \boldsymbol{q} \to \boldsymbol{k}) \\ &- W_{\mathrm{a}\boldsymbol{q}}(\boldsymbol{k} \to \boldsymbol{k} + \boldsymbol{q}) - W_{\mathrm{e}-\boldsymbol{q}}(\boldsymbol{k} \to \boldsymbol{k} + \boldsymbol{q})]\,, \end{aligned} \tag{4.41}$$

i.e., the sum of all scattering probabilities from state $\boldsymbol{k} + \boldsymbol{q}$ into state $\boldsymbol{k}$ with emission of a phonon $\boldsymbol{q}$ or absorption of a phonon $-\boldsymbol{q}$, *minus* all scattering probabilities from the state $\boldsymbol{k}$ into any state $\boldsymbol{k} + \boldsymbol{q}$ with absorption of a phonon $\boldsymbol{q}$ or emission of a phonon $-\boldsymbol{q}$.

The collision term in (4.38) is correspondingly given by the sum of all collision processes involving the emission of a phonon $\boldsymbol{q}$, minus the corresponding absorption processes

$$\left.\frac{\partial g}{\partial t}\right|_{\text{coll}} = \sum_{\boldsymbol{k}} [W_{\mathrm{e}\boldsymbol{q}}(\boldsymbol{k} + \boldsymbol{q} \to \boldsymbol{k}) - W_{\mathrm{a}\boldsymbol{q}}(\boldsymbol{k} \to \boldsymbol{k} + \boldsymbol{q})]\,. \tag{4.42}$$

The principle of microscopic reversibility tells us that the possibility of a process between two states must be independent of the direction in which the process runs. From this we conclude that the factors $W^0_{\mathrm{a}\boldsymbol{q}}$ and $W^0_{\mathrm{e}\boldsymbol{q}}$ in (4.39) and (4.40) are equal. This condition also follows from the requirement that the collision term must vanish in the equilibıium state. We shall return to this below.

First we set down the collision term for the electron system, using (4.39, 40) with $W^0_{\mathrm{a}\boldsymbol{q}} = W^0_{\mathrm{e}\boldsymbol{q}}$, and changing the summation index from $\boldsymbol{q}$ to $-\boldsymbol{q}$ in a few terms. It becomes

$$\begin{aligned} \left.\frac{\partial f}{\partial t}\right|_{\text{coll}} = \sum_{\boldsymbol{q}} W^0_{\boldsymbol{q}}[&\{[1 - f(\boldsymbol{k})]f(\boldsymbol{k} + \boldsymbol{q})[g(\boldsymbol{q}) + 1] - [1 - f(\boldsymbol{k} + \boldsymbol{q})]f(\boldsymbol{k})g(\boldsymbol{q})\} \\ &\times \delta(E(\boldsymbol{k} + \boldsymbol{q}) - E(\boldsymbol{k}) - \hbar\omega_{\boldsymbol{q}}) - \{[1 - f(\boldsymbol{k} - \boldsymbol{q})]f(\boldsymbol{k})[g(\boldsymbol{q}) + 1] \\ &- [1 - f(\boldsymbol{k})]f(\boldsymbol{k} - \boldsymbol{q})g(\boldsymbol{q})\} \cdot \delta(E(\boldsymbol{k} - \boldsymbol{q}) - E(\boldsymbol{k}) + \hbar\omega_{\boldsymbol{q}})]\,. \end{aligned} \tag{4.43}$$

A corresponding equation for the collision term of the phonon system follows from (4.42). Putting the right side of (4.42) first of all into the symmetrized form $(1/2)\sum_{k}[W_{eq}(k + q \to k) + W_{eq}(k \to k - q) - W_{aq}(k \to k + q) - W_{aq}(k - q \to k)]$, the only differences to (4.41) are the factor 1/2, the summation over $\boldsymbol{k}$ instead of $\boldsymbol{q}$, and some changed signs.

In the equilibrium state the left sides of (4.35) and (4.38) become zero. The collision terms must therefore also vanish. If one inserts the Fermi distribution $[\exp\{[E(\boldsymbol{k}) - \mu]/k_{\mathrm{B}}T\} + 1]^{-1}$ and Bose distribution $[\exp(\hbar\omega_{\boldsymbol{q}}/k_{\mathrm{B}}T) - 1]^{-1}$ for the $f(\boldsymbol{k})$ and $g(\boldsymbol{q})$, respectively, in (4.43), the braces vanish due to the energy conservation law expressed by the δ-functions.

One can readily see that both also disappear when one adds to $E(\boldsymbol{k})$ a term $\boldsymbol{\delta}\cdot\boldsymbol{k}$, with any vector $\boldsymbol{\delta}$, in the numerator of the exponent of the Fermi distribution and to $\hbar\omega_{\boldsymbol{q}}$ a term $-\boldsymbol{\delta}\cdot\boldsymbol{q}$ in the Bose distribution. This describes a state in which the thermal equilibrium distributions of electrons and phonons in $\boldsymbol{k}$- and $\boldsymbol{q}$-space, respectively, are displaced relative to the origin in the direction of the vector $\boldsymbol{\delta}$. Such a state is associated with a flow of current. Eq. (4.43) then states that, *under the assumptions leading to this equation*, this state is stationary. When the external fields are removed, it will not revert to the equilibrium state! This, of course, is because we have neglected Umklapp processes and other scattering mechanisms in (4.43). They are needed to return every perturbed system to its equilibrium state. Normal processes are not always sufficient.

The Boltzmann equations for electrons and phonons form a coupled system of differential equations for the distribution functions $f(\boldsymbol{r}, \boldsymbol{k}, t)$ and $g(\boldsymbol{r}, \boldsymbol{q}, t)$. In the usual way one replaces the summation over $\boldsymbol{k}$ and $\boldsymbol{q}$ by integrations in $\boldsymbol{k}$- and $\boldsymbol{q}$-space (in which the integration limits for each term have to be carefully chosen). The Boltzmann equations are then integro-differential equations. We do not want to discuss them further in their general form. Instead we make an assumption which greatly simplifies the solution of the Boltzmann equation for the electron system. We assume that the phonon system sets up an equilibrium state so quickly that we can ignore disturbances in the phonon system (Bloch assumption). We can then replace the $g(\boldsymbol{q})$ in (4.43) by its equilibrium distribution g_0. In this approximation (4.43) can be written in the simpler form

$$\left.\frac{\partial f}{\partial t}\right|_{\mathrm{coll}} = \sum_{\boldsymbol{q}}\{W(\boldsymbol{k}+\boldsymbol{q}, \boldsymbol{k})[1 - f(\boldsymbol{k})]f(\boldsymbol{k}+\boldsymbol{q}) - W(\boldsymbol{k}, \boldsymbol{k}+\boldsymbol{q})[1 - f(\boldsymbol{k}+\boldsymbol{q})]f(\boldsymbol{k})\}\,, \tag{4.44}$$

where the newly introduced $W(\boldsymbol{k}+\boldsymbol{q}, \boldsymbol{k})$ and $W(\boldsymbol{k}, \boldsymbol{k}+\boldsymbol{q})$ contain the $W_{\boldsymbol{q}}^{0}$, g_0-factors and δ-functions from (4.43). We can rewrite (4.44) further by putting $\boldsymbol{k} + \boldsymbol{q} = \boldsymbol{k}'$ and replacing the summation over $\boldsymbol{q}$ by an integral over $\boldsymbol{k}'$

$$\left.\frac{\partial f}{\partial t}\right|_{\mathrm{coll}} = \int d\tau_{\boldsymbol{k}'}\{W(\boldsymbol{k}', \boldsymbol{k})[1 - f(\boldsymbol{k})]f(\boldsymbol{k}') - W(\boldsymbol{k}, \boldsymbol{k}')[1 - f(\boldsymbol{k}')]f(\boldsymbol{k})\}g(\boldsymbol{k}')\,. \tag{4.45}$$

With the assumption that electron spin is unchanged in an interaction, we have to insert for the density of states $g(\boldsymbol{k})$ only the number of states with one spin direction $[1/(2\pi)^3]$.

The $W(\boldsymbol{k}, \boldsymbol{k}')$ can be determined from (4.43). With $g(\boldsymbol{q}) = g(-\boldsymbol{q}) = g_0$ (Bose distribution), we find the following symmetry relationship:

$$W(\boldsymbol{k}', \boldsymbol{k}) \exp[E(\boldsymbol{k})/k_B T] = W(\boldsymbol{k}, \boldsymbol{k}') \exp[E(\boldsymbol{k}')/k_B T]. \tag{4.46}$$

On account of (4.46), the right side of (4.45) vanishes in equilibrium ($f = f_0$).

Splitting the distribution function into its equilibrium value f_0 and a perturbation δf, we see that quadratic terms in δf appear in the integrand in (4.45). These terms can be omitted for small perturbations, the collision term thereby being *linearized*. We obtain a particularly simple form if we use the following definitions:

$$V(\boldsymbol{k}, \boldsymbol{k}') = V(\boldsymbol{k}', \boldsymbol{k}) = W(\boldsymbol{k}', \boldsymbol{k})[1 - f_0(\boldsymbol{k})]f_0(\boldsymbol{k}') = W(\boldsymbol{k}, \boldsymbol{k}')[1 - f_0(\boldsymbol{k}')]f_0(\boldsymbol{k}),$$

$$f = f_0 + \delta f, \qquad \delta f \equiv -\frac{\partial f_0}{\partial E}\delta\Phi = \frac{1}{k_B T} f_0(1 - f_0)\delta\Phi. \tag{4.47}$$

The linearized collision term then becomes

$$\left.\frac{\partial f}{\partial t}\right|_{\text{coll}} = \frac{1}{k_B T}\int d\tau_{\boldsymbol{k}'} V(\boldsymbol{k}', \boldsymbol{k})[\delta\Phi(\boldsymbol{k}') - \delta\Phi(\boldsymbol{k})]g(\boldsymbol{k}'). \tag{4.48}$$

4.2.3 The Relaxation Time Approximation

If an external perturbation drives the electron system into a nonequilibrium state and is then removed, electron-phonon interaction (and other interaction processes not considered here) are responsible for the return of the system to equilibrium. From (4.34) this is described by the differential equation

$$\frac{\partial f}{\partial t} = \left.\frac{\partial f}{\partial t}\right|_{\text{coll}}. \tag{4.49}$$

Relaxation processes of this kind are often exponential in time for small perturbations of the system. The collision term then takes the form of a quotient made up of the deviation of the distribution from equilibrium and a time characteristic of the exponential decay, the *relaxation time*

$$\frac{\partial f}{\partial t} = -\frac{f - f_0}{\tau}, \qquad f = f_0 + C\,\mathrm{e}^{-t/\tau}. \tag{4.50}$$

The introduction of a relaxation time greatly simplifies the solution of the Boltzmann equation. We therefore want to examine under what circumstances this *relaxation time approximation* is possible. To do this, we must attempt to put the collision term in (4.45) into a form in which it is proportional to $\delta f = f - f_0$ with a proportionality factor which is independent of the perturbation.

As a first step we assume *elastic scattering*. The electron energy should then be the same before and after a scattering process; only the direction of $\boldsymbol{k}$ changes. This is of course never rigorously obeyed. However, phonon energies are often negligibly small compared to electron energies if acoustic phonons of long wavelength are emitted or absorbed. This is not the case if optical phonons are involved. However, at low temperatures an electron which absorbs an optical phonon is excited to so high an energy that the probability of its immediate return to a state with the initial energy (but possibly different $\boldsymbol{k}$-direction) with emission of a phonon is very high. This interaction is comparable to elastic scattering by a second-order process with a virtual intermediate state.

We therefore put $E(\boldsymbol{k}') = E(\boldsymbol{k})$. From (4.46) it then follows that $W(\boldsymbol{k}', \boldsymbol{k})$ is already symmetric in $\boldsymbol{k}$ and $\boldsymbol{k}'$. The quadratic terms are removed from the integral in (4.45); the collision term is therefore automatically linearized. We obtain

$$\left.\frac{\partial f(\boldsymbol{k})}{\partial t}\right|_{\text{coll}} = -\delta f(\boldsymbol{k}) \int d\tau_{\boldsymbol{k}'} g(\boldsymbol{k}') W(\boldsymbol{k}', \boldsymbol{k}) \left[1 - \frac{\delta f(\boldsymbol{k}')}{\delta f(\boldsymbol{k})}\right] \delta(E(\boldsymbol{k}') - E(\boldsymbol{k})). \quad (4.51)$$

The δ-function restricts the integration over $\boldsymbol{k}'$ to a surface of constant energy $E = E(\boldsymbol{k})$. With df'_E as an element of this surface, (4.51) becomes

$$\left.\frac{\partial f(\boldsymbol{k})}{\partial t}\right|_{\text{coll}} = -\delta f(\boldsymbol{k}) \int_{E=E(\boldsymbol{k})} W(\boldsymbol{k}', \boldsymbol{k}) \left[1 - \frac{\delta f(\boldsymbol{k}')}{\delta f(\boldsymbol{k})}\right] \frac{g(\boldsymbol{k}') df'_E}{|\nabla_{\boldsymbol{k}'} E|}. \quad (4.52)$$

$g(\boldsymbol{k})$ is again the density of states in $\boldsymbol{k}$-space.

As a *further approximation* we assume that the influence of external fields is on the average only felt by a displacement of the Fermi distribution in $\boldsymbol{k}$-space. If f_0 is displaced in the direction $\boldsymbol{G}$, where the vector $\boldsymbol{G}$ contains the external fields and further only depends on the value of $\boldsymbol{k}$ or $\boldsymbol{k}'$, then we can expand $f(\boldsymbol{k})$ as

$$f(\boldsymbol{k}) = f_0(\boldsymbol{k}) + \frac{\partial f_0}{\partial E} G(\boldsymbol{k}) \cdot \nabla_{\boldsymbol{k}} E + \cdots. \quad (4.53)$$

For this ansatz to lead us to the desired goal, we must further assume *spherical energy surfaces* $E = E(\boldsymbol{k})$. Then $\nabla_{\boldsymbol{k}} E$ is proportional to $\boldsymbol{k}$ and

$$f = f_0(E) - \frac{\partial f_0}{\partial E} \boldsymbol{k} \cdot \mathbf{c}(E). \quad (4.54)$$

We have introduced a minus sign so that (4.54) has the form (4.47) with $\delta\Phi = \boldsymbol{k}\cdot\boldsymbol{c}(E)$.

Finally we call the angle between $\boldsymbol{k}$ and $\boldsymbol{c}$ θ and between $\boldsymbol{k}'$ and $\boldsymbol{c}$ θ'. We then find the value $\cos\theta'/\cos\theta$ for the quotient $\delta f(\boldsymbol{k}')/\delta f(\boldsymbol{k})$. The integral in (4.52) is now independent of δf, and can be defined as the reciprocal relaxation time. Before we present it, we make a last approximation which is intended only to simplify the final equation. The transition probability many depend only on the angle between $\boldsymbol{k}$ and $\boldsymbol{k}'$, not on the individual directions. Eq. (4.52) can then be written as

$$\left.\frac{\partial f}{\partial t}\right|_{\text{coll}} = -\frac{\delta f}{\tau(E)} \tag{4.55}$$

with the energy-dependent relaxation time

$$\frac{1}{\tau(E)} = \int_{E=E(\boldsymbol{k})} W(E,\Theta)(1-\cos\Theta)\frac{g(\boldsymbol{k}')df_E'}{|\nabla_{\boldsymbol{k}'}E|}\,. \tag{4.56}$$

Here Θ is the angle between $\boldsymbol{k}$ and $\boldsymbol{k}'$. In rearranging $\cos\theta'/\cos\theta$, we have used the fact that $\cos\theta' = \cos\theta\cos\Theta + \sin\theta\sin\Theta\cos\varphi$, and that in the integral the second term here disappears as a result the φ-integration.

A further simplification is often used. One assumes that the transition probability does not depend on the scattering angle at all. In this case there is no correlation between the direction of motion of the electron before and after the collision ("memory-extinguishing collisions"). Eq. (4.56) reduces in this case to

$$\frac{1}{\tau(E)} = W(E)\int\frac{g(\boldsymbol{k}')f_E'}{|\nabla_{\boldsymbol{k}'}E|} = W(E)g(E)\,. \tag{4.57}$$

In the approximations (4.56) and (4.57) we can obtain a closed solution of the Boltzmann equation. To this end we first rearrange the left side of the Boltzmann equation (4.35).

In the first term $(\dot{\boldsymbol{k}}\cdot\nabla_{\boldsymbol{k}}f)$, we divide f into f_0 and δf, replace $\nabla_{\boldsymbol{k}}f_0$ with $(\partial f_0/\partial E)\times\nabla_{\boldsymbol{k}}E$ and $\dot{\boldsymbol{k}}$ with (4.36). A term then disappears since $\nabla_{\boldsymbol{k}}E\cdot(\nabla_{\boldsymbol{k}}E\times\boldsymbol{B}) = 0$. Furthermore we neglect $\nabla_{\boldsymbol{k}}\delta f$ compared with $\nabla_{\boldsymbol{k}}f_0$, use (4.47), and obtain

$$\dot{\boldsymbol{k}}\cdot\nabla_{\boldsymbol{k}}f \approx -e\frac{\partial f_0}{\partial E}\left[\boldsymbol{v}\cdot\boldsymbol{E} - \frac{1}{\hbar}(\boldsymbol{v}\times\boldsymbol{B})\cdot\nabla_{\boldsymbol{k}}\delta\Phi\right]. \tag{4.58}$$

In the second term we convert $\nabla_{\boldsymbol{r}}f_0$ as follows:

$$\nabla_{\boldsymbol{r}}f_0 = \frac{\partial f_0}{\partial\dfrac{E-\mu}{k_B T}}\nabla_{\boldsymbol{r}}\frac{E-\mu}{k_B T} = -\frac{\partial f_0}{\partial E}\left(\nabla_{\boldsymbol{r}}\mu + \frac{E-\mu}{T}\nabla_{\boldsymbol{r}}T\right). \tag{4.59}$$

Neglecting $\nabla_r \delta f$ compared with $\nabla_r f_0$ and using (4.47), we finally find

$$\delta\Phi = \tau\left[-e\boldsymbol{v}\cdot\boldsymbol{E} + e(\boldsymbol{v}\times\boldsymbol{B})\cdot\nabla_{\boldsymbol{k}}\delta\Phi - \boldsymbol{v}\cdot\left(\nabla\mu + \frac{E-\mu}{T}\nabla T\right)\right]$$
$$= -\tau\boldsymbol{v}\cdot\left(\nabla\eta + \frac{E-\mu}{T}\nabla T\right) + \frac{e\tau}{\hbar}(\boldsymbol{v}\times\boldsymbol{B})\cdot\nabla_{\boldsymbol{k}}\delta\Phi\,. \tag{4.60}$$

In the second line we have written the electric field as the negative gradient of the electrostatic potential φ, and then combined the chemical potential μ and $-e\varphi$ to form the *electrochemical potential* $\eta = \mu - e\varphi$.

Equation (4.60) already represents the explicit solution of the Boltzmann equation for vanishing magnetic field. When a magnetic field is present, the solution follows by iterative insertion of the right side in $\nabla_{\boldsymbol{k}}\delta\Phi$ as a series expansion in ascending powers of $\boldsymbol{B}$.

As long as we do not know the band structure $E(\boldsymbol{k})$, we can go no further in rearranging (4.60). For the case of free electrons with effective mass m^*, the summation is particularly straightforward. Since $\boldsymbol{v} = \hbar\boldsymbol{k}/m^*$, $\nabla_{\boldsymbol{k}}$ can be converted into $\nabla_{\boldsymbol{v}}$ and one has then, with the abbreviations

$$\boldsymbol{F} = \nabla\eta + \frac{E-\mu}{T}\nabla T\,, \qquad \boldsymbol{s} = \frac{e\tau}{m^*}\boldsymbol{B} \tag{4.61}$$

as the final result

$$\delta\Phi = -\frac{\tau}{1+s^2}[\boldsymbol{v}\cdot\boldsymbol{F} + \boldsymbol{v}\cdot(\boldsymbol{s}\times\boldsymbol{F}) + (\boldsymbol{v}\cdot\boldsymbol{s})(\boldsymbol{s}\cdot\boldsymbol{F})]\,. \tag{4.62}$$

4.2.4 The Variational Method

When the interaction processes do not allow us to introduce a relaxation time in the collision term, the following approach often takes us to our goal. The collision term in the linearized form (4.48) can be taken as an integral operator applied to the function $\delta\Phi(\boldsymbol{k})$ being sought. If we write $L(\delta\Phi)$ or, in short, $L\Phi$ for the (negative) collision term and represent the terms on the left side of the Boltzmann equation (4.35) by $-F$, (4.35) becomes $F = L\Phi$.

We now construct the integral over the product of $L\Phi$ with an as-yet undefined function $\psi(\boldsymbol{k})$

$$(\psi L\Phi) = \frac{1}{k_B T}\frac{1}{(2\pi)^3}\iint V(\boldsymbol{k}',\boldsymbol{k})[\delta\Phi(\boldsymbol{k}) - \delta\Phi(\boldsymbol{k}')]\psi(\boldsymbol{k})d\tau_{\boldsymbol{k}}d\tau_{\boldsymbol{k}'}\,. \tag{4.63}$$

If one exchanges $\boldsymbol{k}, \boldsymbol{k}'$ in the integral, one obtains the same expression except for a reversed sign, and $\psi(\boldsymbol{k}')$ instead of $\psi(\boldsymbol{k})$ in the integrand. One can therefore also write (4.63) as

$$(\psi L\Phi) = \frac{1}{2k_\mathrm{B}T}\frac{1}{(2\pi)^3}\iint V(\boldsymbol{k}', \boldsymbol{k})[\delta\Phi(\boldsymbol{k}) - \delta\Phi(\boldsymbol{k}')][\psi(\boldsymbol{k}) - \psi(\boldsymbol{k}')]d\tau_{\boldsymbol{k}}d\tau_{\boldsymbol{k}'}\,. \tag{4.64}$$

In this form it is clear that

$$(\psi L\Phi) = (\Phi L\psi)\,. \tag{4.65}$$

Furthermore $(\psi L\Phi)$ is positive definite since $V(\boldsymbol{k}', \boldsymbol{k})$, as the transition probability, is always positive.

We now choose ψ such that $(\psi F) = (\psi L\psi)$. Otherwise ψ can be any function. Then

$$([\psi - \Phi]L[\psi - \Phi]) \geq 0\,, \tag{4.66}$$

hence

$$(\psi L\psi) + (\Phi L\Phi) - (\Phi L\psi) - (\psi L\Phi) = (\psi L\psi) + (\Phi L\Phi) - 2(\psi L\Phi) \geq 0\,, \tag{4.67}$$

and since $(\psi L\Phi) = (\psi F) = (\psi L\psi)$,

$$(\Phi L\Phi) \geq (\psi L\psi)\,. \tag{4.68}$$

For all ψ which satisfy the condition $(\psi F) = (\psi L\psi)$, Φ [and therefore the function $\delta\Phi(\boldsymbol{k})$ being sought] maximizes the integral $(\psi L\psi)$. This is the basis for a variational method. One inserts an initial trial function for ψ with unknown parameters, performs a variation with respect to these parameters, and thereby determines their values. Within the limits of the trial function chosen, the result is an optimum approximation to the perturbed distribution function sought.

Which trial functions are chosen depends first on the external fields $\boldsymbol{E}$ and ∇T (an extension of this method to include magnetic fields runs into difficulties) and secondly on the type of interaction, and therefore on the form of the function $V(\boldsymbol{k}', \boldsymbol{k})$. Without knowledge of these two factors it is not possible to discuss this method further. We shall treat one example in Section 4.4.1. We refer the reader to the literature for further material, e.g., the books by *Wilson* [34], *Ziman* [35], and *Haug* [14].

It is however clear that in the case where the relaxation time approximation does not apply it is not possible to present a solution of the Boltzmann equation in closed form that includes all possible external fields. In what follows we shall therefore mostly use this approximation.

4.3 Formal Transport Theory

4.3.1 The Transport Equations

With a knowledge of the electron distribution function, the electric current density and the energy current density can be set down immediately

$$\boldsymbol{i} = \int(-e\boldsymbol{v})f(\boldsymbol{r}, \boldsymbol{k}, t)g(\boldsymbol{k})d\tau_{\boldsymbol{k}} = -\frac{e}{\hbar}\int\nabla_{\boldsymbol{k}}Ef(\boldsymbol{r}, \boldsymbol{k}, t)g(\boldsymbol{k})d\tau_{\boldsymbol{k}}, \tag{4.69}$$

$$\boldsymbol{w} = \int E(\boldsymbol{k})\boldsymbol{v}f(\boldsymbol{r}, \boldsymbol{k}, t)g(\boldsymbol{k})d\tau_{\boldsymbol{k}} = \frac{1}{\hbar}\int\nabla_{\boldsymbol{k}}EE(\boldsymbol{k})f(\boldsymbol{r}, \boldsymbol{k}, t)g(\boldsymbol{k})d\tau_{\boldsymbol{k}}. \tag{4.70}$$

It is however worth deriving these two transport equations in general from the thermodynamics of irreversible processes, thereby gaining a knowledge of the various transport phenomena. This we do in the sections which follow. We can then use the results of the previous sections to calculate quantitatively the transport coefficients which appear, and compare them with experiment. This is the content of Section 4.4.

In Section 2.1.4 we found the condition for equilibrium in homogeneous solids to be the existence of a common chemical potential μ for all electrons. Together with the density of states, the chemical potential determines through the Fermi distribution the energy distribution of the electrons in thermal equilibrium.

We have not so far raised the question of the spatial dependence of any of these parameters. Dependence of the density of states on position may be the result of *inhomogeneity* of the solid [position-dependent band structure in (2.152)]. In the Fermi distribution the energy (band structure), the temperature, and the chemical potential can depend on position. If we limit ourselves as previously to homogeneous solids, $E_n(\boldsymbol{k})$ is independent of position. Internal macroscopic fields can produce a space-dependent electrostatic potential which we can add to the band structure energy as in Section 2.2.10: $E = E_n(\boldsymbol{k}) - e\varphi$. We must then also replace the chemical potential in f_0 by the *electrochemical potential* $\eta = \mu - e\varphi$, as in (4.60)

$$f_0(E, \eta, T) = \{\exp[(E - \eta)/k_B T] + 1\}^{-1}. \tag{4.71}$$

In this formulation the condition for *spatial equilibrium* is that the electrochemical potential and temperature are independent of position. This condition is added to the condition of *local equilibrium*: existence of one chemical potential for all electrons.

The corollary to this statement is the condition for spatial nonequilibrium, i.e., for current flow

$$\nabla\eta \neq 0 \quad \text{or} \quad \nabla T \neq 0 \,. \tag{4.72}$$

Both gradients are the driving forces which lead to current flow.

We shall now set up the basic equations for transport theory using concepts from thermodynamics of irreversible processes. For a more detailed discussion, see, for example, [64]. We start from the relation

$$\rho T\frac{\partial s}{\partial t} = \rho\frac{\partial u}{\partial t} - \mu\frac{\partial n}{\partial t} \quad (Tds = dU - \mu dN \text{ for } dV = 0)\,. \tag{4.73}$$

ρ is the density, s the specific entropy, and u the specific internal energy. μ and n are the chemical potential and concentration of the electrons.

Equation (4.73) can be transformed by taking account of the following laws:

a) Conservation of the number of electrons

$$\frac{\partial n}{\partial t} + \nabla\cdot\boldsymbol{j} = 0 \tag{4.74}$$

where $\boldsymbol{j}$ is the particle current density.

b) Conservation of energy in the absence of external forces

$$\rho\frac{\partial u}{\partial t} + \nabla\cdot\boldsymbol{w} = \boldsymbol{F}\cdot\boldsymbol{j}\,. \tag{4.75}$$

Here $\boldsymbol{w}$ is the energy current density associated with u. Since $\boldsymbol{j}$ is the particle current density, the quantity $\boldsymbol{F}$ in the *local energy production* [right side of (4.75)] must be the external force on an electron.

One can use both equations to formulate a law of conservation of entropy from (4.73). Since entropy can be produced locally in the processes considered here, the right side of the continuity equation for entropy is nonzero, analogous to (4.75). We find

$$\rho\frac{\partial s}{\partial t} + \nabla\cdot\boldsymbol{w}_s = \Sigma\,. \tag{4.76}$$

The entropy current density $\boldsymbol{w}_s$ here is defined by

$$\boldsymbol{w}_s = \frac{1}{T}(\boldsymbol{w} - \mu\boldsymbol{j})\,. \tag{4.77}$$

We thus associate ρs and $\boldsymbol{w}_s$ with each other just as u and $\boldsymbol{w}$, or n and $\boldsymbol{j}$, are associated with each other. Inserting (4.73–75) and (4.77) into (4.76) then yields for the *local entropy production*

$$\Sigma = \frac{1}{T}\left[\boldsymbol{j}\cdot\left(\boldsymbol{F} - T\nabla\frac{\mu}{T}\right) + \boldsymbol{w}\cdot\left(-\frac{\nabla T}{T}\right)\right]. \tag{4.78}$$

The external forces are the Lorentz force of an electromagnetic field: $\boldsymbol{F} = -e(\boldsymbol{E} + \boldsymbol{v}\times\boldsymbol{B})$. We note however that in (4.78) the magnetic term of the Lorentz force disappears due to the scalar product $\boldsymbol{j}\cdot\boldsymbol{F}$ and the velocity $\boldsymbol{v}(\|\boldsymbol{j})$ appearing in $\boldsymbol{F}$. We are left with just $-e\boldsymbol{E}$.

In (4.78) the "fluxes" $\boldsymbol{j}$ (particle current density) and $\boldsymbol{w}$ (energy current density) are related to the "forces" (gradients of the electrostatic potential, the chemical potential, and the temperature). The thermodynamics of irreversible processes now postulates that, when local entropy production is represented as a sum of products of fluxes and forces, both are always linearly related to one another

$$\begin{aligned}\boldsymbol{j} &= L_{11}\left(\boldsymbol{F} - T\nabla\frac{\mu}{T}\right) + L_{12}\left(-\frac{\nabla T}{T}\right),\\ \boldsymbol{w} &= L_{21}\left(\boldsymbol{F} - T\nabla\frac{\mu}{T}\right) + L_{22}\left(-\frac{\nabla T}{T}\right).\end{aligned} \tag{4.79}$$

The L_{ik} are scalar if there are no anisotropies. When the solid considered is anisotropic, or when there is a preferred magnetic field direction, the L_{ik} become *tensors*. In general we have

$$L_{ik}(\boldsymbol{B}) = \tilde{L}_{ki}(-\boldsymbol{B}) \quad (\textit{Onsager relations}). \tag{4.80}$$

Here $\tilde{L}$ is the transposed tensor to L.

These equations can be extended to the case of different types of charge carriers, e.g., electrons and holes in semiconductors. The term $\mu\dot{n}$ in (4.73) must then be replaced by a sum over the respective parameter of the different charge carriers. The equations which follow must also be altered in a corresponding way. For n types of particles, there are then $n + 1$ transport equations with $n + 1$ terms. The L_{ik} form a $(n + 1)\times(n + 1)$-matrix. The L_{ik} are themselves tensors in anisotropic solids. We obtain $3(n + 1)$ transport equations with a correspondingly large number of terms. The Onsager relations apply just the same in all these cases.

The system of (4.79) does not represent the only possible way to formulate the transport equations. By combining equations, other fluxes can be introduced, or by transforming the equations we can go over to other forces. If one chooses fluxes and forces such that the local entropy production remains a sum over products of the new fluxes and forces, then the symmetry of the transport equations given by the Onsager relations is maintained.

First we ask which other fluxes are of interest. The associated forces can then be readily determined from (4.78).

The transition from particle current density $\boldsymbol{j}$ to electric current density $\boldsymbol{i}$ is trivial, involving only multiplication of $\boldsymbol{j}$ by $-e$. On the other hand, the definition of energy current density is not unique. In the equations so far, $\boldsymbol{w}$ is the current density associated with the internal energy u. It does not contain the contribution from the external forces, in our case the electrostatic energy. If we define a *total energy density* $\rho u_{\text{tot}} = \rho u - en\varphi$, the energy current density associated with it is

$$\boldsymbol{w}_{\text{tot}} = \boldsymbol{w} - e\boldsymbol{j}\varphi\,. \tag{4.81}$$

The related continuity equation is

$$\rho\frac{\partial u_{\text{tot}}}{\partial t} + \nabla\cdot\boldsymbol{w}_{\text{tot}} = -e\boldsymbol{E}\cdot\boldsymbol{j} - e\varphi\frac{\partial n}{\partial t} - \nabla\cdot(e\boldsymbol{j}\varphi)\,. \tag{4.82}$$

The right side of this equation vanishes on account of (4.74) and $\boldsymbol{E} = -\nabla\varphi$. It therefore represents the conservation of the total energy.

With $\boldsymbol{w}_{\text{tot}}$, (4.78) takes the form

$$\Sigma = \boldsymbol{i}\cdot\nabla\frac{\eta}{eT} + \boldsymbol{w}_{\text{tot}}\cdot\nabla\frac{1}{T} \tag{4.83}$$

where the electrochemical potential $\eta = \mu - e\varphi$ has again been introduced and the particle current density has been converted into the electric current density.

From this form we can readily go over to a further system in which only the gradients of electrochemical potential and temperature appear as driving forces. To do this we define, analogous to the thermodynamic relation $ds = dq/T$, the *heat current density* $\boldsymbol{w}_q$ by $\boldsymbol{w}_s = \boldsymbol{w}_q/T$. Eq. (4.78) then becomes

$$\Sigma = \frac{1}{T}\left[\boldsymbol{i}\cdot\nabla\frac{\eta}{e} + \boldsymbol{w}_q\cdot\left(-\frac{\nabla T}{T}\right)\right], \tag{4.84}$$

and the associated transport equations are

$$\boldsymbol{i} = N_{11}\nabla\frac{\eta}{e} + N_{12}\left(-\frac{\nabla T}{T}\right), \qquad \boldsymbol{w}_q = N_{21}\nabla\frac{\eta}{e} + N_{22}\left(-\frac{\nabla T}{T}\right). \tag{4.85}$$

with coefficients (tensors) N_{ik} for which the Onsager relations again apply.

We have thus related the electric current density $\boldsymbol{i}$ and heat current density $\boldsymbol{w}_q$ to the driving forces mentioned at the beginning of this section.

Equations of the form (4.85) can easily be obtained from (4.69) and (4.70) if one inserts the perturbed distribution function from the relaxation time approximation (4.82). If one separates f into f_0 and δf in the integrals (4.69) and

(4.70), the contribution from f_0 clearly disappears since in equilibrium the current density must be zero. From (4.47) and (4.62) δf has the form of a sum of terms of the form

$$-\frac{\partial f_0}{\partial E} g(E)\boldsymbol{v}\cdot\boldsymbol{A}, \tag{4.86}$$

where $\boldsymbol{A}$ is one of the vectors $\nabla\eta$, ∇T, $\boldsymbol{B}\times\nabla\eta$, $\boldsymbol{B}\times\nabla T$, $\boldsymbol{B}(\boldsymbol{B}\cdot\nabla\eta)$, or $\boldsymbol{B}(\boldsymbol{B}\cdot\nabla T)$. We thus obtain integrals of the type

$$\int g\boldsymbol{v}(\boldsymbol{v}\cdot\boldsymbol{A})\left(-\frac{\partial f_0}{\partial E}\right)g(\boldsymbol{k})d\tau_{\boldsymbol{k}}. \tag{4.87}$$

In *isotropic* solids, i.e., when $\boldsymbol{v} = (1/\hbar)\nabla_{\boldsymbol{k}}E$ is independent of direction, we can average over the direction of $\boldsymbol{v}$ in the integral. In the integrand this leads instead of $\boldsymbol{v}(\boldsymbol{v}\cdot\boldsymbol{A})$ to $(v^2/3)\boldsymbol{A}$, hence to

$$\int g\frac{v^2}{3}\left(-\frac{\partial f_0}{\partial E}\right)g(\boldsymbol{k})d\tau_{\boldsymbol{k}}\boldsymbol{A}. \tag{4.88}$$

Thus we obtain for $\boldsymbol{i}$ the general form

$$\begin{aligned}\boldsymbol{i} &= \alpha_{11}\nabla\frac{\eta}{e} + \alpha_{12}\nabla T + \beta_{11}\boldsymbol{B}\times\nabla\frac{\eta}{e} + \beta_{12}\boldsymbol{B}\times\nabla T\\ &\quad + \gamma_{11}\boldsymbol{B}\left(\boldsymbol{B}\cdot\nabla\frac{\eta}{e}\right) + \gamma_{12}\boldsymbol{B}(\boldsymbol{B}\cdot\nabla T)\\ &= [\alpha_{11} + \beta_{11}\boldsymbol{B}\times + \gamma_{11}\boldsymbol{BB}\cdot]\nabla\frac{\eta}{e} + [\alpha_{12} + \beta_{12}\boldsymbol{B}\times + \gamma_{12}\boldsymbol{BB}\cdot]\nabla T\end{aligned} \tag{4.89}$$

and a corresponding equation for $\boldsymbol{w}_q = \boldsymbol{w} + (\mu/e)\boldsymbol{i}$. The coefficients α_{ik}, β_{ik}, γ_{ik} are still magnetic field dependent. Eq. (4.89) has precisely the form of the first equation (4.85). The brackets can be identified with the N_{ik}. They are scalar for $\boldsymbol{B} = 0$, and tensors for $\boldsymbol{B} \neq 0$. The form presented in the brackets is the most general tensor relationship between two vectors $\boldsymbol{a}$ and $\boldsymbol{b}$ for a fixed preferred direction $\boldsymbol{c}$, namely the representation of $\boldsymbol{a}$ by three non-coplanar vectors in the directions of $\boldsymbol{b}$, $\boldsymbol{c}$ and perpendicular to both. The most general form of (4.89) is not therefore connected with the relaxation time approximation used. Finally it follows from (4.87) that in nonisotropic media the coefficients associated with the vectors $\boldsymbol{A}$ are always tensors.

4.3.2 Transport Coefficients Without a Magnetic Field

The N_{ik} appearing in (4.85) can be related to parameters which can readily be measured experimentally. To show this we convert (4.85) into two equations for $\nabla(\eta/e)$ and $\boldsymbol{w}_q$

$$\nabla\left(\frac{\eta}{e}\right) = \frac{1}{\sigma}\boldsymbol{i} + \varepsilon\nabla T, \qquad \boldsymbol{w}_q = \Pi\boldsymbol{i} - \kappa\nabla T \tag{4.90}$$

with

$$\begin{aligned} \sigma &= N_{11}, & \varepsilon &= \frac{N_{12}}{TN_{22}}, \\ \kappa &= \frac{N_{11}N_{22} - N_{12}N_{21}}{N_{22}}, & \Pi &= \frac{N_{21}}{N_{22}}. \end{aligned} \tag{4.91}$$

In this form we can compare the coefficients σ, ε, Π, and κ with experimental parameters which can be measured at uniform temperature and at zero current, respectively.

For $\nabla T = 0$ the first equation (4.90) yields

$$\nabla\left(\frac{\eta}{e}\right) = \frac{1}{\sigma}\boldsymbol{i}. \tag{4.92}$$

In homogeneous solids $\nabla(\eta/e) = \nabla(\mu/e - \varphi)$ is identical to the electric field $\boldsymbol{E} = -\nabla\varphi$, since the chemical potential is then independent of position. In this case σ is the constant of proportionality relating $\boldsymbol{i}$ and $\boldsymbol{E}$, i.e., the *electrical conductivity*.

For $\nabla T = 0$, the second equation (4.90) becomes

$$\boldsymbol{w}_q = \Pi\boldsymbol{i}. \tag{4.93}$$

The electric current thus carries with it energy, whose heat component $\boldsymbol{w}_q$ is proportional to $\boldsymbol{i}$ with constant of proportionality Π. Since $\boldsymbol{w}_q = T\boldsymbol{w}_s$, this statement can be reformulated as: Π/T is the constant of proportionality between an electiic current and an accompanying entiopy flow in an isothermal conductor.

Now an energy current cannot be observed directly, only its divergence, i.e., a local production of heat. We must therefore consider the local heat production to obtain the relation between Π and measurable quantities. We start with (4.75).

$$\begin{aligned} \rho\frac{\partial u}{\partial t} &= -\nabla\cdot\boldsymbol{w} - e\boldsymbol{E}\cdot\boldsymbol{j} = -\nabla\cdot\boldsymbol{w}_q - \nabla\cdot(\mu\boldsymbol{j}) + \boldsymbol{i}\cdot\boldsymbol{E} \\ &= \boldsymbol{i}\cdot\boldsymbol{E} - \nabla\cdot(\Pi\boldsymbol{i}_) + \nabla\cdot\frac{\mu}{e}\boldsymbol{i} = \boldsymbol{i}\cdot\nabla\left(\frac{\eta}{e}\right) - \boldsymbol{i}\cdot\nabla\Pi = \frac{i^2}{\sigma} - \boldsymbol{i}\cdot\nabla\Pi. \end{aligned} \tag{4.94}$$

The first term on the right is the Joule heat, the second an additional heat which appears with space-dependent coefficients. Π is constant in a homogeneous material. At the interface between two media, Π will change discontinuously as we go from one material to another. At such an interface (contact), heating or cooling will occur when current flows through it (*Peltier effect*). We can obtain this by integrating (4.94) from a point in conductor A close to the contact to a point in conductor B close to the contact

$$-\int_A^B \boldsymbol{i}\cdot\nabla\Pi ds = i(\Pi_A - \Pi_B). \tag{4.95}$$

Π_A and Π_B are called *Peltier coefficients* of materials A and B, respectively. One should note that the Peltier effect is not limited to contacts but can also appear as a volume effect in inhomogeneous conductors.

For $\nabla T \neq 0$ but $\boldsymbol{i} = 0$, (4.90) becomes

$$\nabla\left(\frac{\eta}{e}\right) = \varepsilon\nabla T, \qquad \boldsymbol{w}_q = -\kappa\nabla T. \tag{4.96}$$

The coefficient κ follows directly here as *thermal conductivity*.

The coefficient ε is seen here to be the constant of proportionality between a temperature gradient and (in the zero current case) the electrochemical potential gradient which it causes (i.e., the voltage appearing along with a gradient of the chemical potential). Here again it is convenient to consider a circuit consisting of two conductors A and B of different material connected at two contacts, but open anywhere within the material B. Let the two contacts be held at temperatures T_1 and T_2. At the location where this "thermo-element" is open, a voltage will appear which can be calculated from (4.96) by integrating around the circuit

$$\oint\nabla\left(\frac{\eta}{e}\right)\cdot d\boldsymbol{s} = \frac{1}{e}\delta\eta = \oint\varepsilon\nabla T\cdot d\boldsymbol{s}$$
$$= \oint\varepsilon dT = \int_{T_1}^{T_2}\varepsilon_A dT + \int_{T_2}^{T_1}\varepsilon_B dT = \int_{T_1}^{T_2}(\varepsilon_A - \varepsilon_B)dT. \tag{4.97}$$

The difference between the electrochemical potentials at both ends $\delta\eta$ is $\delta\mu - e\delta\varphi$. Now $\delta\mu = 0$, since both ends are made of the same material. From (4.97) we are left with

$$-\delta\varphi = \int_{T_1}^{T_2}(\varepsilon_A - \varepsilon_B)dT. \tag{4.98}$$

For a differential temperature difference $T_2 - T_1 = \delta T$, $\varepsilon_A - \varepsilon_B$ is seen to be the constant of proportionality between $\delta\varphi$ and δT. The ε are therefore referred to as absolute differential thermoelectric voltages or *absolute thermoelectric powers*.

As with the Peltier effect, thermoelectric voltages can also appear as a volume effect in inhomogeneous conductors. The appearance of a thermoelectric voltage is often referred to as the *Seebeck effect*.

Other possible transport coefficients are the constants of proportionality between $\boldsymbol{w}_q$ and $\nabla(\eta/e)$ (Thomson heat current coefficient) and between $\boldsymbol{E}$ and ∇T (Thomson potential gradient). Both are of only secondary importance. On the other hand, the Thomson energy coefficient μ_E is important (*Thomson coefficient*). For $\nabla T \neq 0$ and $\boldsymbol{i} \neq 0$, one obtains from (4.75) instead of (4.94)

$$\rho\frac{\partial u}{\partial t} = \boldsymbol{i}\cdot\nabla\left(\frac{\eta}{e}\right) - \nabla\cdot\boldsymbol{w}_q = \frac{i^2}{\sigma} + \varepsilon\boldsymbol{i}\cdot\nabla T - \nabla(\Pi\boldsymbol{i}) + \nabla\cdot(\kappa\nabla T) \qquad (4.99)$$

and for a homogeneous conductor $[\Pi = \Pi(T)]$

$$\rho\frac{\partial u}{\partial t} = \frac{i^2}{\sigma} + \nabla\cdot(\kappa\nabla T) - \mu_E\boldsymbol{i}\cdot\nabla T \qquad (4.100)$$

with $\mu_E = (\partial\Pi/\partial T) - \varepsilon$.

The first term on the right in (4.99) describes the Joule heat, the second term the heat entering the volume by heat conduction, and the third describes heat which is only produced when currents and temperature gradients are simultaneously present (*Thomson heat*).

We see from (4.91) that the coefficients ε and Π are connected through the Onsager relations. We find

$$\Pi = \varepsilon T, \qquad (4.101)$$

and we also find from (4.100)

$$\mu_E = T\frac{\partial\varepsilon}{\partial T}. \qquad (4.102)$$

Eqs. (4.101) and (4.102) are referred to as the *Thomson relations*.

4.3.3 Transport Coefficients with a Magnetic Field

In the presence of a magnetic field, a preferred direction appears even in homogeneous isotropic solids. The coefficients N_{ik} in the transport equations (4.85) therefore become tensors. Since in a homogeneous medium $\nabla(\eta/e)$ is equal to the electric field, we can write (4.85) similar to (4.89) in the following general form:

$$\begin{aligned} \boldsymbol{i} &= \alpha_{11}\boldsymbol{E} + \beta_{11}\boldsymbol{B}\times\boldsymbol{E} + \gamma_{11}\boldsymbol{B}(\boldsymbol{B}\cdot\boldsymbol{E}) + \alpha_{12}\nabla T \\ &\quad + \beta_{12}\boldsymbol{B}\times\nabla T + \gamma_{12}\boldsymbol{B}(\boldsymbol{B}\cdot\nabla T), \\ \boldsymbol{w}_q &= \alpha_{21}\boldsymbol{E} + \beta_{12}\boldsymbol{B}\times\boldsymbol{E} + \gamma_{21}\boldsymbol{B}(\boldsymbol{B}\cdot\boldsymbol{E}) + \alpha_{22}\nabla T \\ &\quad + \beta_{22}\boldsymbol{B}\times\nabla T + \gamma_{22}\boldsymbol{B}(\boldsymbol{B}\cdot\nabla T) \end{aligned} \tag{4.103}$$

with magnetic field-dependent α_{ik}, β_{ik}, γ_{ik}.

These equations allow the description of numerous effects which are just as important as the *thermoelectric effects* discussed in the previous section. One distinguishes here between *galvanomagnetic* and *thermomagnetic effects.* In the first case the primary field, apart of course from the magnetic field itself, is electric and in the second case it is a temperature gradient. By "primary" here we mean that the combined action of the magnetic field and the electric field or the temperature gradient can lead to "secondary" fields or gradients in different directions.

To simplify the discussion we restrict ourselves here to the case where the magnetic field and the other primary field are perpendicular.

If the fields are parallel, the magnetic field has no effect on the charge carriers in a homogeneous isotropic solid. In (4.103) the linear terms in $\boldsymbol{B}$ vanish; the quadratic terms in $\boldsymbol{B}$ just cancel the magnetic field-dependent parts of the terms without $\boldsymbol{B}$—as we shall see explicitly later. In this simple case we therefore find no galvanomaynetic effects. For $\boldsymbol{B} \perp \boldsymbol{E}$, ∇T we want to further arrange that the magnetic field and the primary field lie, respectively, in the z- and x-directions of a Cartesian coordinate system. This means in (4.103) that $\boldsymbol{B}$ has only a z-component while $\boldsymbol{E}$ and ∇T have no z-component. We must however ascribe a y-component to $\boldsymbol{E}$ and ∇T since, due to the action of the Lorentz force, these components can arise as secondary fields.

Equation (4.103) is then

$$\begin{aligned} i_x &= \alpha_{11}E_x - \beta_{11}B_zE_y + \alpha_{12}\frac{\partial T}{\partial x} - \beta_{12}B_z\frac{\partial T}{\partial y}, \\ i_y &= \alpha_{11}E_y + \beta_{11}B_zE_x + \alpha_{12}\frac{\partial T}{\partial y} + \beta_{12}B_z\frac{\partial T}{\partial x}, \\ i_z &= 0 \end{aligned} \tag{4.104}$$

and correspondingly for the w_{qi}.

Choosing the x-direction for the primary fields means that electric current and heat will also flow in the x-direction in the absence of a magnetic field.

With a magnetic field the force on the charge carriers is no longer directed in the x-direction; instead, the Lorentz force has a component perpendicular to the z-direction and to the direction of motion of the charge carriers. In an infinite medium this changes the direction of the current. The (average) direction of motion of the electrons makes an angle θ with their original direction, the so-called *Hall angle*.

We do not wish to make use of this picture however in the following discussion. Experimental facts suggest that we should rather assume that the *geometric shape* of the sample will constrain the current to flow in a given direction. The samples are usually long thin rods. The deflecting force of the magnetic field must then be compensated by an opposing force. Without wishing to go into the physical background at this point, a look at (4.104) will suffice. For $\mathbf{E} = (E_x, 0, 0)$ these equations produce a current component in the y-direction [term with E_x in the second equation (4.104)]. If, on the other hand, $i_y = 0$, the second equation (4.104) (at constant temperature) can only be satisfied if one admits $E_y \neq 0$. One then has to insert this secondary transverse field into the first equation (4.104) to obtain the current i_x as a function of the primary field E_x.

Equation (4.104) shows further that a primary field E_x can also have the secondary effect of setting up a temperature gradient in the y-direction, and that a primary temperature gradient can similarly set up secondary temperature gradients and electric fields. One unites this occurrence of potential and temperature gradients perpendicular to the current and to the magnetic field under the collective heading of *transverse effects* (in a transverse magnetic field). The reactions of these transverse effects on the primary gradients give rise to the *longitudinal effects* (in a transverse magnetic field).

Before we explore all these effects we should point out the following. As a result of the geometry of the sample we do indeed prohibit electrical currents which deviate from the x-direction, since the sample has free surfaces in the y-direction. We can, however, extract the simultaneously produced heat current from the surface, or prevent its extraction by adiabatic conditions. In the first case the temperature in the y-direction remains constant (*isothermal case*). In the second case the heat flow sets up a temperature gradient which produces an opposite flow of heat, so that in equilibrium further flow of heat is prevented (*adiabatic case*).

As in (4.103), we first convert the system of (4.104) into equations in which the quantities sought appear as functions of the readily measurable quantities i_x and $\partial T/\partial x$.

For the *isothermal case* ($\partial T/\partial y = 0$ as well as $i_y = 0$) one finds

$$\begin{aligned} E_x &= (D_0/D_2)i_x + (D_5/D_2)\partial T/\partial x \\ E_y &= -(D_1/D_2)i_x - (D_6/D_2)\partial T/\partial x \\ w_{qx} &= -(D_3/D_2)i_x + (D_7/D_2)\partial T/\partial x\,. \end{aligned} \tag{4.105}$$

For the *adiabatic case* ($w_{qy} = 0$ as well as $i_y = 0$) we have

$$\begin{aligned}
E_x &= (D_0/D_2)(1 - D_4D_6/D_0D_7)i_x + (D_5/D_2)(1 + D_6D_8/D_5D_7)\partial T/\partial x \\
E_y &= -(D_1/D_2)(1 + D_4D_5/D_2D_7)i_x - (D_6/D_2)(1 - D_5D_8/D_6D_7)\partial T/\partial x \\
w_{qx} &= -(D_3/D_2)(1 + D_4D_8/D_3D_7)i_x + (D_7/D_2)[1 + (D_8/D_7)^2]\partial T/\partial x \\
\partial T/\partial y &= -(D_4/D_7)i_x + (D_8/D_7)\partial T/\partial x\,, \qquad (4.106)
\end{aligned}$$

where the D_i are abbreviations for the following determinants:

$$\begin{aligned}
&D_0 = \alpha_{11}, \quad D_1 = \beta_{11}B_z, \quad D_2 = \begin{vmatrix} \alpha_{11} & -\beta_{11}B_z \\ \beta_{11}B_z & \alpha_{11} \end{vmatrix}, \\
&D_3 = \begin{vmatrix} \beta_{11}B_z & \alpha_{11} \\ -\alpha_{21} & \beta_{21}B_z \end{vmatrix}, \quad D_4 = \begin{vmatrix} \alpha_{21} & \alpha_{11} \\ \beta_{21}B_z & \beta_{11}B_z \end{vmatrix}, \\
&D_5 = \begin{vmatrix} -\beta_{11}B_z & \alpha_{12} \\ \alpha_{11} & \beta_{12}B_z \end{vmatrix}, \quad D_6 = \begin{vmatrix} \alpha_{11} & \alpha_{12} \\ \beta_{11}B_z & \beta_{12}B_z \end{vmatrix}, \qquad (4.107) \\
&D_7 = \begin{vmatrix} \alpha_{11} & -\beta_{11}B_z & \alpha_{12} \\ \beta_{11}B_z & \alpha_{11} & \beta_{12}B_z \\ -\alpha_{21} & \beta_{21}B_z & -\alpha_{22} \end{vmatrix}, \quad D_8 = \begin{vmatrix} \alpha_{11} & -\beta_{11}B_z & \alpha_{12} \\ \beta_{11}B_z & \alpha_{11} & \beta_{12}B_z \\ \beta_{21}B_z & \alpha_{21} & \beta_{22}B_z \end{vmatrix}.
\end{aligned}$$

Equations (4.105) and (4.106) contain fourteen coefficients. In the first equations of both systems, for the x-component of the electric field strength, we identify the factors multiplying i_x and $\partial T/\partial x$ as the *reciprocal electric conductivity in a magnetic field*, and the Thomson potential gradient in a magnetic field, respectively. The latter, apart from a μ-part which vanishes in a homogeneous magnetic field, is the *thermoelectric power in a magnetic field.* Correspondingly two coefficients in the equations for w_{qx} give the *Peltier coefficient in a magnetic field*, and the negative coefficient of the *thermal conductivity in a magnetic field*, respectively.

The remaining six coefficients describe the transverse effects. In a transverse magnetic field, as a result of an electric current or a temperature gradient, transverse voltages and (in the adiabatic case) transverse temperature gradients appear.

The appearance of a voltage perpendicular to the electric current and to the magnetic field is known as the *Hall effect*. The associated coefficient is the

$$\text{Hall coefficient} \quad R = \frac{E_y}{i_x B_z}. \qquad (4.108)$$

One has to distinguish between the isothermal and adiabatic Hall coefficients.

The appearance of a voltage perpendicular to both temperature gradient and magnetic field is known as the *Nernst effect*. The associated coefficient is the

$$\text{Nernst coefficient} \quad Q = \frac{E_y}{B_z \partial T/\partial x}. \tag{4.109}$$

Here again we must distinguish between an isothermal and an adiabatic coefficient.

The appearance of a temperature gradient perpendicular to a magnetic field and to an electric current or temperature gradient is known as the *Ettingshausen effect* or the *Righi-Leduc effect*, respectively. The associated coefficients are the

$$\text{Ettingshausen coefficient} \quad P = \frac{\partial T/\partial y}{i_x B_z} \tag{4.110}$$

and the

$$\text{Righi-Leduc coefficient} \quad S = \frac{\partial T/\partial y}{\partial T/\partial x B_z}. \tag{4.111}$$

This is the form in which the galvanomagnetic and thermomagnetic coefficients are mostly given in the literature. We must emphasize however that here they already have acquired a particular form through the restriction to homogeneous solids.

Clearly there is a series of relations between the fourteen coefficients since they are all described by nine parameters D_i. For example it is possible to express the adiabatic coefficients in terms of the isothermal ones. To do this, one must extend the system of (4.105) by the missing equation for w_{qy}, and include the two further isothermal transport coefficients which appear there. However, since these have no great physical significance, one takes the Ettingshausen and Righi-Leduc coefficients along with the six isothermal coefficients and expresses the remaining six adiabatic coefficients in terms of them. The difference between the adiabatic and the isothermal reciprocal electrical conductivity, for example, becomes

$$\sigma_{Ba}^{-1} - \sigma_{Bi}^{-1} = -PQ_i B_z. \tag{4.112}$$

Adiabatic and isothermal thermal conductivity are related by

$$\kappa_{Ba} - \kappa_{Bi} = \kappa_{Bi} S^2 B_z. \tag{4.113}$$

The further relationships can easily be derived from (4.105) and (4.106).

The Onsager relations lead to further relations. We mention here only that the determinants D_6 and D_4 are related through them: $D_6 = -D_4/T$. This leads to the so-called *Bridgman relation*

$$P = Q_i T/\kappa_{Bi}. \tag{4.114}$$

4.4 Transport in Metals and Semiconductors

4.4.1 The Electrical Conductivity

In the first three parts of this chapter we prepared the tools to calculate the transport coefficients. We shall now introduce a few examples. For a systematic discussion we refer the reader to the books by *Smith* et al. [64], *Blatt* [52], and *Ziman* [35], and to the review article by *Blatt* in [101.4].

The electrical conductivity σ relates the electric field $\boldsymbol{E}$ and the current density $\boldsymbol{i}$ produced in a solid by that field. Any theory of conductivity must above all explain the following experimental observations:

1) In most solids σ is field independent for moderate field stiengths. Current density and field are therefore linearly related (Ohm's law).

2) Different solids can have very widely differing conductivities. In metals, σ has values of the order $10^6\ (\Omega\text{cm})^{-1}$, while in semiconductors σ covers a range of 15 to 20 orders of magnitudes below this value.

3) In many metals there is a linear relation between electrical and thermal conductivities $\kappa/\sigma T = \text{const}$ (*Wiedemann-Franz law*).

4) There is a gross difference in the temperature dependence of electrical conductivity in metals and semiconductors. Over a wide temperature range, σ in metals is proportional to T^{-1}. At low temperatures simple metals show a T^{-5}-dependence of conductivity. In undoped semiconductors, the conductivity follows an $\exp(-\alpha/T)$-law. This temperature behaviour relates to samples without lattice imperfections, in which the electron-phonon interaction is the limiting scattering mechanism. An additional scattering mechanism arises in samples with imperfections. This leads to a different temperature dependence of the electrical conductivity which can occur in semiconductors over the entire temperature range, in metals only at very low temperatures.

The electric current density can be written as the product of the charge of an electron e, its mean velocity v, and the electron concentration n. If we define the coefficient relating the mean electron velocity and the electric field as the *electron mobility* $\mu_n = v/E$, σ can be written as the product $\sigma = e\mu_n n$. The effect of the electron-phonon interaction is then only contained in μ_n, while band model statistics determine the electron concentration and its temperature dependence. We are thus in a position to answer parts of the questions posed above, or at least to define them more precisely.

Ohm's law requires that electron concentration and mobility be independent of the field. The former is certainly true in homogeneous solids at moderate field strengths. The latter however remains to be explored.

The difference between metals and semiconductors follows solely from the temperature dependence of the electron concentration. The number of valence electrons involved in electrical conduction in a metal is practically independent of temperature. In semiconductors, however, thermal excitation from the valence to the conduction bands greatly changes the carrier concentrations. One can readily demonstrate that the number of electron-hole pairs in a semiconductor is proportional to $\exp(-E_G/2k_BT)$, where E_G is the energy gap between both bands. In general the temperature dependence of the mobility is given by a power law, and is therefore weak compared to the exponential temperature dependence of the electron concentration.

The first classical theory of conductivity is that of Drude. All electrons are assumed to behave in the same way in an electric field. The interaction with the lattice is by collision processes in which energy and momentum are exchanged. An electron is accelerated by the external field between two collision processes. The interplay of acceleration and collision leads to a constant mean velocity which varies linearly with the electric field (Ohm's law). The Wiedemann-Franz law follows readily too. Nothing can be said about the temperature dependence of the electron concentration. Equally nothing can be said about the temperature dependence of the mobility, or else it is falsely deduced from simple assumptions about the temperature dependence of the parameters involved. These difficulties have not been overcome by later improvements which introduced a velocity distribution for the electrons (Lorentz) and Fermi statistics (Sommerfeld). In spite of a few impressive successes, the Drude-Lorentz-Sommerfeld theory has to be significantly improved by replacing the rather imprecise model of electron collisions with the lattice by electron-phonon interactions. We have already developed the framework required for this in the earlier sections in this chapter.

We start with a calculation of electrical conductivity for the interaction of electrons with longitudinal acoustic phonons. At the outset we make two approximations. We assume the phonon system to be in equilibrium, and we neglect Umklapp processes. The collision term of the Boltzmann equation is then given by (4.43), where the transition probabilities are to be taken from (4.14). We insert the Bose distribution of phonons for the $n_{\boldsymbol{q}}$ in (4.14). For the $n_{\boldsymbol{k}}$ we have to make use of the perturbed distribution function $f(\boldsymbol{k})$. The matrix element which goes into the transition probability is given by (4.9). Usually one introduces the simplifying assumption that the Fourier components $V_{\boldsymbol{q}}$ are independent of $\boldsymbol{q}$. This is needed to be able to evaluate the integrations which follow. As long as we want to calculate only the temperature dependence and not the absolute value of the conductivity, this approximation is quite adequate.

From (4.43) we obtain

$$\left.\frac{\partial f}{\partial t}\right|_{\text{coll}} = \frac{C^2}{NM} \sum_{\boldsymbol{q}} \frac{q^2}{\hbar\omega_{\boldsymbol{q}}} \Big[\{(n_{\boldsymbol{q}} + 1)[1 - f(\boldsymbol{k})]f(\boldsymbol{k}+\boldsymbol{q}) - n_{\boldsymbol{q}}[1 - f(\boldsymbol{k}+\boldsymbol{q})]f(\boldsymbol{k})\}$$
$$\times\, \delta(E(\boldsymbol{k}+\boldsymbol{q}) - E(\boldsymbol{k}) - \hbar\omega_{\boldsymbol{q}}) + \{n_{\boldsymbol{q}}[1 - f(\boldsymbol{k})]f(\boldsymbol{k}+\boldsymbol{q})$$
$$- (n_{\boldsymbol{q}} + 1)[1 - f(\boldsymbol{k}+\boldsymbol{q})]f(\boldsymbol{k})\}\delta(E(\boldsymbol{k}+\boldsymbol{q}) - E(\boldsymbol{k}) + \hbar\omega_{\boldsymbol{q}}) \Big]. \tag{4.115}$$

Introducing the perturbation of the distribution function in accordance with (4.47), replacing the sum over $\boldsymbol{q}$ by an integration, and combining all further constants into C, we find

$$\left.\frac{\partial f}{\partial t}\right|_{\text{coll}} = \frac{C}{T} \int \frac{q^2}{\omega_{\boldsymbol{q}}} n_{\boldsymbol{q}} \{f_0(\boldsymbol{k})[1 - f_0(\boldsymbol{k}+\boldsymbol{q})]\delta(E(\boldsymbol{k}+\boldsymbol{q}) - E(\boldsymbol{k}) - \hbar\omega_{\boldsymbol{q}})$$
$$+ f_0(\boldsymbol{k}+\boldsymbol{q})[1 - f_0(\boldsymbol{k})]\delta(E(\boldsymbol{k}+\boldsymbol{q}) - E(\boldsymbol{k}) + \hbar\omega_{\boldsymbol{q}})\}$$
$$\times\, [\delta\Phi(\boldsymbol{k}+\boldsymbol{q}) - \delta\Phi(\boldsymbol{k})]d\tau_{\boldsymbol{q}}. \tag{4.116}$$

We carry out the integration under two further limiting assumptions. We take $E = \hbar^2k^2/2m$ for $E(\boldsymbol{k})$. We further approximate the phonon dispersion relation $\omega(\boldsymbol{q})$ by a linear dependence $\omega \sim q$, and replace the Brillouin zone by a sphere of radius q_{D} (Debye approximation, Sec. 3.3.4). q_{D} is related to the Debye frequency, the Debye temperature, and the propagation velocity of the longitudinal sound waves by $\hbar\omega_{\mathrm{D}} = \hbar s_l q_{\mathrm{D}} = k_{\mathrm{B}}\Theta_{\mathrm{D}}$. The condition $0 \le q \le q_{\mathrm{D}}$ can then be written

$$\hbar\omega_{\boldsymbol{q}} \le k_{\mathrm{B}}\Theta_{\mathrm{D}}. \tag{4.117}$$

A further condition for q follows from energy conservation $E(\boldsymbol{k}+\boldsymbol{q}) = E(\boldsymbol{k}) \pm \hbar\omega_{\boldsymbol{q}}$. If we insert the quadratic dependence of $E(\boldsymbol{k})$ on $\boldsymbol{k}$ and neglect $\hbar\omega_{\boldsymbol{q}}$ as small compared to the other terms in the energy conservation law, it follows that $q \le 2k$, or

$$\hbar\omega_{\boldsymbol{q}} \le 2\sqrt{\frac{E}{D}}\, k_{\mathrm{B}}\Theta_{\mathrm{D}}, \qquad D = \frac{\hbar^2 q_{\mathrm{D}}^2}{2m^*}. \tag{4.118}$$

Here E is the energy of the scattered electrons. In metals, electrons near the Fermi surface would be the ones involved. E is therefore approximately equal to μ. The second condition becomes important for $\mu < D/4$. This is only the case when the electron density is low. Thus in metals the condition (4.117) applies for the range of integration in (4.116). In semiconductors, on the other hand, in view of the low concentration of conduction electrons, we have to adopt the condition (4.118).

All these restrictions lead to

$$\left.\frac{\partial f}{\partial t}\right|_{\text{coll}} = Ck_x \frac{\partial f_0}{\partial E} E^{-3/2}\left(\frac{T}{\Theta_D}\right)^3 \int_{z_{\min}}^{z_{\max}} \left\{ Ec(\eta) - c(\eta + z) \right.$$
$$\left. \times \left[E + \frac{k_B T}{2} z - \frac{D}{2}\left(\frac{T}{\Theta}\right)^2 z^2\right]\right\} \frac{e^\eta + 1}{e^{\eta+z} + 1} \frac{z^2}{|1 - e^{-z}|} dz \tag{4.119}$$

with D from (4.118) and

$$\delta\Phi(\boldsymbol{k}) \equiv k_x c(E) \text{ [cf. (4.54)]}, \qquad \eta = \frac{E - \mu}{k_B T}, \qquad z = \frac{\hbar\omega_q}{k_B T}.$$

Again we have lumped together all constants appearing into the constant C.

Since in this section we are interested only in the electrical conductivity, we take the external force to be an electric field and we place it along the x-direction. From (4.35) and (4.56) the Boltzmann equation then becomes

$$\left.\frac{\partial f}{\partial t}\right|_{\text{coll}} = -\frac{e\hbar}{m^*}\frac{\partial f_0}{\partial E} k_x E_x . \tag{4.120}$$

Equation (4.120) together with the collision term (4.119) can be approximately solved by iteration. See, for example, *Wilson* [34] for details of the calculation. At moderate temperatures the final result is [cf. (4.54)]

$$f = f_0 - \frac{\partial f_0}{\partial E} k_x c(E), \qquad c \propto -E_x m^{*-3/2} \mu^{3/2} \left(\frac{\Theta_D}{T}\right)^5 J_5^{-1}\left(\frac{\Theta_D}{T}\right) \tag{4.121}$$

with

$$J_5(x) = \int_0^x \frac{z^5 dz}{(e^z - 1)(1 - e^{-z})} \begin{array}{l} = x^4/4 \text{ for small } x, \\ = \text{const for large } x. \end{array} \tag{4.122}$$

Equation (4.121) leads to the conductivity σ through

$$i_x = e\int v_x \frac{\partial f_0}{\partial E} k_x c(E) g(\boldsymbol{k}) d\tau_{\boldsymbol{k}} = -\frac{2e}{3\hbar}\int E\left(-\frac{\partial f_0}{\partial E}\right) c(E) g(E) dE . \tag{4.123}$$

Since from (4.121) $c(E)$ does not depend on E, σ is proportional to c. With the approximations of (4.122), therefore, $\sigma \propto T^{-5}$ for low temperatures, and $\propto T^{-1}$ for high temperatures (compared with the Debye temperature in each case). This result agrees with the observations noted at the beginning of this section. Fig. 4.5 shows the reduced resistivity for several simple metals as a function of temperature.

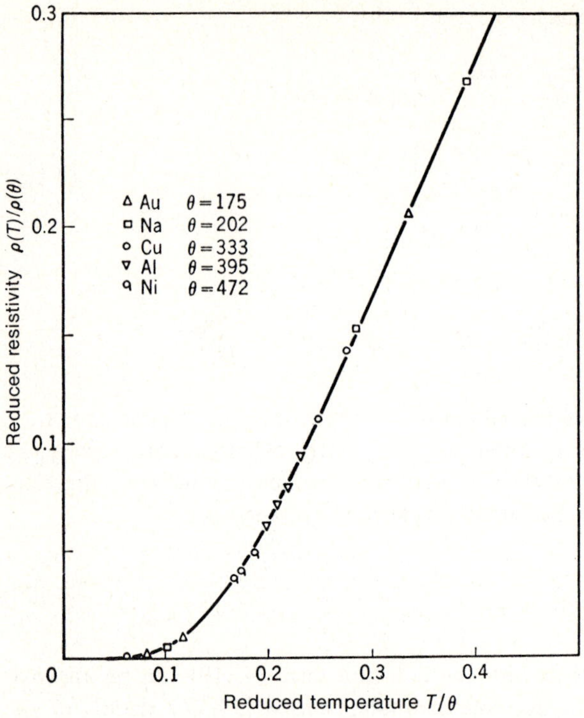

Fig. 4.5. Electrical resistivity of several metals as a function of temperature (in reduced units, Θ = Debye temperature) and theoretical curve obtained from (4.121) and (4.123) (Bloch-Grüneisen relation) (from *Blatt* [101.4]).

The experimental values agree excellently with the theoretical temperature dependence $\sigma \propto T^{-5} J_5^{-1}(T)$ (*Bloch-Grüneisen relation*). In spite of this agreement, these results contribute little to a general theory of the transport properties of metals. Eq. (4.121) gives the distribution function *only* for the case of an external electric field. If temperature gradients or a magnetic field appear with E_x on the right side of (4.120), the iterative technique cannot be carried out (cf. further below).

A relaxation time can be derived formally from (4.121). By putting $c \equiv -(e\hbar/m^*)E_x\tau$, (4.120) becomes

$$\left.\frac{\partial f}{\partial t}\right|_{\text{coll}} = \frac{\partial f_0}{\partial E} k_x \frac{c}{\tau} = -\frac{f - f_0}{\tau}. \tag{4.124}$$

This relaxation time is useless since it can only be used to calculate σ. On the other hand, a generally valid relaxation time can be defined for the temperature region $T \gg \Theta_D$. In this region one can expand the integrand in (4.119) in powers of $z = \Theta_D/T$ and terminate the expansion early. The integration can then be carried out and one finds (details again in *Wilson* [34])

$$\left.\frac{\partial f}{\partial t}\right|_{\text{coll}} = -\frac{f - f_0}{\tau(E)} \quad \text{with} \quad \tau(E) \propto \frac{4m^{*-1/2}E^{3/2}}{e\hbar}\frac{\Theta_D}{T}, \qquad T \gg \Theta_D . \tag{4.125}$$

This expression differs from the τ obtained from (4.121) by the factor $E^{3/2}$ which appears instead of $\mu^{3/2}$ in (4.121). Both expressions lead to the same conductivity since electrons in the vicinity of the Fermi energy preferentially contribute to the electrical conductivity, so that the relevant energy E is almost the same as μ. This follows formally from (4.123). If as in (4.124) above we set c equal to $-(e\hbar/m^*)E_x\tau(E)$, then

$$\sigma = \frac{2e^2}{3m^*}\int E\tau(E)g(E)\left(-\frac{\partial f_0}{\partial E}\right)dE \approx \frac{2e^2}{3m^*}\mu\tau(\mu)g(\mu) . \tag{4.126}$$

Here we have used the fact that the Fermi distribution falls off rapidly at $E = \mu$, so that its negative derivative with respect to energy is a δ-function at $E = \mu$. Using (2.7) and (2.9) we finally find the general expression

$$\sigma = \frac{e^2}{m^*}n\tau(\mu) . \tag{4.127}$$

Separating conductivity into charge, electron concentration, and mobility, the latter is seen to be $\mu_n = e\tau(\mu)/m^*$. From (4.125) the temperature dependence of conductivity, mobility, and relaxation time is equal to T^{-1}.

We find another result for a dilute electron gas where the integration limits of the integral (4.119) are given by (4.118). This case is particularly interesting in semiconductors. There the electron concentration is so small that we can make the further approximation of replacing the Fermi distribution with the Boltzmann distribution [e^{-x} for $1/(e^x + 1)$]. This makes a solution of (4.119) possible for practically the entire temperature range. The result is a relaxation time which is dependent on energy and temperature according to $E^{-1/2}T^{-1}$.

Assuming a Boltzmann distribution, the integral in (4.126) now becomes

$$\sigma = \frac{2e^2}{3m^*k_BT}\exp(\mu/k_BT)\int_0^\infty E\tau(E)g(E)\exp(-E/k_BT)dE . \tag{4.128}$$

On account of the temperature dependence of the relaxation time noted, and since $g(E) \sim E^{1/2}$, the conductivity is proportional to $\exp(\mu/k_BT)$. If we now include the temperature dependence of the concentration of a dilute electron gas found in Section 2.1.4 [$n \sim T^{3/2}\exp(\mu/k_BT)$], it follows that for the electron-LA phonon interaction the mobility in semiconductors obeys a $T^{-3/2}$ law.

We remarked above that the iterative procedure leading to (4.121) is restricted to the case of an electric field as the external force. If other external forces occur along with the electric field, then at low temperatures, i.e., when no relaxation time can be defined from (4.119), the variational method used in Section 4.2.4 has to be employed. We shall indicate the solution for an electric field and temperature gradient which both lie in the x-direction.

We write the collision term (4.119) in the form $-k_x E^{-3/2} Lc(E)$, where L is an integral operator defined by (4.119) which operates on $c(E) = -(f - f_0)/k_x(df_0/dE)$.

With (4.58) and (4.59), the Boltzmann equation then becomes

$$Lc(E) = \frac{\partial f_0}{\partial E}\frac{\hbar}{m^*}\left(eE_x + \frac{\partial \mu}{\partial x} + \frac{E-\mu}{T}\frac{\partial T}{\partial x}\right)E^{3/2}. \tag{4.129}$$

To solve this equation we use as a trial function an expansion of $c(E)$ in powers of E, with unknown coefficients which we determine by the variational method.

Our approach becomes simpler if we divide $c(E)$ into

$$E^{-3/2}c(E) = \frac{\hbar}{m^*}\left(eE_x + \frac{\partial \mu}{\partial x}\right)c_1(E) + \frac{\hbar}{m^*}\frac{d}{dx}(k_B T)c_2(E). \tag{4.130}$$

Because of the linearity of the integral operator (4.129) separates into two integral equations

$$Lc_1(E) = \frac{\partial f_0}{\partial E} \quad \text{and} \quad Lc_2(E) = \eta\frac{\partial f_0}{\partial E} \quad \left(\eta = \frac{E-\mu}{k_B T}\right), \tag{4.131}$$

which we can deal with independently.

Let us consider the first equation (4.131). We take as our trial function the expansion $c_1(\eta) = \sum_r c_r \eta^r$. $c_1(\eta)$ is the trial function designated ψ in Section 4.2.4. We must therefore maximize the value of the integral

$$(c_1(\eta)Lc_1(\eta)) = \sum_{r,s} c_r c_s \int \eta_r L(\eta^s)d\eta \equiv \sum_{r,s} c_r c_s D_{rs} \tag{4.132}$$

subject to the condition

$$\sum_{r,s} c_r c_s D_{rs} = \sum_r c_r \int \eta^r \frac{\partial f_0}{\partial E} d\eta \equiv \sum_r c_r C_r. \tag{4.133}$$

We add this condition, with an undefined Lagrange parameter, to the integral and perform the variation by differentiation with respect to a c_t. This variation must be zero.

$$\frac{d}{dc_t}\left(\sum_{r,s} c_r c_s D_{rs} + \lambda \sum_r c_r C_r\right) = 2\sum_s c_s D_{ts} + \lambda C_t = 0. \tag{4.134}$$

We multiply the right side of this equation by c_t and sum over t

$$2\sum_{t,s} c_t c_s D_{ts} = -\lambda \sum_t c_t C_t . \tag{4.135}$$

Equations (4.133) and (4.135) are identical when $\lambda = -2$. The Lagrange parameter is therefore determined and we obtain the system of equations

$$\sum_s c_s D_{ts} = C_t \tag{4.136}$$

which determine the coefficients c_s sought.

In precisely the same way we can start from the second integral equation (4.131) and obtain

$$c_2(\eta) = \sum_m b_m \eta^m \rightarrow \sum_m b_m D_{tm} = B_t . \tag{4.137}$$

Comparing the two (4.131) and using (4.133) we further find $B_t = C_{t+1}$.

Through (4.136) and (4.137), all the coefficients of the power series expansion for the two components of $c(\eta)$ can be determined. By means of this function we can then determine the electric current density and the heat current density from (4.69) and (4.70).

The execution of the procedure rapidly becomes complicated and consequently one usually includes only the first, or first two, terms in the expansion.

For the calculation of the conductivity in an electric field only the $r = 0$ term in the expansion of $c(\eta)$ is needed. The defining equation (4.136) then becomes $c_0 D_{00} = C_0$ or $c_0 = C_0/D_{00}$. From (4.119) D_{00}^{-1} is found to be proportional to

$$\left(\frac{T}{\Theta_D}\right)^3 \int_{-\infty}^{+\infty} d\eta \int_{-z_{\max}}^{+z_{\max}} dz \left[\frac{k_B T}{2} z - \frac{D}{2}\left(\frac{T}{\Theta_D}\right)^2 z^2\right] \frac{z^2}{|1 - e^{-z}|} \frac{\partial f_0}{\partial \eta} \frac{e^{\eta+1}}{e^{\eta+z}+1} . \tag{4.138}$$

The integral over η can be calculated immediately

$$\int_{-\infty}^{+\infty} d\eta \frac{\partial f_0}{\partial \eta} \frac{e^{\eta+1}}{e^{\eta+z}+1} = -\int_{-\infty}^{+\infty} d\eta \frac{1}{(e^{\eta+z}+1)(e^{-\eta}+1)} = -\frac{z}{e^z - 1} . \tag{4.139}$$

In the remaining integral (4.138), the second term is dominant and yields

$$D_{00}^{-1} \propto \left(\frac{T}{\Theta_D}\right)^5 \int_0^{\Theta/T} \frac{z^5}{(e^z - 1)(1 - e^{-z})} dz = \left(\frac{T}{\Theta_D}\right)^5 J_5\left(\frac{T}{\Theta_D}\right) . \tag{4.140}$$

This is exactly the result (4.121). The iterative method mentioned there is therefore identical with the zero-order approximation of the variational method. The terms with $r = 1$, etc., then give improvements to this approximation.

In this section we have placed the emphasis on the calculation of electrical conductivity in cases where the relaxation time approximation does not apply. Even the zero-order approximation of the variational procedure gave good agreement between the temperature dependence of σ and experimental observations. The absolute value of conductivity is predicted less well. Here the ansatz function for the interaction potential (Bloch, Nordheim, Bardeen, deformation potential, see Sec. 4.1.2) is of considerable importance. Furthermore numerous approximations have to be verified (neglect of Umklapp processes, spherical energy surfaces, etc.). Finally we have restricted ourselves here to the interaction between electrons and longitudinal acoustic phonons.

4.4.2 Transport Coefficients in the Relaxation Time Approximation

The electric current density and heat current density follow from (4.69) and (4.70) with $\boldsymbol{w}_q = \boldsymbol{w} - \mu\boldsymbol{j}$ (Sec. 4.3.2) and $f = f_0 - (\partial f_0/\partial E)\delta\Phi$ as

$$\begin{aligned} \boldsymbol{i} &= -\frac{e}{\hbar}\int \nabla_{\boldsymbol{k}} E\left(-\frac{\partial f_0}{\partial E}\right)\delta\Phi g(\boldsymbol{k})d\tau_{\boldsymbol{k}}\,, \\ \boldsymbol{w}_q &= \frac{1}{\hbar}\int \nabla_{\boldsymbol{k}} E\left(-\frac{\partial f_0}{\partial E}\right)(E-\mu)\delta\Phi g(\boldsymbol{k})d\tau_{\boldsymbol{k}}\,. \end{aligned} \tag{4.141}$$

We have already calculated the perturbation $\delta\Phi$ of the distribution function in the context of the relaxation time approximation. The following approximations are fundamental to the expression given in (4.62): elastic scattering, isotropic scattering probability, free electron gas with effective mass m^*. We shall retain these approximations in this section.

From (4.62) we find, after a few intermediate steps, the coefficients of the equations (4.103) that give $\boldsymbol{i}$ and $\boldsymbol{w}_q$ as functions of the fields $\boldsymbol{E}$ and $\boldsymbol{B}$, and of the temperature gradient. We set down the equations (4.103) again but with different nomenclature

$$\begin{aligned} \boldsymbol{i} &= M_{00}\boldsymbol{E} + M_{10}\boldsymbol{B}\times\boldsymbol{E} + M_{20}\boldsymbol{B}(\boldsymbol{B}\cdot\boldsymbol{E}) \\ &\quad + M_{01}\frac{\nabla T}{T} + M_{11}\boldsymbol{B}\times\frac{\nabla T}{T} + M_{21}\boldsymbol{B}\left(\boldsymbol{B}\cdot\frac{\nabla T}{T}\right), \\ -\boldsymbol{w}_q &= M_{01}\boldsymbol{E} + M_{11}\boldsymbol{B}\times\boldsymbol{E} + M_{21}\boldsymbol{B}(\boldsymbol{B}\cdot\boldsymbol{E}) \\ &\quad + M_{02}\frac{\nabla T}{T} + M_{12}\boldsymbol{B}\times\frac{\nabla T}{T} + M_{22}\boldsymbol{B}\left(\boldsymbol{B}\cdot\frac{\nabla T}{T}\right). \end{aligned} \tag{4.142}$$

We then find for the M_{ik}

$$M_{ik} = -\frac{e}{3\pi^2}\left(\frac{2m^*}{\hbar^2}\right)^{3/2}\int_0^\infty \frac{E^{3/2}}{1+s^2}\frac{\partial f_0}{\partial E}\left(\frac{e\tau}{m^*}\right)^{i+1}\left(\frac{E-\mu}{e}\right)^k dE \tag{4.143}$$

with the abbreviation $\boldsymbol{s} = e\tau\boldsymbol{B}/m^*$. These equations provide all the transport coefficients introduced in Sections 4.3.2 and 4.3.3.

The integral (4.143) can readily be determined in two limiting cases.

1) Strongly Degenerate Electron Gas (*Metals*)
As a first approximation the negative derivative of the Fermi distribution can be replaced by a δ-function $\delta(E-\mu)$. The value of the integral is then the value of the integrand at $E = \mu$. All coefficients with $k = 1$ clearly vanish. This is the complete heat current and all contributions to the electric current due to ∇T. Hence for the thermoelectric and thermomagnetic coefficients one must use the approximation of next higher order

$$\int_0^\infty F(E)\left(-\frac{\partial f_0}{\partial E}\right)dE = F(\mu) + \frac{\pi^2}{6}(k_B T)^2 F''(\mu) + \cdots. \tag{4.144}$$

In the following we consider two transport phenomena which show some of the characteristic features of the relaxation time approximation. We must refer the reader to the literature for a discussion of all other transport phenomena.

First we explore the validity of the Wiedemann-Franz law. Equations (4.141) become

$$\boldsymbol{i} = M_{00}\boldsymbol{E} + M_{01}\frac{\nabla T}{T}, \qquad -\boldsymbol{w}_q = M_{01}\boldsymbol{E} + M_{02}\frac{\nabla T}{T}. \tag{4.145}$$

From (4.90) $\kappa = M_{02}/T$, $\sigma = M_{00}$. The *Wiedemann-Franz law* thus takes the form $\kappa/\sigma T = M_{02}/M_{00}T^2$. The δ-function approximation is good enough for M_{00} and we obtain

$$M_{00} = \frac{e}{3\pi^2}\left(\frac{2m^*}{\hbar^2}\right)^{3/2}\mu^{3/2}\frac{e\tau(\mu)}{m^*}. \tag{4.146}$$

On the other hand, for M_{02} we have to use both terms of the approximation (4.144). We then find

$$\begin{aligned} M_{02} &= \frac{\pi^2}{6}(k_B T)^2\frac{e}{3\pi^2}\left(\frac{2m^*}{\hbar^2}\right)^{3/2}\frac{d^2}{dE^2}\left(E^{3/2}\frac{e\tau(E)}{m^*}\left(\frac{E-\mu}{e}\right)^2\right)_{E=\mu} \\ &= \frac{\pi^2}{3}\frac{(k_B T)^2}{e^2}M_{00}. \end{aligned} \tag{4.147}$$

Division of the two expressions yields for the right side of the Wiedemann-Franz law: $L = (\pi^2/3)(k_B/e)^2$. L is often called *Lorenz number*. This law is therefore always valid wherever the relaxation time approximation applies. When this approximation does not apply, for example in metals at low temperatures, deviations from the law appear. Just as we introduced a formal relaxation time for the electrical conductivity in this temperature range, we can introduce a formal relaxation time for the thermal conductivity. The result above must then be multiplied by the temperature-independent quotient of the two relaxation times. The part played by other scattering mechanisms introduces a further temperature dependence to the Lorenz number.

Galvanomagnetic effects represent another important area for the relaxation time approximation. Since we are then only interested in the electric current density and ∇T is zero, we can use the δ-function approximation for metals. This leads, with the introduction of the *electron mobility* $\mu_n = e\tau(\mu)/m^*$, to

$$M_{i0} = \frac{en\mu_n^{i+1}}{1 + (\mu_n B)^2}. \tag{4.148}$$

Hence

$$\begin{aligned} \boldsymbol{i} &= \frac{en\mu_n}{1 + (\mu_n B)^2}[\boldsymbol{E} + \mu_n \boldsymbol{B} \times \boldsymbol{E} + \mu_n^2 \boldsymbol{B}(\boldsymbol{B} \cdot \boldsymbol{E})] \\ &= en\mu_n + en\mu_n^2 \boldsymbol{B} \times \boldsymbol{E} + en\mu_n^3 \boldsymbol{B} \times (\boldsymbol{B} \times \boldsymbol{E}) + \cdots . \end{aligned} \tag{4.149}$$

This simple result is noteworthy since for weak magnetic fields it is simultaneously a steady-state solution of the equation

$$m^*\left(\dot{\boldsymbol{v}} + \frac{1}{\tau}\boldsymbol{v}\right) = -e(\boldsymbol{E} + \boldsymbol{v} \times \boldsymbol{B}), \qquad \boldsymbol{i} = -en\boldsymbol{v}. \tag{4.150}$$

Equation (4.150) describes the motion of an electron in a viscous medium with friction coefficient $1/\tau(\mu)$ under the action of the Lorentz force of a coupled electric and magnetic field. In this limiting case we can therefore construct a simple classical model for the transport properties of metals, which is closely connected with the Drude-Lorentz-Sommerfeld theory mentioned earlier.

Equation (4.148) is also the result which follows from another approximation. If in (4.143) we assume only that the relaxation time is *independent of energy*, we find that for any form of the Fermi distribution f_0

$$M_{i0} = \frac{e}{3\pi^2}\left(\frac{2m^*}{\hbar^2}\right)^{3/2} \frac{1}{1 + (\mu_n B)^2}\mu_n^{i+1} \int_0^\infty E^{3/2}\left(-\frac{\partial f_0}{\partial E}\right) dE \tag{4.151}$$

and since $n = \int_0^\infty f_0 g(E)\, dE$, we again find precisely (4.148) by partial integration.

The only one of the galvanomagnetic effects we want to consider, briefly, is the *change of resistivity in a magnetic field* (magnetoresistance). We link up here with (4.104). In the approximation (4.148), the coefficients α_{11} and β_{11} which appear are given by

$$\alpha_{11} = \frac{en\mu_n}{1 + (\mu_n B)^2}\,, \qquad \beta_{11} = \mu_n \alpha_{11}\,. \tag{4.152}$$

With the condition $i_y = 0$, the electric current i_x in a magnetic field B_z and electric field E_x is therefore

$$i_x = \alpha_{11} E_x + \frac{\beta_{11}^2}{\alpha_{11}} B_z^2 E_x = \left(\alpha_{11} + \frac{\beta_{11}^2}{\alpha_{11}} B_z^2\right) E_x\,. \tag{4.153}$$

Comparison with (4.152) reveals that $\sigma = en\mu_n$ remains. The conductivity is not therefore affected by the magnetic field. The deflection of the electrons by the Lorentz force is exactly compensated for by the opposing force due to the Hall effect. The assumption involved in this approximation is important here, namely that all electrons which take part in conduction have the energy $E = \mu$, and therefore behave identically under the (velocity-dependent) Lorentz force. A change of resistivity observed in metals is a consequence of *anisotropy*. Our approximation is then no longer valid.

2) Nondegenerate Electron Gas (*Semiconductors*)

If the electron concentration n is small compared to the degeneracy concentration n_0 (2.33), then in (4.143) the Fermi distribution can be replaced with the classical Boltzmann distribution [approximation (2.35)]. If further we describe the relaxation time τ by a power law $\tau(E) = \tau_0 E^r$, and expand (4.143) in increasing powers of the magnetic field, we can reduce all integrals to the form

$$\int_0^\infty E^\alpha \exp\left(-E/k_B T\right) dE = (k_B T)^{\alpha+1}\Gamma(\alpha + 1)\,. \tag{4.154}$$

Since in this approximation the M_{ik} are nonzero, the specific heat conductivity now becomes $\kappa = (1/T)[M_{02} - (M_{10}^2/M_{00})]$. Using (4.154), $\kappa/\sigma T = (r + 5/2)(k_B/e)^2$ follows. Thus the Wiedemann-Franz law is again fulfilled, albeit with a different Lorentz number than for the approximation (4.144). In spite of this the relation has only limited validity in semiconductors. κ only gives the electronic part of the thermal conductivity. In view of the low electron concentration in semiconductors, the heat conductivity κ_l of the lattice must also be considered. Experimentally one measures $(\kappa + \kappa_l)/\sigma T$. The Wiedemann-Franz law is not valid for this quotient. Furthermore in semiconductors, electrons and holes often contribute simultaneously to current flow. This too leads to a more complicated law.

In contrast to (4.149) one finds for isothermal current flow in a magnetic field

$$\boldsymbol{i} = en\mu_n \boldsymbol{E} + A_1 en\mu_n^2 \boldsymbol{E} \times \boldsymbol{B} + A_2 en\mu^3 \boldsymbol{B} \times (\boldsymbol{B} \times \boldsymbol{E}) + \cdots \tag{4.155}$$

with

$$A_i = \left[\Gamma\left(\frac{5}{2}\right)\right]^i \Gamma\left[\frac{5}{2} + (i+1)r\right] \Big/ \left[\Gamma\left(\frac{5}{2} + r\right)\right]^{i+1} . \tag{4.156}$$

This is in agreement with (4.149) only when $r = 0$ (energy-independent relaxation time). The reason for the difference between (4.149) and (4.155) lies in the fact that in the first case, since E was assumed equal to μ, all the electrons possess the same energy and hence the same velocity. In the second case, on the other hand, the electrons have a spectrum of velocities described by the Boltzmann distribution. The A_i are therefore statistical coefficients which stem from an average over different velocity-dependent coefficients.

An important consequence of this difference is that even an isotropic semiconductor exhibits a change of its resistivity in a magnetic field.

Equation (4.155)—extended by a corresponding contribution for holes—forms the basis of the theory of galvanomagnetic effects in semiconductors. We have to refer the reader here to the literature (e.g., *Beer* [102.4]). One reason for the importance of these effects lies in the fact that in (4.155) the magnetic field always appears as $\mu_n B$, and since in many semiconductors the electron mobility is high, these effects are large.

4.4.3 Limits of Validity and Possible Extensions of the Approximations Used

We have made use of numerous approximations in the previous sections. It is therefore appropriate in this final section to explore the limits of validity of a few of these approximations.

The assumption of a *homogeneous solid* need only be rejected in a few cases. We shall discuss inhomogeneous solids (distorted and disordered lattices) in later chapters. It therefore only remains to examine a continuous change of properties, for example the concentration of defects in a semiconductor or the lattice of a mixed crystal. The density of states becomes space dependent in either case. Consequently effects like the Peltier effect or the Seebeck effect, which usually only appear at contacts, can now occur within the crystal.

The assumption of isotropy, on the other hand, demands a closer examination. It includes two approximations: *isotropy of the scattering probability* and *isotropy of the band structure*.

In the relaxation time approximation, isotropic scattering probability leads to a relaxation time which is dependent only on energy, and not on the direction of $\boldsymbol{k}$. Attempts have been made to introduce $\boldsymbol{k}$-dependent relaxation times. However, the validity of such attempts is dubious and their success has been very limited. To this we must add that anisotropic scattering will certainly always be accompanied by an anisotropic band structure, and the latter demands much more significant changes to the results of the isotropic theory.

In *semiconductors*, anisotropic band structure means a direction-dependent effective mass, and possibly equivalent band extrema at different points in the Brillouin zone (for all **k**-vectors of a star, see Fig. 2.34). In *metals*, anisotropy means deviations of the Fermi surface from the spherical form we met, for example, in Fig. 2.27. One of the most important consequences of this anisotropy emerges in the galvanomagnetic effects in metals at high magnetic fields. It is obvious that in weak magnetic fields the electrons will only traverse small regions of the Fermi surface between successive "collisions", while in strong magnetic fields they will execute closed orbits on the Fermi surface. Circulation times are of the order of the reciprocal cyclotron resonance frequency. Hence the boundary between weak and strong magnetic fields is at $\omega_c\tau = 1$ or, since $\omega_c = eB/m^*$ and $\mu_n \approx e\tau/m^*$, at $\mu_n B = 1$.

The fact that the electrons execute closed orbits on the Fermi surface leads to a *saturation* of the magnetoresistance. If there are directions in which the electron can execute *open orbits* in the repeated zone scheme (Fig. 2.30), no saturation is observed when the magnetic field is in this direction. The result is a significant dependence of the magnetoresistance in high magnetic fields on the orientation of the fields relative to the crystal axes. Fig. 4.6 presents an example. In many metals, along with closed and open orbits we also find the coexistence of electron orbits and hole orbits (see again Fig. 2.30). Electrons and holes both contribute to charge transport and we must from the outset consider both contrbiutions together by different transport equations. A saturation of the magnetoresistance cannot appear in this case either. We refer to the literature, particularly to *Smith* et al. [64], to the contribution by *Mackintosh* in [113a], and to a chapter in the book by *Kittel* [16].

The appearance together of different charge carriers, such as we have just referred to, influences all the transport phenomena. The way in which our system of equations has to be extended is evident from the treatment in Section 4.3.1.

The various *scattering mechanisms* which are possible form another complex of possible extensions to the transport theory. We have often emphasized that the coupling between the electrons and the LA-phonons only constitutes one possible interaction. TA-phonons are not involved in normal processes. On the other hand, they are involved in Umklapp processes. There is a further coupling between electrons and LA-phonons in piezoelectric crystals, where acoustic waves are accompanied by polarization. The coupling is then greatly enhanced. We mentioned a similar case earlier when discussing optical phonons. In polar solids (differently charged basis atoms in the Wigner-Seitz cell) the LO-phonons produce a strong polarization. We considered this interaction in Section 4.1.3. If the atoms in the Wigner-Seitz cell are identical (for example in the elements C, Si, Ge which crystallize in the diamond structure), the optical vibrations may be nonpolar. The coupling to the electrons is then weaker.

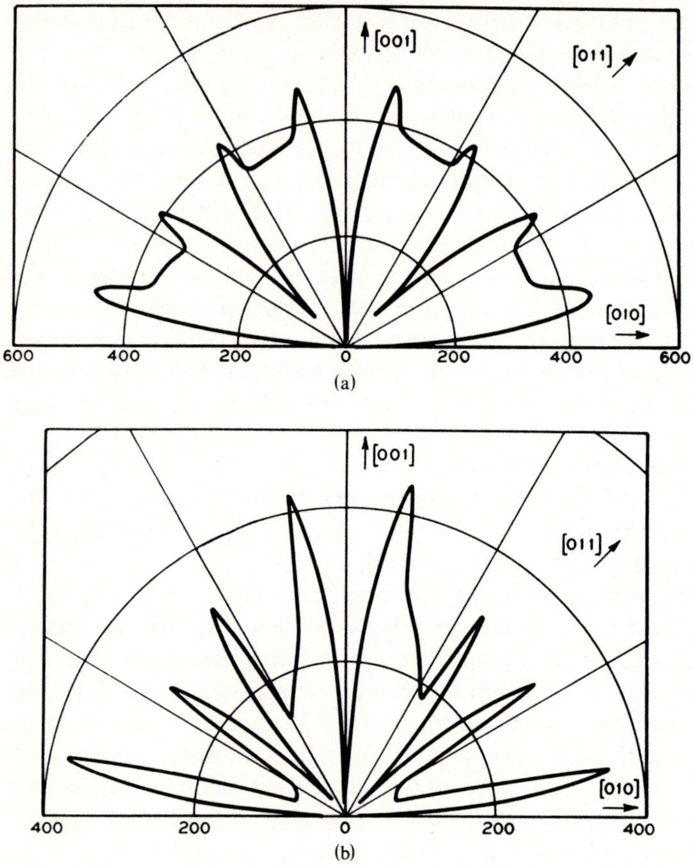

Fig. 4.6a and b. Transverse magnetoresistance $\Delta R/R_0$ in copper at 18 000 gauss and 4.2 K, as a function of the orientation of the current direction relative to the crystal axes. (a) Magnetic field in [100] direction, (b) magnetic field inclined to the [100] direction by less that 3° [from J. R. Klauder, J. E. Kunzler: in *The Fermi Surface*, ed. by W. A. Harrison, M. B. Webb (John Wiley & Sons, New York 1960)].

Other mechanisms to take account of in scattering processes are the transition of an electron between different equivalent minima in the conduction band of semiconductors (intervalley scattering, as distinct from intravalley scattering), the elastic scattering of charge carriers at charged impurities, and the scattering at surfaces and inner boundaries. We cannot go into all these scattering mechanisms here and therefore refer the reader to the literature. Within the relaxation time approximation, one finds the difference between the various possibilities to be a different energy dependence of the relaxation time $\tau(E) = \tau_0 E^r$. The contributions from the various scattering mechanisms add to the collision term in the Boltzmann equation. Each term added is proportional to a reciprocal relaxation time. The relaxation times of the individual scattering mechanisms thus add together reciprocally to give an effective relaxation time. Since the resistivity is proportional to the reciprocal relaxation time in a first approximation, the resistivities to be expected from the separate action of the individual scattering mechanisms also add together. This rule (*Mathiessen's rule*) is only valid when the relaxation time approximation applies. Otherwise it can be broken (see, for example *Ziman* [35]). Fig. 4.7 shows the resistivity which remains in sodium at low temperatures due to scattering at lattice defects (residual resistivity).

We close our discussion of scattering mechanisms with a few remarks on the *polar optical interaction*. The individual process is clearly nonelastic. In view of the advantages of a relaxation time approximation, there have been many attempts to introduce quasi-relaxation times here too. Fig. 4.8 shows the quasi-relaxation times for the electrical conductivity, thermoelectric power, and Hall coefficient in semiconductors (Boltzmann statistics) which have been deduced by the variational method. One can see that at low and at high temperature the quasi-relaxation times agree, so that the use of a unified relaxation time with exponent 1/2 and 0 respectively in the power law $\tau \propto E^r$ appears to be a justifiable approximation.

In polar crystals, following Section 4.1.3, we can take the electron together with its polarization cloud as a *polaron*. For weak coupling the polaron is a quasi-particle which differs from a crystal electron only in its effective mass. For

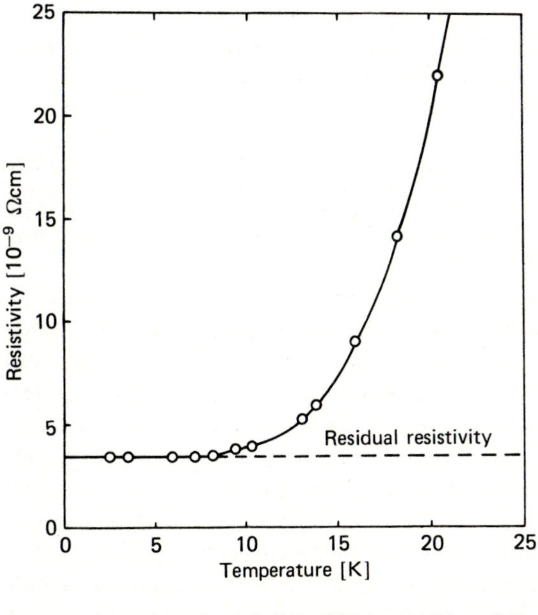

Fig. 4.7. Superposition of the residual resistivity of Na due to scattering at lattice defects and the resistivity due to electron-phonon interaction (Mathiessen's rule) (from *Blakemore* [4]).

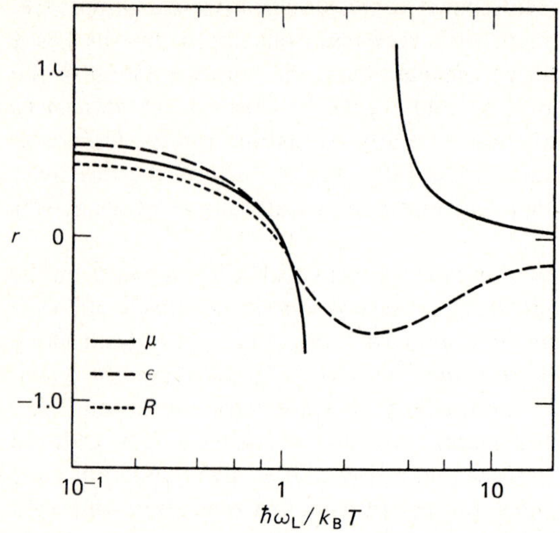

Fig. 4.8. Exponent r in the power law $\tau \propto E^r$ of effective relaxation times for electron-LO-phonon interaction. The effective relaxation times calculated for mobility μ_n, thermopower ε, and Hall coefficient R by the variational method approximately agree at high and at low temperatures. The assumption of a unified relaxation time thus appears to be justified there [from H. Ehrenreich: J. Appl. Phys. **32,** 2155 (1961)].

stronger coupling, other difficulties appear. In such crystals the mobility of electrons (polarons) is mostly very low. If one calculates from the mobility the *mean free path* between two interaction processes, one finds values of the order of the atomic separation in the lattice. It is then clearly no longer sensible to describe the quasi-particle "crystal electron" by the Boltzmann equation. We should rather examine the microscopic motion of the electrons in the presence of both the lattice potential and the external field. This motion consists of individual steps in which the electron hops from one localized state to the next (*hopping processes*). For this we need other theoretical aids which we shall not be concerned with until Section 8.3.6.

The last approximations we want to mention here are the assumptions we have made about the *phonon system*. There is no fundamental difficulty in using the correct phonon dispersion spectrum instead of the Debye approximation. More critical is the assumption that the phonon system should be in equilibrium. Each interaction process between electrons and phonons disturbs *both* systems. The approximation then only implies that we can neglect the feedback to the electron system due to the perturbation of the phonon system. This is certainly justified in most cases. It may be, however, that the phonon system is already perturbed by external forces. In a temperature gradient, for example, lattice heat conduction means a flow of phonons from the hotter to the colder end of the sample. Through scattering processes the electrons will then be made aware of the preferred direction of the phonons; they will be carried along (*phonon-drag*).

Correspondingly in the isothermal case the electrons flowing in an electric field will carry phonons along with them (*electron-drag*). The first effect makes a contribution to the thermopower, the second effect a contribution to the Peltier effect.

5. Electron-Electron Interaction by Exchange of Virtual Phonons: Superconductivity

5.1 Introduction

Figure 4.1 contains the graph of an interaction which we have not so far taken a close look at: an electron emits a virtual phonon, and this phonon is absorbed by another electron. This results in an *effective, additional electron-electron interaction.* The physical background is simple: the emission of virtual phonons by electrons means a deformation (polarization) of the lattice in the vicinity of the electron. If a second electron is near this polarization cloud, it experiences a force of attraction or repulsion which has nothing to do with the Coulomb interaction between the two electrons. It is with this interaction that we now want to concern ourselves.

For the electron-phonon coupling we take (4.9)

$$H_{\text{el-ph}} = \sum_{\boldsymbol{k}\boldsymbol{q}\sigma} M_{\boldsymbol{q}}(a^{+}_{-\boldsymbol{q}} + a_{\boldsymbol{q}})c^{+}_{\boldsymbol{k}+\boldsymbol{q},\sigma}c_{\boldsymbol{k},\sigma}\,. \tag{5.1}$$

We therefore restrict ourselves to LA-phonons. From (4.8) the matrix element for free electrons depends only on $\boldsymbol{q}$; Umklapp processes do not appear. The direction of electron spin is not changed by the interaction.

The graph in Fig. 4.1 describes a process which proceeds via a virtual intermediate state. Two intermediate states are possible: either electron $\boldsymbol{k}'$ emits a phonon $\boldsymbol{q}$ which is absorbed by electron $\boldsymbol{k}$, or electron $\boldsymbol{k}$ emits a phonon $-\boldsymbol{q}$ which is absorbed by electron $\boldsymbol{k}'$. If one represents the initial state by $|i\rangle$, the virtual state by $|m\rangle$ and the final state by $|f\rangle$, the interaction is described by the expression

$$\frac{1}{2}\sum_{\substack{\boldsymbol{k}\boldsymbol{k}'\boldsymbol{q}\\ \sigma\sigma'}}\Bigg[\frac{\langle f|c^{+}_{\boldsymbol{k}+\boldsymbol{q},\sigma}c_{\boldsymbol{k},\sigma}a_{\boldsymbol{q}}|m_1\rangle\langle m_1|c^{+}_{\boldsymbol{k}'-\boldsymbol{q},\sigma'}c_{\boldsymbol{k}',\sigma'}a^{+}_{\boldsymbol{q}}|i\rangle}{E(\boldsymbol{k}') - E(\boldsymbol{k}'-\boldsymbol{q}) - \hbar\omega_{\boldsymbol{q}}} + \frac{\langle f|c^{+}_{\boldsymbol{k}'-\boldsymbol{q},\sigma'}c_{\boldsymbol{k}',\sigma'}a_{-\boldsymbol{q}}|m_2\rangle\langle m_2|c^{+}_{\boldsymbol{k}+\boldsymbol{q},\sigma}c_{\boldsymbol{k},\boldsymbol{q}}a^{+}_{-\boldsymbol{q}}|i\rangle}{E(\boldsymbol{k}) - E(\boldsymbol{k}+\boldsymbol{q}) - \hbar\omega_{\boldsymbol{q}}}\Bigg]. \tag{5.2}$$

From energy conservation one can replace $E(\boldsymbol{k}') - E(\boldsymbol{k}'-\boldsymbol{q})$ by $E(\boldsymbol{k}+\boldsymbol{q}) - E(\boldsymbol{k})$ in the first denominator. The factor 1/2 is needed since, because of the summation over $\boldsymbol{k}$, $\boldsymbol{k}'$ and $\boldsymbol{q}$, one term in the brackets already covers both intermediate states.

We do not want to use (5.2) as a starting point for our examination of the effective electron-electron interaction. It suits us better to put the Hamiltonian into a form in which the interaction appears explicitly. This can be achieved through the canonical transformation described in Section 3.1.5. The Hamiltonian (3.34)—extended by spin indices—just describes our problem. The canonical transformation (3.36) leads to an equivalent Hamiltonian which among others contains a term of the type (3.39). This is the interaction we seek. From (3.35) and (3.36) we directly find for this term the explicit form

$$H_{\text{eff}} = \frac{1}{2} \sum_{\substack{\boldsymbol{k}\boldsymbol{k}'\boldsymbol{q} \\ \sigma\sigma'}} |M_{\boldsymbol{q}}|^2 (\alpha - \beta) c^+_{\boldsymbol{k}+\boldsymbol{q},\sigma} c^+_{\boldsymbol{k}'-\boldsymbol{q},\sigma'} c_{\boldsymbol{k}',\sigma'} c_{\boldsymbol{k},\sigma} . \tag{5.3}$$

The matrix element for the transition corresponding to (5.2) is then, making use of (3.37)

$$\langle f | H_{\text{eff}} | i \rangle = \frac{1}{2} \sum_{\substack{\boldsymbol{k}\boldsymbol{k}'\boldsymbol{q} \\ \sigma\sigma'}} \langle f | V_{\boldsymbol{k}\boldsymbol{q}} c^+_{\boldsymbol{k}+\boldsymbol{q},\sigma} c^+_{\boldsymbol{k}'-\boldsymbol{q},\sigma'} c_{\boldsymbol{k}',\sigma'} c_{\boldsymbol{k},\sigma} | i \rangle \tag{5.4}$$

with

$$\begin{aligned} V_{\boldsymbol{k}\boldsymbol{q}} &= |M_{\boldsymbol{q}}|^2 \left[\frac{1}{E(\boldsymbol{k}) - E(\boldsymbol{k}+\boldsymbol{q}) - \hbar\omega_{\boldsymbol{q}}} - \frac{1}{E(\boldsymbol{k}) - E(\boldsymbol{k}+\boldsymbol{q}) + \hbar\omega_{\boldsymbol{q}}} \right] \\ &= \frac{2|M_{\boldsymbol{q}}|^2 \hbar\omega_{\boldsymbol{q}}}{[E(\boldsymbol{k}+\boldsymbol{q}) - E(\boldsymbol{k})]^2 - (\hbar\omega_{\boldsymbol{q}})^2} . \end{aligned} \tag{5.5}$$

One can readily see the equivalence of the two expressions (5.2) and (5.5). $V_{\boldsymbol{k}\boldsymbol{q}}$ has the significance of a Fourier coefficient of the effective interaction. If $V_{\boldsymbol{k}\boldsymbol{q}}$ is positive the interaction is repulsive, if negative it is attractive. The latter is the case when $|E(\boldsymbol{k}+\boldsymbol{q}) - E(\boldsymbol{k})| < \hbar\omega_{\boldsymbol{q}}$.

To examine the effective interaction we omit all the terms in the transformed Hamiltonian which contain phonon creation or annihilation operators. We are left with the two terms

$$\begin{aligned} H_s = \sum_{\boldsymbol{k}\sigma} E(\boldsymbol{k}) c^+_{\boldsymbol{k},\sigma} c_{\boldsymbol{k},\sigma} + \sum_{\substack{\boldsymbol{k}\boldsymbol{k}'\boldsymbol{q} \\ \sigma\sigma'}} |M_{\boldsymbol{q}}|^2 \frac{\hbar\omega_{\boldsymbol{q}}}{[E(\boldsymbol{k}+\boldsymbol{q}) - E(\boldsymbol{k})]^2 - (\hbar\omega_{\boldsymbol{q}})^2} \\ \times c^+_{\boldsymbol{k}+\boldsymbol{q},\sigma} c^+_{\boldsymbol{k}'-\boldsymbol{q},\sigma'} c_{\boldsymbol{k}',\sigma'} c_{\boldsymbol{k},\sigma} . \end{aligned} \tag{5.6}$$

As long as we can restrict ourselves to these two terms the Hamiltonian operates only on electron variables. We can therefore completely ignore the boson system.

In the sections which follow we shall study the interaction more closely. It will emerge that, with particular assumptions, an attractive effective interaction leads to a *correlation* of the electrons, which results in a reduction in energy of the ground state. The correlation takes place predominantly in pairs between electrons of opposite spin and opposite wave vector (*Cooper pairs*). In Section 5.2 we shall look at individual Cooper pairs, and in Section 5.3 we shall treat the ground state of an electron gas with attractive interaction. In the treatment of excited states in Section 5.4 we shall find indications of the justification for using this interacting electron gas to help explain the phenomenon of *superconductivity*. A simple explanation for the appearance of undamped current flow in superconductors follows from the existence of an energy gap between ground and first excited states.

Then in Section 5.5 we present a review of the most important properties of superconductors. We shall find that a number of these properties can be interpreted with the aid of the theory presented in the previous sections. As an extension of this discussion, we present in Section 5.6 an explanation for the Meissner-Ochsenfeld effect, i.e., the expulsion of a magnetic field out of a superconductor. Finally in Section 5.7 we briefly go into further experimental facts and their interpretation. In so doing we discuss the limits of the theory presented here, and point to other theoretical treatments.

The topic of this chapter is not primarily the theory of superconductivity in all its aspects but the virtual exchange of phonons by electrons and its consequences on physical phenomea. To a large extent the subject of this chapter is contained in the original work of *Bardeen* et al.[1] The theory is called BSC theory after its originators. Here however we shall often take a different path to the derivation of the results, one which is more suited to the concept of elementary excitations. The method used is based on work by *Bogoliubov*.

Superconductivity is a self-contained subject on which there are numerous monographs. We particularly recommend the books mentioned in the reference list [71–77].

5.2 Cooper Pairs

To grasp the significant aspects of the new interaction, we consider an idealized case: a noninteracting electron gas fills the Fermi sphere in $\boldsymbol{k}$-space. All the states below k_F, E_F are occupied, and all above it empty. Into this system we introduce two electrons $[\boldsymbol{k}_1, E(\boldsymbol{k}_1)]$ and $[\boldsymbol{k}_2, E(\boldsymbol{k}_2)]$. We take the positive part of $V_{\boldsymbol{kq}}$ (5.5) as the interaction between these two electrons. Interaction processes involving phonon exchange will therefore only occur for $|E(\boldsymbol{k}+\boldsymbol{q}) - E(\boldsymbol{k})| \leqslant \hbar\omega_{\boldsymbol{q}}$.

We construct the wave function for the electron pair by application of two creation operators to the ground state $|G\rangle$ (filled Fermi sphere), summing over all possible $\boldsymbol{k}_1$ and $\boldsymbol{k}_2$ ($k_i > k_F$) and over the electron spins

$$\psi_{12} = \sum_{\boldsymbol{k}_1\boldsymbol{k}_2\sigma_1\sigma_2} a_{\sigma_1\sigma_2}(\boldsymbol{k}_1, \boldsymbol{k}_2) c^+_{\boldsymbol{k}_1\sigma_1} c^+_{\boldsymbol{k}_2\sigma_2} |G\rangle . \tag{5.7}$$

[1] J. Bardeen, L. N. Cooper, J. R. Schrieffer: Phys. Rev. **108**, 1175 (1957).

To form a state with defined total momentum, we carry out the summation in (5.7) subject to the condition $\boldsymbol{K} = \boldsymbol{k}_1 + \boldsymbol{k}_2 = \text{const.}$

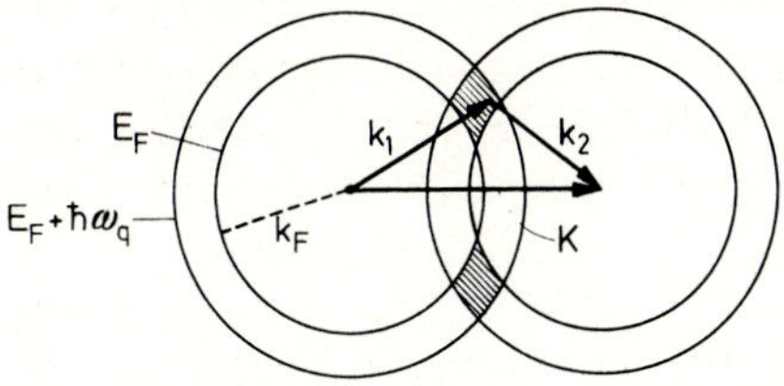

Fig. 5.1. Figure to assist the determination of the $\boldsymbol{k}$-vectors of two interacting electrons, when both should lie in states within a shell of thickness $\hbar\omega_q$ above E_F, and their total wave vector $\boldsymbol{K} = \boldsymbol{k}_1 + \boldsymbol{k}_2$ is given.

The energy of the electron pair is made up of the individual energies of the electrons, and their interaction energy ΔE. Our intention is to calculate this interaction energy. The greater the number of sum terms contributing to (5.7), the larger it will be. ΔE will be largest if we choose $\boldsymbol{K} = 0$. One can see this by considering two electrons just above E_F. An interaction should occur only when the electrons are in states outside the Fermi sphere, with energies $E(\boldsymbol{k}_i) \leqslant E_F + \hbar\omega_{\boldsymbol{q}}$. Since further $\boldsymbol{k}_1 + \boldsymbol{k}_2 = \boldsymbol{K}$, the regions of $\boldsymbol{k}$-space involved in the sum in (5.7) are given by the hatched areas in Fig. 5.1. They are at their largest when $\boldsymbol{K} = 0$. In what follows we restrict ourselves to this case, and we also assume that both electron spins are antiparallel. Eq. (5.7) then becomes

$$\psi_{12} = \sum_{\boldsymbol{k}} a(\boldsymbol{k}) c^+_{\boldsymbol{k},\sigma} c^+_{-\boldsymbol{k},-\sigma} |G\rangle\,. \tag{5.8}$$

With the wave function (5.8) we now wish to calculate the energy of the electron pair. We must first introduce a further approximation to make the calculation possible. We take the $V_{\boldsymbol{kq}}$ to be constant in the range of the attractive interaction ($V_{\boldsymbol{kq}} = -V$), and zero elsewhere. Hence we take the Hamiltonian

$$\begin{aligned} H &= \sum_{\boldsymbol{k}\sigma} E(\boldsymbol{k}) c^+_{\boldsymbol{k},\sigma} c_{\boldsymbol{k},\sigma} - \frac{V}{2} \sum_{\boldsymbol{kq}\sigma} c^+_{\boldsymbol{k}+\boldsymbol{q},\sigma} c^+_{-\boldsymbol{k}-\boldsymbol{q},-\sigma} c_{-\boldsymbol{k},-\sigma} c_{\boldsymbol{k}\sigma}\,. \\ V &\neq 0 \quad \text{only for} \quad |E(\boldsymbol{k}+\boldsymbol{q}) - E(\boldsymbol{k})| \leq \hbar\omega_{\boldsymbol{q}}\,. \end{aligned} \tag{5.9}$$

For $\omega_{\boldsymbol{q}}$ we choose a characteristic frequency of the phonon spectrum, e.g., the Debye frequency ω_D (Sec. 3.3.4), as the maximum value of $\omega_{\boldsymbol{q}}$ in the Debye approximation.

We then find for the energy

$$E = \langle\psi|H|\psi\rangle = 2\sum_{\boldsymbol{k}} E(\boldsymbol{k})|a(\boldsymbol{k})|^2 - V\sum_{\boldsymbol{kq}} a^*(\boldsymbol{k}+\boldsymbol{q})a(\boldsymbol{k})\,. \tag{5.10}$$

We determine the $a(\boldsymbol{k})$ by varying E subject to the condition $\sum_{\boldsymbol{k}} |a(\boldsymbol{k})|^2 = 1$

$$\frac{\partial}{\partial a^*_{\boldsymbol{k}'}}\left[E - \lambda \sum_{\boldsymbol{k}''} |a(\boldsymbol{k}'')|^2\right] = 2E(\boldsymbol{k}')a(\boldsymbol{k}') - V\sum_{\boldsymbol{q}} a(\boldsymbol{k}' - \boldsymbol{q}) - \lambda a(\boldsymbol{k}') = 0 \tag{5.11}$$

or

$$[2E(\boldsymbol{k}) - \lambda]a(\boldsymbol{k}) = V\sum_{\boldsymbol{k}'} a(\boldsymbol{k}'). \tag{5.12}$$

We satisfy the restriction on the interaction by taking V only nonzero for energies in the range E_F to $E_F + \hbar\omega_{\boldsymbol{q}}$. The same is also true for the $a(\boldsymbol{k})$, and the sum in (5.12) runs over a finite number of $\boldsymbol{k}'$. Calling this sum C we find

$$a(\boldsymbol{k}) = \frac{VC}{2E(\boldsymbol{k}) - \lambda}, \qquad \sum_{\boldsymbol{k}} a(\boldsymbol{k}) = C = \sum_{E(\boldsymbol{k})} \frac{VC}{2E(\boldsymbol{k}) - \lambda}. \tag{5.13}$$

The sum here runs over all states between E_F and $E_F + \hbar\omega_{\boldsymbol{q}}$.

The final step is to return to (5.12). If we take the complex conjugate of this equation, multiply by $a(\boldsymbol{k})$, and sum over $\boldsymbol{k}$, an equation follows which agrees with (5.10) if one puts $\lambda = E$. Thus we have determined the Lagrange parameter and can write the second equation (5.13) in the following form

$$1 = \sum_{E(\boldsymbol{k})} \frac{V}{2E(\boldsymbol{k}) - E} = V\int_{E_F}^{E_F + \hbar\omega_{\boldsymbol{q}}} \frac{g(x)dx}{2x - E}. \tag{5.14}$$

In transforming the sum into an integration, the density of states $g(E)$ has been introduced. In view of the narrow range of integration we put $g(E) \approx g(E_F)$. The integral can be evaluated and one obtains

$$E = 2E_F - \frac{2\hbar\omega_{\boldsymbol{q}} \exp[-2/g(E_F)V]}{1 - \exp[-2/g(E_F)V]} \approx 2E_F - 2\hbar\omega_{\boldsymbol{q}} \exp[-2/g(E_F)V] \tag{5.15}$$

where the expression on the extreme right is valid for small V (weak interaction).

The energy of the electron pair is therefore less than its minimum energy $2E_F$ in the absence of the interaction. One can show that all other solutions of (5.10) lead to energies $>2E_F$. The lowest energy state of the electron pair is thus a *bound* state. One calls bound electron pairs *Cooper pairs*. The electrons in a pair have opposite wave vector and opposite spin. We shall not prove here that the state with opposite spins is preferred. It must however be emphasized that the derivation of (5.15) assumed antiparallel spins. If we had assumed parallel spins in (5.8), the constant C in (5.13) would have been zero because of the antisymmetry of the spatial part of the wave function. The assumption of antiparallel spin was therefore necessary to get rid of the factor C in the transition from (5.13) to (5.14).

The appearance of a bound state for the additional electron pair considered here means that the excitation of two electrons from states directly below the Fermi surface into states above it—in the context of our idealized interaction model—leads to a lower total energy. The filled Fermi sphere is then unstable and one can gain energy by combining electrons into Cooper pairs. This is the basic idea from which we can explain the ground state of the superconducting electron gas, and it is to this that we now turn our attention.

5.3 The Ground State of the Superconducting Electron Gas

From the consideration of a single Cooper pair we go now to the calculation of the energy of the ground state of a superconducting electron gas. We write the Hamiltonian (5.9) once again in a simpler form. In the following a positive spin is to be associated with the index $\boldsymbol{k}$, and a negative spin with $-\boldsymbol{k}$. Eq. (5.9) then becomes

$$H = \sum_{\boldsymbol{k}} E(\boldsymbol{k})(c_{\boldsymbol{k}}^{+} c_{\boldsymbol{k}} + c_{-\boldsymbol{k}}^{+} c_{-\boldsymbol{k}}) - V \sum_{\boldsymbol{k}\boldsymbol{k}'} c_{\boldsymbol{k}'}^{+} c_{-\boldsymbol{k}'}^{+} c_{-\boldsymbol{k}} c_{\boldsymbol{k}} \,. \tag{5.16}$$

We rearrange this equation further by introducing creation and annihilation operators, analogous to (2.17–20). In our current representation the defining equations for these operators (2.17) are

$$\begin{aligned} \alpha_{\boldsymbol{k}} &= u_{\boldsymbol{k}} c_{\boldsymbol{k}} - v_{\boldsymbol{k}} c_{-\boldsymbol{k}}^{+} \qquad & \alpha_{-\boldsymbol{k}} &= u_{\boldsymbol{k}} c_{-\boldsymbol{k}} + v_{\boldsymbol{k}} c_{\boldsymbol{k}}^{+} \\ \alpha_{\boldsymbol{k}}^{+} &= u_{\boldsymbol{k}} c_{\boldsymbol{k}}^{+} - v_{\boldsymbol{k}} c_{-\boldsymbol{k}} \qquad & \alpha_{-\boldsymbol{k}}^{+} &= u_{\boldsymbol{k}} c_{-\boldsymbol{k}}^{+} + v_{\boldsymbol{k}} c_{\boldsymbol{k}} \,. \end{aligned} \tag{5.17}$$

For the moment we neglect the conditions on the $u_{\boldsymbol{k}}$ and $v_{\boldsymbol{k}}$ contained in (2.17). The conditions (2.18) are however taken to apply: $u_{\boldsymbol{k}}^2 + v_{\boldsymbol{k}}^2 = 1$, $u_{\boldsymbol{k}} = u_{-\boldsymbol{k}}$, $v_{\boldsymbol{k}} = -v_{-\boldsymbol{k}}$. Their validity guarantees that the same commutation relations apply for the α-operators as for the c-operators.

We further introduce—as in Section 2.1.3.—the operator $\bar{H} = H - E_{\mathrm{F}} N_{\mathrm{op}}$, and the energy $\varepsilon(\boldsymbol{k}) = E(\boldsymbol{k}) - E_{\mathrm{F}}$. The transition from the c- to the α-operators, known as the *Bogoliubov-Valatin transformation*, then leads to

$$\begin{aligned} \bar{H} = {} & \sum_{\boldsymbol{k}} \varepsilon(\boldsymbol{k})[2v_{\boldsymbol{k}}^2 + (u_{\boldsymbol{k}}^2 - v_{\boldsymbol{k}}^2)(\alpha_{\boldsymbol{k}}^{+}\alpha_{\boldsymbol{k}} + \alpha_{-\boldsymbol{k}}^{+}\alpha_{-\boldsymbol{k}}) + 2u_{\boldsymbol{k}}v_{\boldsymbol{k}}(\alpha_{\boldsymbol{k}}^{+}\alpha_{-\boldsymbol{k}}^{+} + \alpha_{-\boldsymbol{k}}\alpha_{\boldsymbol{k}})] \\ & - V \sum_{\boldsymbol{k}\boldsymbol{k}'} [u_{\boldsymbol{k}}v_{\boldsymbol{k}}u_{\boldsymbol{k}'}v_{\boldsymbol{k}'}(1 - \alpha_{-\boldsymbol{k}'}^{+}\alpha_{-\boldsymbol{k}'} - \alpha_{\boldsymbol{k}'}^{+}\alpha_{\boldsymbol{k}'})(1 - \alpha_{-\boldsymbol{k}}^{+}\alpha_{-\boldsymbol{k}} - \alpha_{\boldsymbol{k}}^{+}\alpha_{\boldsymbol{k}}) \\ & + (u_{\boldsymbol{k}}^2 - v_{\boldsymbol{k}}^2)u_{\boldsymbol{k}'}v_{\boldsymbol{k}'}(1 - \alpha_{-\boldsymbol{k}'}^{+}\alpha_{-\boldsymbol{k}'} - \alpha_{\boldsymbol{k}'}^{+}\alpha_{\boldsymbol{k}'})(\alpha_{-\boldsymbol{k}}\alpha_{\boldsymbol{k}} + \alpha_{\boldsymbol{k}}^{+}\alpha_{-\boldsymbol{k}}^{+}) \\ & + (u_{\boldsymbol{k}}^2\alpha_{-\boldsymbol{k}}\alpha_{\boldsymbol{k}} - v_{\boldsymbol{k}}^2\alpha_{\boldsymbol{k}}^{+}\alpha_{-\boldsymbol{k}}^{+})(u_{\boldsymbol{k}'}^2\alpha_{\boldsymbol{k}'}^{+}\alpha_{-\boldsymbol{k}'}^{+} - v_{\boldsymbol{k}'}^2\alpha_{-\boldsymbol{k}'}\alpha_{\boldsymbol{k}'})] \,. \end{aligned} \tag{5.18}$$

We examine initially the first line in (5.18). It contains the free electron gas contribution to $\bar{H}$. Through the condition that u_k and v_k should only be nonzero outside and inside the Fermi sphere, respectively, we found in Section 2.1.3 that the last term on the first line vanished. Only the first term (energy of the filled Fermi sphere) and the second term (energy of the quasi-particles defined by the particle number operators $\alpha_k^+ \alpha_k$) remain. If we include the other lines in (5.18), we have to choose different u_k and v_k to eliminate all terms containing the products $\alpha_k^+ \alpha_{-k}^+$ or $\alpha_{-k}\alpha_k$.

We can simplify the procedure in the present case. Since we are only interested in this section in the ground state, we can set all particle number operators $\alpha_k^+ \alpha_k$ in (5.18) equal to zero. Finally we can neglect the last line in (5.18). It contains fourfold products of α-operators. As one can readily show, it makes only a small contribution to the final result.

We are left with

$$\bar{H} = 2\sum_k \varepsilon(k) v_k^2 - V \sum_{kk'} u_k v_k u_{k'} v_{k'}$$
$$+ \sum_k \left[2u_k v_k \varepsilon(k) - (u_k^2 - v_k^2) V \sum_{k'} u_{k'} v_{k'} \right] (\alpha_k^+ \alpha_{-k}^+ + \alpha_{-k}\alpha_k) . \tag{5.19}$$

We determine the u_k and v_k from the requirement that the bracket in (5.19) vanish. As in (5.13), we put $\sum_{k'} u_{k'} v_{k'}$ equal to a constant, which we call Δ/V. For the bracket in (5.19) to be zero, we must have

$$2u_k v_k \varepsilon(k) = \Delta(u_k^2 - v_k^2) . \tag{5.20}$$

Since the relation $u_k^2 + v_k^2 = 1$ must also apply, it follows that

$$u_k^2 = \frac{1}{2}(1 + \xi_k) , \qquad v_k^2 = \frac{1}{2}(1 - \xi_k) ; \qquad \xi_k = \frac{\varepsilon(k)}{\sqrt{\varepsilon^2(k) + \Delta^2}} . \tag{5.21}$$

Eq. (5.21) reduces to the condition (2.17) when the interaction V and hence Δ are zero. With the interaction the u_k and v_k change smoothly from value 1 to value 0 in the vicinity of k_F. In this region the excitation described by the α-operators is neither an electron nor a hole, but a complicated mixture of both.

We can calculate Δ by inserting (5.21) into the definition for Δ

$$\Delta = V \sum_k u_k v_k = \frac{V}{2} \sum_k \sqrt{1 - \xi_k^2} = \frac{V}{2} \sum_k \frac{\Delta}{\sqrt{\varepsilon^2(k) + \Delta^2}} . \tag{5.22}$$

We are not interested in the solution $\Delta = 0$ (vanishing interaction); we can therefore divide both sides of (5.22) by Δ. An equation of the type (5.14) remains. We must watch two things in converting the sum into an integral. In contrast to (5.14), the sum over $\boldsymbol{k}$ here, in accordance with our convention concerning the spin indices, means a sum over one spin direction only. In the integral we therefore take $g(E)/2\,[\approx g(E_\mathrm{F})/2]$. The integration limits are the energies at which V disappears. As in the last section we restrict V to the range $|\varepsilon(\boldsymbol{k})| \leqslant \hbar\omega_{\boldsymbol{q}}$. We then find

$$1 = \frac{V}{4}\int_{-\hbar\omega_q}^{+\hbar\omega_q} \frac{g(\varepsilon)d\varepsilon}{\sqrt{\varepsilon^2 + \Delta^2}} \approx \frac{Vg(E_\mathrm{F})}{4}\int_{-\hbar\omega_q}^{+\hbar\omega_q} \frac{d\varepsilon}{\sqrt{\varepsilon^2 + \Delta^2}} \tag{5.23}$$

and this leads to

$$\Delta = 2\hbar\omega_{\boldsymbol{q}} \exp\left[-2/g(E_\mathrm{F})V\right]. \tag{5.24}$$

Δ agrees with the binding energy of a Cooper pair (5.15). The energy of the ground state can now be readily found from (5.19). Subtracting from (5.19) the energy of the filled Fermi sphere, we find for the difference between the Hamiltonians with and without interaction

$$\bar{H} - \bar{H}^0 = 2\sum_{\boldsymbol{k}} \varepsilon(\boldsymbol{k})v_{\boldsymbol{k}}^2 - 2\sum_{k<k_\mathrm{F}} \varepsilon(\boldsymbol{k}) - V\sum_{\boldsymbol{k}\boldsymbol{k}'} u_{\boldsymbol{k}}v_{\boldsymbol{k}}u_{\boldsymbol{k}'}v_{\boldsymbol{k}'}\,. \tag{5.25}$$

Since there are no longer operators in this equation, this is the desired energy by which the ground state of the interacting electron gas is separated from the ground state of the noninteracting gas. Inserting (5.21) yields

$$E = \sum_{k<k_\mathrm{F}} |\varepsilon|\left(1 - \frac{|\varepsilon|}{\sqrt{\varepsilon^2 + \Delta^2}}\right) + \sum_{k>k_\mathrm{F}} \varepsilon\left(1 - \frac{\varepsilon}{\sqrt{\varepsilon^2 + \Delta^2}}\right) - \sum_{\boldsymbol{k}} \frac{\Delta^2}{2\sqrt{\varepsilon^2 + \Delta^2}} \tag{5.26}$$

and converting to an integration (again with only one spin direction), one obtains

$$E = g(E_\mathrm{F})\int_0^{\hbar\omega_q}\left(\varepsilon - \frac{1}{2}\frac{2\varepsilon^2 + \Delta^2}{\sqrt{\varepsilon^2 + \Delta^2}}\right)d\varepsilon\,. \tag{5.27}$$

This can be easily integrated and leads to

$$E = \frac{g(E_\mathrm{F})}{2}(\hbar\omega_{\boldsymbol{q}})^2\left[1 - \sqrt{1 + \left(\frac{\Delta}{\hbar\omega_{\boldsymbol{q}}}\right)^2}\,\right] \approx -\frac{g(E_\mathrm{F})\Delta^2}{4}\,. \tag{5.28}$$

The final form here is only valid for weak interactions ($\Delta \ll \hbar\omega_{\boldsymbol{q}}$).

Equation (5.28) is the *condensation energy* of the new ground state.

We have not used the wave function of the ground state in this derivation.

We can obtain this wave function, which we represent by $|0\rangle$, by the application of α-operators to the vacuum state "empty Fermi sphere" ($|\mathrm{vac}\rangle$).

For a *noninteracting* electron gas we can construct the filled Fermi sphere by "annihilating holes" in all states with $k < k_\mathrm{F}$. Using (5.17), we find

$$\prod_k \alpha_k \alpha_{-k} |\mathrm{vac}\rangle = \prod_k (u_k c_k - v_k c_{-k}^+)(u_k c_{-k} + v_k c_k^+)|\mathrm{vac}\rangle$$
$$= \prod_{k<k_\mathrm{F}} c_k^+ c_{-k}^+ |\mathrm{vac}\rangle = |0\rangle \,. \tag{5.29}$$

For an *interacting* electron gas we need only take the other meaning of the u_k, v_k (5.21)

$$\prod_k \alpha_k \alpha_{-k} |\mathrm{vac}\rangle = \prod_k [u_k^2 c_k c_{-k} + u_k v_k (c_k c_k^+ - c_{-k}^+ c_{-k}) + v_k^2 c_k^+ c_{-k}^+]|\mathrm{vac}\rangle$$
$$= \prod_k (u_k v_k + v_k^2 c_k^+ c_{-k}^+)|\mathrm{vac}\rangle \,. \tag{5.30}$$

Eq. (5.30) is not yet normalized. Since

$$\langle \mathrm{vac}| \prod_k (u_k v_k + v_k^2 c_{-k} c_k)(u_k v_k + v_k^2 c_k^+ c_{-k}^+)|\mathrm{vac}\rangle$$
$$= \langle \mathrm{vac}| \prod_k (u_k^2 v_k^2 + v_k^4 c_{-k} c_k c_k^+ c_{-k}^+)|\mathrm{vac}\rangle = \prod_k (u_k^2 v_k^2 + v_k^4) = \prod_k v_k^2$$

the normalized wave function is

$$|0\rangle = \prod_k (u_k + v_k c_k^+ c_{-k}^+)|\mathrm{vac}\rangle \,. \tag{5.31}$$

We note first of all that in the ground state only Cooper pairs $(\boldsymbol{k}\uparrow, -\boldsymbol{k}\downarrow)$ appear. u_k^2 is the probability that a pair of states with opposite $\boldsymbol{k}$ and σ is unoccupied, v_k^2 is the probability that it is occupied. If we carry out the multiplication (5.31), terms having a different number of pair-creation operators appear. Thus (5.31) is not a state with a definite particle number. However, we could have regarded (5.31) as a trial function for the wave function, and determined the u_k and v_k by a variation which minimizes the energy. This is the path originally followed by *Bardeen* et al. One then obtains precisely the results derived above by a different method. The variation has to be performed for fixed particle number. We have therefore to add the condition $N = \mathrm{const}$. This is done by adding a term—λN_op to the Hamiltonian before the variation. The Lagrange parameter turns out to be the chemical potential, here the Fermi energy E_F. This is the real reason why we went over from H to $\overline{H}$ in (5.18).

The results so far tell us that there is a reduction in the energy of the ground state. We can only find out that the phenomenon of superconductivity is associated with this when we turn our attention to the excited states. This we do now.

5.4 Excited States

We again describe the excited states above the ground state in terms of quasi-particles. They are created and annihilated by operators α_k^+ and α_k. These operators were introduced by the Bogoliubov-Valatin transformation (5.17). The associated quasi-particles are different in several respects from quasi-particles previously introduced. In particular we remember the results of the last section, by which such a quasi-particle is an electron in state $\boldsymbol{k}\uparrow$ for energies $E \gg E_F + \Delta$, and a hole in state $-\boldsymbol{k}\downarrow$ for energies $E \ll E_F - \Delta$. In a region of width 2Δ about E_F, it is a complicated mixture of the two possibilities.

The energy of one of these quasi-particles can readily be obtained from (5.18). In addition to terms contributing to the ground state and terms of the fourth order in the α which we neglected in the last section, this equation contains other terms with α-operators in the combination $\alpha_k^+\alpha_k + \alpha_{-k}^+\alpha_{-k}$. Both terms are particle number operators for the quasi-particles considered here. From (5.18) the contributions from the Hamiltonian $\bar{H}$ include the terms for the ground state and the terms

$$\bar{H} = \cdots + \sum_k [\varepsilon(\boldsymbol{k})(u_k^2 - v_k^2) + V\sum_{k'} 2u_k v_k u_{k'} v_{k'}](\alpha_k^+\alpha_k + \alpha_{-k}^+\alpha_{-k}) + \cdots . \tag{5.32}$$

The difference between the excited and the ground state is then equal to the terms listed in (5.32), where the particle number operators are replaced by the particle numbers themselves

$$E - E_0 = \sum_k [\varepsilon(\boldsymbol{k})(u_k^2 - v_k^2) + 2\Delta u_k v_k](n_{k\uparrow} + n_{-k\downarrow}) = \sum_k \sqrt{\varepsilon^2 + \Delta^2}\, n_k . \tag{5.33}$$

The energy of an individual quasi-particle is then

$$\bar{\varepsilon}(\boldsymbol{k}) = \sqrt{\varepsilon^2 + \Delta^2} . \tag{5.34}$$

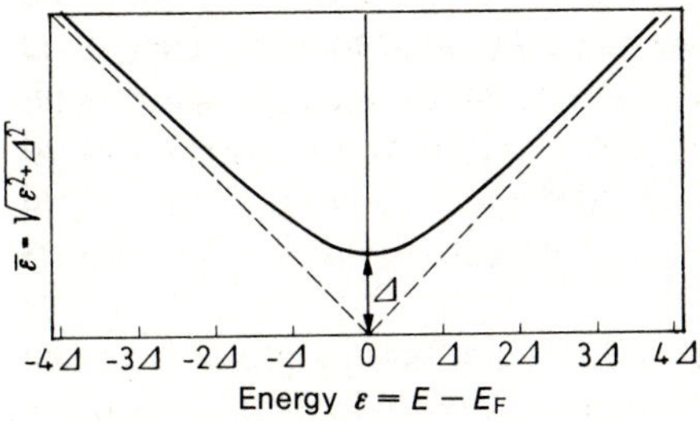

Fig. 5.2. Energy $\bar{\varepsilon}$ of the quasi-particles which describe the excited states of a superconducting electron gas.

This function is shown in Fig. 5.2. $\varepsilon(\boldsymbol{k})$ is the energy as measured from the Fermi surface. While it is possible in the noninteracting gas to create electrons just above E_F by the input of an infinitesimal amount of energy, here a threshold energy Δ is needed. Ground and first excited states are thus separated by an *energy gap*. Moreover, in a scattering process (energy transfer to the electron system) a quasi-particle is never created alone, always in pairs—corresponding to the electron-hole pair of the noninteracting electron gas. The threshold energy of an excitation out of the ground state is therefore 2Δ!

This result allows a qualitative interpretation of the superconducting properties of the electron gas. From Fig. 2.5 we know to represent a current-carrying state of the electron gas by displacing the Fermi sphere in $\boldsymbol{k}$-space. When the driving field is removed, equilibrium is set up again by scattering processes which scatter the electrons back into the original Fermi sphere with absorption or emission of phonons (Fig. 5.3). Such processes readily occur in the noninteracting electron gas. For the superconducting electron gas, energy conservation demands that in this process the initial energy must lie at least 2Δ above the final energy. If this is not satisfied, i.e., if the displacement of the Fermi sphere is so little that no electron from the left hatched area in Fig. 5.3 has an energy 2Δ higher than the states in the right hatched area, then a relaxation from the current-carrying state cannot occur—at least not by mechanisms known to the transport theory for normal conductors. Current flows free of resistance.

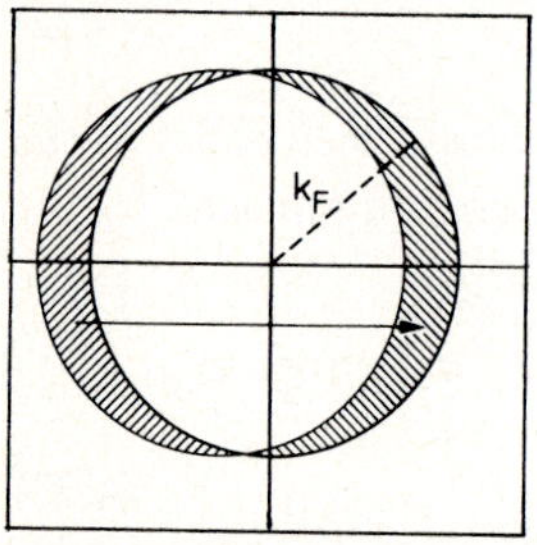

Fig. 5.3. In an electric field all electrons within the Fermi sphere are "displaced" in $\boldsymbol{k}$-space. When the field is removed, the equilibrium state is regained by transitions between the two hatched areas outside and inside the Fermi sphere. In a superconducting electron gas the energy gap will also be displaced. Scattering processes can then only occur if the energy 2Δ can be delivered.

If the displacement of the Fermi sphere is $\delta \boldsymbol{k} = (m/\hbar)\delta \boldsymbol{v} = (m/\hbar e n)\boldsymbol{i}$, an electron from the hatched area in Fig. 5.3 can only be scattered into the original Fermi sphere if $(\hbar^2/2m)(k_F + \delta k)^2 - (\hbar^2/2m)(k_F - \delta k)^2 \geqslant 2\Delta$. Inserting values of the right order of magnitude ($n = 3\cdot 10^{22}\,\mathrm{cm}^{-3}$, $\Delta = 10^{-16}$ erg, $k_F = 10^8\,\mathrm{cm}^{-1}$) yields for the current density below which there is no resistance the value $i = 2ne\Delta/\hbar k_F \approx 10^7\,\mathrm{Acm}^{-2}$.

The need to provide a minimum energy 2Δ in a scattering process can also be attributed to the fact that the Cooper pairs must be broken up in the process. This leads to the picture that Cooper pairs and single quasi-particles are present together in an excited state. Cooper pairs lead to a current which flows without resistance; individual particles are scattered (*two fluid model*).

We now leave this consideration of individual excitations to examine an excited state at a temperature $T \neq 0$. At temperatures above absolute zero we have to note that states $\boldsymbol{k}\uparrow$ or $-\boldsymbol{k}\downarrow$ are occupied statistically. We do this by replacing the particle numbers n_k by their statistical averages

$$n_{\boldsymbol{k}} \to \langle n_{\boldsymbol{k}} \rangle \equiv f_{\boldsymbol{k}} = \left[\exp \frac{\bar{\varepsilon}(\boldsymbol{k})}{k_{\mathrm{B}} T} + 1\right]^{-1}. \tag{5.35}$$

Here we have taken the Fermi distribution (2.30) as the occupation probability, and have inserted the energy of the quasi-particles (5.34) (referred to the Fermi surface) for the energy difference $E - \mu$.

Equation (5.34) contains the energy gap Δ. We calculated this parameter from (5.23). Eq. (5.23) was the result of a diagonalization procedure for the Hamiltonian carried out under the assumption $n_{\boldsymbol{k}} = 0$ (ground state, $T = 0$). To obtain the correct value of Δ in (5.34) we have therefore to repeat the diagonalization procedure for $n_{\boldsymbol{k}} \neq 0$ ($T \neq 0$). We obtain a different value of Δ for each temperature; the energy gap becomes temperature dependent: $\Delta = \Delta(T)$!

To calculate $\Delta(T)$ extend (5.20) by the particle number operators associated with the relevant term of (5.18), and replace them immediately by the mean occupation numbers (5.35)

$$2u_{\boldsymbol{k}}v_{\boldsymbol{k}}\varepsilon(\boldsymbol{k}) - (u_{\boldsymbol{k}}^2 - v_{\boldsymbol{k}}^2)V \sum_{\boldsymbol{k}'} u_{\boldsymbol{k}'}v_{\boldsymbol{k}'}(1 - 2f_{\boldsymbol{k}'}) = 0\,. \tag{5.36}$$

Corresponding to (5.20), we now define $\Delta(T)$ by

$$\Delta(T) = V \sum_{\boldsymbol{k}'} u_{\boldsymbol{k}'}v_{\boldsymbol{k}'}(1 - 2f_{\boldsymbol{k}'})\,. \tag{5.37}$$

With this we find instead of (5.23)

$$1 = \frac{V_{\mathrm{g}}(E_{\mathrm{F}})}{4} \int_{-\hbar\omega_q}^{+\hbar\omega_q} \frac{d\varepsilon}{\sqrt{\varepsilon^2 + \Delta^2(T)}} \left\{1 - 2f\left(\frac{\sqrt{\varepsilon^2 + \Delta^2(T)}}{k_{\mathrm{B}}T}\right)\right\}. \tag{5.38}$$

This equation yields $\Delta(T)$. A closed solution is not possible. From (5.38) one can see first of all that when $T = 0$ the equation becomes identical to (5.23). With increasing temperature $\Delta(T)$ decreases continuously. At a given temperature T_c $\Delta(T)$ becomes zero. We interpret this temperature as the *critical temperature* above which superconductivity disappears. We shall return to this. Above T_c we must set $\Delta(T)$ equal to zero. We can do this since a factor $\Delta(T)$ has been removed from (5.38) just as from (5.23), i.e., $\Delta = 0$ is also a solution of the original equation.

For our discussion we rearrange (5.38). After division by $Vg(E_{\mathrm{F}})/4$, the use of (5.24) for $\Delta(0)$, and the separation of the integral on the right into two partial integrals, we find for the left side $2 \ln [2\hbar\omega_q/\Delta(0)]$, for the first integral on the right side 2 ln $[2\hbar\omega_q/\Delta(T)]$, and hence with the approximation $\Delta(0) \ll \hbar\omega_q$

$$\ln\frac{\Delta(T)}{\Delta(0)} = -2\int_0^\infty \frac{dx}{\sqrt{x^2+1}} f\left(\sqrt{x^2+1}\,\frac{\Delta(T)}{\Delta(0)}\left[\frac{k_BT}{\Delta(0)}\right]^{-1}\right) = F\left(\frac{\Delta(T)}{\Delta(0)}, \frac{k_BT}{\Delta(0)}\right). \tag{5.39}$$

This equation now only contains the reduced energy gap $\Delta(T)/\Delta(0)$ and a reduced temperature $k_BT/\Delta(0)$. From this it follows that the transition temperature T_c calculated from the condition $\Delta(T_c) = 0$ is linearly dependent on $\Delta(0)$. Numerical integration gives

$$k_BT_c \approx 0.57\Delta(0). \tag{5.40}$$

Hence we can also replace $k_BT/\Delta(0)$ by T/T_c in (5.39).

Equation (5.39) presents in reduced units a relationship between energy gap and temperature which contains no other parameter (see Fig. 5.7). In this form it applies to all superconductors for which the condition of weak coupling is satisfied $[\Delta(0) \ll \hbar\omega_q = \hbar\omega_D]$.

5.5 Comparison with Experiment

One of the most important characteristics of superconductivity is its limitation to temperatures below a critical temperature T_c. From (5.40), k_BT_c is approximately half the binding energy of a Cooper pair at $T = 0$. From (5.24), this again is proportional to the energy of the virtual phonon times an exponential factor containing the interaction constant V and the density of states at the Fermi surface. Since weak coupling V is assumed, $\Delta(0)$ and hence k_BT_c are small compared with $\hbar\omega_q$. If one inserts for ω_q the maximum value ω_D of the Debye approximation, it follows that the transition temperature T_c of a superconductor is small compared to its Debye temperature. This is verified experimentally. Typical values for T_c lie between 20 K (for several niobium alloys) and the lowest detected limit (0.01 K for tungsten). The Debye temperatures of metals on the other hand almost always lie above 100 K.

One can obtain a second statement about the transition temperature from the linear relationship between T_c and ω_q. ω_q is the frequency of an acoustic phonon which, in the Debye approximation, is seen from (3.178) to be proportional to $\rho^{-1/2}$, and therefore also to $M^{-1/2}$ (M = ion mass). For the isotopes of a superconducting metal we therefore expect a law of the form $M^aT_c =$ const, with $a = 1/2$. Indeed this *isotopy effect* was the first experimental pointer to the fact that lattice vibrations are involved in the interaction relevant to superconductivity. In general the values for a differ somewhat from 0.5, a sign that the neglect of all other interactions apart from the simplified BCS interaction is too coarse an approximation.

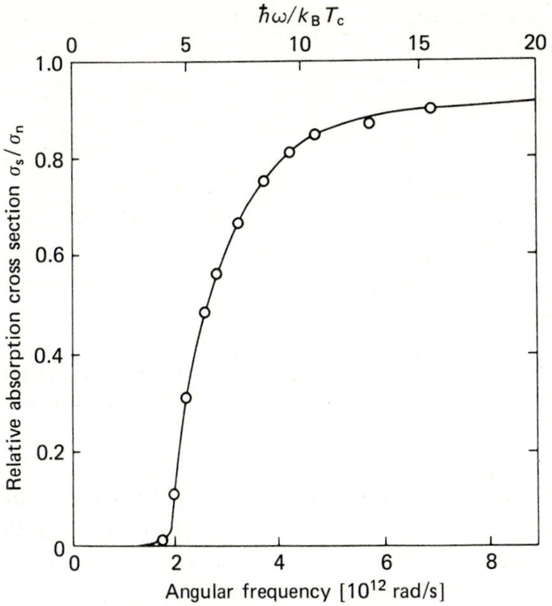

Fig. 5.4. Absorption edge of a superconductor (indium). Photons can only be absorbed when they have the energy 2Δ needed to break up a Cooper pair (from *Blakemore* [4]).

One expects to find evidence of the energy gap $2\Delta(T)$ in a superconductor in its optical absorption. As in semiconductors and insulators, absorption should only take place above a threshold energy which is given here by the energy $2\Delta(T)$ needed to break up a Cooper pair. This is verified experimentally as exemplified in Fig. 5.4. The experimental difficulties associated with such measurements are considerable, however, since the frequency $\omega = 2\Delta(0)/\hbar \approx 4k_BT_c/\hbar$ lies in the range 10^{10} to 10^{12} Hz, i.e., in the far infrared.

A method which can yield more information is based on *tunneling experiments,* in which the current flowing from one metal through a thin oxide layer into another metal is measured. For an understanding of this phenomenon we first present the density of states in a superconductor. The density of states, described by the function $g_n(E)\,dE$ in a normal conductor, is altered in superconductors by the appearance of an energy gap. Since no states are lost, $g_n(E)\,dE = g_s(\bar{\varepsilon})\,d\bar{\varepsilon}$. Here $\bar{\varepsilon}$ is the energy of the quasi-particles given by (5.34). From this equation E and $\bar{\varepsilon}$ are related by $E = E_F + \sqrt{\bar{\varepsilon}^2 + \Delta^2}$. We have therefore

$$g_s(\bar{\varepsilon}) = g_n(E)\frac{dE}{d\bar{\varepsilon}} = g_n(E)\frac{\bar{\varepsilon}}{\sqrt{\bar{\varepsilon}^2 + \Delta^2}} \tag{5.41}$$

for $|\bar{\varepsilon}| > \Delta$, while for $|\bar{\varepsilon}| < \Delta$ the density of states is zero (cf. Fig. 5.5).

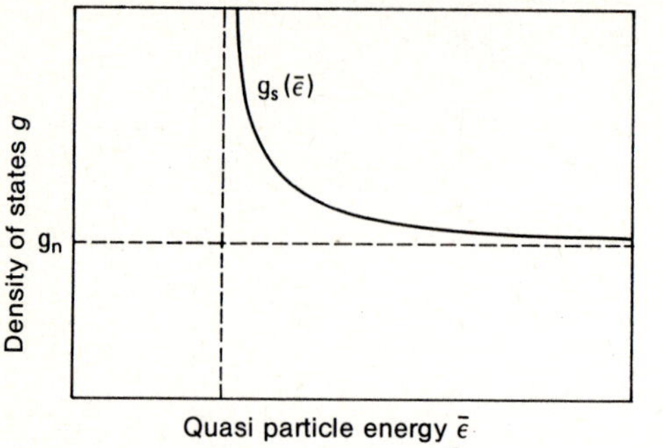

Fig. 5.5. Density of states $g_s(\bar{\varepsilon})$ for a superconducting electron gas. The appearance of an energy gap changes the density of states of the normal conductor. Well above Δ, g_s is equal to g_n.

In Fig. 5.6 we compare the three following possibilites: (a) The oxide layer separates two normal conducting metals. By applying a voltage to this system, the chemical potentials (Fermi energies) of the two metals are displaced relative to one another. Current then flows since there are now occupied states in one metal at the same energy as unoccupied states in the other. One can readily see that this current depends linearly on the applied voltage.

(b) The oxide layer separates a normal conducting metal and a superconducting metal. In accordance with Fig. 5.6 (second row), current can then only flow from the normal conductor to the superconductor when the applied voltage reaches the value Δ/e. The current will then rapidly increase since, as we see in Fig. 5.5, there are a great many states available for occupation just above Δ. This line of argument is only strictly valid at $T = 0$ when the Fermi distribution is a step function. The flow of current will commence earlier for $T \neq 0$.

(c) The oxide layer separates two superconductors with different widths of energy gap. From Fig. 5.6, similar to the case just considered, a current will start to flow at $V = (\Delta_1 + \Delta_2)/e$. At $T \neq 0$ this tunnel current is preceded by a weak current which reaches its maximum when the upper edges of the two energy gaps are at the same energy ($V = (\Delta_1 - \Delta_2)/e$).

In addition to these currents due to the motion of individual quasi-particles (*Giaever tunneling*) there is also the possibility in the last case examined that Cooper pairs tunnel through the oxide layer from one superconductor to the other (*Josephson tunneling*). One can show that this current can flow in the absence of an applied voltage if the phases of the wave functions describing the Cooper pairs on both sides of the junction are different [see (5.70) and Problem 5.4]. When $V > 0$, current oscillations occur.

A further method of measuring $\Delta(T)$ is by *ultrasonic attenuation*. The energy of ultrasonic phonons is so small that they cannot split Cooper pairs. They are only absorbed by excited quasi-particles. Since their number is set by $\Delta(T)$, the energy gap can then be determined. In Fig. 5.7 we compare the variation

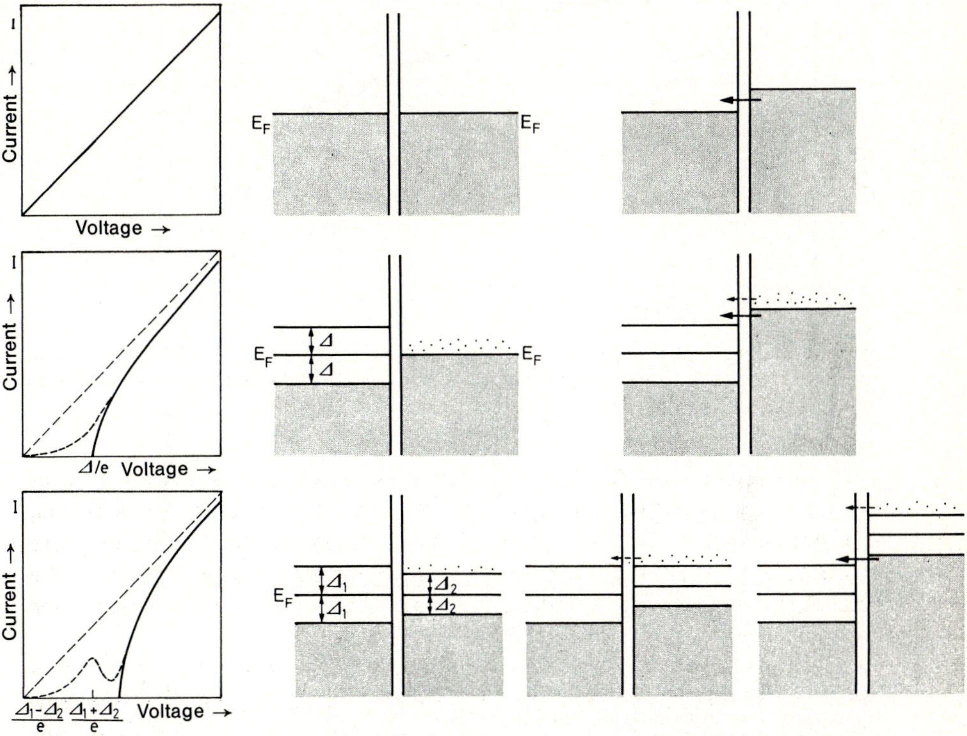

Fig. 5.6. Tunneling through an oxide layer between two normal or superconducting solids. For the density of states of both sides of the layer see Fig. 5.5. Upper row: If both solids are normal conductors, conduction band levels up to the energy E_F are occupied in equilibrium. If a voltage is applied, the chemical potential (Fermi energy) of one conductor will be raised relative to that of the other. Current will flow. The current-voltage characteristic is linear. Middle row: If one of the solids is a superconductor, near $T = 0$ current will only flow when the chemical potential of the normal conductor is raised by Δ. For $T \neq 0$, thermal excitation of individual electrons results in a weak current appearing at lower voltages. Bottom row: If two superconductors are involved, near $T = 0$ current does not flow before the chemical potential is raised by $\Delta_1 + \Delta_2$. When $T \neq 0$, an additional current appears which has its maximum at $\Delta_1 - \Delta_2$.

of the energy gap with temperature as determined by ultrasonic measurements in tin and the function $\Delta(T)$ from (5.39).

In a superconductor the electronic part of the *specific heat* is carried by the excited particles. From the statistics the number of particles excited across the energy gap is proportional to $\exp(-E_G/2k_BT)$, where E_G is the width of the gap. We therefore expect here an $\exp(-\Delta(T)/k_B)$-law. Consequently the specific heat of a superconductor will also have a temperature dependence completely different from that of a normal conductor. Initially it increases exponentially; as a result of the temperature dependence of the energy gap, deviations from the exponential law then appear. At T_c the specific heat changes discontinuously to the value for the normal conducting metal.

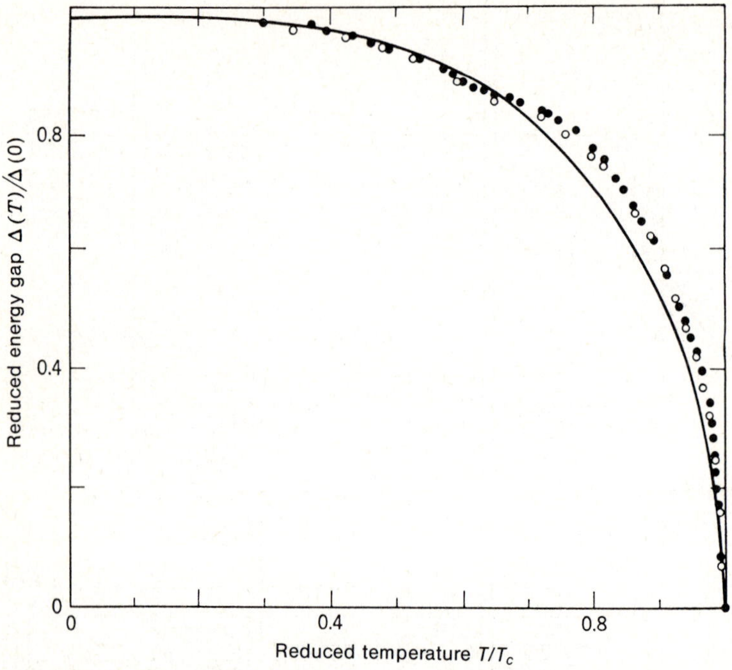

Fig. 5.7. Temperature dependence of the energy gap according to the BCS theory, and comparison with experimental results for tin (from *Rickayzen* [74]).

The BCS theory in the main verifies these qualitative considerations of the behaviour of the specific heat with temperature. For a more exact calculation we have to determine the free energy and the entropy of a superconductor. We choose not to do this and for the thermodynamic characteristics of superconductors we again refer the reader to the literature.

Knowledge of the free energy of the superconducting and normal states also permits the calculation of the *critical magnetic field* needed to destroy superconductivity at a given temperature below T_c.

In the previous section we presented a qualitative explanation for the appearance of persistent currents in superconductors. A second test for any theory of superconductivity is the explanation of the *Meissner-Ochsenfeld effect*. i.e., the fact that when a superconductor in a magnetic field is cooled down below the transition temperature the magnetic field is expelled from within it. In view of the fundamental significance of this effect we devote the next section to it.

The last phenomenon connected with superconductivity that we shall mention is *flux quantization*. London's phenomenological theory predicts the quantization of the magnetic flux through a superconducting ring. It gives a flux quantum of magnitude $h/2e^*$, where e^* is the charge of the charge carriers of the persistent current flowing in the ring. Experimentally one finds $e^* = 2e$, i.e., double the electron charge. This can be explained by the result from the BCS theory that the persistent current is carried by Cooper pairs (see also Sec. 5.7).

Up till now we have restricted ourselves to the discussion of phenomena in superconductors which can be explained by the BCS theory. In the main these are the characteristics of the so-called type I superconductors. We shall look at more comprehensive possibilities for describing superconductors in the final Section 5.7.

5.6 The Meissner-Ochsenfeld Effect

To explain the Meissner-Ochsenfeld effect, i.e., the expulsion of a magnetic field from a superconductor, we look at the general case of a superconducting electron gas in a magnetic field. Let the field be described by a vector potential $\boldsymbol{A}$. We choose $\boldsymbol{A}$ so that $\nabla \cdot \boldsymbol{A} = 0$. We have then to extend the Hamiltonian by an additional term of the form

$$H' = \frac{1}{2m}(\boldsymbol{p} + e\boldsymbol{A})^2 - \frac{1}{2m}p^2 \approx \frac{e}{2m}(\boldsymbol{p}\cdot\boldsymbol{A} + \boldsymbol{A}\cdot\boldsymbol{p}) . \tag{5.42}$$

Since we want to restrict our considerations to weak magnetic fields we have omitted a term of order A^2 in (5.42).

We are mostly interested in the current density induced by the magnetic field

$$\boldsymbol{i} = \frac{\mathrm{i}e\hbar}{2m}(\psi^*\nabla\psi - \psi\nabla\psi^*) - \frac{e^2}{m}\boldsymbol{A}\psi^*\psi . \tag{5.43}$$

First we transform (5.42) and (5.43) into the occupation number representation. We use a procedure which we have not applied so far but which follows logically from the Appendix. We replace the wave functions in (5.43) by *field operators* in accordance with

$$\psi^* \to \frac{1}{\sqrt{V_g}}\sum_{\boldsymbol{k}\sigma} \exp(-\mathrm{i}\boldsymbol{k}\cdot\boldsymbol{r})c^+_{\boldsymbol{k}\sigma} , \qquad \psi \to \frac{1}{\sqrt{V_g}}\sum_{\boldsymbol{k}'\sigma'} \exp(\mathrm{i}\boldsymbol{k}'\cdot\boldsymbol{r})c_{\boldsymbol{k}'\sigma'} . \tag{5.44}$$

The $c^+_{\boldsymbol{k}}$ and $c_{\boldsymbol{k}}$ create and annihilate electrons. $\boldsymbol{i}$ then becomes the following current density *operator*:

$$\begin{aligned}
\boldsymbol{i} &= \sum_{\boldsymbol{k}\boldsymbol{k}'\sigma\sigma'}\left[-\frac{e\hbar}{2mV_g}(\boldsymbol{k}+\boldsymbol{k}') - \frac{e^2\boldsymbol{A}}{mV_g}\right]\exp[\mathrm{i}(\boldsymbol{k}'-\boldsymbol{k})\cdot\boldsymbol{r}]c^+_{\boldsymbol{k}\sigma}c_{\boldsymbol{k}'\sigma'} \\
&= \sum_{\boldsymbol{q}}\exp(\mathrm{i}\boldsymbol{q}\cdot\boldsymbol{r})\sum_{\boldsymbol{k}\sigma\sigma'}\left[-\frac{e\hbar}{2mV_g}(2\boldsymbol{k}-\boldsymbol{q}) - \frac{e^2\boldsymbol{A}}{mV_g}\right]c^+_{\boldsymbol{k}-\boldsymbol{q},\sigma}c_{\boldsymbol{k}\sigma'} \\
&= \sum_{\boldsymbol{q}}\exp(\mathrm{i}\boldsymbol{q}\cdot\boldsymbol{r})\boldsymbol{i}_{\boldsymbol{q}} .
\end{aligned} \tag{5.45}$$

The right-hand side of this equation can be interpreted as the Fourier expansion of the operator.

Since we only need products $c_{k'}^{+}c_{k}$ with fixed spin for both electrons, here and in the following equations we combine the sum over σ, σ' into a single sum.

Equation (5.42) can be transformed in the same way. We first form the interaction energy

$$E' = \frac{e}{2m}\langle\psi|\boldsymbol{p}\cdot\boldsymbol{A} + \boldsymbol{A}\cdot\boldsymbol{p}|\psi\rangle\,. \tag{5.46}$$

Using (5.44), this becomes an operator

$$H' = -\frac{ie\hbar}{2mV_g}\sum_{\boldsymbol{k}\boldsymbol{k}'\sigma\sigma'}\int \exp\left[i(\boldsymbol{k}' - \boldsymbol{k})\cdot\boldsymbol{r}\right]\boldsymbol{A}\cdot\boldsymbol{i}(\boldsymbol{k} + \boldsymbol{k}')d\tau c_{\boldsymbol{k}\sigma}^{+}c_{\boldsymbol{k}'\sigma'}\,, \tag{5.47}$$

or, if one introduces the Fourier expansion for the vector potential and restricts oneself to one spin summation

$$\begin{aligned} H' &= \frac{e\hbar}{2mV_g}\sum_{\boldsymbol{k}\boldsymbol{k}'\boldsymbol{q}\sigma}\int \exp\left[i(\boldsymbol{k}' + \boldsymbol{q} - \boldsymbol{k})\cdot\boldsymbol{r}\right]\boldsymbol{A}_{\boldsymbol{q}}\cdot(\boldsymbol{k} + \boldsymbol{k}')d\tau c_{\boldsymbol{k}\sigma}^{+}c_{\boldsymbol{k}'\sigma} \\ &= \frac{e\hbar}{2m}\sum_{\boldsymbol{k}\boldsymbol{q}\sigma}\boldsymbol{A}_{\boldsymbol{q}}\cdot(2\boldsymbol{k} - \boldsymbol{q})c_{\boldsymbol{k}\sigma}^{+}c_{\boldsymbol{k}-\boldsymbol{q},\sigma}\,. \end{aligned} \tag{5.48}$$

We now form the expectation value for the current density. We divide $\boldsymbol{i}$ into two parts given by the sum terms in (5.45). We call the first and second parts $\boldsymbol{i}_1$, $\boldsymbol{i}_2$, respectively. The expectation value of $\boldsymbol{i}_2$ is

$$\begin{aligned} \langle\boldsymbol{i}_2\rangle &= -\frac{e^2\boldsymbol{A}}{mV_g}\langle\psi|\sum_{\boldsymbol{q}}\exp(i\boldsymbol{q}\cdot\boldsymbol{r})\sum_{\boldsymbol{k}\sigma}c_{\boldsymbol{k}-\boldsymbol{q},\sigma}^{+}c_{\boldsymbol{k}\sigma}|\psi\rangle \\ &= -\frac{e^2\boldsymbol{A}}{mV_g}\langle\psi|\sum_{\boldsymbol{k}\sigma}c_{\boldsymbol{k}\sigma}^{+}c_{\boldsymbol{k}\sigma}|\psi\rangle = -\frac{e^2 n}{m}\boldsymbol{A}\,. \end{aligned} \tag{5.49}$$

Here we have made use of the fact that the matrix element on the right gives the particle number, regardless of whether $|\psi\rangle$ describes a state of a normal conductor (to which the c^{+}, c have to be applied) or a superconductor (where the c^{+}, c must first be transformed into the α^{+}, α).

For the expectation value of $\boldsymbol{i}_1$ we have first to undertake the transformation to the α^{+}, α. This requires the spin summation to be carried out. We write

$$\boldsymbol{i}_1 = \sum_{\boldsymbol{k}\boldsymbol{q}}\left[-\frac{e\hbar}{2mV_g}(2\boldsymbol{k} - \boldsymbol{q})\exp(i\boldsymbol{q}\cdot\boldsymbol{r})\right](c_{(\boldsymbol{k}-\boldsymbol{q})\uparrow}^{+}c_{\boldsymbol{k}\uparrow} + c_{(\boldsymbol{k}-\boldsymbol{q})\downarrow}^{+}c_{\boldsymbol{k}\downarrow}) \tag{5.50}$$

and in the second summation replace $\boldsymbol{k} - \boldsymbol{q}$ by $-\boldsymbol{k}$ and $\boldsymbol{k}$ by $-(\boldsymbol{k} - \boldsymbol{q})$.

This only leads to a change of one sign

$$\boldsymbol{i}_1 = \sum_{\boldsymbol{kq}} \left[\frac{e\hbar}{2mV_g}(2\boldsymbol{k} - \boldsymbol{q}) \exp(i\boldsymbol{q}\cdot\boldsymbol{r}) \right] (c^+_{(\boldsymbol{k}-\boldsymbol{q})\uparrow} c_{\boldsymbol{k}\uparrow} - c^+_{-\boldsymbol{k}\downarrow} c_{-(\boldsymbol{k}-\boldsymbol{q})\downarrow}) . \tag{5.51}$$

Using (5.17), we then find

$$\begin{aligned} \boldsymbol{i}_1 = \sum_{\boldsymbol{kq}} & \left[-\frac{e\hbar}{2mV_g}(2\boldsymbol{k} - \boldsymbol{q}) \exp(i\boldsymbol{q}\cdot\boldsymbol{r}) \right] \\ & \times [(u_{\boldsymbol{k}-\boldsymbol{q}}u_{\boldsymbol{k}} + v_{\boldsymbol{k}-\boldsymbol{q}}v_{\boldsymbol{k}})(\alpha^+_{(\boldsymbol{k}-\boldsymbol{q})\uparrow}\alpha_{\boldsymbol{k}\uparrow} - \alpha^+_{-\boldsymbol{k}\downarrow}\alpha_{-(\boldsymbol{k}-\boldsymbol{q})\downarrow}) \\ & + (u_{\boldsymbol{k}-\boldsymbol{q}}v_{\boldsymbol{k}} - u_{\boldsymbol{k}}v_{\boldsymbol{k}-\boldsymbol{q}})(\alpha^+_{(\boldsymbol{k}-\boldsymbol{q})\uparrow}\alpha^+_{-\boldsymbol{k}\uparrow} - \alpha_{\boldsymbol{k}\downarrow}\alpha_{-(\boldsymbol{k}-\boldsymbol{q})\downarrow})] . \end{aligned} \tag{5.52}$$

We now make an approximation. At the end of the calculation we shall restrict ourselves to the limiting case $\boldsymbol{q} \to 0$. Then $u_{\boldsymbol{k}-\boldsymbol{q}} = u_{\boldsymbol{k}}$, $v_{\boldsymbol{k}-\boldsymbol{q}} = v_{\boldsymbol{k}}$. If we use these relations already now, (5.52) simplifies to

$$\boldsymbol{i}_1 = \sum_{\boldsymbol{kq}} \left[-\frac{e\hbar}{2mV_g}(2\boldsymbol{k} - \boldsymbol{q}) \exp(i\boldsymbol{q}\cdot\boldsymbol{r}) \right] (\alpha^+_{(\boldsymbol{k}-\boldsymbol{q})\uparrow}\alpha_{\boldsymbol{k}\uparrow} - \alpha^+_{-\boldsymbol{k}\downarrow}\alpha_{-(\boldsymbol{k}-\boldsymbol{q})\downarrow}) . \tag{5.53}$$

Thus, compared to (5.51), only the nomenclature of the operators changes.

We can apply the same argument for H' in (5.47). We find

$$\begin{aligned} H' &= \frac{e\hbar}{m} \sum_{\boldsymbol{kq}} \boldsymbol{A}_{\boldsymbol{q}}\cdot(2\boldsymbol{k} - \boldsymbol{q})(c^+_{\boldsymbol{k}\uparrow}c_{(\boldsymbol{k}-\boldsymbol{q})\uparrow} - c^+_{-(\boldsymbol{k}-\boldsymbol{q})\downarrow}c_{-\boldsymbol{k}\downarrow}) \\ &\approx \frac{e\hbar}{m} \sum_{\boldsymbol{kq}} \boldsymbol{A}_{\boldsymbol{q}}\cdot(2\boldsymbol{k} - \boldsymbol{q})(\alpha^+_{\boldsymbol{k}\uparrow}\alpha_{(\boldsymbol{k}-\boldsymbol{q})\uparrow} - \alpha^+_{-(\boldsymbol{k}-\boldsymbol{q})\downarrow}\alpha_{-\boldsymbol{k}\downarrow}) . \end{aligned} \tag{5.54}$$

To form the expectation value of $\boldsymbol{i}$ we use the wave function [cf. (4.23)]

$$|n\rangle_1 = |n\rangle_0 + \sum_{m(\neq 0)} \frac{\langle m|H'|n\rangle_0}{E_n - E_m}|m\rangle_0 + \cdots , \tag{5.55}$$

where the $|n\rangle_0$ are wave functions of the magnetic field-free case. In contrast to $\boldsymbol{i}_2$, where the matrix element between wave functions of zero order would lead to (5.49), $\langle n|\boldsymbol{i}_1|n\rangle_0$ vanishes. The first nonvanishing contribution to $\langle \boldsymbol{i}_1\rangle$ comes from the linear terms in $\boldsymbol{A}$

$$\langle \boldsymbol{i}_1 \rangle = \sum_{m(\neq 0)} \frac{\langle n|\boldsymbol{i}_1|m\rangle\langle m|H'|n\rangle}{E_n - E_m} + \sum_{m'(\neq 0)} \frac{\langle n|H'|m'\rangle\langle m'|\boldsymbol{i}_1|n\rangle}{E_n - E_{m'}} . \tag{5.56}$$

In accordance with (5.53) and (5.54), the energy of the intermediate states $|m\rangle$ lies $|(\bar{\varepsilon}(\boldsymbol{k} - \boldsymbol{q}) - \bar{\varepsilon}(\boldsymbol{k})|$ above the initial state. All the sum terms in (5.56) can thus be brought to a single denominator. Inserting (5.53) and (5.54) into

(5.56), a Fourier coefficient of $\boldsymbol{i}_1$ is

$$\langle \boldsymbol{i}_{1q} \rangle = \sum_{\boldsymbol{k}} \left(-\frac{e^2\hbar^2}{2m^2 V_g} \right) \boldsymbol{A}_q \cdot (\boldsymbol{k} - \boldsymbol{q})(2\boldsymbol{k} - \boldsymbol{q}) 2 \frac{n_{\boldsymbol{k}-\boldsymbol{q}} - n_{\boldsymbol{k}}}{\bar{\varepsilon}(\boldsymbol{k} - \boldsymbol{q}) - \bar{\varepsilon}(\boldsymbol{k})} . \tag{5.57}$$

We are interested in the temperature dependence of the current. We therefore replace the occupation numbers $n_{\boldsymbol{k}}$ by the occupation probabilities $f_{\boldsymbol{k}}$, as in (5.35). If we go to the limit $\boldsymbol{q}$ approaching zero the difference quotient in (5.57) becomes the differential quotient $\partial f_{\boldsymbol{k}}/\partial \bar{\varepsilon}$. We can set all other $\boldsymbol{q}$ appearing in this equation to zero. When we replace the sum over $\boldsymbol{k}$ (one spin direction) by an integration, we obtain

$$\langle \boldsymbol{i}_{10} \rangle = \frac{1}{V_g} \sum_{\boldsymbol{k}} \left(-\frac{e^2\hbar^2}{2m^2} \right) (\boldsymbol{A}_0 \cdot \boldsymbol{k}) 4\boldsymbol{k} \frac{\partial f_{\boldsymbol{k}}}{\partial \bar{\varepsilon}} = \frac{2e\hbar^2}{3m^2} \frac{\boldsymbol{A}_0}{(2\pi)^3} \int k^2 d\tau_{\boldsymbol{k}} \frac{\partial f_{\boldsymbol{k}}}{\partial \bar{\varepsilon}} . \tag{5.58}$$

We combine this expression with (5.49) to give the following equation which is valid for the limiting case $\boldsymbol{q} = 0$:

$$\langle \boldsymbol{i}_0 \rangle = -\frac{e^2 n}{m} \boldsymbol{A}_0 \left[1 - \frac{2E_F}{k_F^5} \int_0^\infty k^4 dk \left(-\frac{\partial f_{\boldsymbol{k}}}{\partial \bar{\varepsilon}} \right) \right] . \tag{5.59}$$

Here (2.8) and (2.9) have also been used.

The integral can be solved easily in two cases. a) For a normal conducting electron gas ($\bar{\varepsilon} = E - E_F$) $-\partial f_{\boldsymbol{k}}/\partial E$ becomes a δ-function and the two terms in the brackets cancel. b) For $T = 0$, $\partial f_{\boldsymbol{k}}/\partial \bar{\varepsilon}$ is zero since $\bar{\varepsilon} \neq 0$. The brackets then take the value one.

Equation (5.59) is a linear relation between the current density and the vector potential of the magnetic field, and is thus identical to one of the two *London equations* in the phenomenological theory of superconductivity. It is usually written in the form

$$\boldsymbol{i} = -\frac{1}{\mu_0 \lambda^2} \boldsymbol{A} . \tag{5.60}$$

To show that this equation contains the Meissner-Ochsenfeld effect, we examine the boundary between the superconductor and a vacuum. Let the superconductor fill the half-space $z < 0$ and the vacuum the half-space $z > 0$. Let there also be a magnetic field $\boldsymbol{B} = (B_x, 0, 0)$ in the outer space. The following equations then apply in the superconductor:

$$\nabla \times \boldsymbol{i} = -\frac{1}{\mu_0 \lambda^2} \boldsymbol{B} , \qquad \nabla \times \boldsymbol{B} = \mu_0 \boldsymbol{i} , \qquad \nabla \cdot \boldsymbol{B} = 0 . \tag{5.61}$$

We can arrange these into

$$\nabla^2 \boldsymbol{i} = \frac{1}{\lambda^2} \boldsymbol{i} , \qquad \nabla^2 \boldsymbol{B} = \frac{1}{\lambda^2} \boldsymbol{B} . \tag{5.62}$$

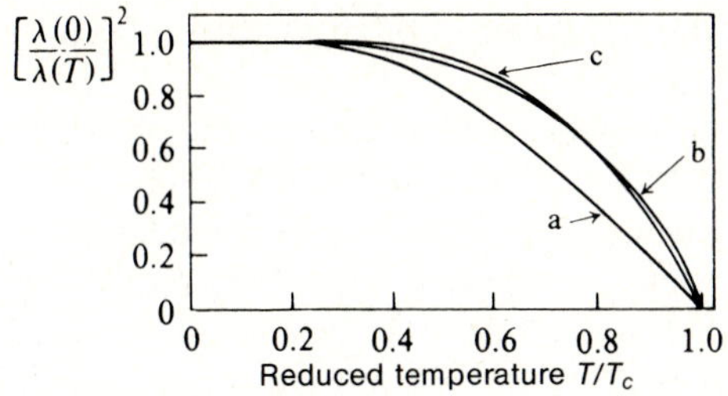

Fig. 5.8a-c. Penetration depth of a magnetic field according to the BCS theory: a) local approximation, b) nonlocal approximation, and c) according to the empirical law $\lambda^{-2} \propto 1 - (T/T_c)^4$ (from *Fetter* and *Walecka* [37]).

For the boundary conditions given here the solutions for $\boldsymbol{B} = (B_x, 0, 0)$ and $\boldsymbol{i} = (0, i_y, 0)$ are clearly proportional to the boundary values at $z = 0$ times a factor $\exp(-z/\lambda)$. The external magnetic field therefore induces in a surface layer of thickness λ a supercurrent which prevents the magnetic field from penetrating into the interior of the superconductor. The parameter λ defined in (5.60) is the *penetration depth* of the magnetic field. With (5.60) and (5.59) we can write

$$\lambda(T) = \lambda(0)\left[1 - \frac{2E_F}{k_F^5}\int_0^\infty k^4 dk\left(-\frac{\partial f_k}{\partial \bar{\varepsilon}}\right)\right]^{-1/2}. \tag{5.63}$$

The integral can be evaluated numerically when one knows the energy gap Δ as a function of temperature (Fig. 5.7). One finds the result presented in Fig. 5.8. At the transition temperature the penetration depth becomes infinite and the Meissner-Ochsenfeld effect disappears.

Figure 5.8 presents two further curves, an empirical law for the penetration depth $\lambda(T) = \lambda(0)[1 - (T/T_c)^4]^{-1/2}$, and the result of a "nonlocal BCS theory." By this we mean the following: If we do not restrict ourselves in the derivation of (5.60) to the limiting case $q = 0$, the relation between $\boldsymbol{i}$ and $\boldsymbol{A}$ becomes "nonlocal"

$$\langle \boldsymbol{i}_{\boldsymbol{q}} \rangle = f(\boldsymbol{q})\boldsymbol{A}_{\boldsymbol{q}}, \qquad \boldsymbol{i}(\boldsymbol{r}) = \int f(\boldsymbol{r} - \boldsymbol{r}')\boldsymbol{A}(\boldsymbol{r}')d\tau' \tag{5.64}$$

where $f(\boldsymbol{r})$ is the Fourier transform of $f(\boldsymbol{q})$. If $f(\boldsymbol{q}) = \text{const}, f(\boldsymbol{r} - \boldsymbol{r}') = \delta(\boldsymbol{r} - \boldsymbol{r}')$. In this case $\boldsymbol{i}(\boldsymbol{r})$ depends only on the local value of $\boldsymbol{A}(\boldsymbol{r})$. If $f(\boldsymbol{q})$ varies with $\boldsymbol{q}$ the current density at a particular position $\boldsymbol{r}$ depends on all values of $\boldsymbol{A}(\boldsymbol{r}')$ within a certain range given by a *coherence length* ξ. We return briefly to this in the next section.

5.7 Further Theoretical Concepts

In this chapter we have limited ourselves to *one* aspect of the theory of superconductivity: the presentation of the BCS theory formulated in terms of elementary excitations. There are, however, a number of other theories, which we can only mention briefly.

We mention first the *phenomenological theory* which describes the properties of superconductors by modifying the Maxwell equations of electrodynamics. The most important of these is the *London theory*. Since superconductivity is regarded as another state of the matter, the Maxwell equations remain untouched and only the material equations are modified. The change only involves the current density which is put in terms of a sum of a normal current density and a current density due to superconductivity. Ohm's law continues to apply for the normal current. For that due to superconductivity the London equations are introduced

$$\nabla \times \boldsymbol{i}_s = -\Lambda \boldsymbol{B}, \qquad \frac{\partial \boldsymbol{i}_s}{\partial t} = \Lambda \boldsymbol{E}, \qquad \Lambda^{-1} = \mu_0 \lambda^2 . \tag{5.65}$$

The first equation (5.65) was derived from the BCS theory in the last section.

Pippard presented an extension of these equations which showed that (particularly for problems involving spatial parameter variation) the spatial coherence of the wave functions has to be considered. This leads to the nonlocal London equation already referred to above, in which the current density at a particular position is related to the value of the vector potential in its vicinity. The *coherence length* ξ can be determined from the BCS theory when one does not restrict the calculation, as we did in the last section, to the limiting case $\boldsymbol{q} = 0$. The curves shown in Fig. 5.8 apply to the two limiting cases of large and small coherence length (compared with the penetration depth).

A later phenomenological theory by Ginzburg and Landau starts out from a different concept. The states of the electrons in a superconductor are divided into normal and superconducting states (two fluid model). An *order parameter* $\Psi(\boldsymbol{r})$ (macroscopic wave function) is introduced to identify the density of Cooper pairs condensed in the superconducting states: $n_s^* = |\Psi|^2$. The order parameter is determined by the nonlinear *Ginzburg-Landau equations*

$$\frac{1}{2m^*}(-i\hbar\nabla + e^*\boldsymbol{A})^2\Psi - \alpha(T_c - T)\Psi + \beta|\Psi|^2\Psi = 0 \tag{5.66}$$

and the supercurrent is given by

$$\boldsymbol{i}_s = \frac{ie^*\hbar}{2m^*}(\Psi^*\nabla\Psi - \Psi\nabla\Psi^*) - \frac{e^{*2}}{m^*}|\Psi|^2\boldsymbol{A} . \tag{5.67}$$

Since Ψ describes Cooper pairs, e^* is twice the electronic charge, m^* is twice the electronic mass, and n_s^* is half the electron densities in the superconducting state.

For a spatially uniform order parameter (5.67) becomes

$$\boldsymbol{i}_s = -\frac{e^{*2}}{m^*}|\Psi|^2\boldsymbol{A} = -\frac{e^2 n_s}{m}\boldsymbol{A}\,. \tag{5.68}$$

This is identical to (5.59) if one identifies the expression in brackets in (5.59) with the fraction of superconducting electrons n_s/n.

Comparing (5.59), (5.63) and (5.68), we find the relation $n_s = (\lambda(T)/\lambda(0))^2 n$. Thus Fig. 5.7 also shows the variation of n_s between $T = 0$ and $T = T_c$.

The features of the Ginzburg-Landau equations are revealed in the treatment of systems in which the order parameter changes with position. Introducing dimensionless quantities by $\Phi = \Psi/\Psi_0$, $\nabla_\xi = \xi\nabla$, $\Psi_0 = [\alpha/\beta\cdot(T_c - T)]^{1/2}$, $\xi = [\hbar^2/\alpha(T_c - T)]^{1/2}$ (5.66) becomes

$$-\nabla_\xi^2\Phi - (1 - |\Phi|^2)\Phi = 0\,. \tag{5.69}$$

From this we see that ξ is the characteristic length for spatial variations of the order parameter. It is the *coherence length* mentioned at the end of the last section.

For several applications it is useful to write the order parameter as

$$\Psi = |\Psi|\,\mathrm{e}^{\mathrm{i}\delta} = n_s^{1/2}\,\mathrm{e}^{\mathrm{i}\delta}\,. \tag{5.70}$$

Flux quantization and Josephson effect mentioned in Section 5.5 can be handled very easily by using (5.70) in the Ginzburg-Landau formalism (see Problem 5.4).

A further important parameter contained in the theory is the relation between penetration depth and coherence length $\kappa = \lambda/\xi$. It can be shown that superconductors behave in a fundamentally different way depending upon whether the *Ginzburg-Landau parameter* κ is larger or smaller than unity. In *type* I *superconductors* it is smaller than unity, in *type* II *superconductors* larger than unity. In the first case one finds a full Meissner-Ochsenfeld effect below a critical magnetic field, whereas above this field strength superconductivity disappears. In type II superconductors there is an intermediate state (mixed state) between two critical magnetic fields in which, in homogeneous conductors, normal and superconducting regions appear together. The regions of normal conduction extend in the direction of the magnetic field and contain a defined (quantized) magnetic flux. The requirement that they must contain at least one flux quantum means that there is a lower limit to their extent.

The phenomenologically derived Ginzburg-Landau equations were reduced later by Gorkov to the microscopic theory. In its present form (having been developed further by Abrikosov) the theory is called *GLAG theory* after Ginzburg, Landau, Abrikosov, and Gorkov. For a deeper discussion of this important theory and its applications we refer the reader to [71–77].

6. Interaction with Photons: Optics

6.1 Fundamentals

6.1.1 Introduction

The interaction between a solid and electromagnetic waves differs in many respects from the behaviour of the solid under the influence of static electric and magnetic fields. The optical phenomena of absorption, reflection, and dispersion of incident radiation appear in place of the transport of charge and energy. These processes can be described as the action of a high-frequency macroscopic field on the solid, or as the interaction between the elementary excitations of the solid and the quanta of the electromagnetic field—the *photons*.

In Sections 6.1.2 to 6.1.4 we lay the foundation for the theoretical description of optical effects in solids. We begin with a short discussion of the concept of the photon as an elementary excitation (Sec. 6.1.2). If the photons in the solid are very strongly coupled to other elementary excitations (optical phonons, excitons) the interaction can no longer be treated by perturbation theory. Photon and phonon (exciton) are then inextricably linked and a new elementary excitation must be introduced to describe the combination. This special case of *polaritons* is examined in Section 6.1.3. In Section 6.1.4 we introduce the complex dielectric constant, which connects the microscopic interaction processes between the elementary excitations with photons and the macroscopic effects of absorption, reflection, and dispersion.

In Sections 6.2 and 6.3, we concern ourselves with the electron-photon interaction and the phonon-photon interaction, respectively. We particularly refer to the following literature as relevant to the content of this chapter: the monographs, conference volumes, and collected papers [84–88, 112.34, 114b,c], also individual contributions in [112.52, 117] and the appropriate chapters from the introductions to solid-state theory mentioned in the reference list.

6.1.2 Photons

The Maxwell equations for the electromagnetic field in vacuum are (for vanishing current and charge density)

$$\nabla \times \boldsymbol{H} = \dot{\boldsymbol{D}}, \qquad \nabla \times \boldsymbol{E} = -\dot{\boldsymbol{B}}, \qquad \nabla \cdot \boldsymbol{B} = 0, \qquad \nabla \cdot \boldsymbol{D} = 0. \tag{6.1}$$

In the system of units employed here, $\boldsymbol{D} = \varepsilon_0 \boldsymbol{E}$ and $\boldsymbol{B} = \mu_0 \boldsymbol{H}$. This leads to the wave equations

$$\nabla^2 X - \frac{1}{c^2}\frac{\partial^2 X}{\partial t^2} = 0\,, \tag{6.2}$$

where X is a component of the electric or the magnetic field. Both fields can be expressed in terms of a vector potential $\boldsymbol{A}$. Since the Maxwell equations are homogeneous, the scalar potential can be put equal to zero. Hence

$$\boldsymbol{E} = -\dot{\boldsymbol{A}}\,, \qquad \boldsymbol{H} = \frac{1}{\mu_0}\nabla \times \boldsymbol{A}\,, \qquad \nabla \cdot \boldsymbol{A} = 0\,. \tag{6.3}$$

From this we find directly that the wave equation (6.2) also applies for the components of $\boldsymbol{A}$. Particular solutions of the wave equations are the monochromatic plane waves

$$X = X_0 \exp\left[\mathrm{i}(\boldsymbol{\kappa}\cdot\boldsymbol{r} - \omega_{\boldsymbol{\kappa}} t)\right], \qquad \omega_{\boldsymbol{\kappa}} = \kappa c \tag{6.4}$$

and from these we can construct general solutions by superposition.

We enclose the field in a finite volume V_{g}, so that it can be represented as a Fourier series

$$\boldsymbol{A} = \sum_{\boldsymbol{\kappa}} \left[\boldsymbol{a}_{\boldsymbol{\kappa}} \exp(\mathrm{i}\boldsymbol{\kappa}\cdot\boldsymbol{r}) + \boldsymbol{a}_{\boldsymbol{\kappa}}^* \exp(-\mathrm{i}\boldsymbol{\kappa}\cdot\boldsymbol{r})\right]. \tag{6.5}$$

The time dependence of the $\boldsymbol{a}_{\boldsymbol{\kappa}}$ in (6.5) is $\exp(-i\omega_{\boldsymbol{\kappa}} t)$. Our goal is the quantization of this field. As in Section 3.1.4 for the electric field of the collective oscillations of the electron gas, we introduce canonical field variables $\boldsymbol{Q}_{\boldsymbol{\kappa}}$ and $\boldsymbol{P}_{\boldsymbol{\kappa}}$. We set

$$\boldsymbol{Q}_{\boldsymbol{\kappa}} = \sqrt{V_{\mathrm{g}}\varepsilon_0}(\boldsymbol{a}_{\boldsymbol{\kappa}} + \boldsymbol{a}_{\boldsymbol{\kappa}}^*)\,, \qquad \boldsymbol{P}_{\boldsymbol{\kappa}} = \dot{\boldsymbol{Q}}_{\boldsymbol{\kappa}} = -\mathrm{i}\omega_{\boldsymbol{\kappa}}\sqrt{V_{\mathrm{g}}\varepsilon_0}(\boldsymbol{a}_{\boldsymbol{\kappa}} - \boldsymbol{a}_{\boldsymbol{\kappa}}^*)\,. \tag{6.6}$$

Since $\nabla\cdot\boldsymbol{A} = 0$, the vectors $\boldsymbol{a}_{\boldsymbol{\kappa}}$, and hence also the $\boldsymbol{Q}_{\boldsymbol{\kappa}}$ and $\boldsymbol{P}_{\boldsymbol{\kappa}}$, are perpendicular to $\boldsymbol{\kappa}$. In the plane perpendicular to $\boldsymbol{\kappa}$ they have therefore two components, which we distinguish by the index α. The Hamiltonian then follows after a simple calculation

$$H = \frac{1}{2}\int(\varepsilon_0 E^2 + \mu_0 H^2)d\tau = \frac{1}{2}\sum_{\boldsymbol{\kappa}\alpha}(P_{\boldsymbol{\kappa}\alpha}^2 + \omega_{\boldsymbol{\kappa}\alpha}^2 Q_{\boldsymbol{\kappa}\alpha}^2)\,. \tag{6.7}$$

We can now quantize exactly in accordance with Appendix A.

Equation (6.7) has the form of (A.1) with real $Q_{\boldsymbol{\kappa}\alpha}$ and $P_{\boldsymbol{\kappa}\alpha}$. We therefore introduce the commutation relations

$$[Q_{\boldsymbol{\kappa}\alpha}, P_{\boldsymbol{\kappa}'\alpha'}] = i\hbar\delta_{\boldsymbol{\kappa}\boldsymbol{\kappa}'}\delta_{\alpha\alpha'} \tag{6.8}$$

for the operators $Q_{\kappa\alpha}$ and $P_{\kappa\alpha}$, and combine the operators into new *photon creation and annihilation operators*

$$c^+_{\kappa\alpha} = \frac{1}{\sqrt{2\hbar\omega_{\kappa\alpha}}}(\omega_{\kappa\alpha}Q_{\kappa\alpha} - \mathrm{i}P_{\kappa\alpha}), \qquad c_{\kappa\alpha} = \frac{1}{\sqrt{2\hbar\omega_{\kappa\alpha}}}(\omega_{\kappa\alpha}Q_{\kappa\alpha} + \mathrm{i}P_{\kappa\alpha}). \tag{6.9}$$

The Hamiltonian then takes the form

$$H = \sum_{\kappa\alpha} \hbar\omega_{\kappa\alpha}\left(c^+_{\kappa\alpha}c_{\kappa\alpha} + \frac{1}{2}\right). \tag{6.10}$$

6.1.3 Polaritons

Absorption of electromagnetic radiation means energy transfer from the electromagnetic field to the solid, or put another way, absorption of photons with creation of elementary excitations. For the energy to indeed be extracted from the radiation field the coupling between elementary excitations and photons should not be stronger than other interactions in the solid which further dissipate the energy absorbed. Otherwise the probability that photons are reemitted is too high, and the energy absorbed will be returned to the radiation field (Fig. 6.1). A typical example is the interaction of the transverse electromagnetic field

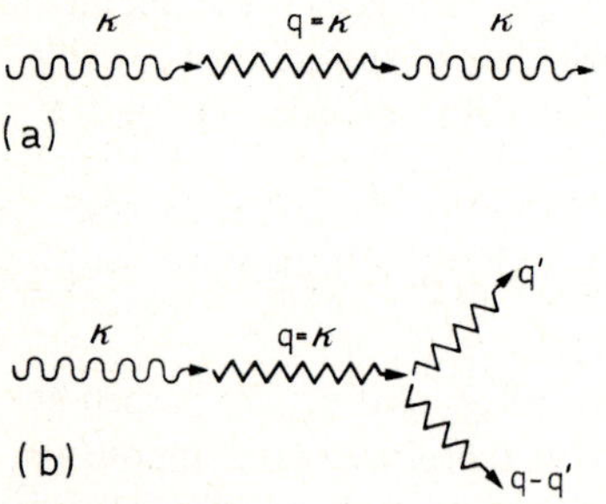

Fig. 6.1a and b

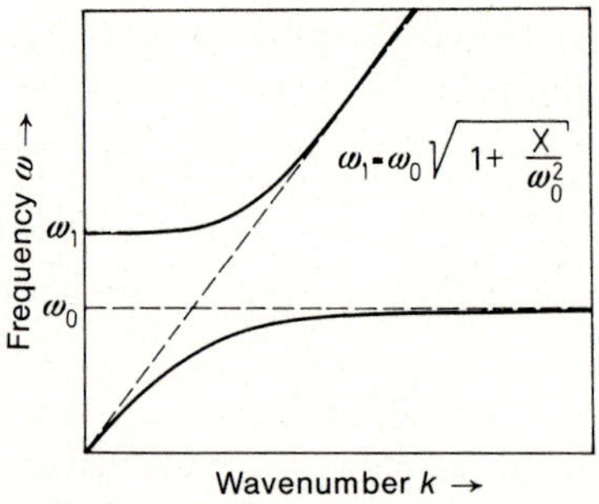

Fig. 6.2

Fig. 6.1a and b. Absorption of a photon with emission of a phonon. (a) In a subsequent process the photon is reemitted. (b) In a subsequent process the phonon decays into two phonons of lower energy; only then has the photon energy finally been extracted from the radiation field.

Fig. 6.2. At the crossover point of two dispersion curves (here the dispersion of photons $\omega = ck$, and polarization quanta $\omega = \omega_0$), energy and wave vector conservation allow a transformation of the one excitation into the other in accordance with Fig. 6.1. The strong interaction leads to a mixed form, the *polariton*. The dispersion relation of the polariton has two branches. Well above the crossover, the polariton of one of the branches becomes a photon, and that of the other becomes a polarization quantum. The same applies below crossover ($k \to 0$), only there the velocity of the photon and the frequency of the polarization quantum are changed.

with the transverse polarization waves of the solid. The associated elementary excitations are the TO-phonons, for the polarization of the ion lattice, and the excitons, for the polarization of the electron system. Both are Bose particles. We designate both in the following as polarization quanta. The interaction of the electromagnetic field then consists of photon-polarization quantum transformation and vice versa. Absorption takes place only if the polarization quantum decays into (one or more) other elementary excitations instead of reemitting a photon.

The dispersion of polarization quanta is very weak and the dispersion curves of photons and polarization quanta intersect at low wave vector (Fig. 6.2). Near the point of intersection—where energy and momentum are the same for both excitations—the coupling becomes so strong that neither can continue to be regarded as an independent elementary excitation. Photon and polarization quantum rather exist together as a single quantity which can be interpreted as a new elementary excitation, and one whose interaction with the other elementary excitations in the solid represents the real absorption process. It is called a *polariton.*

To examine the polariton concept more clearly we introduce, along with the vector potential of light, the polarization of the dielectric. We express both vectors in terms of creation and annihilation operators

$$\boldsymbol{A} = \sum_{\kappa\alpha} A_0(c_{\kappa\alpha} + c^+_{-\kappa\alpha}) \exp(\mathrm{i}\boldsymbol{\kappa}\cdot\boldsymbol{r}), \tag{6.11}$$

$$\boldsymbol{P} = \sum_{k\alpha} \boldsymbol{P}_0(b_{k\alpha} + b^+_{-k\alpha}) \exp(\mathrm{i}\boldsymbol{k}\cdot\boldsymbol{r}). \tag{6.12}$$

The summations are over the wave vectors $\boldsymbol{\kappa}$ and $\boldsymbol{k}$, respectively, and the two directions of polarization in the plane perpendicular to it. For simplicity we shall in future use the index $\boldsymbol{k}$ for the combination of summation indices in both cases.

For the Hamiltonian of the photon-polarization quantum interaction we put

$$H = \sum_{k}\left[E_{1k}\left(c_k^+c_k + \frac{1}{2}\right) + E_{2k}\left(b_k^+b_k + \frac{1}{2}\right) + E_{3k}(c_k^+b_k - c_kb_k^+ - c_kb_{-k} + c^+_{-k}b_k^+)\right]. \tag{6.13}$$

The first term is the Hamiltonian for the photon field, the second that for the polarization field. The third term describes the interaction, the signs having been fixed in accordance with the following argument. The interaction depends on the product of the two fields $\boldsymbol{E}\cdot\boldsymbol{P}$. Since $\boldsymbol{E}\cdot\boldsymbol{P} = -\dot{\boldsymbol{A}}\cdot\boldsymbol{P}$, following (6.11) and (6.12) the interaction term contains products of the form $-(c_k - c^+_{-k}) \times (b_k + b^+_{-k})$ or, after exchanging the summation index $\boldsymbol{k}$ for $-\boldsymbol{k}$ in the b_k, the four combinations in the last term in (6.13). To find the energies E_{ik} we would have to derive the Hamiltonian explicitly. We want to avoid this laborious approach. Instead, we obtain the E_{ik} by a different method.

For this we return to the classical problem. We describe the electromagnetic field by Maxwell's equations and the polarization field by an equation of motion in which free oscillators of eigenfrequency ω_0 are coupled with the electric field through a susceptibility χ

$$\nabla \times \boldsymbol{H} = \varepsilon_0 \dot{\boldsymbol{E}} + \dot{\boldsymbol{P}}, \qquad \nabla \times \boldsymbol{E} = -\mu_0 \dot{\boldsymbol{H}}, \qquad \ddot{\boldsymbol{P}} + \omega_0^2 \boldsymbol{P} = \chi \varepsilon_0 \boldsymbol{E}. \tag{6.14}$$

These are the three defining equations for the fields $\boldsymbol{E}$, $\boldsymbol{H}$, and $\boldsymbol{P}$. We take these fields to be transverse plane waves which propagate in the z-direction

$$\begin{aligned} E_x &= E_{x0} \exp[\mathrm{i}(kz - \omega t)], \qquad H_y = H_{y0} \exp[\mathrm{i}(kz - \omega t)], \\ P_x &= P_{x0} \exp[\mathrm{i}(kz - \omega t)]. \end{aligned} \tag{6.15}$$

Inserting (6.15) into (6.14) leads to

$$\begin{aligned} &\omega \varepsilon_0 E_x + \omega P_x - k H_y = 0, \qquad k E_x - \omega \mu_0 H_y = 0, \\ &\chi \varepsilon_0 E_x + (\omega^2 - \omega_0^2) P_x = 0. \end{aligned} \tag{6.16}$$

This system of equations only has a solution if the determinant

$$\begin{vmatrix} \omega \varepsilon_0 & \omega & -k \\ k & 0 & -\omega \mu_0 \\ \chi \varepsilon_0 & \omega^2 - \omega_0^2 & 0 \end{vmatrix} \tag{6.17}$$

vanishes. Thus

$$\omega^4 - \omega^2(\omega_0^2 + \chi + c^2 k^2) + \omega_0^2 c^2 k^2 = 0 \tag{6.18}$$

from which the dispersion relation $\omega(k)$ for polaritons can be calculated. We discuss this relation (shown in Fig. 6.2) further below.

First we attempt to obtain the same relation from the Hamiltonian (6.13). Eq. (6.13) contains creation and annihilation operators for photons and polarization quanta. By linear combination we introduce creators and annihilators for *polaritons*

$$\alpha_{\boldsymbol{k}} = w c_{\boldsymbol{k}} + x b_{\boldsymbol{k}} + y c_{-\boldsymbol{k}}^{+} + z b_{-\boldsymbol{k}}^{+}. \tag{6.19}$$

We define the $\alpha_{\boldsymbol{k}}^{+}$ correspondingly. The coefficients w, x, y, z can be obtained, similar to the antiferromagnetic magnon case in Section 3.4.3, by requiring that the Hamiltonian expressed in terms of the α_k should take the form

$$H = \sum_k \left[E_k^{(1)}\left(\alpha_{1k}^+\alpha_{1k} + \frac{1}{2}\right) + E_k^{(2)}\left(\alpha_{2k}^+\alpha_{2k} + \frac{1}{2}\right)\right]. \tag{6.20}$$

In this account has been taken of the fact that, for a given $\boldsymbol{k}$, a creation and an annihilation operator exist for each of the two polariton branches (Fig. 6.2).

From (6.19) and (6.20) we find for the commutator of one of the new operators with H

$$[\alpha_k, H] = E_k\alpha_k. \tag{6.21}$$

We have omitted the index 1, 2 here. This relation must also apply between the operator defined in (6.19) and the Hamiltonian (6.13). If we write it explicitly, we obtain an operator equation of the form

$$A_1 c_k + A_2 b_k + A_3 c_{-k}^+ + A_4 b_{-k}^+ = 0 \tag{6.22}$$

which reduces to the equations $A_i = 0$ or $a_i w + b_i x + c_i y + d_i z = 0$ by forming suitable matrix elements. These equations for w, x, y, z can only be solved when their determinant vanishes. One obtains here

$$\begin{vmatrix} E_{1k} - E_k & -E_{3k} & 0 & -E_{3k} \\ E_{3k} & E_{2k} - E_k & -E_{3k} & 0 \\ 0 & -E_{3k} & -E_{1k} - E_k & -E_{3k} \\ -E_{3k} & 0 & E_{3k} & -E_{2k} - E_k \end{vmatrix} = 0. \tag{6.23}$$

This gives the secular equation

$$E_k^4 - E_k^2(E_{1k}^2 + E_{2k}^2) + E_{1k}^2E_{2k}^2 + 4E_{1k}E_{2k}E_{3k}^2 = 0. \tag{6.24}$$

By comparing with (6.18) one therefore finds for the E_{ik} of the Hamiltonian (6.13)

$$E_{1k} = \hbar ck, \qquad E_{2k} = \hbar\omega_0\sqrt{1 + \chi/\omega_0^2}, \qquad E_{3k} = \mathrm{i}\left(\frac{\chi ck\hbar^2}{4\omega_0\sqrt{1 + \chi/\omega_0^2}}\right)^{1/2}. \tag{6.25}$$

We now examine Fig. 6.2. For large k the two branches of the dispersion spectrum become identical to the branches $\omega = ck$ for photons, and $\omega = \omega_0$ for polarization quanta. For small k, on the other hand, the spectrum exhibits gross deviations. The one branch admittedly increases linearly from zero but its slope is not c but $c/[1 + (\chi/\omega_0^2)]^{1/2}$. The other branch starts at $\omega = \omega_0[1 + (\chi/\omega_0^2)]^{1/2}$.

Now we consider the two possible polarization quanta, the optical phonons and the excitons.

In Section 3.3.8 we found that the frequency at $\boldsymbol{q} = 0$ of the longitudinal optical phonons is larger than that of the transverse phonons (Lyddane-Sachs-Teller relation). This follows from the fact that the longitudinal vibrations of charged ions are connected with the appearance of an internal macroscopic electric field, while the transverse on the other hand are not. Similarly, the shift of the frequency of the upper polariton branch at $\boldsymbol{q} = 0$ to higher frequencies is caused by the electric field of the light.

To determine the dispersion spectrum of the *phonon-polaritons* (as distinct from the *exciton-polaritons*), we put the two equations (3.186) in place of the third equation (6.14). Along with $\boldsymbol{E}$ and $\boldsymbol{H}$, the fields $\boldsymbol{P}$ and $\boldsymbol{w}$ therefore appear. Eq. (6.17) then becomes a 4×4-determinant with the secular equation

$$\omega^4 - \omega^2\left[\omega_L^2 + \frac{c^2k^2}{\varepsilon(\infty)}\right] + \omega_T^2\frac{c^2k^2}{\varepsilon(\infty)} = 0\,. \tag{6.26}$$

The corresponding spectrum is shown in Fig. 6.3. The frequency shift of the upper branch at $\boldsymbol{k} = 0$ is exactly the Lyddane-Sachs-Teller shift. The only difference from Fig. 6.2 is the fact that for large k the slope of the upper branch is no longer c but $c/\sqrt{\varepsilon(\infty)}$. The lower branch initially has the slope $c/\sqrt{\varepsilon(0)}$ and later approaches the value ω_T.

We find very similar relations for the exciton polariton. Here we must point out something we have not mentioned so far, namely that there are longitudinal and transverse excitons. There are, therefore, exciton states which are associated with polarization perpendicular or parallel to the exciton wave vector $\boldsymbol{K}$. For further details we refer to the book by *Knox* [102.5].

Optical phonons and excitons possess many common characteristics. Both are Bose particles. In nonpolar cubic solids the three phonon branches are degenerate at $\boldsymbol{q} = 0$. Excitons with $\boldsymbol{K} = 0$, produced by an allowed dipole transition between two bands (e.g., an s-p-transition), also occupy a threefold degenerate state (degeneracy of the three p-functions). If one goes to small $\boldsymbol{q}$

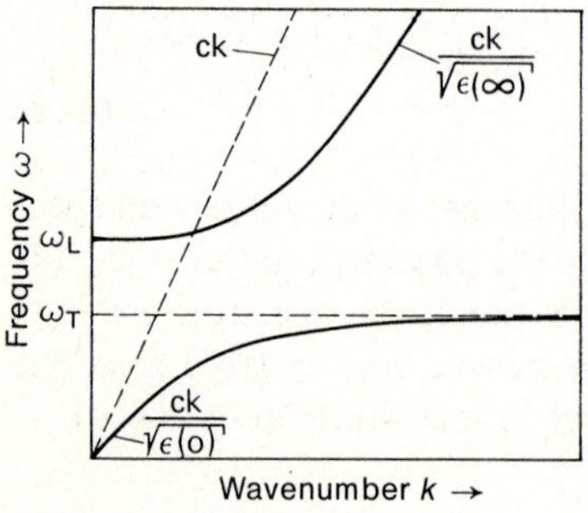

Fig. 6.3. Dispersion curves for the phonon-polariton. In contrast to Fig. 6.2, the slope of the photonlike branch above the crossover point is less than the velocity of light.

or $\boldsymbol{K}$ values, the threefold degeneracy splits up into a single and a double branch, which can be identified with the longitudinal and transverse branches, respectively. We saw this for the phonons in Fig. 3.11. For polar solids the threefold degeneracy is already lifted at $\boldsymbol{q} = 0$ and $\boldsymbol{K} = 0$.

Our earlier discussion of phonons however also shows that in both cases the strict division into longitudinal and transverse polarization quanta is valid only along individual axes of symmetry. It is not valid at a general point in the Brillouin zone. This is of secondary importance for the polariton problem since in the immediate vicinity of $\boldsymbol{k} = 0$ (the $\boldsymbol{\kappa}$-vector of light is about 10^{-4} times smaller than the dimensions of the Brillouin zone) it is always possible to separate the branches into longitudinal and transverse.

"Mixed forms" of elementary excitations such as the polaritons treated here also arise in other cases, for example in the strong coupling between LO-phonons and plasmons (e.g., *Mooradian* [103.IX]). For further discussion on the polariton problem see, in particular, the article by *Hopfield* in [117], *Mills* and *Burstein* in [111c.36], and the conference proceedings [119].

6.1.4 The Complex Dielectric Constant

The macroscopic interaction of electromagnetic fields with matter is already contained in the Maxwell equations. The equations (6.1) also apply to matter, with the first equation extended for conductors to include a current term, and with the possible appearance of space charges as sources of the displacement field. We do not have to make use of either extension if we restrict ourselves to optical effects in nonconductors.

The fields $\boldsymbol{E}$ and $\boldsymbol{D}$, $\boldsymbol{B}$ and $\boldsymbol{H}$ are now related by the equations

$$\boldsymbol{D} = \varepsilon_0 \varepsilon \boldsymbol{E}, \qquad \boldsymbol{B} = \mu_0 \mu \boldsymbol{H}. \tag{6.27}$$

In what follows we shall put $\mu = 1$, thereby restricting ourselves to nonmagnetic solids. The constant ε then contains all information about the interaction of the fields with the matter.

The two equations (6.27) (together with Ohm's law $\boldsymbol{i} = \sigma \boldsymbol{E}$ in conductors) express the empirical fact that, in electrostatics and magnetostatics, the "excitations" are linearly related to the exciting field. It is not obvious that these equations are also valid for high-frequency fields. It is therefore desirable, without explicit recourse to the Maxwell equations, to formulate the first equation (6.27) anew from some general assumptions about the nature of the interaction.

In optics there is also a relationship between the electric field of the light wave and the displacement field in the solid. In (6.27) the relationship is *scalar, linear, local,* and *synchronous.* The scalar form means the assumption of an isotropic medium, an assumption which we wish to adopt in the following. The change to anisotropic media by introducing a tensor relationship does not encounter any fundamental difficulties. The linearity of the relationship means a restriction to weak (compared with the internal electric fields) external fields. A nonlinear connection cannot of course be excluded, but *linear optics* is adequate—as long as high-intensity excitation, e.g., by a laser, is excluded. Eq. (6.27) also contains the assumption that the displacement field $\boldsymbol{D}(\boldsymbol{r}, t)$ is given by the value of the macroscopic field at the same location and the same time. We shall also maintain the requirement of locality although the possibility of nonlocality (*spatial dispersion*) cannot be altogether discounted (see, for example, *Cardona* in [112.52]). On the other hand, for rapidly changing fields the excitation is not necessarily synchronous with the exciting field. It is only necessary that the coupling be *causal*, i.e., only values of $\boldsymbol{E}$ before time t should contribute to the excitation at time t.

As a result of these assumptions we define the *dielectric constant* as follows (for further details see, for example, the contribution by *Tauc* in [112.34]): We put

$$\boldsymbol{D} = \varepsilon_0 \boldsymbol{E} + \boldsymbol{P} = \varepsilon_0 \left(\boldsymbol{E} + \int_{-\infty}^{t} f(t - t')\boldsymbol{E}(t')dt' \right), \tag{6.28}$$

where we have also introduced the polarization $\boldsymbol{P}$ and linked it causally with the electric field. If one interprets (6.28) as a linear relationship between $\boldsymbol{D}$ and $\boldsymbol{E}$ through a "dielectric constant ε," then ε is a linear integral operator. Representing $\boldsymbol{E}$ as a Fourier sum of plane waves of different frequency, the relationship between the individual terms becomes frequency dependent

$$\begin{aligned} \boldsymbol{P}(t) &= \int_{-\infty}^{t} f(t - t')\varepsilon_0 \boldsymbol{E}_0 \exp(-\mathrm{i}\omega t')dt' \\ &= \exp(-\mathrm{i}\omega t) \int_0^{\infty} f(t'') \exp(\mathrm{i}\omega t'')dt''\varepsilon_0 \boldsymbol{E}_0 = \exp(-\mathrm{i}\omega t)\boldsymbol{P}_0 . \end{aligned} \tag{6.29}$$

This defines the *complex, frequency-dependent dielectric constant*

$$\varepsilon(\omega) = 1 + \int_0^{\infty} f(t) \exp(\mathrm{i}\omega t)dt . \tag{6.30}$$

In view of the assumption of locality this quantity is not wave-number dependent like the dielectric constant of the electron gas defined in Section 3.1.6. The two quantities also differ in other respects. Thus the constant introduced here describes the response of the solid to a *transverse* disturbance, while that in Section 3.1.6 was concerned with a *longitudinal* disturbance (associated with density fluctuations of the electron gas).

The expression (6.28) is a special case of the general relationship between a disturbance and the excitation induced according to *linear response theory*. The integral operator involved in this relationship, the *generalized susceptibility*, possesses some remarkable properties as, for example, a relation between its real and imaginary parts (dispersion relation, Kramers-Kronig relation). This relation is a consequence of causality, i.e., the fact that there can be no response before a field is applied.

In the present case (transverse dielectric constant in a nonconductor), the dispersion relation is given by (see the literature given in Sec. 6.1.1 for a proof)

$$\varepsilon(\omega) - 1 = \frac{1}{\mathrm{i}\pi} P \int_{-\infty}^{+\infty} \frac{\varepsilon(\xi) - 1}{\xi - \omega} d\xi \tag{6.31}$$

where the principal value of the integral has to be taken. Separating the real and imaginary parts of the dielectric constant

$$\varepsilon(\omega) = \varepsilon_1(\omega) + \mathrm{i}\varepsilon_2(\omega) \tag{6.32}$$

the two relations below follow from (6.31)

$$\varepsilon_1(\omega) - 1 = \frac{1}{\pi} P \int_{-\infty}^{+\infty} \frac{\varepsilon_2(\xi)}{\xi - \omega} d\xi = \frac{2}{\pi} P \int_0^{\infty} \frac{\xi \varepsilon_2(\xi)}{\xi^2 - \omega^2} d\xi \,, \tag{6.33}$$

$$\varepsilon_2(\omega) = -\frac{1}{\pi} P \int_{-\infty}^{+\infty} \frac{\varepsilon_1(\xi) - 1}{\xi - \omega} d\xi = -\frac{2\omega}{\pi} P \int_0^{\infty} \frac{\varepsilon_1(\xi)}{\xi^2 - \omega^2} d\xi \,. \tag{6.34}$$

The real part can thus be calculated when the imaginary part is known for all frequencies, and vice versa.

In what follows we shall reduce practically all theoretical results to the two functions $\varepsilon_1(\omega)$ and $\varepsilon_2(\omega)$. We must therefore determine first of all how they are related to parameters which can be measured experimentally.

We consider a monochromatic plane wave in an absorbing medium: $\boldsymbol{E} = \boldsymbol{E}_0 \exp[\mathrm{i}(\kappa z - \omega t)]$. Let the dielectric constant of the medium be $\neq 1$, but let $\mu = 1$ ($\boldsymbol{D} = \varepsilon\varepsilon_0 \boldsymbol{E}$, $\boldsymbol{B} = \mu_0 \boldsymbol{H}$). Instead of the relation $\kappa = \omega/c$ (6.4) we then have $\kappa^2 = (\omega/c)^2\varepsilon$. If we introduce the *complex refractive index* N by $N^2 = \varepsilon$, $N = n + \mathrm{i}k$, then $\kappa = (\omega/c)N$. The two *optical constants* n (real index of refraction) and k (extinction coefficient) are related to ε_1 and ε_2 by

$$\varepsilon_1 = n^2 - k^2 \,, \qquad \varepsilon_2 = 2nk \,. \tag{6.35}$$

If one puts this into the equation for a plane wave propagating in the z-direction one finds

$$\boldsymbol{E} = \boldsymbol{E}_0 \exp\left(-\frac{\omega k}{c} z\right) \exp\left[\mathrm{i}\left(\frac{n\omega}{c} z - \omega t\right)\right]. \tag{6.36}$$

The wave is damped, k describes the *absorption* of the wave in the medium, and n its *dispersion*. From (6.35), ε_1 determines the dispersion (through the factor n) and ε_2 the absorption (through k).

The optical constants are often determined by measuring *transmission* and *reflection*. For normal incidence on a layer of thickness d, the intensities of transmission and reflection, I_t and I_r are given by

$$I_t = I_0 \frac{(1 - R^2)\exp(-Kd)}{1 - R^2 \exp(-2Kd)}\left(1 - \frac{k^2}{n^2}\right), \qquad I_r = I_0 R \frac{1 - \exp(-2Kd)}{1 - R^2 \exp(-2Kd)}, \tag{6.37}$$

where K is the *absorption coefficient*

$$K = \frac{2\omega k}{c} \tag{6.38}$$

and R is the *reflection coefficient*

$$R = \frac{(n-1)^2 + k^2}{(n+1)^2 + k^2}. \tag{6.39}$$

6.2 Electron-Photon Interaction

6.2.1 Introduction

In the sections which follow we discuss those interaction processes in which electrons are excited into higher one-electron states of the band model by photon absorption. We assume first of all that the initial state belongs to an occupied band (valence band) and that the final state lies in an empty band (conduction band). We shall therefore be considering optical absorption in semiconductors and insulators.

Figure 6.4 presents some typical experimental results. Fig. 6.4a shows the reflection spectrum of germanium in the region below 20 eV; Fig. 6.4b, c show the functions $\varepsilon_1(\omega)$ and $\varepsilon_2(\omega)$ determined from the reflection spectrum. The fact that real and imaginary parts of the complex dielectric constant can be obtained from a single experimental spectrum follows from the Kramers-Kronig relations (6.33, 34), which allow us to find the one function from a knowledge of the other over the entire spectral region. On this point see [85], for example.

For us the most important curve here is Fig. 6.4c, since ε_2 determines the absorption. One can see that above a threshold energy there is a region of high absorption with significant structure, after which the absorption decreases continuously. The spectrum in Fig. 6.4c is typical of all semiconductors and insulators. The threshold energy corresponds to the energy gap between the maximum of the valence band and the minimum of the conduction band. It is the smallest energy

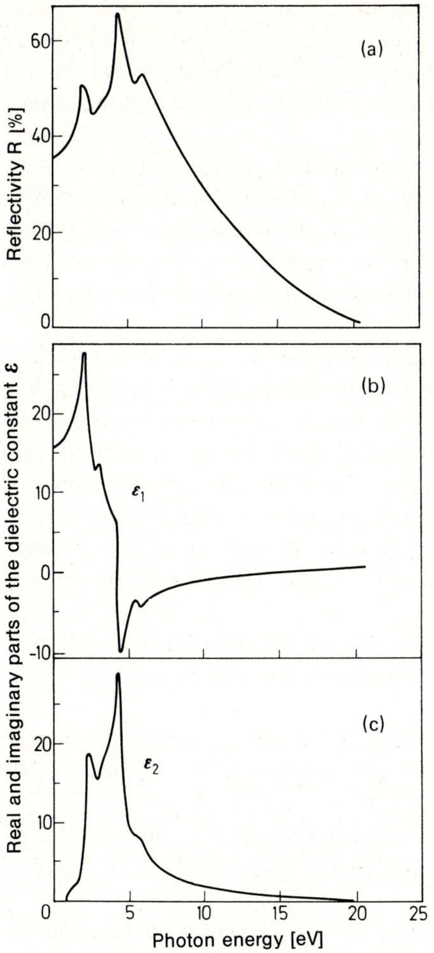

Fig. 6.4a-c. The reflectivity of germanium between 0 and 20 eV and the spectra derived from it for the real and imaginary parts of the complex dielectric constant.

an electron must receive in order to make a transition into the conduction band. The structure above the threshold concerns transitions from states in the valence band into higher conduction bands. Their contribution to ε_2 is strongly dependent on where the initial and final states lie in the Brillouin zone, how large the densities of states are, and whether the transitions are allowed or forbidden by the symmetry of the states concerned.

In the simplest case, apart from the photon and the electron, no other elementary excitations are involved in the interaction (direct transitions).

The laws of conservation of energy and wave vector require that the excitation energy $E_f(\boldsymbol{k}_f) - E_i(\boldsymbol{k}_i)$ be equal to the energy of the photon absorbed

$\hbar\omega$, and that $\boldsymbol{k}_f - \boldsymbol{k}_i$ be equal to the photon wave vector $\boldsymbol{\kappa}$. Since for the energies involved (of order eV) the photon wave numbers are several orders of magnitude smaller than the dimensions of the Brillouin zone, $\boldsymbol{\kappa}$ can be neglected. We can therefore assume conservation of $\boldsymbol{k}$ in the transition. We shall be looking at such transitions in Section 6.2.2.

As in Fig. 4.1, it is convenient to represent the possible interactions by graphs (Fig. 6.5). The individual processes are the absorption of a photon with the transition of an electron from the valence to the conduction band (at approximately the same $\boldsymbol{k}$), and the relaxation of the electron with emission of a photon (radiative transition). These processes are indicated at the top of Fig. 6.5.

The complete process which we want to examine in the following sections is the transition of an electron from state $\boldsymbol{k}$ in the valence band into state $\boldsymbol{k} + \boldsymbol{q}$ in the conduction band. Three conditions should be considered; a) The transition must be either one of the individual processes or be composed of individual processes with virtual intermediate states, b) the individual processes must be allowed, i.e., the transition matrix elements should not vanish on symmetry grounds, and c) momentum must be conserved in each individual process, whereas energy conservation is required only for the complete transition, not for the individual steps.

If $\boldsymbol{q} = 0$, *direct processes* are possible. Energy conservation and selection rules can require the successive absorption of two photons for certain transitions. We discuss this case in Section 6.2.4.

If $\boldsymbol{q} \neq 0$, momentum can only be conserved if an additional phonon of momentum $\boldsymbol{q}$ is absorbed or a phonon $-\boldsymbol{q}$ is emitted (Fig. 6.5, below). We treat these *indirect transitions* in Section 6.2.3.

Section 6.2.5 considers an interaction neglected up till then, namely that of the electron with the hole it leaves behind in the valence band, i.e., *exciton effects* in the absorption spectrum.

In Sections 6.2.2 to 6.2.5 we shall not be able to present a comparison between theory and experiment, since all the processes mentioned collectively contribute to the shape of the absorption and reflection spectra. Section 6.2.6 therefore is devoted to such a comparison.

In considering only *interband transitions,* we have ignored the possibility of an electron transition within the same band. This process can occur in metals and semiconductors, but requires other methods of treatment. We treat it as a transport problem—the behaviour of an electron gas in a high-frequency electric field. We relate the high-frequency conductivity obtained to the dielectric constant. This is done in Section 6.2.7.

Optical spectra can be affected by external parameters. One of the most marked effects is the change in electron transitions due to a change of band structure in a magnetic field (Sec. 2.1.6). We treat the phenomenon of *magneto-absorption* and *magneto-reflection* in Section 6.2.8, and the *magneto-optics of free charge carriers* in Section 6.2.9.

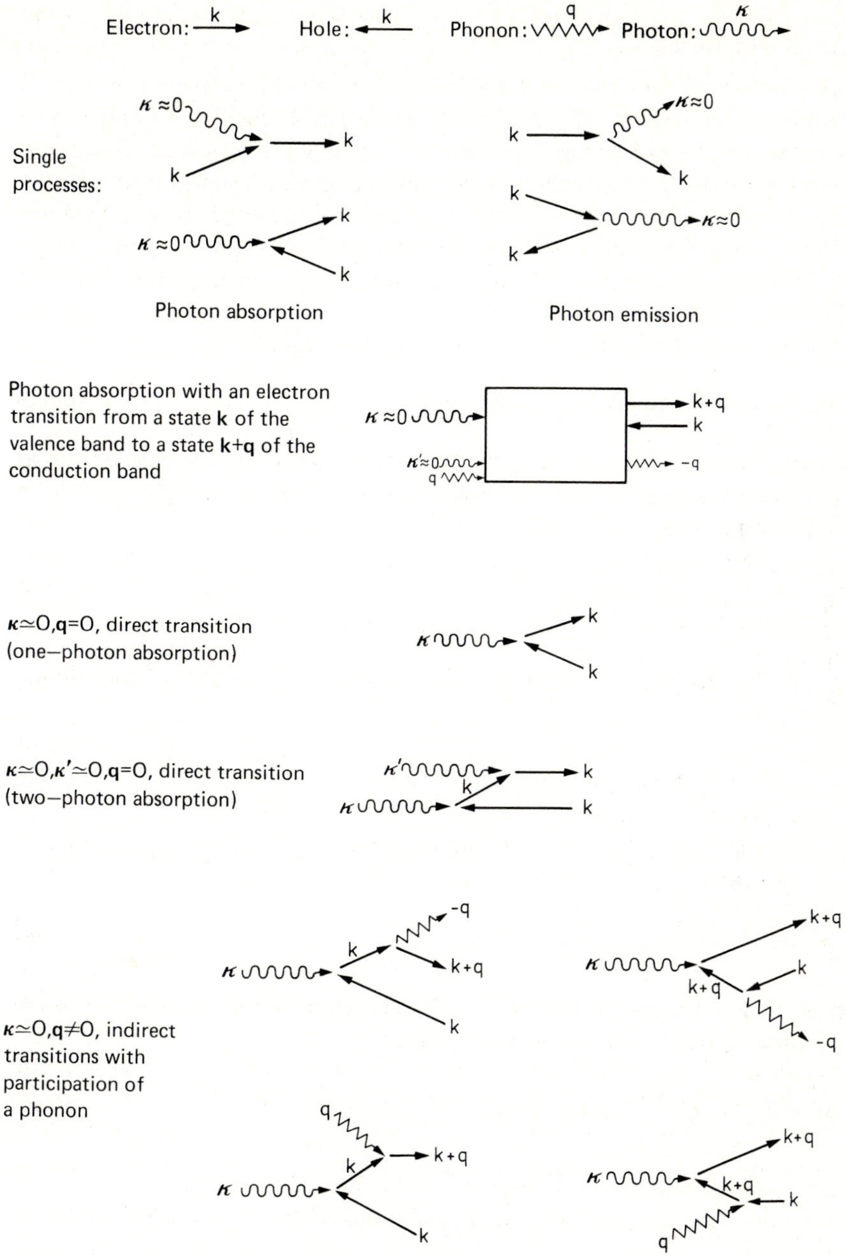

Fig. 6.5. Graphs for the electron-photon interaction.

6.2.2 Direct Transitions

In this section we examine the simplest interaction process, the absorption of a photon by an electron. In this process the electron changes its energy and momentum by the energy and momentum of the photon absorbed. Our aim is the calculation of the absorption coefficient due to such elementary processes.

The absorption coefficient is defined as the energy absorbed (per unit volume and time) divided by the incident energy (per unit area and time). If $W(\omega)$ is the number of photons absorbed per unit volume per unit time, then $\hbar\omega W(\omega)$ is the energy absorbed.

We represent the incident light by its vector potential

$$A(r, t) = A_0 e \exp[\mathrm{i}(\kappa \cdot r - \omega t)] + \text{c.c.} \tag{6.40}$$

The energy incident per unit area and time then follows as the product of the energy density $u = (1/2)(\varepsilon\varepsilon_0 E^2 + \mu_0 H^2)$, and the velocity of light $v = c/n$ in the medium. Hence

$$K = \frac{2\hbar}{\varepsilon_0 n c \omega A_0^2} W(\omega) . \tag{6.41}$$

From (6.35) and (6.38) we find for the imaginary part of the complex dielectric constant

$$\varepsilon_2 = \frac{2\hbar}{\varepsilon_0 \omega^2 A_0^2} W(\omega) . \tag{6.42}$$

To determine the function $W(\omega)$ we start from the Schrödinger equation

$$\left[\frac{1}{2m}(p + eA)^2 + V(r)\right]\psi = \mathrm{i}\hbar\dot{\psi} . \tag{6.43}$$

Here p is the momentum operator $-\mathrm{i}\hbar\nabla$. The Hamiltonian is made up of the unperturbed part H_0 and the perturbation H'

$$H = H_0 + H' , \qquad H' = \frac{e}{m} p \cdot A \tag{6.44}$$

where we have omitted the A^2-term as small compared with H'.

We now expand ψ in terms of the solutions of the unperturbed problem: $H_0\psi_0 = \mathrm{i}\hbar\dot{\psi}_0$.

$$\psi = \sum_{nk} a_n(k, t) \exp\left[-\frac{\mathrm{i}}{\hbar} E_n(k) t\right] |n, k\rangle . \tag{6.45}$$

$|n, k\rangle$ means the Bloch function $\psi_n(k, r)$.

Inserting (6.45) into (6.43) and multiplying by $\exp[(\mathrm{i}/\hbar)E_m(\boldsymbol{k}')t]\,\langle m, \boldsymbol{k}'|$ gives

$$\dot{a}_m(\boldsymbol{k}', t) = \frac{1}{\mathrm{i}\hbar}\sum_{n\boldsymbol{k}} a_n(\boldsymbol{k}, t)\exp\left(\frac{\mathrm{i}}{\hbar}[E_m(\boldsymbol{k}') - E_n(\boldsymbol{k})]t\right)\langle m,\boldsymbol{k}'|H'|n,\boldsymbol{k}\rangle\,. \tag{6.46}$$

Now let $a_j(\boldsymbol{k}, 0) = 1$ and all other $a_n(\boldsymbol{k}, 0) = 0$, i.e., we assume that at time $t = 0$ the electron described by (6.43) is in state $\boldsymbol{k}$ of the jth band. The probability $W(j, \boldsymbol{k}, j', \boldsymbol{k}', \omega, t)$ that at time t it will be in state $|j', \boldsymbol{k}'\rangle$ is then equal to $|a_{j'}(\boldsymbol{k}', t)|^2$. In the first approximation we find

$$\dot{a}_{j'}^{(1)}(\boldsymbol{k}', t) = \frac{1}{\mathrm{i}\hbar}\exp\left(\frac{\mathrm{i}}{\hbar}[E_{j'}(\boldsymbol{k}') - E_j(\boldsymbol{k})]t\right)\langle j', \boldsymbol{k}'|H'|j, \boldsymbol{k}\rangle \tag{6.47}$$

and

$$W(j, \boldsymbol{k}, j', \boldsymbol{k}', \omega, t) = -\frac{1}{\hbar^2}\left|\int_0^t \exp\left(\frac{\mathrm{i}}{\hbar}[E_{j'}(\boldsymbol{k}') - E_j(\boldsymbol{k})]t'\right)\right.$$
$$\left.\times\,\langle j', \boldsymbol{k}'|H'|j, \boldsymbol{k}\rangle dt'\right|^2\,. \tag{6.48}$$

Next we examine more closely the matrix element which appears in (6.48)

$$\langle j', \boldsymbol{k}'|H'|j, \boldsymbol{k}\rangle = \frac{e}{m}\frac{\hbar}{\mathrm{i}}A_0[\exp(-\mathrm{i}\omega t)\langle j', \boldsymbol{k}'|\exp(\mathrm{i}\boldsymbol{\kappa}\cdot\boldsymbol{r})\boldsymbol{e}\cdot\nabla|j, \boldsymbol{k}\rangle$$
$$+ \exp(\mathrm{i}\omega t)\langle j', \boldsymbol{k}'|\exp(-\mathrm{i}\boldsymbol{\kappa}\cdot\boldsymbol{r})\boldsymbol{e}\cdot\nabla|j, \boldsymbol{k}\rangle]\,. \tag{6.49}$$

We shall see further below that only the first term contributes to the absorption. We therefore neglect the second term for the moment. In the first term we divide the integration into partial integrations over the individual Wigner-Seitz cells. Since the value of a Bloch function at equivalent points in two Wigner-Seitz cells differs only by a factor $\exp(\mathrm{i}\boldsymbol{k}\cdot\boldsymbol{R}_l)$, we have

$$\langle\ldots|\ldots|\ldots\rangle = \langle\ldots|\ldots|\ldots\rangle_{\mathrm{WSC}}\sum_l \exp[\mathrm{i}(\boldsymbol{k} + \boldsymbol{\kappa} - \boldsymbol{k}')\cdot\boldsymbol{R}_l]. \tag{6.50}$$

The last sum is only nonzero when $\boldsymbol{k} + \boldsymbol{\kappa} - \boldsymbol{k}'$ is equal to a lattice vector in the reciprocal lattice. Since $\boldsymbol{k}$ and $\boldsymbol{k}'$ lie in the Brillouin zone and $\boldsymbol{\kappa}$ is very small compared with each vector $\boldsymbol{K}_{j'}$, it follows that

$$\boldsymbol{k}' = \boldsymbol{k} + \boldsymbol{\kappa}. \tag{6.51}$$

This is the expression for wave number conservation in the transition.

The change of electron momentum from $\boldsymbol{k}$ to $\boldsymbol{k}'$ is small. As we emphasized earlier, the $\boldsymbol{\kappa}$-vector is several orders of magnitude less than the dimensions of the Brillouin zone. Therefore to a good approximation we can put

$$\boldsymbol{\kappa} = 0\,, \qquad \boldsymbol{k} = \boldsymbol{k}'\,. \tag{6.52}$$

The transition takes place with conservation of the electron $\boldsymbol{k}$-vector (direct transition). Eq. (6.48) then becomes

$$W(j, j', \boldsymbol{k}, \omega, t) = -\frac{e^2 A_0^2}{m^2}\left| M_{jj'}(\boldsymbol{k}) \int_0^t \exp\left(\frac{\mathrm{i}}{\hbar}[E_{j'}(\boldsymbol{k}) - E_j(\boldsymbol{k}) - \hbar\omega]t'\right) dt'\right|^2 \tag{6.53}$$

where $M_{jj'}(\boldsymbol{k})$ is the matrix element (6.50).

We find for the integral in (6.53)

$$-\left|\int_0^t \cdots dt'\right|^2 = \frac{\sin^2 xt}{x^2} \approx \pi t\delta(x) = 2\pi\hbar t\delta(2\hbar x)\,; \qquad x = \frac{E_{j'} - E_j - \hbar\omega}{2\hbar}\,. \tag{6.54}$$

The last terms on the right-hand side (including the δ-function) are valid for large t only. We have then

$$W(j, j', \boldsymbol{k}, \omega, t) = \frac{e^2 A_0^2}{m^2}|M_{jj'}(\boldsymbol{k})|^2 2\pi\hbar t\delta(E_{j'}(\boldsymbol{k}) - E_j(\boldsymbol{k}) - \hbar\omega)\,. \tag{6.55}$$

The δ-function in (6.55) means that energy as well as momentum is conserved in the transition

$$E_{j'}(\boldsymbol{k}) = E_j(\boldsymbol{k}) + \hbar\omega\,. \tag{6.56}$$

If we had considered the second term in (6.49) we would have had $-\hbar\omega$ instead of $\hbar\omega$ in (6.55). This term therefore describes the induced emission of a photon in the transition.

We find the total number of transitions per unit volume and time from (6.55) by dividing by t, summing over all j (occupied bands) and j' (unoccupied bands), and integrating throughout the Brillouin zone

$$W(\omega) = \sum_{jj'} \frac{1}{t} \frac{2}{(2\pi)^3} \int W(j, j', \boldsymbol{k}, \omega, t) d\tau_{\boldsymbol{k}}\,. \tag{6.57}$$

We find for the final result

$$\varepsilon_2 = \frac{4\pi e^2\hbar^2}{\varepsilon_0 m^2\omega^2} \sum_{jj'} \frac{2}{(2\pi)^3} \int |\langle j', \boldsymbol{k}|\boldsymbol{e}\cdot\nabla|j, \boldsymbol{k}\rangle|^2 \delta(E_{j'}(\boldsymbol{k}) - E_j(\boldsymbol{k}) - \hbar\omega) d\tau_{\boldsymbol{k}}\,. \tag{6.58}$$

We rearrange this further using (2.152):

$$\varepsilon_2 = \frac{4\pi e^2\hbar^2}{\varepsilon_0 m^2\omega^2}\sum_{jj'}\frac{2}{(2\pi)^3}\int_{E_{j'}-E_j=\hbar\omega}|\langle j', \boldsymbol{k}|\boldsymbol{e}\cdot\nabla|j, \boldsymbol{k}\rangle|^2\cdot\frac{df}{|\nabla_{\boldsymbol{k}}[E_{j'}(\boldsymbol{k}) - E_j(\boldsymbol{k})]|}\,. \tag{6.59}$$

An approximation often used starts with the assumption that the matrix element does not change much with $\boldsymbol{k}$. One can then put it in front of the integral. An integral remains which differs from the density of states only in that, instead of the band structure function $E_j(\boldsymbol{k})$, one has the difference of two such functions. One designates this integal the *combined density of states* $g_{jj'}(\omega)$. It gives the number of pairs of states in bands j and j' separated by a given energy $\hbar\omega$. With this approximation one obtains

$$\varepsilon_2 = \frac{4\pi e^2\hbar^2}{\varepsilon_0 m^2\omega^2}\sum_{jj'}|\langle j', \boldsymbol{k}|\boldsymbol{e}\cdot\nabla|j, \boldsymbol{k}\rangle|^2 g_{jj'}(\omega)\,. \tag{6.60}$$

With regard to the range of validity of (6.60), we should point out that we have assumed all states in band j to be occupied, and all in band j' to be empty. If this is not so we have to include the occupation probabilities of the states of both bands in a suitable form.

Direct transitions in semiconductors are particularly important if the valence band maximum and conduction band minimum lie at the same $\boldsymbol{k}$. Then they are the lowest transitions of the absorption spectrum and determine the shape of the *absorption edge*.

We assume that we can describe the extrema by an effective mass

$$\hbar\omega = E_{j'}(\boldsymbol{k}) - E_j(\boldsymbol{k}) = E_G + \frac{\hbar^2k^2}{2m_{j'}} + \frac{\hbar^2k^2}{2m_j} = E_G + \frac{\hbar^2k^2}{2m_{\text{comb}}}\,. \tag{6.61}$$

In accordance with (2.152) and (2.7), we then find for the combined density of states

$$g_{jj'}(\omega) = \frac{1}{2\pi^2}\left(\frac{2m_{\text{comb}}}{\hbar^2}\right)^{3/2}(\hbar\omega - E_G)^{1/2}\,. \tag{6.62}$$

To examine the energy dependence of the matrix element in (6.60), we insert Bloch functions

$$\begin{aligned}\langle j', \boldsymbol{k} + \boldsymbol{\kappa}|\boldsymbol{e}\cdot\nabla|j, \boldsymbol{k}\rangle &= \int u_{j'}^*(\boldsymbol{k} + \boldsymbol{\kappa}, \boldsymbol{r})\exp[-\mathrm{i}(\boldsymbol{k} + \boldsymbol{\kappa})\cdot\boldsymbol{r}]\boldsymbol{e}\cdot\nabla[u_j(\boldsymbol{k}, \boldsymbol{r})\exp(\mathrm{i}\boldsymbol{k}\cdot\boldsymbol{r})]d\tau \\ &= \int u_{j'}^*(\boldsymbol{k} + \boldsymbol{\kappa}, \boldsymbol{r})\exp(-\mathrm{i}\boldsymbol{\kappa}\cdot\boldsymbol{r})\boldsymbol{e}\cdot\nabla u_j(\boldsymbol{k}, \boldsymbol{r})d\tau \\ &\quad + \mathrm{i}\boldsymbol{e}\cdot\boldsymbol{k}\int u_{j'}^*(\boldsymbol{k} + \boldsymbol{\kappa}, \boldsymbol{r})u_j(\boldsymbol{k}, \boldsymbol{r})d\tau \\ &\equiv M_{j'j}^{\mathrm{a}} + \mathrm{i}\boldsymbol{e}\cdot\boldsymbol{k}M_{j'j}^{\mathrm{f}}\,.\end{aligned} \tag{6.63}$$

The second term of the sum can in general be neglected compared with the first since $\kappa \approx 0$. It would be zero if the $\boldsymbol{k}$-vector of the electron really were conserved in the transition (orthogonality of Bloch functions with the same $\boldsymbol{k}$ and different j).

We have not yet asked whether the symmetries of the initial and final states will at all allow the transitions associated with the matrix element $M^{\mathrm{a}}_{j'j}$, or whether the $M^{\mathrm{a}}_{j'j}$ vanish for reasons of symmetry. In the latter case (forbidden transitions), the matrix element $M^{\mathrm{f}}_{j'j}$ determines the transition intensity.

The energy dependence of ε_2 can readily be found for allowed and forbidden transitions. For forbidden transitions we replace $\boldsymbol{k}$ in accordance with (6.61) by $[(2m_{\mathrm{comb}}/\hbar^2)(\hbar\omega - E_{\mathrm{G}})]^{1/2}$. We find

$$\text{for } \textit{allowed transitions} \quad \varepsilon_2 \propto (\hbar\omega)^{-2}|M^{\mathrm{a}}_{j'j}|^2(\hbar\omega - E_{\mathrm{G}})^{1/2}\,, \tag{6.64}$$

$$\text{for } \textit{forbidden transitions} \quad \varepsilon_2 \propto (\hbar\omega)^{-2}|M^{\mathrm{f}}_{j'j}|^2(\hbar\omega - E_{\mathrm{G}})^{3/2}\,. \tag{6.65}$$

In addition to the appearance of an absorption edge, i.e., a threshold energy for absorption due to direct transitions, the absorption spectrum of a semiconductor or insulator is characterized by the appearance of a distinct structure. Fig. 6.4 presents an example. This structure can certainly not be attributed to a strong energy dependence of the matrix element $M_{j'j}$ (though this factor will always be energy dependent). When the initial or final state of the transition lies in an energy range where subbands of the valence or conduction band overlap, i.e., where the density of states has a distinct structure, the combined density of states will depend strongly on the transition energy. Eq. (6.59) shows that the combined density of states always has singularities in its integrand when $\nabla_{\boldsymbol{k}}[E_{j'}(\boldsymbol{k}) - E_j(\boldsymbol{k})] = 0$. This is the case for transitions between band (or subband) extrema, for which $\nabla_{\boldsymbol{k}}E_{j'}$ and $\nabla_{\boldsymbol{k}}E_j$ both vanish. The condition is also satisfied when the gradients of the two states are equal: $\nabla_{\boldsymbol{k}}E_{j'}(\boldsymbol{k}) = \nabla_{\boldsymbol{k}}E_j(\boldsymbol{k})$. This does not mean that the combined density of states becomes infinite at these *critical points*. One can show that at each critical point kinks appear in the spectrum, i.e., discontinuities in the first derivative of $g_{jj'}(\omega)$.

Inserting a Taylor series for $E_{j'}(\boldsymbol{k}) - E_j(\boldsymbol{k})$ about the critical point, the first two nonvanishing terms in the expansion are

$$E_{j'}(\boldsymbol{k}) - E_j(\boldsymbol{k}) = E_0 + \sum_i a_i(k_i - k_{0i})^2\,, \tag{6.66}$$

where the values and sign of the a_i depend on band structure in the vicinity of the critical point. If all a_i are positive and equal, (6.66) takes the form (6.61). The absorption edge is therefore a possible critical point with the combined density of states $g = 0$ below E_0, and $g \propto (E - E_0)^{1/2}$ above E_0. The same energy dependence follows for positive but unequal a_i.

Other critical points appear when all a_i are negative [$g \propto (E_0 - E)^{1/2}$ below and $g = 0$ above E_0], when two a_i are negative and one is positive

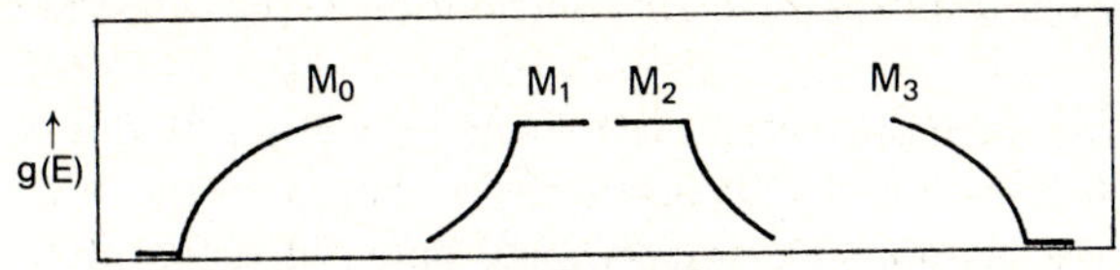

Fig. 6.6. Behaviour of the combined density of states near a critical point.

[$g \propto$ const $- (E_0 - E)^{1/2}$ below and $g =$ const above E_0], and when two a_i are positive and one a_i is negative [$g =$ const below and $g \propto$ const $- (E - E_0)^{1/2}$ above E_0]. These possibilities, together with the designations of the different critical points (M_0 to M_3), are shown in Fig. 6.6.

The succession of numerous critical points—superimposed on a weaker energy-dependent background—gives rise to the detailed structure of an absorption spectrum. We look at examples in Section 6.2.6.

6.2.3 Indirect Transitions

Direct transitions only determine the shape of the absorption edge of an insulator or semiconductor when the valence band maximum and the conduction band minimum lie at the same $\boldsymbol{k}$-vector. In many solids this, however, is not the case. An example is the band structure of silicon shown in Fig. 2.31. There the valence band maximum lies at Γ and the conduction band minima lie on the Δ-axes.

Transitions can also occur between these extrema, so that in the absorption spectrum the direct absorption edge is preceded by a forerunner. Following the results of the last section, the transitions responsible for the latter absorption cannot be one-step photon absorption processes since then momentum would not be conserved. The change of momentum of the electron has to be taken up by a *phonon* which is absorbed or emitted in the transition process. If $\boldsymbol{q}$ is the wave number and $\hbar\omega_{\boldsymbol{q}}$ the phonon energy, momentum and energy conservation yield

$$\boldsymbol{k}' = \boldsymbol{k} + \boldsymbol{\kappa} \pm \boldsymbol{q}, \qquad E(\boldsymbol{k}') = E(\boldsymbol{k}) + \hbar\omega \pm \hbar\omega_{\boldsymbol{q}}. \tag{6.67}$$

The plus sign means absorption, the minus sign emission of a phonon.

We connect up with the formalism of the last section. The perturbation operator H' now contains two terms

$$H' = H'_{\text{photon}} + H'_{\text{phonon}} \tag{6.68}$$

where the first contribution is given by (6.44). For the second we use (4.9), thus limiting ourselves to LA-phonons. The main contribution to the transition energy is then carried by the photon, and the main contribution to the transition momentum by the phonon.

In what follows we shall concern ourselves with two-step processes. In one of the two steps the electron changes state by absorption of a photon, in the other by absorption or emission of a phonon. The intermediate state is a virtual state. Energy is only conserved in the complete process. In contrast, momentum must be conserved in each of the two steps since otherwise the transition matrix elements vanish.

We start with the general formalism. H'_{photon} has the form

$$H'_{\text{photon}} = \frac{e\hbar}{\mathrm{i}m} A_0 \exp(-\mathrm{i}\omega t) \boldsymbol{e} \cdot \nabla . \tag{6.69}$$

This follows from (6.44) if we allow only photon *absorption* and set $\boldsymbol{\kappa}$ to zero. H'_{phonon} becomes

$$H'_{\text{phonon}} = M_{\boldsymbol{kq}}[a^+_{-\boldsymbol{q}0} \exp(\mathrm{i}\omega_{\boldsymbol{q}} t) + a_{\boldsymbol{q}0} \exp(-\mathrm{i}\omega_{\boldsymbol{q}} t)] c^+_{\boldsymbol{k}+\boldsymbol{q}} c_{\boldsymbol{k}} . \tag{6.70}$$

This follows from (4.9) with the time dependence of the a^+ and a explicitly expressed. This is necessary in view of the use of H'_{phonon} in the time-dependent Schrödinger equation. The remaining time-independent parts a_0^+ and a_0 operate on the state vectors of the phonons in the same way as the a^+ and a, and lead likewise to the matrix elements (4.12) and (4.13).

For the description of the two-step process we have to use a perturbation calculation of second order. We start with (6.46). Once again, at $t = 0$, let $a_j = 1$ and all other $a_n = 0$. For time t we put

$$a_j(\boldsymbol{k}, t) = 1 + a_j^{(1)} + a_j^{(2)} + \cdots , \qquad a_n(\boldsymbol{k}, t) = a_n^{(1)} + a_n^{(2)} + \cdots \quad (n \neq j) . \tag{6.71}$$

Each of the terms in the sum here is of order H' smaller than the term in front of it. From (6.69, 70) we can write H' in (6.68) in the form $H' = \sum_\alpha H'_\alpha \exp(-\mathrm{i}\omega_\alpha t)$. If we insert (6.71) into (6.46) and respectively compare terms of given order on the left and on the right, we obtain

$$\dot{a}_n^{(1)}(\boldsymbol{k}'', t) = \frac{1}{\mathrm{i}\hbar} \sum_\alpha \exp\left(\frac{\mathrm{i}}{\hbar}[E_n(\boldsymbol{k}'') - E_j(\boldsymbol{k}) - \hbar\omega_\alpha] t\right) \langle n, \boldsymbol{k}'' | H'_\alpha | j, \boldsymbol{k} \rangle , \tag{6.72}$$

$$\dot{a}_{j'}^{(2)}(\boldsymbol{k}', t) = \frac{1}{\mathrm{i}\hbar} \sum_{\alpha'} \sum_{n\boldsymbol{k}''} a_n^{(1)}(\boldsymbol{k}'', t) \times \exp\left(\frac{\mathrm{i}}{\hbar}[E_{j'}(\boldsymbol{k}') - E_n(\boldsymbol{k}'') - \hbar\omega_{\alpha'}] t\right) \langle j', \boldsymbol{k}' | H | n, \boldsymbol{k}'' \rangle . \tag{6.73}$$

From (6.72), $a_n^{(1)}$ can be found by integration. If we put the result into (6.73), we obtain on the right side an expression with two terms, in each of which the time dependence is given by $\exp[(i/\hbar)\Delta Et]$. The time integration is again straightforward and one finds

$$a_f^{(2)}(\boldsymbol{k}_f, t) = \sum_{m,\alpha,\alpha'} \frac{\langle f|H'_{\alpha'}|m\rangle\langle m|H'_{\alpha}|i\rangle}{E_m - E_i - E_\alpha} \times \left\{ \frac{\exp[(i/\hbar)(E_f - E_i - E_\alpha - E_{\alpha'})t] - 1}{E_f - E_i - E_\alpha - E_{\alpha'}} - \frac{\exp[(i/\hbar)(E_f - E_m - E_{\alpha'})t] - 1}{E_f - E_m - E_{\alpha'}} \right\}. \quad (6.74)$$

E_i and $|i\rangle$ are the energy and the wave function of the initial state of the electron, E_f and $|f\rangle$ are the corresponding quantities for the final state, and E_m and $|m\rangle$ those of the virtual intermediate states. E_α and $E_{\alpha'}$ are the energies $\hbar\omega_{\alpha(')}$ associated with the frequencies $\omega_{\alpha(')}$.

The summation in (6.74) is over all possible processes which involve an intermediate state, and over all intermediate states. The following are the possibilities which have to be considered (Fig. 6.7):

1) The electron goes from its initial state in the valence band into an inter-

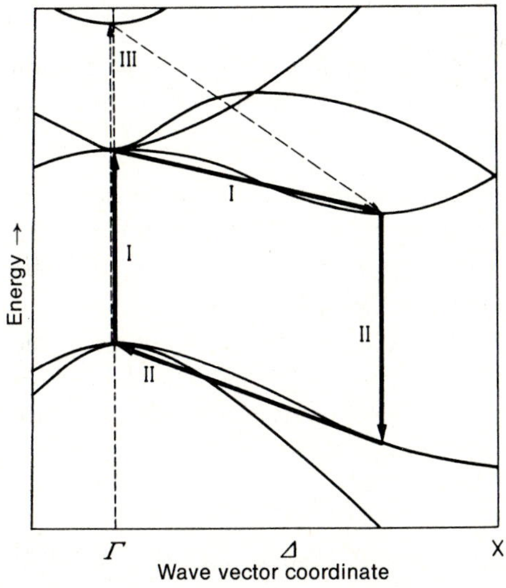

Fig. 6.7. Indirect transitions in the band structure of silicon (see Fig. 2.31). I: Electron transition from valence band into the conduction band with absorption of a photon, followed by a transition within the conduction band with absorption or emission of a phonon. II: Hole transition from conduction to valence band with absorption of a photon, followed by a transition within the valence band with absorption or emission of a phonon. Transitions involving higher bands (III) are possible, but contribute little to absorption.

mediate state with the same $\boldsymbol{k}_i$ in the lowest conduction band or in one of the higher lying conduction bands (direct transition with absorption of a photon). From there it makes the second transition with the absorption or emission of a phonon of wave number $\boldsymbol{q} = \pm(\boldsymbol{k}_f - \boldsymbol{k}_i)$ into the final state. Of these possibilities the most probable transition will be that for which the virtual intermediate state is nearest in energy to the final state. In future we shall therefore only consider the lowest conduction band state with $\boldsymbol{k}_i$ as the possible intermediate state.

2) Photon absorption takes place at the wave number $\boldsymbol{k}_f$ of the final state. Initial and virtual intermediate states are coupled by phonon absorption or emission. The most likely process will be that for which the virtual state, like the initial state, lies in the uppermost valence band.

These considerations reduce the number of processes to be considered in (6.74) to just a few. We compare first the two terms in the brackets on the right. From conservation of energy the denominator of the first term is zero. The square of this term—as in (6.54)—leads to a δ-function. In view of the resonant denominator this term always dominates and the second term can be omitted.

The square of the right side of (6.74) then gives the transition probability. The terms associated with the different processes can be squared separately since the individual processes are independent of each other. One can see this from the fact that the mixed terms do not make any contributions to the transition probability which increase with t. Finally we find

$$|a_f^{(2)}(\boldsymbol{k}_f, t)|^2 = \sum_{m\alpha\alpha'} \frac{|\langle f|H'_{\alpha'}|m\rangle|^2 |\langle m|H'_\alpha|i\rangle|^2}{(E_m - E_i - E_\alpha)^2} 2\pi\hbar t \delta(E_f - E_i - E_\alpha - E_{\alpha'}) . \tag{6.75}$$

One obtains the total transition rate as in (6.57) by dividing by t and summing over all initial and final states. Here we must only watch the fact that now $\boldsymbol{k}_f \neq \boldsymbol{k}_i$, so that we have to integrate separately over $\boldsymbol{k}_f$ and $\boldsymbol{k}_i$. The imaginary part of the complex dielectric constant then follows as

$$\begin{aligned} \varepsilon_2 = \sum_{\mathrm{I\,II}} \int g(\boldsymbol{k}_i) d\tau_{\boldsymbol{k}_i} \int g(\boldsymbol{k}_f) d\tau_{\boldsymbol{k}_f} \\ \times \left\{ \frac{C_{if}^{(\mathrm{abs})}}{\omega^2} \delta(E_f(\boldsymbol{k}_f) - E_i(\boldsymbol{k}_i) - \hbar\omega - \hbar\omega_{\boldsymbol{q}}) \right. \\ \left. + \frac{C_{if}^{(\mathrm{em})}}{\omega^2} \delta(E_f(\boldsymbol{k}_f) - E_i(\boldsymbol{k}_i) - \hbar\omega + \hbar\omega_{\boldsymbol{q}}) \right\} . \end{aligned} \tag{6.76}$$

The sum is over both the possibilities mentioned above (direct transitions at $\boldsymbol{k}_i$ or $\boldsymbol{k}_f$). The factors C_{if} contain

1) *the square of the matrix element for a phonon transition.* From (4.12) and (4.13) it is proportional to $(1 - \bar{n}_{\boldsymbol{k}+\boldsymbol{q}})\bar{n}_{\boldsymbol{k}}\bar{n}_{\boldsymbol{q}}$ for phonon absorption and to

$(1 - \bar{n}_{k+q})\bar{n}_k(\bar{n}_q + 1)$ for phonon emission. If we start with a filled valence band and empty conduction band, the factors $(1 - \bar{n}_{k+q})\bar{n}_k$ allow only two transitions. In the one an electron first makes a vertical transition at $\boldsymbol{k}_i$ from valence to conduction band, and then a transition within the conduction band to $\boldsymbol{k}_f$ with phonon absorption or emission. In the other case an electron first makes a vertical transition at $\boldsymbol{k}_f$ into the conduction band, then another electron in the valence band is scattered from $\boldsymbol{k}_i$ to $\boldsymbol{k}_f$. One can also interpret this latter process as a hole transition from $\boldsymbol{k}_f$ in the conduction band, via $\boldsymbol{k}_f$ in the valence band, to $\boldsymbol{k}_i$ in the valence band. It is then exactly the inverse process to the first possibility (Fig. 6.7).

2) *an energy denominator* $(E_m - E_i - \hbar\omega_\alpha)$. Since the transition $i \to m$ is always a direct transition, $E_m - E_i$ is the energy of the "direct absorption edge" at $\boldsymbol{k}_i$ and $\hbar\omega_\alpha$ the photon energy $\hbar\omega$. Because of their comparatively low intensity, indirect transitions are only of interest where direct transitions are not yet possible, i.e., below the direct absorption edge. In this region, however, $\hbar\omega$ is small compared with $E_m - E_i$. The energy denominator does therefore not influence the shape of the indirect absorption edge.

We have further to pay attention to the factor $\bar{n}_q$ and $\bar{n}_q + 1$ in $C_{if}^{(\mathrm{abs})}$ and $C_{if}^{(\mathrm{em})}$, respectively. This is important at low temperatures since there $C_{if}^{(\mathrm{abs})}$ also approaches zero as $\bar{n}_q \to 0$. Emission processes remain possible, however.

The integrations in (6.76) can be carried out if, as in (6.61), we assume parabolic energy surfaces with scalar effective mass in the vicinity of the extrema of valence and conduction bands. If, in addition, direct transitions are allowed, all the matrix elements can be brought in front of the integral as $\boldsymbol{k}$-independent. If they are forbidden, then following (6.63) a factor k^2 is added. Including all constant factors in the C_{if}, we find for allowed transitions

$$\begin{aligned}\varepsilon_2 &= \sum_{\mathrm{I\,II}} \frac{C_{if}}{\omega^2} \int dE_i \int dE_f \sqrt{(E_V - E_i)(E_f - E_C)}\,\delta(E_f - E_i - \hbar\omega \pm \hbar\omega_q) \\ &= \sum_{\mathrm{I\,II}} \frac{C_{if}}{\omega^2} \int_{E_C}^{E_V + \hbar\omega \pm \hbar\omega_q} dE_f \sqrt{(E_f - E_C)(E_V + \hbar\omega \pm \hbar\omega_q - E_f)} \\ &= \sum_{\mathrm{I\,II}} \frac{C_{if}}{\omega^2} (\hbar\omega \pm \hbar\omega_q - E_G)^2 . \end{aligned} \tag{6.77}$$

This leads to

$$\varepsilon_2 = C^{(\mathrm{abs})}\omega^{-2}(\hbar\omega + \hbar\omega_q - E_G)^2 + C^{(\mathrm{em})}\omega^{-2}(\hbar\omega - \hbar\omega_q - E_G)^2 . \tag{6.78}$$

For forbidden transitions we would have the third power of the energy differences in (6.78) instead of the square. $E_G = E_C - E_V$ is the difference in energy between the two band extrema (the energy gap). We would finally point out that the terms in (6.78) are only nonzero when the quantities in parentheses are positive, i.e., when $\hbar\omega \pm \hbar\omega_q > E_G$. We present examples of indirect transitions in Section 6.2.6.

6.2.4 Two-Photon Absorption

A further important two-step process is the absorption of two photons. In the first step an electron is raised into a virtual state by absorption of a photon of frequency ω_1 and polarization $\boldsymbol{e}_1$; in the second it transfers into the final state with absorption of a second photon of frequency ω_2 and polarization $\boldsymbol{e}_2$. The probability of such processes is very small, so that light sources of high intensity (lasers) are needed. Their significance therefore is not so much the fact that they represent a further absorption mechanism to direct and indirect transitions, but rather that for two-photon processes other selection rules apply than for one-photon processes. Transitions which cannot be observed in the normal absorption spectrum can thus be measured by two-photon absorption.

The Hamiltonian in this case is

$$H = \frac{1}{2m}(\boldsymbol{p} + e\boldsymbol{A}_1 + e\boldsymbol{A}_2)^2 + V(\boldsymbol{r}). \tag{6.79}$$

Neglecting terms with A_i^2 compared with $\boldsymbol{p}\cdot\boldsymbol{A}_i$, as in (6.44), we find for the perturbation operator H'

$$H' = \frac{e}{m}(\boldsymbol{p}\cdot\boldsymbol{A}_1 + \boldsymbol{p}\cdot\boldsymbol{A}_2) + \frac{e^2}{m}\boldsymbol{A}_1\cdot\boldsymbol{A}_2. \tag{6.80}$$

We look first at the terms linear in $\boldsymbol{A}$. In a first-order perturbation calculation one finds direct transitions with absorption of one of the two photons. In a second-order calculation one finds the transition probability from (6.75) with $i, i' = 1, 2$ and $\boldsymbol{k}_i = \boldsymbol{k}_m = \boldsymbol{k}_f$. It follows that

$$W(i, f, \boldsymbol{k}, \omega, t) = \frac{e^4 A_{10}^2 A_{20}^2}{m^4} 2\pi\hbar t\delta(E_f(\boldsymbol{k}) - E_i(\boldsymbol{k}) - \hbar\omega_1 - \hbar\omega_2)$$
$$\times \left[\frac{|\langle f|\boldsymbol{e}_2\cdot\boldsymbol{p}|m\rangle\langle m|\boldsymbol{e}_1\cdot\boldsymbol{p}|i\rangle|^2}{(E_m - E_i - \hbar\omega_1)^2} + \frac{|\langle f|\boldsymbol{e}_1\cdot\boldsymbol{p}|m\rangle\langle m|\boldsymbol{e}_2\cdot\boldsymbol{p}|i\rangle|^2}{(E_m - E_i - \hbar\omega_2)^2}\right]. \tag{6.81}$$

With absorption of a photon, the electron is first raised into an intermediate state, from which it progresses to the final state by absorption of a second photon. Two processes contribute to (6.81), the order of the two absorption processes being different in the two cases. Again energy is conserved only in the complete process; momentum is conserved in each individual process.

The second term on the right in (6.80) already leads to two-photon processes in a first-order perturbation calculation, if we do not assume $\boldsymbol{\kappa}_1 = 0$ and $\boldsymbol{\kappa}_2 = 0$. We find then

$$H' = \frac{e^2}{m} A_{10} A_{20} \boldsymbol{e}_1 \cdot \boldsymbol{e}_2 \exp(\mathrm{i}[(\boldsymbol{\kappa}_1 + \boldsymbol{\kappa}_2)\cdot \boldsymbol{r} - (\omega_1 + \omega_2)t])$$

$$= \frac{e^2}{m} A_{10} A_{20} \boldsymbol{e}_1 \cdot \boldsymbol{e}_2 \exp(-\mathrm{i}(\omega_1 + \omega_2)t)[1 + \mathrm{i}(\boldsymbol{\kappa}_1 + \boldsymbol{\kappa}_2)\cdot \boldsymbol{r} + \cdots]. \quad (6.82)$$

If with this operator one forms matrix elements between Bloch states, they vanish in the dipole approximation ($\boldsymbol{\kappa}_i = 0$). The second term in the expansion on the right (quadrupole approximation) provides contributions to the transition probability, and hence to the absorption, proportional to

$$W^{A^2}(f, i, \boldsymbol{k}, \omega, t) \propto \boldsymbol{e}_1 \cdot \boldsymbol{e}_2 |\langle f|(\boldsymbol{\kappa}_1 + \boldsymbol{\kappa}_2)\cdot \boldsymbol{r}|i\rangle|^2 . \quad (6.83)$$

One can show that this contribution is small compared with (6.81) and therefore contributes little to two-photon absorption. Consequently only the first contribution is treated in the literature.

Different selection rules apply for direct transitions involving one photon, and for those involving two photons. This can readily be seen qualitatively by a comparison with transitions in the free atom. For a corresponding transition, the selection rule which applies is that the quantum number l must change by ± 1. Two-photon transitions are made up of two one-photon transitions. The selection rules are then $\Delta l = 0$ or $\Delta l = 2$. For the same initial state the two possibilities thus lead to different final states. Corresponding selection rules apply between levels in the band model.

In addition to such symmetry considerations of the wave functions, the polarization of the two photons can be different. For the transitions in (6.82), which we shall not pursue here, one can see immediately that a polarization dependence will arise from the factor $\boldsymbol{e}_1 \cdot \boldsymbol{e}_2$. The polarization dependence is more complicated for the transitions of (6.81). It emerges, for example, that, for a $\Gamma_1 \to \Gamma_1$ transition, polarization is taken account of by a factor $(\boldsymbol{e}_1 \cdot \boldsymbol{e}_2)^2$, and for a $\Gamma_1 \to \Gamma_{15}$ transition by the factor $(\boldsymbol{e}_1 \times \boldsymbol{e}_2)^2$ (see for example, Fig. 2.31 for the nomenclature of the transitions). The one transition is thus forbidden for $\boldsymbol{e}_1 \perp \boldsymbol{e}_2$, the other for $\boldsymbol{e}_1 \parallel \boldsymbol{e}_2$. The symmetries of two-photon transitions can thus be found by using light of different polarization. For further details and a summary of the literature on two-photon transitions see, for example, *Fröhlich* [103, X]. We shall discuss one experimental result in Section 6.2.6.

In this section we have limited ourselves to the case where both photons in the two-photon process are absorbed. Processes in which one photon is absorbed and a second emitted are, however, also of interest. This is a process involving inelastic scattering of light by matter (*Raman effect*). We shall not treat these two-photon processes until Section 6.2.9.

6.2.5 Exciton Absorption

The absorption processes considered in the last two sections lead to the formation of an electron-hole pair. So far we have always assumed that there is no interaction between electron and hole. That may be so when the energy delivered is considerably larger than the threshold energy E_G, since then the electron and the hole acquire sufficient kinetic energy. However in the neighbourhood of the absorption edge E_G, the electron-hole interaction can play an important role.

Furthermore, in our consideration of direct and indirect transitions, we have not so far taken account of the fact that immediately after their creation electron and hole are at the same location.

The mutual interaction and spatial correlation of the electron-hole pair lead to the formation of an exciton (Sec. 3.2). We therefore examine in this section exciton effects in the absorption spectrum of a solid. We shall neglect the polariton aspects treated in Section 6.1.3, and assume that the excitons decay so rapidly due to coupling with phonons or lattice defects that exciton creation is a true absorptive mechanism.

We link up with the formalism of Section 3.2. There we described the excitons by introducing Slater determinants $\Phi_{c\boldsymbol{k}_e v\boldsymbol{k}_h}$ which described a situation in which there is a a state $|v\boldsymbol{k}_h\rangle$ free in an otherwise full valence band (v), and hence a state $|c\boldsymbol{k}_e\rangle$ occupied in an otherwise empty conduction band (c). We constructed the exciton wave function by a superposition of such determinants, taking account of the fact that only determinants with given $\boldsymbol{K} = \boldsymbol{k}_e - \boldsymbol{k}_h$ should be used

$$\Psi_{\boldsymbol{K}} = \sum_{\boldsymbol{k}_e} A_{\boldsymbol{k}_e,\boldsymbol{k}_e-\boldsymbol{K}} \Phi_{\boldsymbol{k}_e,\boldsymbol{k}_e-\boldsymbol{K}} \,. \tag{6.84}$$

By introducing the Fourier transform of A

$$U_{\boldsymbol{K}}(\boldsymbol{\beta}) = \frac{1}{\sqrt{N}} \sum_{\boldsymbol{k}_e} A_{\boldsymbol{k}_e,\boldsymbol{k}_e-\boldsymbol{K}} \exp(\mathrm{i}\boldsymbol{k}_e \cdot \boldsymbol{\beta}) \tag{6.85}$$

one arrives at the exciton wave function of (3.76) and (3.74)

$$\Psi_{\boldsymbol{K}} = \frac{1}{\sqrt{N}} \sum_{\boldsymbol{k}_e} \sum_{\boldsymbol{\beta}} U_{\boldsymbol{K}}(\boldsymbol{\beta}) \exp(-\mathrm{i}\boldsymbol{k}_e \cdot \boldsymbol{\beta}) \Phi(\boldsymbol{k}_e - \boldsymbol{K}, \boldsymbol{k}_e) \tag{6.86}$$

or finally with $\Phi(\boldsymbol{K}, \boldsymbol{\beta}) = (1/\sqrt{N}) \sum_{\boldsymbol{k}_e} \exp(-\mathrm{i}\boldsymbol{k}_e \cdot \boldsymbol{\beta}) \Phi(\boldsymbol{k}_e - \boldsymbol{K}, \boldsymbol{k}_e)$ at

$$\Psi_{\boldsymbol{K}} = \sum_{\boldsymbol{\beta}} U_{\boldsymbol{K}}(\boldsymbol{\beta}) \Phi(\boldsymbol{K}, \boldsymbol{\beta}) \,. \tag{6.87}$$

Further transformations were involved for Wannier excitons, leading to the Schrödinger equation (3.86) for the relative motion of electron and hole.

If we consider first *direct transitions* only, the wave vectors of electron and hole are equal ($\boldsymbol{K} = 0$). The additional transformations in Section 3.2.4, leading from U through F to the function F', are then superfluous. All three functions are identical. The wave function (6.87) and the Schrödinger equation (3.86) then become

$$\Psi_0 = \sum_{\boldsymbol{\beta}} U_0(\boldsymbol{\beta})\Phi(0, \boldsymbol{\beta}), \qquad \left(-\frac{\hbar^2}{2\mu}\nabla^2 - \frac{e^2}{4\pi\varepsilon_0\beta}\right)U_0 = (E - E_0)U_0. \tag{6.88}$$

The difference compared to the theoretical description of absorption processes without electron-hole interaction lies only in the fact that, instead of the matrix elements $\langle c\boldsymbol{k}|\boldsymbol{e}\cdot\boldsymbol{p}|v\boldsymbol{k}\rangle$ for the direct transition, we must now consider matrix elements of the form

$$\int \Psi_0 \boldsymbol{e}\cdot\boldsymbol{p}_i\Phi_0 d\tau_1 \ldots d\tau_N. \tag{6.89}$$

Here Ψ_0 is a wave function of an exciton with $\boldsymbol{K} = 0$, given by (6.88) and Φ_0 the wave function of the ground state (filled valence band, free conduction band). Then

$$\begin{aligned}\int \Psi_0 \boldsymbol{e}\cdot\boldsymbol{p}\Phi_0 d\tau_1 \ldots d\tau_N &= \sum_{\boldsymbol{k}} A_{\boldsymbol{k},\boldsymbol{k}}\langle c\boldsymbol{k}|\boldsymbol{e}\cdot\boldsymbol{p}|v\boldsymbol{k}\rangle \\ &= \sum_{\boldsymbol{k}} A_{\boldsymbol{k},\boldsymbol{k}}(M_{cv}^{\mathrm{a}} + \mathrm{i}\boldsymbol{e}\cdot\boldsymbol{k}M_{cv}^{f}),\end{aligned} \tag{6.90}$$

since, because of the normalization of the eigenfunctions, only the integral over the coordinates of one electron remains in the multiple integration. The last step in (6.90) used (6.63).

We restrict ourselves now to transitions from the vicinity of the maximum of a parabolic isotropic valence band into a corresponding conduction band. We assume that only $\boldsymbol{k}$-vectors from the immediate vicinity of $\boldsymbol{k} = 0$ contribute to the sum in (6.90). We can then remove the matrix elements from the sum. (We can of course also do this without the above assumption if we can show that the matrix elements in general are $\boldsymbol{k}$-independent). We then find

$$\int \Psi_0 \boldsymbol{e}\cdot\boldsymbol{p}\Phi_0 d\tau_1 \ldots d\tau_N = M_{cv}^{\mathrm{a}}\sum_{\boldsymbol{k}} A_{\boldsymbol{k},\boldsymbol{k}} + \mathrm{i}M_{cv}^{f}\sum_{\boldsymbol{k}} A_{\boldsymbol{k},\boldsymbol{k}}\boldsymbol{e}\cdot\boldsymbol{k}, \tag{6.91}$$

or using (6.85)

$$\frac{1}{\sqrt{N}}\int \Psi_0 \boldsymbol{e}\cdot\boldsymbol{p}\Phi_0 d\tau_1 \ldots d\tau_N = M_{cv}^{a}U_0(0) + M_{cv}^{f}\boldsymbol{e}\cdot\nabla_{\boldsymbol{\beta}}U_0(\boldsymbol{\beta})|_{\boldsymbol{\beta}=0}. \tag{6.92}$$

This is different from (6.63) only in the two factors containing the function $U(\boldsymbol{\beta})$. We saw in (6.63) that the first term describes the *allowed*, and the second the *forbidden transitions.* Since the square of the matrix element enters the final expression, we can immediately write for the imaginary part of the complex dielectric constant

$$\begin{aligned} \varepsilon_2 &= \varepsilon_2^{0a}|U_0(0)|^2 && \text{for allowed transitions} \\ &= \varepsilon_2^{0f}|\boldsymbol{e}\cdot\nabla_{\boldsymbol{\beta}}U_0(\boldsymbol{\beta})|_{\boldsymbol{\beta}=0}|^2 && \text{for forbidden transitions.} \end{aligned} \tag{6.93}$$

For the two correction factors we need the function $U_0(\boldsymbol{\beta})$ and its derivative at $\boldsymbol{\beta} = 0$. U is the solution of the Schrödinger equation (6.88) which we know from the treatment of the H-atom. From there we adopt for the *discrete spectrum* of the exciton

$$|U_0(0)|^2 = \frac{1}{a_0^3 n^3 \pi}, \qquad |\boldsymbol{e}\cdot\nabla_{\boldsymbol{\beta}}U_0(\boldsymbol{\beta})|_{\boldsymbol{\beta}=0}|^2 = \frac{(n^2-1)}{a_0^5 n^5 \pi} \tag{6.94}$$

with a_0 = Bohr's radius (in a medium of dielectric constant ε).

For allowed transitions, the correction factor is only nonzero for *s*-excitons; for forbidden transitions it is only nonzero for *p*-excitons Thus on the low-energy side of the absorption edge a line spectrum appears which in one case shows all *s*-transitions, in the other all *p*-transitions beginning with $n = 2$(!).

The correction factors also change the absorption spectrum above the absorption edge. Here we have to take for the U_0 the eigenfunctions of the H-atom in the *continuous spectrum.* We obtain

$$\begin{aligned} |U_0(0)|^2 &= \frac{\gamma e^\gamma}{\sinh\gamma}, \qquad |\boldsymbol{e}\cdot\nabla_{\boldsymbol{\beta}}U_0(\boldsymbol{\beta})|_{\boldsymbol{\beta}=0}|^2 = \frac{\gamma e^\gamma}{\sinh\gamma}\left(1+\frac{\gamma^2}{\pi^2}\right) \\ \gamma &= \sqrt{\frac{\pi^2 R}{E-E_G}}, \qquad R = \frac{\mu e^4}{32\pi^2\hbar^2\varepsilon^2\varepsilon_0^2}. \end{aligned} \tag{6.95}$$

When we talk here of *s*- and *p*-excitons we refer to the symmetry of the function U_0.

If one inserts (6.95) into (6.93), one finds a large increase in ε_2 at $E \approx E_G$. With increasing E the correction factors (6.95) approach the value 1, in agreement with the fact that the interaction between electron and hole diminishes in importance with increasing kinetic energy of the charge carriers. One also finds exciton effects in the ε_2-spectrum at energies $E \gg E_G$. These effects are always most marked however at critical points in the ε_2 spectrum.

We now look briefly at indirect exciton transitions, contenting ourselves, however, with just a few remarks.

Let the energy of an indirect exciton (electron transition from a parabolic maximum of the valence band at $\boldsymbol{k} = 0$ into a parabolic minimum of the conduction band at $\boldsymbol{k} = \boldsymbol{k}_0$) be

$$E = E_G - \frac{R}{n^2} + \frac{\hbar^2}{2M}(\boldsymbol{K} - \boldsymbol{k}_0)^2 . \tag{6.96}$$

Here E_G is the indirect energy gap. M and R are defined by the effective masses at both extrema. Similar to Section 6.2.3, we now have to average over all possible $\boldsymbol{K}$, taking energy conservation into account. The absorption is then proportional to

$$\begin{aligned} \int g(\boldsymbol{K})\delta\left(\hbar\omega - E_G + \frac{R}{n^2} - \frac{\hbar^2}{2M}(\boldsymbol{K} - \boldsymbol{k}_0)^2 \pm \hbar\omega_{\boldsymbol{q}}\right)d\tau_{\boldsymbol{k}} \\ \propto \int \left[E - \left(E_G - \frac{R}{n^2}\right)\right]^{1/2} \delta(\hbar\omega - E \pm \hbar\omega_{\boldsymbol{q}})dE \\ \propto \left(\hbar\omega - E_G + \frac{R}{n^2} \pm \hbar\omega_{\boldsymbol{q}}\right)^{1/2} . \end{aligned} \tag{6.97}$$

Here too we have to average the transition probability with a factor giving the probability of finding electron and hole at the same place. This produces a factor n^{-3}, as in the first equation (6.94). Finally summing over all n, one finds for indirect allowed transitions into *discrete exciton states*

$$\varepsilon_2 \propto \sum_{n=1}^{\infty} \frac{1}{n^3}\left(\hbar\omega - E_G + \frac{R}{n^2} \pm \hbar\omega_{\boldsymbol{q}}\right)^{1/2} . \tag{6.98}$$

This differs fundamentally from the corresponding spectrum for direct transitions. Instead of discrete exciton lines in the spectrum, a series of square root shaped steps appears, each of which corresponds to the threshold energy of the transition into a discrete exciton band, with absorption or emission of a phonon.

The spectrum is again also different above E_G. The most significant result here is that, instead of the quadratic dependence of energy in (6.78), an exponent 3/2 appears. Moreover the factors C in (6.78) are changed by the addition of a correction factor as in (6.93).

One eventually finds for *forbidden transitions of the indirect exciton*

$$\varepsilon_2 \propto \sum_{n=1}^{\infty} \frac{n^2 - 1}{n^5}\left(\hbar\omega - E_G + \frac{R}{n^2} \pm \hbar\omega_{\boldsymbol{q}}\right)^{3/2} \tag{6.99}$$

for the discrete spectrum and an $E^{5/2}$-law for the continuum.

In the section following we present a few experimentally obtained exciton spectra. For further details we refer to the book by *Knox* [102.5] and the articles by *Elliott* in [118] and *Dimmock* in [104.3].

6.2.6 Comparison with Experimental Absorption and Reflection Spectra

The most important details of an absorption spectrum occur in the region of the *absorption edge*. We begin with a discussion of this region and discuss later what happens above the absorption edge.

Absorption due to band to band transitions sets in above the threshold energy $E_G - \hbar\omega_q - E_{ex}$, where E_G is the energy gap between valence and conduction band of a semiconductor or insulator, $\hbar\omega_q$ the energy of a phonon which may be absorbed in this process, and E_{ex} the binding energy of an exciton formed in the transition. All these phenomena occur in the absorption spectrum of germanium. As an instructive example, Fig. 6.8 shows the absorption coefficient in the region 0.62 to 0.86 eV for various temperatures. This is the region where indirect transitions can be observed. Since the energy gap E_G is temperature dependent, the threshold energy shifts with temperature. At 4.2 K, two threshold energies at 0.75 and 0.77 eV can be seen, above which absorption increases strongly. At 77 K one can distinguish four components with threshold energies at 0.705, 0.725, 0.745, and 0.760 eV. Above 195 K an exact analysis shows six components. Two other components are presumably not resolved so that altogether eight components exist. They are associated in pairs with indirect transitions with phonon emission or absorption. Hence four different phonons are involved in the transitions.

In germanium, indirect transitions take place from the highest level in the valence band at Γ into the lowest level in the conduction band at L (see Fig.

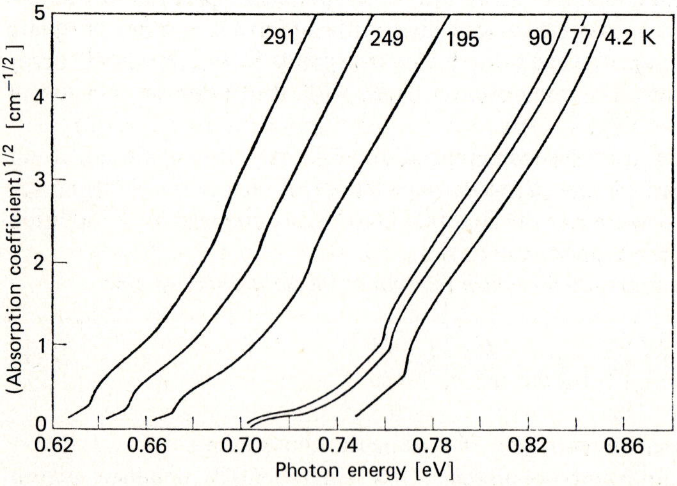

Fig. 6.8. The absorption edge of germanium in the region of indirect transitions, for various temperatures [from G. G. Macfarlane et al.: Phys. Rev. **108**, 1377 (1957)].

2.33). The phonons involved must therefore have $\boldsymbol{q}$-vectors in the $\omega_j(\boldsymbol{q})$-spectrum (e.g., Fig. 3.11 for diamond) which lie at point L. The four transition pairs are each coupled with a TA(L)-, LA(L)-, TO(L)-, and LO(L)-phonon.

The form of the individual components decides the assignment of the individual phonons to the transitions, and the question of the participation of excitons. The lowest transition at 20 K begins with the slope $(\hbar\omega - E')^{3/2}$; the second has the complicated form $a(\hbar\omega - E' + 0.0027\,\mathrm{eV})^{1/2} + b(\hbar\omega - E' + 0.0017\,\mathrm{eV})^{1/2} + c(\hbar\omega - E')^{3/2}$. Here the value of E' is the energy gap plus the energy of an LA(L)-phonon at the temperature concerned. The two additional components at 77 K each have the same form and only a different E'-value, so that the first two transitions take place with phonon emission and the two later ones with phonon absorption. This assignment is obvious since at low temperatures phonons are not available for absorption but can be emitted. From Fig. 3.11, the two phonons of lowest energy are a TA(L)- and an LA(L)-phonon.

The above-mentioned form of the one 20 K component has the shape of the indirect exciton absorption treated in the previous section. Transitions occur into the discrete ground state (binding energy 0.0027 eV) and into the first excited state (binding energy 0.0017 eV), then into the continuum.

Consideration of the TA- and LA-transitions shows that the former are forbidden, the latter allowed. This explains the exponents of the energy dependence of the individual components. The discrete steps are evidently not resolved in the TA-transitions.

The structure and threshold energies of the additional components at higher temperatures can also be explained by indirect (allowed or forbidden) exciton transitions with absorption or emission of an optical phonon.

At higher phonon energies, transitions take place in germanium between the valence band maximum and the lowest conduction band at Γ. These are the direct transitions of lowest threshold energy. Fig. 6.9 shows the shape of the direct absorption edge at various temperatures. At higher temperature the absorption edge has approximately the $(\hbar\omega - E_G)^{1/2}$-dependence required of direct transitions. With falling temperature, exciton effects appear, and these alter the shape in accordance with (6.93, 95). The discrete exciton states are not individually resolved. They reveal their presence by an absorption peak.

Before we turn to the absorption above the edge, we look at the absorption spectrum of Cu_2O in Fig. 6.10 as an example of a clearly resolved discrete spectrum of direct excitons. Here we can recognize several lines associated with different transitions. The best known "yellow" line $17.5 \cdot 10^3\,\mathrm{cm}^{-1}$ (5700 to 5800 Å) is shown again in expanded form and different representation in the bottom figure. Since these transitions are forbidden one finds p-excitons from $n = 2$ on [(6.93, 95)]. The s-exciton transitions do not appear. If one applies an electric field, these transitions become weakly allowed (lower part of Fig. 6.10).

Another possibility for measuring otherwise forbidden exciton transitions

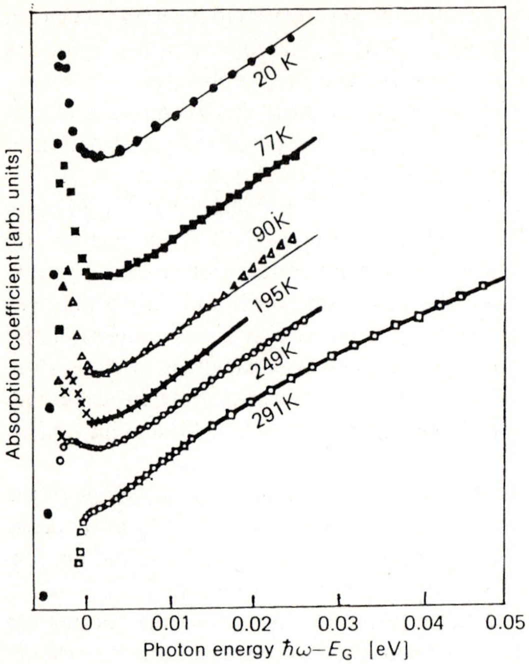

Fig. 6.9. Exciton lines near the direct absorption edge of germanium. The lines merge into a continuum which is very marked at low temperatures. At high temperatures the exciton contributions disappear (from *McLean* [118]).

is two-photon absorption. Fig. 6.11 shows the one-photon and the two-photon spectrum of ZnO. Direct allowed transitions are involved here.

We now turn to higher transitions. We already presented the ε_2-spectrum of germanium in Fig. 6.4. In Section 6.2.2 we saw that peaks and edges appear in the absorption spectrum when the photon energy equals the energy of a critical point in the combined density of states. Fig. 6.12 shows a theoretical ε_2-spectrum. The transitions at the critical points, which almost always lie at points of symmetry in the Brillouin zone, and an experimental curve are also shown. Exact analyses of this type contributed a lot to the understanding of the band structure of different semiconductors. We cannot go into this any further here, and refer the reader, for example, to *Greeneway* and *Harbeke* [85], *Phillips* [101.18], and the literature cited in Section 6.1.1.

6.2.7 Absorption by Free Charge Carriers

In addition to the transitions between different bands which we have considered so far, there can also be optical transitions in which the state of the absorbing electron belongs to the same band before and after the transition. Initial and final states are linked by a continuous succession of stationary states.

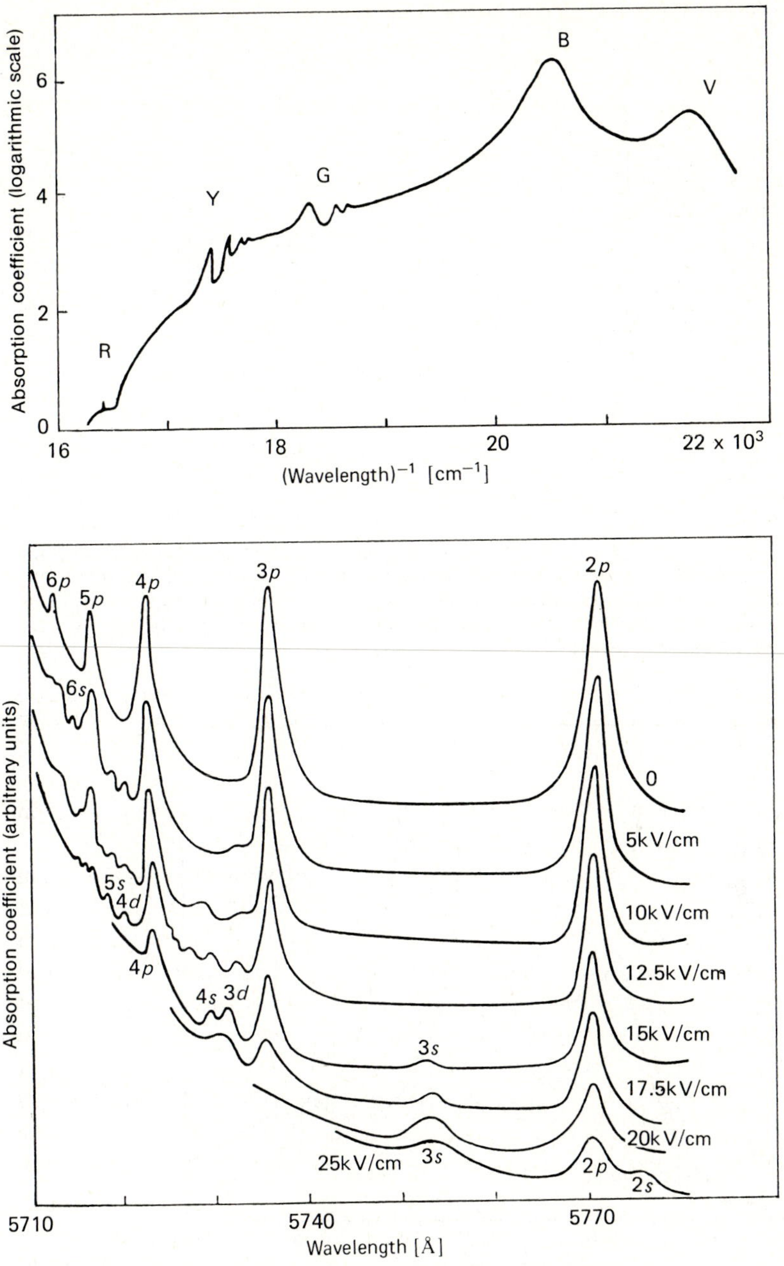

Fig. 6.10. Exciton lines in Cu_2O. The discrete spectrum indicated by Y in the upper part is shown again in another representation in the lower part. Because of selection rules, only p-exciton lines appear. If one disturbs the symmetry with an electric field, s-exciton transitions also become possible (from *Grosmann* [118]).

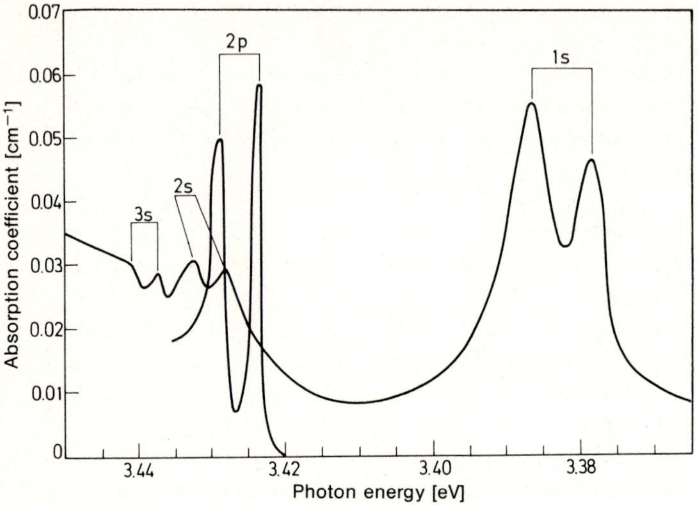

Fig. 6.11. One-photon and two-photon absorption in ZnO. For one-photon absorption, selection rules allow the appearance of only *s*-exciton lines, for two-photon absorption only *p*-exciton lines. The splitting into pairs of lines comes from the splitting of the valence band in ZnO (from H. Fröhlich: Proc. 10th Conference on the Physics of Semiconductors, Cambridge, Mass. 1970).

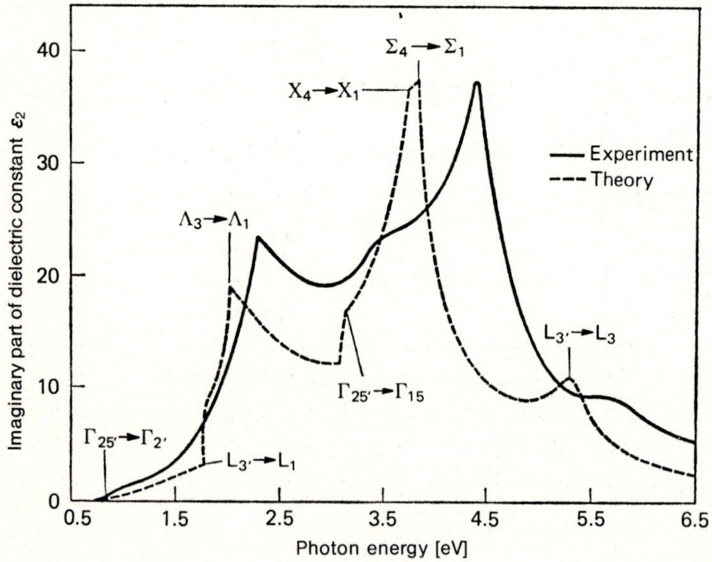

Fig. 6.12. Theoretical ε_2-spectrum for germanium with assignment of critical points to transitions in the band model (from *Phillips* [101.18]).

Such transitions are important when occupied and unoccupied states are present in one band (metals and semiconductors). We treat this case as an interaction between light and a gas of free electrons, characterized by an effective mass. For metals, the density of the gas is very high, for semiconductors low. In the following we neglect all effects due to nonspherical Fermi surfaces, or nonscalar or anisotropic effective masses.

In a free electron gas, direct transitions between two states $|\boldsymbol{k}\rangle$ and $|\boldsymbol{k}'\rangle$ are not possible since the matrix element $\langle \boldsymbol{k}'|\boldsymbol{e}\cdot\nabla|\boldsymbol{k}\rangle$ vanishes when one inserts plane waves for the $|\boldsymbol{k}\rangle$. The electron-phonon interaction must therefore be involved in the transitions. The continuous succession of states which can be traversed by an electron suggests that the "transition" can be interpreted as an acceleration of an electron by the high-frequency electric field of the light wave. This *absorption* of light *by quasi-free charge carriers* can consequently be treated as a transport problem.

We do not want to adopt the Boltzmann equation route to the solution of this problem, but rather limit ourselves to a simple approximation which reveals all the important features.

In (4.150) we saw that in the relaxation time approximation with energy-independent relaxation time, the motion of a charge carrier with effective mass m^* under an external force could be described as though it were in a viscous medium with friction constant $1/\tau$, where τ is the relaxation time. This model will now be used to calculate the *high-frequency conductivity*. To compare its order of magnitude with the angular frequency of the light, we write the damping constant as $1/\tau = \omega_0$. From (4.150) we put

$$m^*(\dot{\boldsymbol{v}} + \omega_0\boldsymbol{v}) = -e\boldsymbol{E} = -e\boldsymbol{e}E_0 \exp\left[\mathrm{i}(\boldsymbol{\kappa}\cdot\boldsymbol{r} - \omega t)\right]. \tag{6.100}$$

We solve this equation by putting the same time dependence $\exp(-\mathrm{i}\omega t)$ for the velocity $\boldsymbol{v}$ as for the electric field $\boldsymbol{E}$. We find

$$\boldsymbol{i} = \sigma\boldsymbol{E} = -en\boldsymbol{v} = \sigma_0\omega_0\frac{\omega_0 + \mathrm{i}\omega}{\omega_0^2 + \omega^2}\boldsymbol{E}, \qquad \sigma_0 = \frac{en^2\tau}{m^*}. \tag{6.101}$$

The complex conductivity given by (6.101) can be used to determine the optical constants. In Section 6.1.4 we avoided connecting the complex dielectric constant with the material constants in the Maxwell equations and chose a more general approach. It is necessary here to return to the wave equation following from the Maxwell equations

$$\frac{\varepsilon_\mathrm{L}}{c^2}\ddot{\boldsymbol{E}} + \frac{\sigma}{\varepsilon_0 c^2}\dot{\boldsymbol{E}} = -\nabla\times(\nabla\times\boldsymbol{E}) \tag{6.102}$$

where ε_L is the "static" dielectric constant which emerges from the Maxwell equations. The index L refers to the fact that the main contribution to ε_L comes from the lattice (see further below). Inserting (6.101) yields for isotropic media

$$\left[\left(\frac{\omega}{c}\right)^2 \varepsilon_L + \frac{i\omega\sigma}{\varepsilon_0 c^2}\right]\boldsymbol{E} = \kappa^2 \boldsymbol{E} - \boldsymbol{\kappa}(\boldsymbol{\kappa}\cdot\boldsymbol{E})\,. \tag{6.103}$$

By comparing both sides it follows from (6.103) that $\boldsymbol{E}$ is perpendicular to $\boldsymbol{\kappa}$. Since $\kappa = (\omega/c)(n + ik)$ then

$$(n + ik)^2 = \varepsilon_L + i\frac{\sigma}{\omega\varepsilon_0} \tag{6.104}$$

and we find

$$\varepsilon_1 = n^2 - k^2 = \varepsilon_L - \frac{1}{\varepsilon_0\omega}\,\mathrm{Im}\,(\sigma) = \varepsilon_L\left(1 - \frac{\omega_p^2}{\omega_0^2 + \omega^2}\right), \tag{6.105}$$

$$\varepsilon_2 = 2nk = \frac{1}{\varepsilon_0\omega}\,\mathrm{Re}(\sigma) = \varepsilon_L\frac{\omega_0}{\omega}\frac{\omega_p^2}{\omega_0^2 + \omega^2}, \tag{6.106}$$

where ω_p is the plasma frequency in a medium of dielectric constant ε_L: $\omega_p^2 = ne^2/m^*\varepsilon_L\varepsilon_0$.

We look first of all at ε_1. Along with the term ε_L, we have an additional (negative) term proportional to the electron concentration. ε_L is the contribution of all polarization mechanisms other than the free carrier contribution in the region under consideration.

The free carrier contribution is negligible if the electron concentration (and thus ω_p) is small, if the electron-phonon coupling (ω_0) is large, or if the frequency of the light (ω) is high. The contribution from the electron plasma becomes important if ω and ω_p are of the same order of magnitude.

ε_1 determines the reflection. From (6.39) the reflection coefficient for weak absorption (semiconductor) is equal to $(n - 1)^2/(n + 1)^2$. If $\varepsilon_2 \approx 0$, we have total reflection for $n = 0$ ($\varepsilon_1 = 0$) and no reflection at all for $n = 1$ ($\varepsilon_1 = 1$). Eq. (6.105) shows that for $\omega_0 = 0$ these cases occur when $\omega = \omega_p$ and when $\omega = [\varepsilon_L/(\varepsilon_L - 1)]^{1/2}\omega_p$, respectively. When the dielectric constant ε_L is large, as in semiconductors, the values $R = 1$ and $R = 0$ can lie very close together (*plasma reflection edge*, see Fig. 6.18).

The imaginary part of the dielectric constant ε_2 determines the absorption of the electron gas and also the minimum reflectivity at the plasma edge. ε_2 is clearly proportional to the electron concentration but also to the electron-phonon coupling ω_0. The most interesting region in the absorption spectrum is that where ω is large compared to ω_0. Then $\varepsilon_2 \propto \omega^{-3}$. The frequency dependence of the absorption coefficient $K = 2\omega k/c$ depends on whether $\omega \gg \omega_p$ or $\omega \ll \omega_p$. If we go to such high frequencies that ω is larger than ω_p, then $k \propto \omega^{-3}$ and hence $K \propto \omega^{-2} \propto \lambda^2$. We then find that absorption increases quadratically with increasing optical wavelength. This is shown in Fig. 6.13

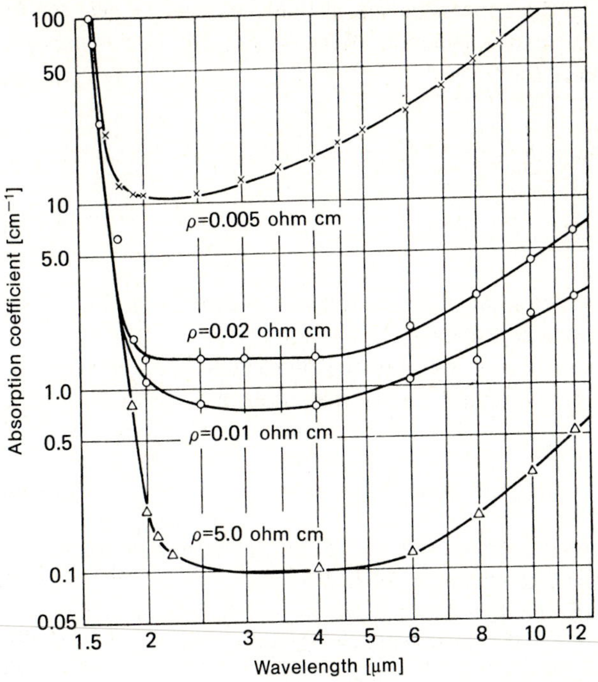

Fig. 6.13. Absorption constant of *n*-type germanium for wavelengths above the absorption edge. Initially the absorption falls rapidly with increasing wavelength when the photon energy is no longer sufficient to excite an electron from the valence into the conduction band. The absorption constant then increases approximately quadratically as a result of absorption by free electrons in the conduction band. The absolute value of K is proportional to the electron concentration, and hence inversely proportional to the resistivity ρ (from H. Y. Fan, M. Becker: Proc. of the International Conference on Semiconductors, Reading, 1951).

(again for germanium). For wavelengths above the absorption edge, K exhibits a λ^2-behaviour in which the absolute value of K increases with decreasing resistivity (increasing electron concentration).

All these statements are results from the simplest approximation. Deviations arise when one considers the detailed scattering mechanism which determines the relaxation time τ, usually reflected in a frequency dependence $\tau \propto \omega^{-r}$, which alters the frequency dependence of the absorption coefficient discussed above.

6.2.8 Absorption and Reflection in a Magnetic Field

In Section 2.1.6 we treated the change in the energy spectrum, and hence in the free electron density of states, due to a magnetic field. We now use these results to study the effect of a magnetic field on interband transitions (*magneto-absorption*).

We restrict ourselves to the approximation of (6.60), i.e., we neglect the $\boldsymbol{k}$-dependence of the interband matrix elements. When a magnetic field is present, the Hamiltonian (6.44) takes the form

$$H = H_0 + H', \qquad H_0 = \frac{1}{2m}(\boldsymbol{p} + e\boldsymbol{A}_{\text{magn}})^2 + V(\boldsymbol{r}),$$

$$H' = \frac{e}{m}\boldsymbol{A}\cdot(\boldsymbol{p} + e\boldsymbol{A}_{\text{magn}}), \tag{6.107}$$

where $\boldsymbol{A}_{\text{magn}}$ is the vector potential of the external magnetic field. This form follows from (6.43) when one considers $\boldsymbol{A}_{\text{magn}}$ as well as $\boldsymbol{A}$.

To calculate the matrix elements we need the Bloch functions for electrons in a magnetic field. In (2.57) we found for the free electron wave functions the form

$$F_{\nu k_y k_z}(x) = (L_y L_z)^{-1/2} \exp[\mathrm{i}(k_y y + k_z z)]\varphi_\nu(x - x_0). \tag{6.108}$$

A normalization factor has been included here. The $\varphi_\nu(x)$ are the oscillator wave functions of (2.58) with the eigenvalues E_ν of (2.59).

If one looks particularly at direct transitions at $\boldsymbol{k} = 0$, one must multiply F by a factor $u_n(0, \boldsymbol{r})$ to obtain the desired Bloch functions. We shall not prove this obvious assumption here.

In place of the matrix elements (6.49) for direct transitions we therefore have the integrals

$$\langle u_{n'} F_{\nu' k_y' k_z'} | H' | u_n F_{\nu k_y k_z} \rangle. \tag{6.109}$$

In the wave functions, F is a slowly varying, and u a rapidly varying factor. If as in (6.50) one divides the integration into partial integrals over individual Wigner-Seitz cells, one can put all slowly varying functions as constant in one Wigner-Seitz cell. The final sum over the cells can be transformed into an integration over the slowly varying parts. Then we find from (6.109)

$$\langle \ldots | H' | \ldots \rangle = \langle F_{\nu' k_y' k_z'} | H' | F_{\nu k_y k_z} \rangle \langle u_{n'} | u_n \rangle_{\text{WSC}} + \langle F_{\nu' k_y' k_z'} | F_{\nu k_y k_z} \rangle \left\langle u_{n'} \left| \frac{e}{m}\boldsymbol{A}\cdot\boldsymbol{p} \right| u_n \right\rangle_{\text{WSC}}. \tag{6.110}$$

For $n' \neq n$, the first term is again very small compared with the second [cf. 6.63]. One finds for the integrals in (6.110) which contain F

$$\langle F_{\nu' k_y' k_z'} | F_{\nu k_y k_z} \rangle = C \delta_{k_z k_z'} \delta_{\nu\nu'} \tag{6.111}$$

and

$$\langle F_{\nu' k_y' k_z'} | H' | F_{\nu k_y k_z} \rangle = C' \delta_{k_z k_z'} \delta_{\nu,\nu' \pm 1}. \tag{6.112}$$

Associated with these equations there will be selection rules for the k_y, but we do not require these here.

For direct allowed transitions we can restrict our interest in (6.110) to the second term. The selection rules (6.111) then say that the transitions take place with conservation of wave number k_z and subband index ν.

In the approximation of (6.60), we can put for the imaginary part of the dielectric constant

$$\varepsilon_2 \propto \sum_\nu \int \frac{dk_y}{2\pi} \int \frac{dk_z}{2\pi} \delta\left(\hbar\omega - \left[E_{n'} + \left(\nu' + \frac{1}{2}\right)\hbar\omega_{cn'} + \frac{\hbar^2 k_z^2}{2m_{n'}}\right] + \left[E_n - \left(\nu + \frac{1}{2}\right)\hbar\omega_{cn} - \frac{\hbar^2 k_z^2}{2m}\right]\right)$$

$$= \sum_\nu \int \frac{dk_y}{2\pi} \int \frac{dk_z}{2\pi} \delta\left(\hbar\omega - E_G - \left(\nu + \frac{1}{2}\right)\hbar\omega_c^* - \frac{\hbar^2 k_z^2}{2m^*}\right) \tag{6.113}$$

with

$$\omega_c^* = \frac{eB}{m^*}, \qquad \frac{1}{m^*} = \frac{1}{m_{n'}} + \frac{1}{m_n}.$$

The two integrations in (6.113) are independent. We obtain the integration limits from Section 2.1.6. There k_y was related to x_0, the reference point in the x-direction, by $k_y = -(eB/\hbar)x_0$. Since x_0 must lie in the volume V_g (and can therefore only vary between two constant values), k_y should also lie between two limiting values, both linear in $\boldsymbol{B}$. The integral over k_y is then proportional to $\boldsymbol{B}$, and hence to ω_c^*. The integral over k_z can readily be evaluated. Since

$$\delta(f(x)) = \frac{\delta(x - x_0)}{|f'(x_0)|} \tag{6.114}$$

we find

$$\int dk_z \delta\left(\alpha - \frac{\hbar^2 k_z^2}{2m^*}\right) = \left[\left|\frac{d}{dk_z}\left(\alpha - \frac{\hbar^2 k_z^2}{2m^*}\right)\right|_{\alpha = \hbar^2 k_z^2/2m^*}\right]^{-1} \propto \alpha^{-1/2} \tag{6.115}$$

and finally

$$\varepsilon_2(\omega, \omega_c^*) \propto \hbar\omega_c \sum_\nu \left[\hbar\omega - E_G - \left(\nu + \frac{1}{2}\right)\hbar\omega_c^*\right]^{-1/2}. \tag{6.116}$$

Before discussing this equation we give the corresponding equations for the indirect and for the direct forbidden transitions. From (6.76) we have to carry out a double integration over k_y' and k_y, and over k_z' and k_z, and we have to sum over ν' and ν. An additive term $\hbar\omega_{\boldsymbol{q}}$ appears in the δ-function of (6.113). The double integration over k_y' and k_y yields a factor $(\hbar\omega_c)^{1/2}$, corresponding to (6.116). The first integration over k_z' leads to

$$\varepsilon_2 \propto (\hbar\omega_c^*)^2 \sum_{\nu\nu'} \int dk_z \left[\hbar\omega - E_G - \left(\nu' + \frac{1}{2}\right)\hbar\omega_{cn'} - \left(\nu + \frac{1}{2}\right)\hbar\omega_{cn} \pm \hbar\omega_{\boldsymbol{q}} - \frac{\hbar^2 k_z^2}{2m^*} \right]^{-1/2} . \tag{6.117}$$

The integral extends over the region for which the root is real. Then

$$\varepsilon_2 \propto (\hbar\omega_c^*)^2 \sum_{\nu\nu'} S\left(\hbar\omega - E_G - \left(\nu' + \frac{1}{2}\right)\hbar\omega_{cn'} - \left(\nu + \frac{1}{2}\right)\hbar\omega_{cn} \pm \hbar\omega_{\boldsymbol{q}} \right) . \tag{6.118}$$

$S(x)$ is a step function which vanishes for $x < 0$ and is unity for $x > 0$.

For *forbidden direct transitions* we state the corresponding expressions without proof. Depending on the polarization of the incident radiation we have to distinguish between

$$\boldsymbol{E} \parallel \boldsymbol{B}\colon \varepsilon_2 \propto \hbar\omega_c^* \sum_{\nu} \left[\hbar\omega - E_G - \left(\nu + \frac{1}{2}\right)\hbar\omega_c^* \right]^{1/2} , \tag{6.119}$$

and

$$\boldsymbol{E} \perp \boldsymbol{B}\colon \varepsilon_2 \propto (\hbar\omega_c^*)^2 \sum_{\nu} \left\{ c_1(\nu + 1)\left[\hbar\omega - E_G - \left(\nu + \frac{1}{2}\right)\hbar\omega_c^* - \hbar\omega_{cn} \right]^{-1/2} + c_2\nu\left[\hbar\omega - E_G - \left(\nu + \frac{1}{2}\right)\hbar\omega_c^* - \hbar\omega_{cn'} \right]^{-1/2} \right\} . \tag{6.120}$$

The sum in (6.119, 120) as also in (6.116) extends over all ν for which the roots are real. In addition to the selection rule $k_z' = k_z$, in (6.118) $\nu' = \nu$ applies, and in (6.120) $\nu' = \nu + 1$ for the first term and $\nu' = \nu - 1$ for the second term.

All the expressions given so far have been based on the effective mass approximation without spin. When spin is taken into account, additional terms of the general form $\pm(g/2)\mu_B B$ appear in the root expressions and in $S(x)$. The g-factor here depends on the type of transition and on the polarization of the incident radiation (cf. Problem 6.7).

From (6.116–120) we can expect two different types of absorption spectra. For allowed direct transitions and for forbidden direct transitions with $\boldsymbol{E} \parallel \boldsymbol{B}$, the roots occur in the denominator of the sum terms. We therefore obtain absorption peaks when the energy of the incident radiation equals the minimum energy required for the transition between two subbands. For allowed direct transitions and within the two series (6.120), the separations between two absorption peaks are $\Delta\omega = \omega_c$. The separation between these two series is $\omega_{cn'} - \omega_{cn}$. If one takes account of spin splitting, the series (6.116) splits into four series (two each for $\boldsymbol{E} \parallel \boldsymbol{B}$ and $\boldsymbol{E} \perp \boldsymbol{B}$); each of the two series (6.120) split likewise into two.

The behaviour for indirect transitions, and for forbidden direct transitions with $\boldsymbol{E} \perp \boldsymbol{B}$, is different. Here the roots occur in the numerator of the sum terms. The absorption spectrum then consists of steps at the energies where the roots become zero, or at which the step function $S(x)$ jumps from 0 to 1. The separation between steps is again ω_c.

The equations just discussed only apply of course for the ideal case $T = 0$. Lattice vibrations cause a broadening of the spectra. The steps are then blurred and the absorption peaks lead to oscillatory behaviour of the absorption.

All the details of the absorption spectrum in a magnetic field discussed are confirmed experimentally. We find corresponding phenomena in the reflection spectrum. We restrict ourselves to one example: the *direct* transitions in the magneto-absorption spectrum of germanium. The final states of these transitions lie in a higher conduction band (cf. Fig. 2.33). It is approximately isotropic so that, following Section 2.1.6, we can expect two ladders of levels for the two spin possibilities, each with equidistant spacing at $\boldsymbol{k} = 0$. The relationships for the initial states are more complicated. The top of the valence band is fourfold degenerate and splits at $\boldsymbol{k} \neq 0$ into two subbands, the "heavy-hole" band and the "light-hole" band. In a magnetic field this leads to four ladders for $\boldsymbol{k} = 0$, whose levels are not equidistant because of the interaction between the subbands. For the ladder of terms for the "light holes" moreover no subbands with $\nu = 0$ and $\nu = 1$ exist. This is shown in Fig. 6.14. The four ladders of the valence band levels are given by

$$\begin{aligned} E_1^{\pm}(\nu) &= \Bigg(a\nu - \left(\frac{a}{2} + b - \frac{c}{2}\right) \\ &\qquad \pm \left\{\left[b\nu - \left(a - c + \frac{b}{2}\right)\right]^2 + 3d\nu(\nu - 1)\right\}^{1/2}\Bigg)\hbar\omega_c\,, \\ E_2^{\pm}(\nu) &= \Bigg(a\nu - \left(\frac{a}{2} - b + \frac{c}{2}\right) \\ &\qquad \pm \left\{\left[b\nu + \left(a - c - \frac{b}{2}\right)\right]^2 + 3d\nu(\nu - 1)\right\}^{1/2}\Bigg)\hbar\omega_c \end{aligned} \tag{6.121}$$

in which the constants follow from the structure of the valence band in the absence of a magnetic field.

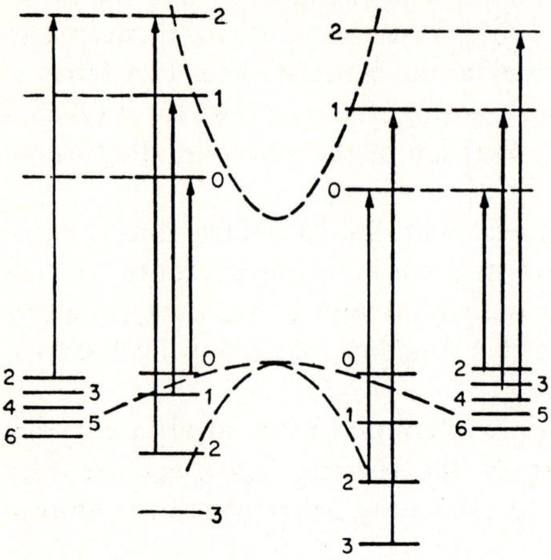

Fig. 6.14. Magnetic field induced splitting of the levels in the conduction band and valence band of germanium at point Γ in the Brillouin zone. See the text for the transitions indicated [from L. M. Roth, B. Lax, S. Zwerdling: Phys. Rev. **114**, 90 (1959)].

The direct transitions are described by (6.116). If we distinguish the ladders in the conduction band by $c^+(v)$ and $c^-(v)$, we find as selection rules

$$\left.\begin{aligned} &E_1^{\pm}(v) \to c^-(v) \\ &E_2^{\pm}(v) \to c^+(v-2) \end{aligned}\right\} \boldsymbol{E} \parallel \boldsymbol{B}$$
$$\left.\begin{aligned} &E_1^{\pm}(v) \to c^+(v) \quad \text{and} \quad c^+(v-2) \\ &E_2^{\pm}(v) \to c^-(v) \quad \text{and} \quad c^-(v-2) \end{aligned}\right\} \boldsymbol{E} \perp \boldsymbol{B}\,. \tag{6.122}$$

The transitions for $\boldsymbol{E} \parallel \boldsymbol{B}$ are indicated in Fig. 6.14.

Figure 6.15 illustrates the fine structure of magneto-absorption in Ge. The different behaviour for $\boldsymbol{E} \parallel \boldsymbol{B}$ and $\boldsymbol{E} \perp \boldsymbol{B}$ can be clearly seen. Fig. 6.16 gives a comparison between theory and experiment for $\boldsymbol{E} \parallel \boldsymbol{B}$. Almost all details in the spectrum can be explained theoretically.

6.2.9 Magneto-Optics of Free Charge Carriers

For a description of the optical properties of free charge carriers in a static magnetic field, we add to the right-hand side of (6.100) the term $-e\boldsymbol{v} \times \boldsymbol{B}$. We choose $\boldsymbol{B} = (0, 0, B)$, i.e., the magnetic field lies along the z-direction of our coordinate system. Introducing the cyclotron resonance frequency $\omega_c = eB/m^*$ then

$$\sigma = \begin{vmatrix} \sigma_{xx} & \sigma_{xy} & 0 \\ -\sigma_{xy} & \sigma_{xx} & 0 \\ 0 & 0 & \sigma_{zz} \end{vmatrix} \tag{6.123}$$

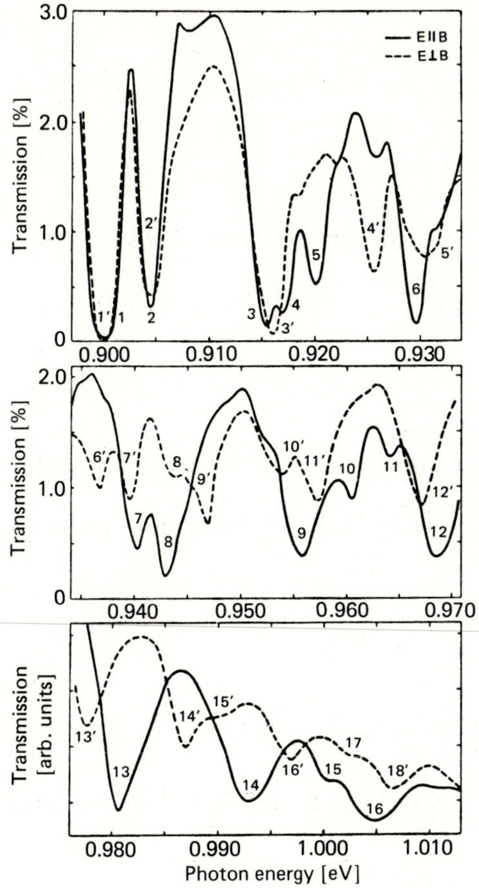

Fig. 6.15. Fine structure in the magneto-absorption of Ge at 38.9 kilogauss and 4.2 K [from S. Zwerdling, L. M. Roth, B. Lax: Phys. Rev. **109**, 2207 (1958)].

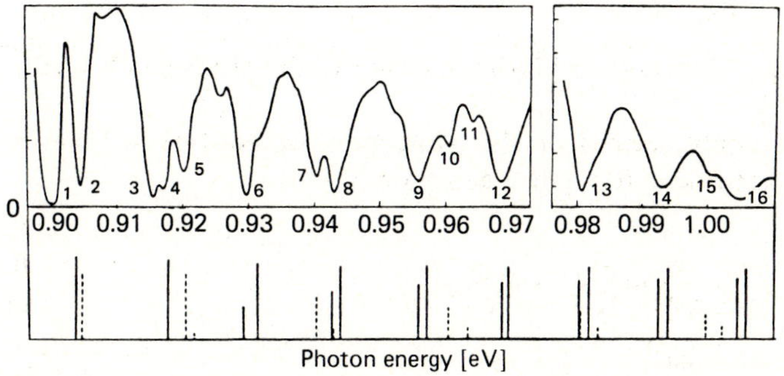

Fig. 6.16. Comparison of the absorption spectrum of Fig. 6.15 with theory for $\boldsymbol{E} \| \boldsymbol{B}$. Above: experimental spectrum. Below: calculated transition energies. Solid lines: transitions from "heavy hole" levels (upper valence band), dotted lines: transitions from "light hole" levels (lower valence band). The height of the lines gives the theoretical intensity [from L. M. Roth, B. Lax, S. Zwerdling: Phys. Rev. **114**, 90 (1959)].

with

$$\sigma_{xx} = \sigma_0\omega_0 \frac{\omega_0 - i\omega}{(\omega_0 - i\omega)^2 + \omega_c^2}, \qquad \sigma_{xy} = \sigma_0\omega_0 \frac{\omega_c}{(\omega_0 + i\omega)^2 + \omega_c^2},$$

$$\sigma_{zz} = \sigma_0\omega_0 \frac{1}{\omega_0 - i\omega}. \tag{6.124}$$

The dispersion equation (6.103) is unchanged except that the conductivity is now a tensor

$$\kappa^2 \boldsymbol{E} - \boldsymbol{\kappa}(\boldsymbol{\kappa}\cdot\boldsymbol{E}) = \left(\frac{\omega}{c}\right)^2 \varepsilon_L \boldsymbol{E} + \frac{i}{\varepsilon_0}\frac{\omega}{c^2}\sigma\boldsymbol{E}. \tag{6.125}$$

We examine this equation in two limiting cases.

1) *Longitudinal case.* Magnetic field parallel to the propagation vector of the light.
Then the vector $\boldsymbol{\kappa}$ has a z-component only. From (6.125) it follows that, as well as $E_z = 0$,

$$\begin{aligned} \kappa^2 E_x &= \left(\frac{\omega}{c}\right)^2 \varepsilon_L E_x + \frac{i}{\varepsilon_0}\frac{\omega}{c^2}(\sigma_{xx}E_x + \sigma_{xy}E_y), \\ \kappa^2 E_y &= \left(\frac{\omega}{c}\right)^2 \varepsilon_L E_y + \frac{i}{\varepsilon_0}\frac{\omega}{c^2}(-\sigma_{xy}E_x + \sigma_{xx}E_y) \end{aligned} \tag{6.126}$$

or, for circularly polarized light, by combining E_x and E_y into $E_\pm = E_x \pm iE_y$ and $\sigma_\pm = \sigma_{xx} \pm i\sigma_{xy}$ finally

$$(n + ik)_\pm^2 = \varepsilon_L + \frac{i}{\varepsilon_0\omega}\sigma_\mp. \tag{6.127}$$

2) *Transverse case.* Magnetic field perpendicular to the propagation vector of light.
Since we have so far only decided on the z-direction, we are free to fix $\boldsymbol{\kappa}$ in the x-y plane. We choose $\boldsymbol{\kappa} = (0, \kappa, 0)$. Then from (6.125)

$$\begin{aligned} \kappa^2 E_x &= \left(\frac{\omega}{c}\right)^2 \varepsilon_L E_x + \frac{i\omega}{\varepsilon_0 c^2}(\sigma_{xx}E_x + \sigma_{xy}E_y), \\ 0 &= \left(\frac{\omega}{c}\right)^2 \varepsilon_L E_y + \frac{i\omega}{\varepsilon_0 c^2}(-\sigma_{xy}E_x + \sigma_{xx}E_y), \\ \kappa^2 E_z &= \left(\frac{\omega}{c}\right)^2 \varepsilon_L E_z + \frac{i\omega}{\varepsilon_0 c^2}\sigma_{zz}E_z. \end{aligned} \tag{6.128}$$

From the second equation we see first of all that there is now a field component in the direction of propagation of the light. Since the direction of the electric field vector has not yet been fixed, we distinguish two further cases.

a) $\boldsymbol{E} \parallel \boldsymbol{B}$, i.e., $E_x = E_y = 0$. The third equation (6.128) then gives

$$(n + \mathrm{i}k)_{\parallel}^2 = \varepsilon_{\mathrm{L}} + \frac{\mathrm{i}}{\varepsilon_0 \omega} \sigma_{zz} . \tag{6.129}$$

b) $\boldsymbol{E} \perp \boldsymbol{B}$, i.e., $E_z = 0$. From the first two equations (6.128) it follows that

$$(n + \mathrm{i}k)_{\perp}^2 = \varepsilon_{\mathrm{L}} + \frac{\mathrm{i}}{\varepsilon_0 \omega}\left(\sigma_{xx} + \frac{\mathrm{i}\sigma_{xy}^2}{\varepsilon_0 \varepsilon_{\mathrm{L}} \omega + \mathrm{i}\sigma_{xx}}\right) . \tag{6.130}$$

The optical constants follow from the equations (6.127), (6.129), and (6.130) by separating real and imaginary parts.

1) *Longitudinal case.*

$$\varepsilon_1 = \varepsilon_{\mathrm{L}}\left[1 - \frac{\omega_0^2}{\omega} \frac{\omega \pm \omega_{\mathrm{c}}}{(\omega \pm \omega_{\mathrm{c}})^2 + \omega_0^2}\right] , \tag{6.131}$$

$$\varepsilon_2 = \varepsilon_{\mathrm{L}} \frac{\omega_0}{\omega} \omega_{\mathrm{p}}^2 \frac{1}{(\omega \pm \omega_{\mathrm{c}})^2 + \omega_0^2} \tag{6.132}$$

where the two signs refer to the two directions of circular polarization of the light.

2a) *Transverse case,* $\boldsymbol{E} \parallel \boldsymbol{B}$.

$$\varepsilon_1 = \varepsilon_{\mathrm{L}}\left(1 - \frac{\omega_{\mathrm{p}}^2}{\omega^2 + \omega_0^2}\right) , \tag{6.133}$$

$$\varepsilon_2 = \varepsilon_{\mathrm{L}} \frac{\omega_0}{\omega} \omega_{\mathrm{p}}^2 \frac{1}{\omega^2 + \omega_0^2} . \tag{6.134}$$

2b) *Transverse case,* $\boldsymbol{E} \perp \boldsymbol{B}$.

$$\varepsilon_1 = \varepsilon_{\mathrm{L}}\left(1 - \frac{\omega_{\mathrm{p}}^2 \beta}{\omega^2 \beta^2 + \omega_0^2 \alpha^2}\right) , \tag{6.135}$$

$$\varepsilon_2 = \varepsilon_{\mathrm{L}} \frac{\omega_0}{\omega} \omega_{\mathrm{p}}^2 \frac{\alpha}{\omega^2 \beta^2 + \omega_0^2 \alpha^2} \tag{6.136}$$

with

$$\alpha = 1 + \frac{\omega^2 \omega_{\mathrm{c}}^2}{\omega^2 \omega_0^2 + (\omega^2 - \omega_{\mathrm{p}}^2)^2} , \tag{6.137}$$

$$\beta = 1 - \frac{\omega_{\mathrm{c}}^2 (\omega^2 - \omega_{\mathrm{p}}^2)}{\omega^2 \omega_0^2 + (\omega^2 - \omega_{\mathrm{p}}^2)^2} . \tag{6.138}$$

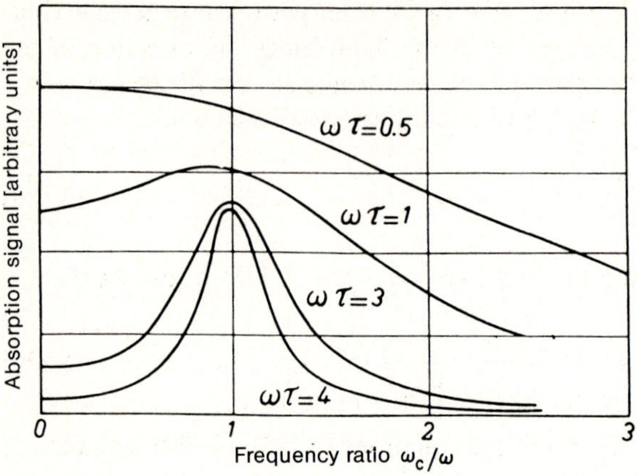

Fig. 6.17. Absorption in the neighbourhood of the cyclotron resonance frequency ω_c. In the absence of electron-phonon interaction ($\omega_0 = 1/\tau = 0$) the absorption curve is a δ-function at $\omega = \omega_c$. With increasing electron-phonon interaction the absorption peak will become increasingly flat in accordance with (6.140).

Equations (6.131–138) contain four characteristic frequencies: ω, ω_p, ω_c, ω_0, which represent the four parameters light, electron concentration, magnetic field, and electron-phonon coupling. The interrelation of these parameters determines the appearance of different phenomena. We shall treat one phenomenon each from absorption, reflection, and dispersion: cyclotron resonance, magneto-plasma reflection, and Faraday and Voigt effects.

Cyclotron Resonance

We consider the longitudinal case but in contrast to (6.127) we choose linearly polarized light ($E_x \neq 0$, $E_y = 0$). It then follows from the first equation (6.126) that

$$\kappa^2 = \left(\frac{\omega}{c}\right)^2 (\varepsilon_1 + i\varepsilon_2) = \left(\frac{\omega}{c}\right)^2 \varepsilon_L + \frac{i\omega}{\varepsilon_0 c^2}\sigma_{xx}. \tag{6.139}$$

ε_2 and hence the absorption constant K become proportional to the real part of σ_{xx}

$$K \propto \frac{\omega_0^2 + \omega_c^2 + \omega^2}{(\omega_0^2 + \omega_c^2 - \omega^2)^2 + 4\omega_0^2\omega^2}. \tag{6.140}$$

As a result of this, an absorption maximum appears near $\omega = \omega_c$, which is more marked the smaller ω_0 is (Fig. 6.17). This phenomenon can readily be understood both classically and quantum mechanically. In a magnetic field the electrons, which have a thermal velocity v_{th}, will be accelerated in orbits about the direction of the magnetic field. Their angular frequency is the cyclotron resonance frequency ω_c; the radius of the orbit is $r = v_{th}/\omega_c$. A high-frequency electric field, whose $\boldsymbol{E}$-vector is oscillating in the orbital plane, will then be most strongly absorbed when its frequency is exactly equal to the angular frequency ω_c. In this case the periodic motion of the electrons is in phase with the changes in the electric field, and the electrons draw most energy out of the field. We can also readily understand cyclotron resonance from the splitting of the band into magnetic subbands. In the theory of direct transitions in a magnetic field we omitted the first term on the right in (6.110), since the matrix element $\langle u_{n'}|u_n\rangle$ is very small for $n' \neq n$. For $n' = n$ (direct transitions between magnetic subbands of a band n), the matrix element, on the other hand, approaches unity. From (6.112) the selection rules for this term are $\nu' = \nu \pm 1$, where ν' and ν index the subbands. Transitions between adjacent subbands, however, require the energy $\hbar\omega_c$, which must be delivered from the photons. The resonance condition $\omega = \omega_c$ thus follows directly.

Measurements of cyclotron resonance are important for the determination of the frequency ω_c and hence the effective mass of the charge carriers in semiconductors. There are a number of obstacles to the observation of this resonance in metals. Electromagnetic waves of frequency ω_c only penetrate into a thin surface layer much smaller than the orbit radius v_{th}/ω_c. In spite of this a signal can be detected when the magnetic field is applied parallel to the surface (*Azbel-Kaner resonance*). Only those fractions of the orbits reaching into the surface layer contribute. Related to this is the further peculiarity that not only the frequency ω_c appears in the signal but also all integer multiples $n\omega_c$, where the electron only enters the surface layer at every nth oscillation of the field. A difficulty arises in metals from the structure of the Fermi surfaces, which often deviates considerably from a sphere. The assumption of a scalar, isotropic effective mass is then no longer justified. This complication, on the other hand, allows the use of resonance measurements to determine the shape of Fermi surfaces.

Magneto-Plasma Reflection

From (6.131), one finds for circularly polarized light, if $\omega \approx \omega_p \gg \omega_c, \omega_0$

$$R = 1 \quad \text{for} \quad \omega = \omega_p \pm \frac{1}{2}\omega_c + \cdots. \tag{6.141}$$

In a magnetic field the plasma edge (cf. Sec. 6.2.7) is subject to a displacement by $\pm\omega_c/2$, depending on the polarization. In unpolarized light both possibilities are superimposed, so that the plasma edge is split into two components, separated by ω_c (Fig. 6.18). This can be used for a determination of ω_c and hence m^*.

Faraday and Voigt Effect

As indicated in (6.131), right- and left-circularly polarized waves have different

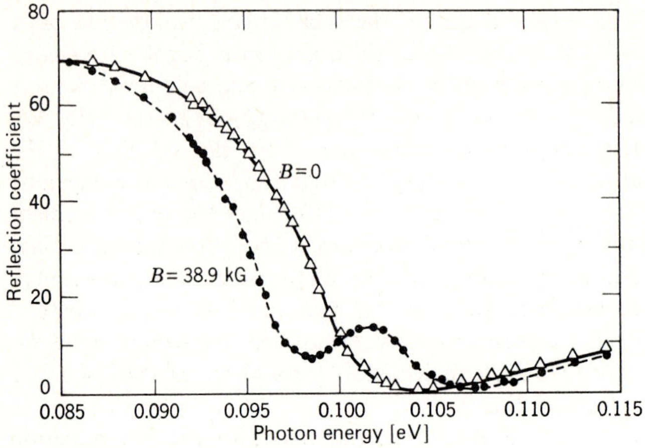

Fig. 6.18. Transverse magneto-plasma reflection in *n*-InAs at room temperature. Also shown is the plasma edge in the absence of a magnetic field [from B. Wright, B. Lax: J. Appl. Phys. **32**, 2113 (1961)].

propagation velocities. The phase difference which arises after propagating a distance d is $\delta = (\omega/c)(n_+ - n_-)d$. If (for small magnetic fields) n_+ and n_- are only slightly different, this can be approximated by

$$\delta = \frac{\omega}{2nc}(n_+^2 - n_-^2)d = \frac{\omega}{2nc}(\varepsilon_{1+} - \varepsilon_{1-})d\,. \tag{6.142}$$

Instead of circularly polarized light we consider now a linearly polarized wave which we take to consist of two circularly polarized components with opposite senses of polarization. From (6.142) the polarization direction will then change by an angle θ, which amounts to half the phase difference δ (*Faraday effect*).

In the transverse case too the magnetic field influences the polarization of the light, since the dispersion is different for $\boldsymbol{E} \parallel \boldsymbol{B}$ and $\boldsymbol{E} \perp \boldsymbol{B}$. Linearly polarized light, with electric vector inclined by 45° to the direction of the magnetic field, will consequently be converted into elliptically polarized light (*Voigt effect*).

As an example of the Faraday effect, Fig. 6.19 shows measurements on GaAs. From (6.142) and (6.131) we find a Faraday angle $\theta \propto \omega_p^2 \omega_c/\omega^2$. The effect is thus linear in B, and quadratic in the wavelength. The λ^2-law is closely followed by the experimental data over a wide range of wavelengths.

In many semiconductors, deviations both to higher and to lower angles have been found at short wavelengths. They are ascribed to contributions from interband transitions. Since we have not treated these here, let us give a brief explanation of the effect. We return to the contents of Section 6.2.8. There we

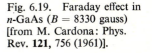

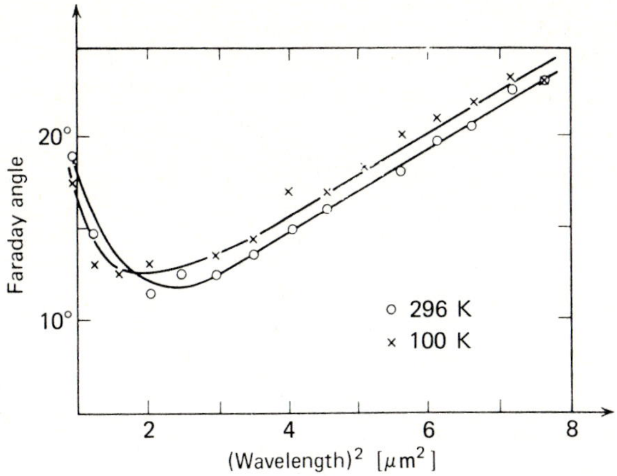

Fig. 6.19. Faraday effect in n-GaAs ($B = 8330$ gauss) [from M. Cardona: Phys. Rev. **121**, 756 (1961)].

saw that for direct allowed transitions between two bands, the selection rule $\nu' = \nu$ applies. If one considers the spin-splitting of the individual subbands, the transition energy $\hbar\omega = E_G + (\nu + 1/2)\hbar\omega_c$ is extended by additive terms $\pm(g/2)\mu_B B$, where the g-factor depends on the band structure and is therefore different for valence band and conduction band (see also Problem 6.7). A further selection rule now arises, namely that a transition is only possible between a subband $\pm(g_V/2)\mu_B B$ of the valence band and a subband $\mp(g_C/2)\mu_B B$ of the conduction band, where the upper or lower signs apply to the two possible directions of circular polarization. Hence for both polarization directions the energy difference of equivalent interband transitions is $(g_C + g_V)\mu_B B$. g-factors in semiconductors can deviate from the value 2 for free electrons both in magnitude and in sign. One can therefore appreciate that the contribution to dispersion from interband transitions can be both positive and negative, and that it can also vanish for different signs of g_C and g_V.

6.3 Phonon-Photon Interaction

6.3.1 Introduction

Following the electron-photon interaction, we want to close this chapter with a treatment of the interaction of light with the lattice vibrations. We again consider the possibilities with the aid of graphs (Fig. 6.20, cf. Figs. 4.1, 6.5). Individual processes are the transformation of a photon into a phonon of identical energy and wave vector. The selection rules and conservation laws restrict these processes to the formation of a TO-phonon and to polar solids. We have therefore in addition to consider second-order processes, in which a photon decays into two phonons or into a photon and a phonon. Moreover, lattice anharmonicity can lead to a further process in which one phonon decays

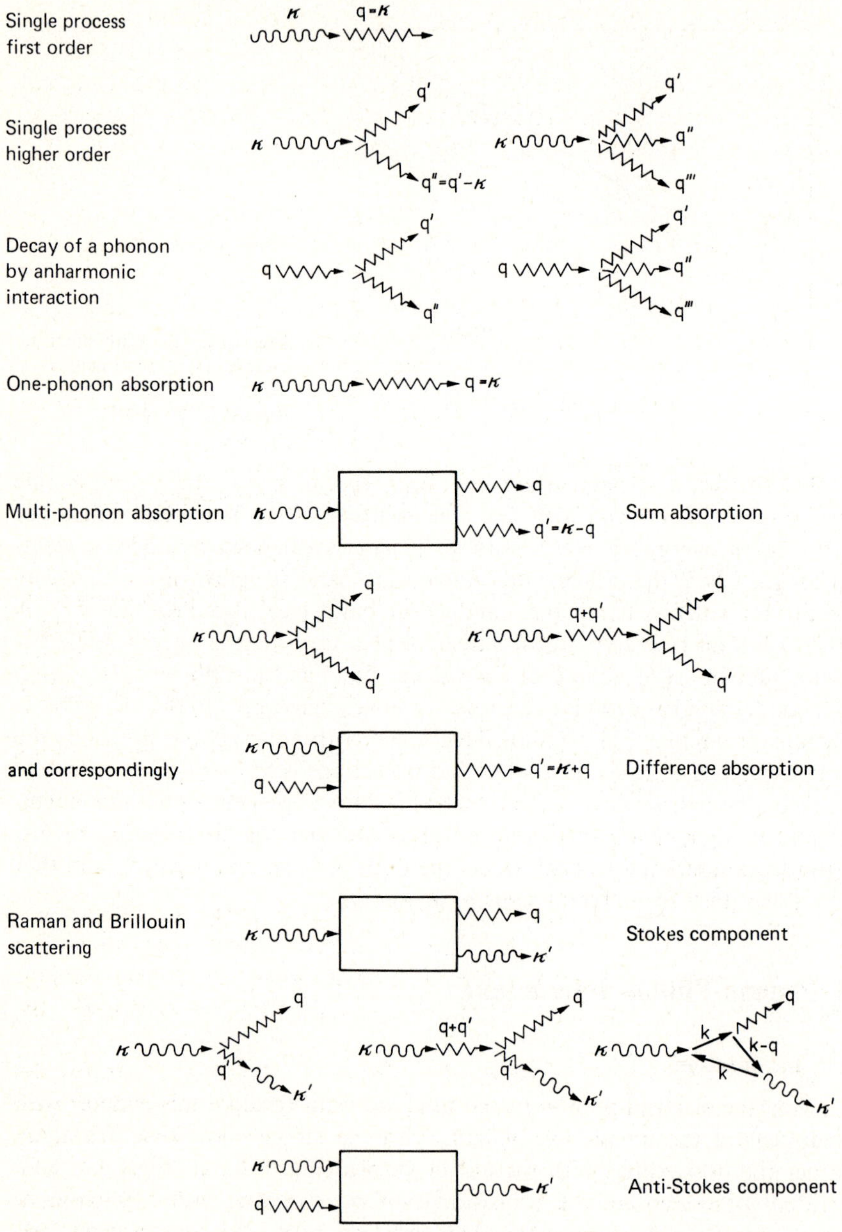

Fig. 6.20. Graphs for the phonon-photon interaction.

into two phonons. These processes are also important for a different reason. When introducing the polariton in Section 6.1.3 we saw that the absorption of a photon with emission of a TO-phonon is only an absorption mechanism if the phonon decays rapidly enough for its energy not to be returned to the radiation field. In view of this, the following processes are important (where a check is still required in each case to see if the selection rules and conservation laws allow the one or other process).

One-phonon absorption, i.e., transformation of a photon into a phonon of identical energy and wave vector. The optical spectra associated with this will be discussed in Section 6.3.2.

Multi-phonon absorption, i.e., absorption of a photon with formation of two or more phonons. Various concurrent processes are possible here. We shall discuss this in Section 6.3.3.

Scattering of light, i.e., absorption of a photon, emission of another photon of different energy with absorption, or emission of one or more phonons. If the phonons involved in such processes belong to the optical branch, one speaks of *Raman scattering*, if they belong to the acoustic branch one speaks of *Brillouin scattering*.

In this case too there are several possibilities. First the phonon can be emitted (Stokes scattering). The secondary photon then has lower energy. In the other case (phonon absorption, anti-Stokes scattering) it has higher energy. In each of the two cases the complete process can be a single process of higher order, or be based on successive individual processes with virtual intermediate states. We examine these possibilities in Section 6.3.4.

We shall give references to literature on the photon-phonon interaction in the course of the following sections.

6.3.2 One-Phonon Absorption

In the absorption of a photon with emission of a phonon, energy and momentum conservation require (as long as Umklapp processes are not involved) that the energy and wave vector of the two elementary excitations be the same. Phonon energies lie below 0.1 eV, hence one-photon absorption takes place in the infrared.

The conservation laws tell us further that only optical phonons can be created. The velocity of light is greater than the propagation velocity of acoustic waves in solids by a factor of 10^3 to 10^5. There are therefore no acoustic phonons with the same energy and wave vector as photons.

Along with the laws of conservation we have also to consider the possible mechanisms by which radiation field and lattice vibrations can be coupled. In the dipole approximation one only finds nonvanishing matrix elements for one-quantum transitions in the case of optical phonons. We start with a classical derivation of optical absorption by excitation of transverse optical vibrations of the lattice.

We link up with the limiting case of long wavelength optical vibrations which we treated in Section 3.3.8. We consider a lattice with two oppositely charged ions in each Wigner-Seitz cell. Associated with the relative displacement $\boldsymbol{s} = \boldsymbol{s}_+ - \boldsymbol{s}_-$ there is a time-dependent dipole moment $e^*\boldsymbol{s}$, where e^* is the effective charge on the ions. The equations of motion for the $\boldsymbol{s}$ and for the reduced displacements $\boldsymbol{w} = \sqrt{NM/V_g}\boldsymbol{s}$ are given in (3.186). We insert the values for the coefficients b_{ik} into the first equation (3.186) and extend it with a damping term

$$\ddot{\boldsymbol{w}} + \gamma\dot{\boldsymbol{w}} + \omega_T^2\boldsymbol{w} = \omega_T\sqrt{[\varepsilon(0) - \varepsilon(\infty)]\varepsilon_0}\,\boldsymbol{E}\,. \tag{6.143}$$

The polarization $\boldsymbol{P}$ is related to the reduced displacements and to the macroscopic electric field through the second equation (3.186)

$$\boldsymbol{P} = \omega_T\sqrt{[\varepsilon(0) - \varepsilon(\infty)]\varepsilon_0}\,\boldsymbol{w} + [\varepsilon(\infty) - 1]\varepsilon_0\boldsymbol{E}\,. \tag{6.144}$$

Furthermore polarization and electric field are related to each other by

$$\varepsilon\boldsymbol{E} = \varepsilon_0\boldsymbol{E} + \boldsymbol{P}\,. \tag{6.145}$$

If we put $\exp(i\omega t)$ for the time dependence of the vectors $\boldsymbol{w}$, $\boldsymbol{E}$, and $\boldsymbol{P}$, from (6.143) to (6.145) we find the dielectric constant to be given by

$$\varepsilon = \varepsilon(\infty) + [\varepsilon(0) - \varepsilon(\infty)]\frac{\omega_T^2}{\omega_T^2 - \omega^2 - i\gamma\omega} \tag{6.146}$$

or, separated into real and imaginary parts

$$\varepsilon_1 = \varepsilon(\infty) + [\varepsilon(0) - \varepsilon(\infty)]\frac{\omega_T^2(\omega_T^2 - \omega^2)}{(\omega_T^2 - \omega^2)^2 + \gamma^2\omega^2}\,, \tag{6.147}$$

$$\varepsilon_2 = [\varepsilon(0) - \varepsilon(\infty)]\frac{\omega_T^2\omega\gamma}{(\omega_T^2 - \omega^2)^2 + \gamma^2\omega^2}\,. \tag{6.148}$$

One can deduce the dispersion and absorption directly from these two equations and determine the reflectivity using (6.39) (Fig. 6.21). For zero damping one finds total reflection between $\omega = \omega_T$ and $\omega = \omega_L$, and just above ω_L a frequency where the reflection coefficient is equal to zero. Damping smooths out the idealized reflection spectrum.

The one-phonon absorption can accordingly be represented as a classical interaction between an electromagnetic wave and damped "harmonic oscillators" of frequency ω_T. As an example, Fig. 6.22 shows the reflectivity of GaAs. The spectrum is completely defined by $\varepsilon(0)$, $\varepsilon(\infty)$, ω_T, and γ, so that one can determine these parameters by matching. The Lyddane-Sachs-Teller relation (3.194) then yields ω_L.

We have introduced the macroscopic quantities $\varepsilon(0)$ and $\varepsilon(\infty)$ as parameters here. The quantities fundamental to the model—effective charge on the ions

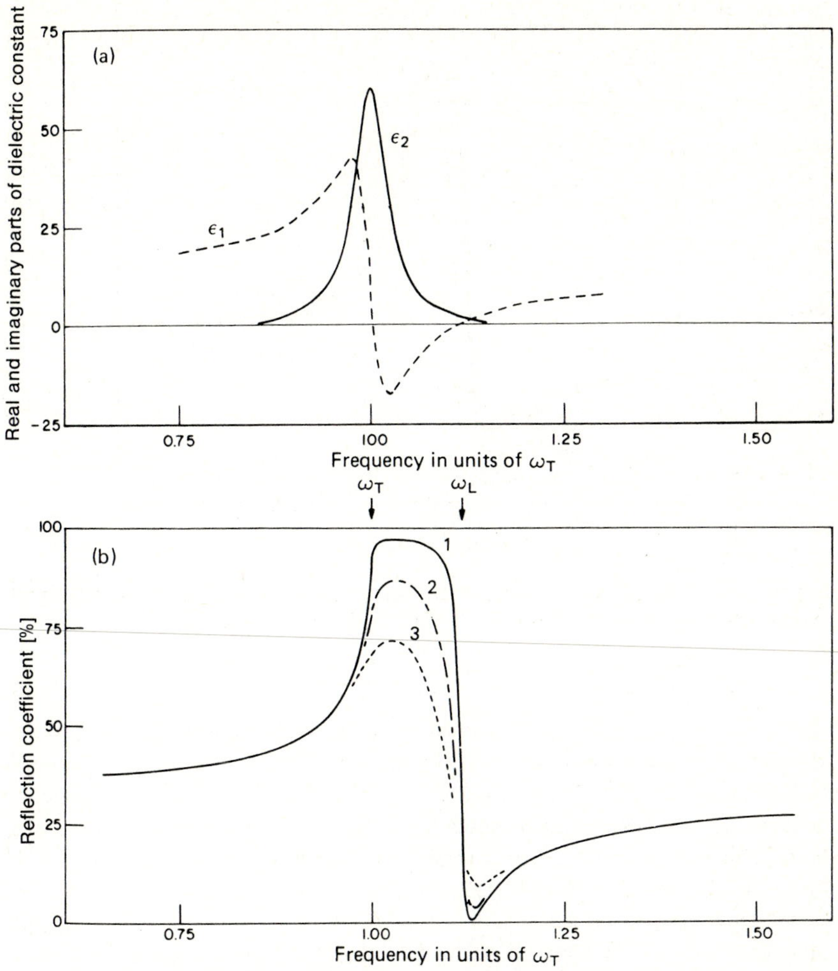

Fig. 6.21. (a) Real and imaginary parts of the complex dielectric constant from (6.147) and (6.148) for $\varepsilon(0) = 15$, $\varepsilon(\infty) = 12$, $\gamma/\omega_T = 0.05$. (b) Reflectivity for $\gamma/\omega_T = 0.004$ (curve *1*), $= 0.02$ (curve *2*), $= 0.05$ (curve *3*) (from *Hass* [104.3]).

and polarizability of the ions, for example—do not explicitly appear. One often defines an *effective ionic charge* e^* by the equation

$$\boldsymbol{P} = \frac{N}{V_g} e^* \boldsymbol{s} . \tag{6.149}$$

This definition is not unique. For transverse vibrations $\boldsymbol{E} = 0$ and from (3.186)

$$\boldsymbol{P} = b_{21}\boldsymbol{w} = \frac{N}{V_g}(\omega_T\{[\varepsilon(0) - \varepsilon(\infty)]\varepsilon_0 V_g \overline{M}/N\}^{1/2})\boldsymbol{s} . \tag{6.150}$$

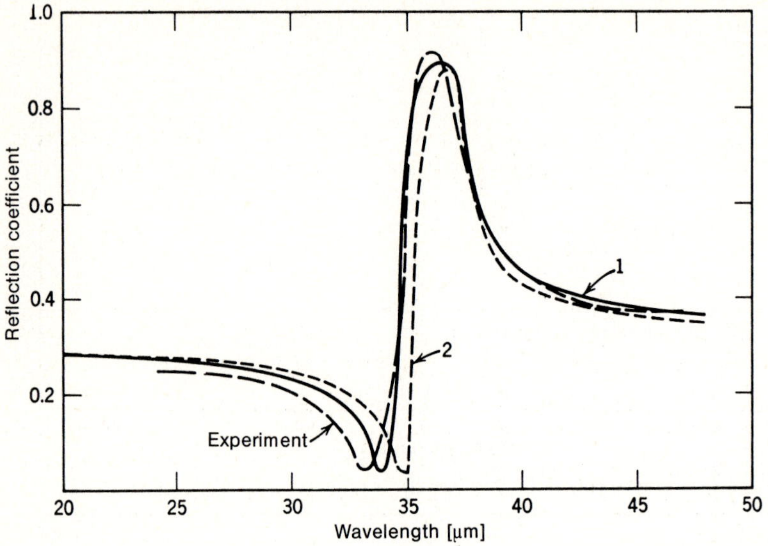

Fig. 6.22. Reflectivity of GaAs and two theoretical curves obtained by matching of the parameters $\varepsilon(0)$, $\varepsilon(\infty)$, ω_T and γ [from H. Ehrenreich: Phys. Rev. **120**, 1951 (1960)].

The quantity in the brackets is designated the effective ion charge e_B^* (index B stands for Born). If one starts from the longitudinal vibrations, relates $\boldsymbol{E}$ with $\boldsymbol{s}$ according to (3.190) and puts $\boldsymbol{P} = -\varepsilon_0\boldsymbol{E}$, (6.149) leads to the effective ion charge $e_C^* = e_B^*/\varepsilon(\infty)$ introduced by Callen. Following Szigeti, it is also customary to use an effective ion charge e_S^*, which appears in (6.149) if (3.186) is used and the effective field [(3.182)] is set to zero for transverse vibrations: $e_S^* = 3e_B^*/[\varepsilon(\infty) + 2]$.

We shall only give an indication here of the quantum mechanical formulation of one-phonon absorption. Classically the interaction energy is the negative of the product of the polarization vector and the electric field: $U = -\boldsymbol{P}\cdot\boldsymbol{E}$. We have related the polarization vector to lattice ion displacements through an effective charge, so we can write in general

$$\boldsymbol{P} = \frac{1}{V_g}\sum_{\alpha n} e_\alpha^* s_{n\alpha} = \sqrt{\frac{\hbar}{2\omega_{TO}N}}\,\frac{1}{V_g}\sum_{\alpha n q}\frac{e_\alpha^*}{\sqrt{M_\alpha}}\,\boldsymbol{e}_\alpha(\boldsymbol{q})(a_{-\boldsymbol{q}}^+ + a_{\boldsymbol{q}})\exp(\mathrm{i}\boldsymbol{q}\cdot\boldsymbol{R}_n)\,. \tag{6.151}$$

e_α^* is the effective charge on the αth ion. The electric field is given by a plane wave with transverse polarization $\boldsymbol{e}_\kappa$ (parallel to the polarization vector of the TO-phonon) and with space dependence $\exp(\mathrm{i}\boldsymbol{\kappa}\cdot\boldsymbol{r})$. The Hamiltonian for the photon-phonon interaction is then

$$H' = -\boldsymbol{P}\cdot\boldsymbol{E} \propto \left(\sum_{\alpha} \frac{e_{\alpha}^{*}}{\sqrt{M_{\alpha}}} \boldsymbol{e}_{\alpha}\cdot\boldsymbol{e}_{\kappa}\right) \frac{1}{\sqrt{\omega_{\mathrm{TO}}}} \sum_{qn} (a_{-q}^{+} + a_{q}) \exp[\mathrm{i}(\boldsymbol{q}\cdot\boldsymbol{R}_n + \boldsymbol{\kappa}\cdot\boldsymbol{r})]. \tag{6.152}$$

Replacing $\boldsymbol{r}$ by $\boldsymbol{R}_n$ in each term of the summation over n, the summation over $\exp[\mathrm{i}(\boldsymbol{q} + \boldsymbol{\kappa})\cdot\boldsymbol{R}_n]$ yields the condition $\boldsymbol{q} = -\boldsymbol{\kappa}$. We are left with

$$H' \propto (\cdots)\frac{1}{\sqrt{\omega_{\mathrm{TO}}}}(a_{\kappa}^{+} + a_{-\kappa}). \tag{6.153}$$

The two phonon operators describe the two possibilities associated with the absorption of the photon of momentum $\boldsymbol{\kappa}$: creation of a phonon $\boldsymbol{\kappa}$ or annihilation of a phonon $-\boldsymbol{\kappa}$. We must exclude the second term since it contravenes energy conservation. We obtain for the probability of absorption an expression proportional to the square of the matrix element formed by H', to which the law of energy conservation must then be added. We find

$$\begin{aligned}\text{Absorption} &\propto |\langle f|H'|i\rangle|^2\delta(\omega_{\mathrm{TO}} - \omega)\\ &\propto \left(\sum_{\alpha} \frac{e_{\alpha}^{*}}{\sqrt{M_{\alpha}}} \boldsymbol{e}_{\alpha}\cdot\boldsymbol{e}_{\kappa}\right)^2 \frac{1}{\omega_{\mathrm{TO}}}(n_{\mathrm{TO}} + 1)\delta(\omega_{\mathrm{TO}} - \omega).\end{aligned} \tag{6.154}$$

In addition to the δ-function which produces a sharp absorption line at $\omega = \omega_{\mathrm{TO}}$, the factor $n_{\mathrm{TO}} + 1$ (stimulated and spontaneous emission of a TO-phonon) and the factor containing the ion charge and mass are important. The latter shows that for transverse optical vibrations with no associated electrical moment, there is no photon-phonon coupling of the type considered here.

In calculating the total absorption due to one-phonon processes, one has also to note that phonon-photon transitions take place as well as photon-phonon transitions. From the photon-to-phonon transformation rate (proportional to $n_{\mathrm{TO}} + 1$), one has therefore to subtract those processes in which, with the same exchange of energy, phonons transform into photons. The latter process is proportional to n_{TO}. Since $(n_{\mathrm{TO}} + 1) - n_{\mathrm{TO}} = 1$, the total transition rate is therefore independent of n_{TO}.

For further reading we refer the reader to the book by *Born* and *Huang* [67] and the article by *Hass* in [104.3].

6.3.3 Multi-Phonon Absorption

We consider now the absorption of a photon with formation of two or more phonons. In this process, two coupling mechanisms come into the picture (cf. Fig. 6.20).

1) *Dipole moments of higher order*. The linear relation between polarization and displacement (6.151) leads to processes in which one phonon is emitted

or absorbed. The interaction operator is linear in $\boldsymbol{P}$, hence also in $\boldsymbol{s}$, and finally also in $a_{\boldsymbol{q}}^{+}$ and $a_{\boldsymbol{q}}$. This linear relation however only describes the vibration of rigid ions whose effective charge we imagine to be at the centre of gravity. The displacement of neighbouring ions, however, distorts the electron shells of an ion, and thereby induces an additional effective charge Δe^*. This charge produces an extra coupling to the electric field. The corresponding term in the interaction operator must depend on two displacements, that of the inducing ion and that of the ion induced

$$H_{\text{int}} \propto \sum_{\substack{n\alpha \\ n'\alpha'}} \boldsymbol{s}_{n\alpha} \cdot f_{n\alpha}^{n'\alpha'} \boldsymbol{s}_{n'\alpha'}, \tag{6.155}$$

where $f_{n\alpha}^{n'\alpha'}$ can be a tensor. Since there are two displacements in (6.155), the interaction operator contains products of two phonon creation or annihilation operators. In absorbing a photon, therefore, two phonons can be emitted, or one phonon emitted and one absorbed. One can interpret (6.155) as the quadratic term of an expansion of the interaction operator in increasing powers of the displacements. The term of nth order is then connected with an n-phonon process.

We presented the graph of the two-phonon processes in Fig. 6.20. We have not, however, decided whether this three-quantum process contains virtual intermediate states. An example would be the creation of a virtual electron-hole pair by the photon, the subsequent emission of a phonon $\boldsymbol{q}$ by the electron or hole, and finally the recombination of the pair with emission of the second phonon $-\boldsymbol{q}$. In what follows we shall only consider the single process and shall not discuss the case of intermediate electronic states.

2) *Lattice anharmonicity.* Just as photons can decay into several phonons, as in the process just considered, a phonon can also decay into several phonons. Since the primary particle is also a phonon, such processes must be of higher order in the displacements, for example, the decay into two phonons must be a third order effect in the $\boldsymbol{s}_{n\alpha}$. This and terms of higher order are just the anharmonic terms left out in the expansion (3.96) of the lattice potential. They cause an interaction between the phonons. We shall treat them in more detail in the next chapter. They are important here since the mechanism shown in Fig. 6.20 (conversion of a photon into one phonon and subsequent decay of the phonon) also leads to multi-phonon absorption. In contrast to coupling through higher order dipole moments, coupling through lattice anharmonicity can only be effective in polar solids since one subprocess is forbidden in non-polar solids.

Two-phonon absorption is therefore determined by the following factors.

Momentum conservation demands $\boldsymbol{\kappa} = \boldsymbol{q} \pm \boldsymbol{q}'$. Since $\boldsymbol{\kappa}$ is negligibly small, $\boldsymbol{q} \pm \boldsymbol{q}' = 0$ follows (sum or difference absorption).

Energy conservation demands $\omega = \omega_j(\boldsymbol{q}) \pm \omega_{j'}(\boldsymbol{q})$ where the two signs apply to emission of two phonons (sum absorption), and emission and absorption of a phonon (difference absorption). The two phonons can belong to different branches but the length of the wave vector $\boldsymbol{q}$ must be the same.

Selection rules result from the symmetries of the matrix elements. Since $\boldsymbol{q}$ is not restricted by $\boldsymbol{\kappa}$ to the vicinity of $\boldsymbol{q} = 0$, phonons with any $\boldsymbol{q}$ of the Brillouin zone can take part in the process. For each pair of phonons from the two branches j and j', we have a range of transition energies given by the various possible $\boldsymbol{q}$. The multi-phonon spectrum is therefore continuous. The contribution from each phonon combination has an upper energy limit, given by the number of phonons involved times the highest energy $\hbar\omega_j$ in the $\omega_j(\boldsymbol{q})$-spectrum. Each of these contributions also has a different temperature dependence: If one regards the temperature dependence as only determined by the $n_{\boldsymbol{q}}$ (Bose distributions), then, following the remarks at the end of the last section, the one-phonon absorption is temperature independent. In accordance with these considerations the absorption rate for two-phonon processes is proportional to $(1 + n_{q_1})(1 + n_{q_2}) - n_{q_1}n_{q_2} = 1 + n_{q_1} + n_{q_2}$ for sum processes and $(1 + n_{q_1})n_{q_2} - n_{q_1}(1 + n_{q_2}) = n_{q_2} - n_{q_1}$ for difference processes. Corresponding factors can be found for processes of higher order.

All the above considerations lead to an absorption according to

$$\varepsilon_2 \propto \sum_{\boldsymbol{q}jj'} \frac{|M(\boldsymbol{q},j,j')|^2}{\omega_j(\boldsymbol{q})\omega_{j'}(\boldsymbol{q})} \begin{Bmatrix} 1 + n_{j\boldsymbol{q}} + n_{j'\boldsymbol{q}} \\ n_{j'\boldsymbol{q}} - n_{j\boldsymbol{q}} \end{Bmatrix} \delta(\omega - \omega_j(\boldsymbol{q}) \mp \omega_{j'}(\boldsymbol{q})) . \qquad (6.156)$$

Within the limits of our discussion we can say nothing about the matrix element.

In treating direct transitions in Section 6.2.2, we simplified a similar equation by removing from the sum over $\boldsymbol{q}$ all the factors except the δ-functions, assuming them to be approximately constant. The sum over the δ-functions gave a combined density of states whose physical content was easy to discuss. We can proceed similarly here. We can consequently expect a strong structure in the continuous multi-phonon absorption spectrum, which we can associate with phonon processes at points $\boldsymbol{q}$ of high symmetry. While critical points in the electron transitions are defined by the condition $\nabla_{\boldsymbol{k}}E_{j'}(\boldsymbol{k}) = \nabla_{\boldsymbol{k}}E_j(\boldsymbol{k})$ (Sec. 6.2.2), for two-phonon processes here the two conditions $\nabla_{\boldsymbol{q}}\omega_{j'}(\boldsymbol{q}) = \pm\nabla_{\boldsymbol{q}}\omega_j(\boldsymbol{q})$ appear.

As an example we show in Fig. 6.23 multi-phonon spectra of AlSb. The structure of Fig. 6.23a can be satisfactorily explained by two-phonon processes, that of Fig. 6.23b by three- and four-phonon processes, respectively, with the sole involvement of the two optical phonons at Γ, X, and L and the two acoustic phonons at X and L.

As references on the content of this section we mention *Balkanski* [117], *Johnson* [112.34], and *Spitzer* [104.3].

6.3.4 Raman and Brillouin Scattering

One refers to inelastic scattering of light by emission or absorption of a phonon as Raman scattering, if the phonon belongs to the optical branch, and as Bril-

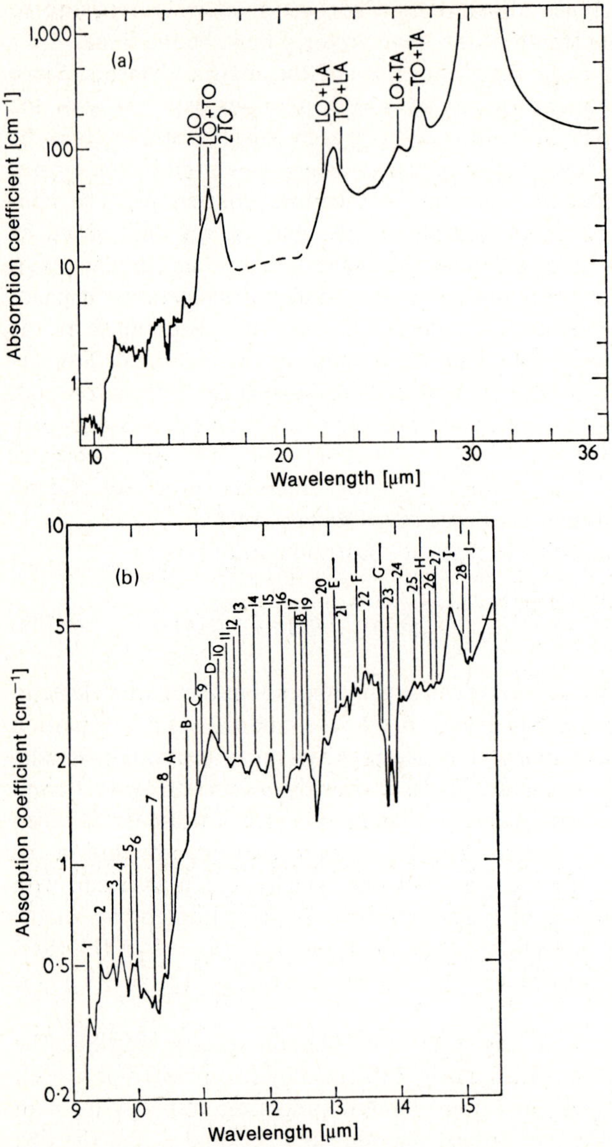

Fig. 6.23a and b. Multi-phonon spectra of AlSb and association of the structure with two-phonon processes (a) and three- and four-phonon processes (b) [from F. A. Johnson: Progress in Semiconductors, Vol. 9 (Heywood, London 1966)].

louin scattering, if the phonon belongs to the acoustic branch. In Fig. 6.20 we presented three interaction processes.

1) A photon decays into two phonons by coupling with dipole moments of higher order; one of these phonons changes into a photon.
2) First the conversion photon-phonon takes place. This decays into two phonons as a result of anharmonic coupling and one of these phonons then changes into a photon.
3) The photon creates a virtual electron-hole pair. The electron (or hole) emits a phonon and the secondary photon arises in the recombination of the pair.

All three processes describe the emission of a phonon (Stokes scattering). Corresponding processes can be stated for the absorption of a phonon (anti-Stokes scattering). It can be shown that 3) is the most likely process. It is the only process which is possible in nonpolar solids, since there the subprocess "photon-phonon conversion" or vice versa is forbidden. In the following we treat Raman scattering through process 3) and restrict ourselves to Stokes scattering.

The scattering probability is clearly proportional to the square of the following expression (perturbation calculation of third order):

$$\sum_{m_1 m_2} \frac{\langle f|\boldsymbol{e}_2\cdot\boldsymbol{p}|m_2\rangle\langle m_2|\boldsymbol{s}\cdot\nabla V|m_1\rangle\langle m_1|\boldsymbol{e}_1\cdot\boldsymbol{p}|i\rangle}{(\omega_{m_2}-\omega)(\omega_{m_1}-\omega)} + \cdots$$

$$\equiv \sum_{m_1 m_2} \frac{p_{fm_2}\boldsymbol{e}\cdot\boldsymbol{m}p_{m_1 i}}{(\omega_{m_2}-\omega)(\omega_{m_1}-\omega)} + \cdots \equiv \sum_i e_i R_{12}^{(i)} . \qquad (6.157)$$

$|i\rangle$, $|m_1\rangle$, $|m_2\rangle$, $|f\rangle$ are wave functions of the initial state (one photon), of the first intermediate state (an electron-hole pair), of the second intermediate state (electron-hole pair + emitted phonon), and of the final state (phonon + secondary photon). Two of the matrix elements describe the electron-photon interaction, one the electron-phonon interaction. $\boldsymbol{e}_1$, $\boldsymbol{e}_2$, and $\boldsymbol{e}$ are the polarization vectors; $\hbar\omega$, $\hbar\omega'$, and $\hbar\omega_{\boldsymbol{q}}$ the energies of the primary photon, the secondary photon, and the phonon. The abbreviations on the second line then follow. The plus signs + ... indicate that five further terms appear in the perturbation calculation, in each of which the order of the three individual processes is different. One refers to the tensor $R_{12}^{(i)}$ as the *Raman tensor*. The indices 1,2 indicate the directions of polarization of the two photons.

The form of (6.157) suggests a classical interpretation of the interaction processes. We already found two of the three matrix elements together with an energy denominator in (6.81), when studying two-photon absorption. The only difference is that now a photon absorption is coupled with a photon emission. The light polarizes the solid (formation of virtual electron-hole pairs) and the lattice vibrations couple to this polarization. So just as the phonon absorption is associated with the dipole moment, the Raman effect is associated with the polarizability tensor. The *first-order Raman effect* considered here stems from the first terms of an expansion of this tensor in powers of the lattice vibrations. The quadratic term in the $s_{n\alpha}$ of such an expansion—which is not contained in (6.157)!—leads to the *second-order Raman effect*. It involves the emission of two phonons, or the emission of one phonon and absorption of a second, or finally the absorption of two phonons. With this, two first-order processes can be connected with each other through a virtual photon, or both phonons can be emitted (absorbed) by one virtual electron-hole pair. In the first case a line spectrum results, with a difference in energy between primary and secondary photon which is the sum or difference of the first-order Raman energies. In the second case both phonons together have to satisfy energy and momentum conservation; each phonon however can have a $\boldsymbol{q}$-vector from anywhere in the Brillouin zone. The associated spectrum is therefore a continuum. A discussion of the transition matrix elements leads to selection rules, i.e., statements about which optical phonons can be involved in Raman scattering. Since optical absorption and Raman scattering are the result of different interactions, the selection rules are different for the two phenomena. Individual lattice vibrations can be "Raman active" but not "infrared active" and vice versa. Group theoretical considerations must be used to decide these questions. In contrast to infrared absorption, the Raman effect can involve LO-phonons.

The fact that energy and wave vector of the emitted photon and phonon are related to the angle of emission is also important. If ψ is the angle between the direction of the primary and secondary photons, it follows from energy conservation $\hbar\omega = \hbar\omega' + \hbar\omega_{\boldsymbol{q}}$ and from wave vector conservation $\boldsymbol{\kappa} = \boldsymbol{\kappa}' + \boldsymbol{q}$, that

$$\frac{q^2c^2}{\varepsilon(\infty)} = 4\omega(\omega - \omega_{\boldsymbol{q}}) \sin^2\frac{\psi}{2} + \omega_{\boldsymbol{q}}^2 . \tag{6.158}$$

In general the phonon energy is small. If we neglect it in (6.158) we simplify the equation to

$$q = 2\kappa \sin\frac{\psi}{2} . \tag{6.159}$$

The theory of Brillouin scattering is not significantly different from that for Raman scattering. The phonons now belong to the acoustic branch. Their energy is considerably less than that of the optical phonons. We refer to the literature for details.

As an example Fig. 6.24 shows Raman scattering in InP and AlSb. One can see the lines of the first-order Raman effect with TO- and LO-phonons, and the continuum of the second-order Raman effect. The strong structure stems—as with multi-phonon absorption—from the critical points in the combined density of states.

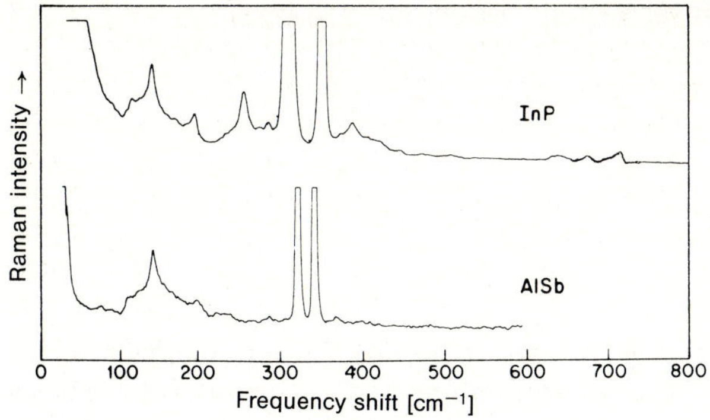

Fig. 6.24. Frequency shift in the Raman spectrum of first and second order for InP and AlSb at room temperature (from *Mooradian* [103.IX]).

In addition to the Raman effect involving phonons, the Raman effect involving plasmons or magnons is also important. We refer to the literature mentioned below.

With decreasing scattering angle ψ, the wave number of the phonons decreases in accordance with (6.158). Polariton effects can then appear. Instead of the phonon frequency ω_{TO} one measures the polariton frequency, which decreases with decreasing q (cf. Fig. 6.3). The Raman effect therefore assists in the measurement of the lower branch of the polariton spectrum.

One can determine the upper branch of an (exciton-) polariton spectrum with two-photon spectroscopy which is similar to the Raman effect (Sec. 6.2.4). In this one allows two rays of photons (in the solid polaritons from the lower branch of the spectrum) to impinge at such an angle that the two-polariton absorption process produces a polariton belonging to the upper branch which can then be detected as a photon in leaving the solid.

Further literature on Raman scattering, and other topics in this section, is to be found in *Cardona* [111b.8] and in the contributions by *Balkanski* and by *Burstein* in [117], by *Cummins* and by *Loudon* in [112.31], by *Loudon* in [111a.13], by *Mooradian* in [103.IX], and *Richter* in [111d.78]. On general questions we also refer the reader to the book by *Born* and *Huang* [67].

7. Phonon-Phonon Interaction: Thermal Properties

7.1 Introduction

The theory of lattice vibrations treated in Section 3.3 was restricted to the harmonic approximation. The potential energy of a lattice ion was expanded in powers of the displacements of the ions from their equilibrium positions, and the expansion was terminated after the first nonvanishing (quadratic) term.

The most significant result from this approximation was the possibility of decoupling the lattice vibrations by a transformation to normal coordinates. This led when the normal vibrations were quantized to the concept of *phonons* as noninteracting excitations of the lattice.

This decoupling is not possible when we take higher terms in the expansion of the potential. If one nevertheless wants to retain the concept of elementary excitations, then these higher terms mean a mutual *interaction* between the phonons. As a result of this interaction a phonon from a given state $\boldsymbol{q}, j$ will disappear after a finite time through a multi-phonon process, e.g., decay into other phonons. The phonons then have a finite *lifetime*. At the same time the higher terms in the expansion contribute to the phonon energy: they cause a renormalization of the phonon frequency $\omega_j(\boldsymbol{q})$. It is with this energy shift, and with the lifetime, that we want to concern ourselves in the following section.

The anharmonic terms play an important role in the *thermodynamics* of crystals. Phenomena such as thermal expansion, the difference between adiabatic and isothermal processes, between specific heat at constant volume and at constant pressure, are not contained in the harmonic approximation. We present an introduction to this set of questions in Section 7.3.

A second important area in which phonon-phonon interaction plays a part is *lattice heat conduction*. We treat this in Section 7.4. In the transport theory we only took account of energy transport through the electron system. The energy flux in the phonon system, however, is also important, particularly in insulators. For this we shall be making use of the Boltzmann equation for phonons which we established in Section 4.2.2.

There are a number of good comprehensive reviews on the phonon-phonon interaction. We refer particularly to the presentations by *Leibfried* and *Ludwig* in [101.12, 111d.43] and by *Cowley* and *Cochran* [111c.31/1; 106, XXV/2a] and additionally to the articles by *Krumhansl* in [117], *Klemens* in [101.7] and *Mendelssohn* and *Rosenberg* in [101.12]. We would further point to the corresponding chapters in the books by *Peierls* [31] and *Ziman* [35].

7.2 Frequency Shift and Lifetime of Phonons

In the harmonic approximation the phonons are noninteracting elementary excitations characterized by the eigenfrequencies $\omega_j(\boldsymbol{q})$. By including anharmonic terms in the Hamiltonian, their frequencies are shifted. At the same time a lifetime can be defined as the mean time a phonon remains in a state $\boldsymbol{q}, j$ before it disappears due to phonon-phonon interaction.

The first term left out of the Hamiltonian in the harmonic approximation has the form

$$H_3 = \frac{1}{3!} \sum_{ii'i''} \sum_{\substack{nn'n'' \\ \alpha\alpha'\alpha''}} \Phi \begin{pmatrix} n & \alpha & i \\ n' & \alpha' & i' \\ n'' & \alpha'' & i'' \end{pmatrix} s_{n\alpha i} s_{n'\alpha' i'} s_{n''\alpha'' i''} . \tag{7.1}$$

The $s_{n\alpha i}$ are again the ith component of the displacement of the αth basis atom of the nth Wigner-Seitz cell.

Here too a number of symmetry relations apply to the Φ. We shall mention only that, as a result of the translation invariance of the lattice, the Φ remain unaltered when one adds to each of the $\boldsymbol{R}_n$, $\boldsymbol{R}_{n'}$, $\boldsymbol{R}_{n''}$ a primitive translation $\boldsymbol{R}_m$ (displacement of the origin to an equivalent point in another Wigner-Seitz cell). From this invariance it follows that (7.1) contains a factor $\delta_{\boldsymbol{q}+\boldsymbol{q}'+\boldsymbol{q}'',\boldsymbol{K}_m}$ which is equal to unity if the sum of the $\boldsymbol{q}$ is equal to zero or to a primitive translation in the reciprocal lattice, and otherwise disappears. If one notes that from (3.121) the $s_{n\alpha i}$ contain a factor $\exp(\mathrm{i}\boldsymbol{q}\cdot\boldsymbol{R}_n)$, one can extract from (7.1) a factor

$$\begin{aligned} &\sum_{nn'n''} \Phi \begin{pmatrix} n & \alpha & i \\ n' & \alpha' & i' \\ n'' & \alpha'' & i'' \end{pmatrix} \exp[\mathrm{i}(\boldsymbol{q}\cdot\boldsymbol{R}_n + \boldsymbol{q}'\cdot\boldsymbol{R}_{n'} + \boldsymbol{q}''\cdot\boldsymbol{R}_{n''})] \\ &\quad = \sum_n \exp[\mathrm{i}(\boldsymbol{q} + \boldsymbol{q}' + \boldsymbol{q}'')\cdot\boldsymbol{R}_n] \sum_{n'n''} \Phi \begin{pmatrix} n & \alpha & i \\ n' & \alpha' & i' \\ n'' & \alpha'' & i'' \end{pmatrix} \\ &\qquad \times \exp[\mathrm{i}(\boldsymbol{q}'\cdot(\boldsymbol{R}_{n'} - \boldsymbol{R}_n) + \boldsymbol{q}''\cdot(\boldsymbol{R}_{n''} - \boldsymbol{R}_n))] . \end{aligned} \tag{7.2}$$

The second sum in this expression is however independent of n, as one can readily see when before the summation over $\boldsymbol{R}_{n'}$ and $\boldsymbol{R}_{n''}$ one goes over to a summation over $\boldsymbol{R}_{n'} - \boldsymbol{R}_n$ and $\boldsymbol{R}_{n''} - \boldsymbol{R}_n$ and takes account of the translation invariance of the Φ. Using (4.105), the first sum gives the factor mentioned.

If we now transfer to a description in terms of creation and annihilation operators for phonons [in the following we write $\omega_{\boldsymbol{q}j}$ instead of $\omega_j(\boldsymbol{q})$]

$$s_{n\alpha i} = \frac{1}{\sqrt{NM_\alpha}} \sum_{j\boldsymbol{q}} \left(\frac{\hbar}{2\omega_{\boldsymbol{q}j}}\right)^{1/2} e_{\alpha i}^{(j)}(\boldsymbol{q}) \exp(\mathrm{i}\boldsymbol{q}\cdot\boldsymbol{R}_n)(a^+_{-\boldsymbol{q}j} + a_{\boldsymbol{q}j}) \tag{7.3}$$

then it follows for H_3 that

$$H_3 = \frac{1}{3!} \sum_{qq'q''} \sum_{jj'j''} \frac{\hbar^{3/2}}{2^{3/2}N^{1/2}} \frac{\Phi(qj, q'j', q''j'')}{\sqrt{\omega_{qj}\omega_{q'j'}\omega_{q''j''}}} \delta_{q+q'+q'',K_m}$$
$$\times (a^+_{-qj} + a_{qj})(a^+_{-q'j'} + a_{q'j'})(a^+_{-q''j''} + a_{q''j''}) . \qquad (7.4)$$

Correspondingly we find for the fourth order term

$$H_4 = \frac{1}{4!} \sum_{qq'q''q'''} \sum_{jj'j''j'''} \frac{\hbar^2}{4N} \frac{\Phi(qj, q'j', q''j'', q'''j''')}{\sqrt{\omega_{qj}\omega_{q'j'}\omega_{q''j''}\omega_{q'''j'''}}} \delta_{q+q'+q''+q''',K_m}$$
$$\times (a^+_{-qj} + a_{qj})(a^+_{-q'j'} + a_{q'j'})(a^+_{-q''j''} + a_{q''j''})(^+_{-}a_{q'''j'''} + a_{q'''j'''}) . \qquad (7.5)$$

We shall not require terms of fifth and higher orders. The coefficients $\Phi(qj, q'j', q''j'')$ and $\Phi(qj, q'j', q''j'', q'''j''')$ follow directly from (7.1) to (7.3) and from the corresponding equations for H_4.

The Hamiltonian (7.4) describes three-phonon interactions, the operator (7.5) four-phonon interactions. In three-phonon processes the number of phonons changes. There are four basic processes: absorption of a phonon with formation of two other phonons, absorption of two phonons with formation of another, simultaneous disappearance and simultaneous creation of three phonons. The two last possibilities mentioned clearly violate energy conservation. We must however include them in our considerations as possible virtual subprocesses of multiple-step interactions. In Fig. 7.1, all the possible interactions are shown by which a given phonon qj can appear or disappear as a result of a three-phonon process.

We define the lifetime of a phonon as the reciprocal value of the probability

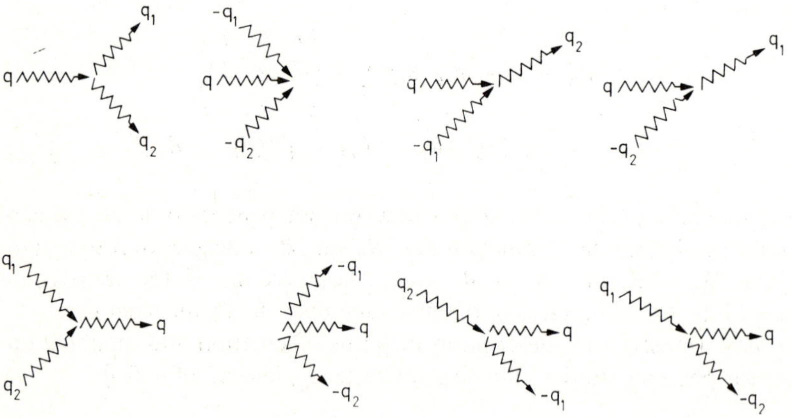

Fig. 7.1. Graphs for three-phonon interaction processes. In the process in the first row a phonon q is destroyed, in the second row a phonon q is created.

that a phonon $\boldsymbol{q}j$ will vanish as a result of one of the processes mentioned. This probability is the sum of the probabilities for the first four processes in Fig. 7.1, minus the probabilities for the four further processes by which phonons $\boldsymbol{q}j$ are created.

If we are not interested in numerical factors, we can readily give the expression for the lifetime. Each partial probability contains the square of the transition matrix element times a delta function which guarantees energy conservation. The second conservation law is given by the factor $\delta_{\boldsymbol{q}+\boldsymbol{q}'+\boldsymbol{q}'',\boldsymbol{K}_m}$. It states that the sum of the wave vectors of the phonons involved, apart from a primitive translation in the reciprocal lattice, is conserved. The often-used expression "momentum conservation" has to be used with care.

The squares of the matrix elements for the eight subprocesses differ only in the occupation numbers of the phonons n_1 and n_2. For the phonon considered we put $n_{\boldsymbol{q}} = 1$ in the first four subprocesses in Fig. 7.1, for the four others $n_{\boldsymbol{q}} = 0$. Four of the subprocesses can be combined to pairs with a factor $(n_1 + 1) \times (n_2 + 1) - n_1 n_2 = n_1 + n_2 + 1$, the other four to pairs with the factor $n_1(n_2 + 1) - n_2(n_1 + 1) = n_1 - n_2$. Altogether (including the numerical factors not calculated here) it follows that the *reciprocal lifetime* $\Gamma(\boldsymbol{q}j)$ is given by

$$\begin{aligned}\Gamma(\boldsymbol{q}j) = \frac{\pi\hbar}{16N\omega_{\boldsymbol{q}j}} \sum_{\boldsymbol{q}_1 j_1 \boldsymbol{q}_2 j_2} & \frac{|\Phi(-\boldsymbol{q}j, \boldsymbol{q}_1 j_1, \boldsymbol{q}_2 j_2)|^2}{\omega_{\boldsymbol{q}_1 j_1}\omega_{\boldsymbol{q}_2 j_2}} \delta_{\boldsymbol{q}_1+\boldsymbol{q}_2-\boldsymbol{q},\boldsymbol{K}_m} \\ & \times \{(n_1 + n_2 + 1)[\delta(\omega - \omega_1 - \omega_2) - \delta(\omega + \omega_1 + \omega_2)] \\ & + (n_1 - n_2)[\delta(\omega + \omega_1 - \omega_2) - \delta(\omega - \omega_1 + \omega_2)]\} \,. \end{aligned} \tag{7.6}$$

The ω_i are the frequencies associated with the $\boldsymbol{q}_i j_i$.

A second route to the definition of the lifetime proceeds by way of a calculation of the contribution of the anharmonic terms H_3 and H_4 to the "one-particle energy" $\omega_{\boldsymbol{q}j}$. By including these terms, the frequency $\omega_{\boldsymbol{q}j}$ is shifted by an amount $\Delta(\boldsymbol{q}j)$. One calculates this amount best by a method which uses Green's functions. Since we have not discussed this method here we content ourselves with a few remarks.

The vibrational energy of the lattice also includes contributions from all three- and more-phonon processes for which the phonon distribution is the same in the initial and final states. Only two-step processes can contribute here from H_3, since each individual process causes a change in the numbers of phonons. These processes are illustrated in Fig. 7.2. In each case the first process, leading to the virtual intermediate state, is identical with one of the eight three-phonon processes shown in Fig. 7.1.

The contribution from two-step processes to the energy is given according to (4.24) by terms of the form

$$\sum_{m(\neq n)} \frac{|\langle m|H_3|n\rangle|^2}{E_n - E_m} \,. \tag{7.7}$$

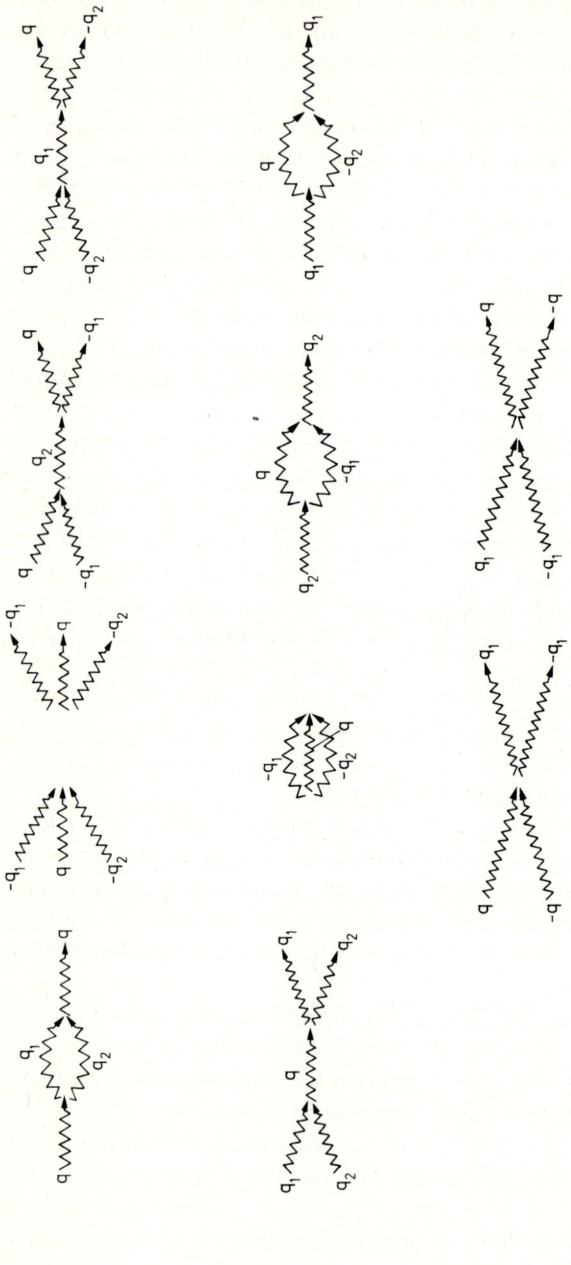

Fig. 7.2. Phonon-phonon processes which contribute to the renormalization of the phonon energy.

The sum is over all eight intermediate states represented in Fig. 7.2. One thus obtains a relation similar in its structure to (7.6). Instead of the δ-functions there are energy denominators $E_n - E_m$. Moreover the sum is over all eight processes, while in (7.6), four processes were added and the other four subtracted. This latter fact however leads to identical signs in the sum terms of both expressions, since the energy denominator of the last four processes in Fig. 7.2 has the opposite sign to that of the first four.

A further comment on the energy denominator is important. If one makes the transition from the normalization volume V_g to a medium of infinite extent, the energies E_n form a continuum and the sum in (7.7) becomes an integral. We can avoid the pole at $E_n = E_m$ in the integration if we add an imaginary additional term $i\delta$ to the denominator $E_n - E_m$, and subsequently let δ go to zero. We can then make use of (3.59) and we obtain

$$\lim_{\delta\to 0} \frac{1}{E_n - E_m + i\delta} = P\left(\frac{1}{E_n - E_m}\right) - i\pi\delta(E_n - E_m)\,, \tag{7.8}$$

where P indicates the principal part. Thus (7.7) provides a contribution $\Delta(\boldsymbol{q}j) + i\Gamma(\boldsymbol{q}j)$ to the energy, where $\Gamma(\boldsymbol{q}j)$ is the reciprocal lifetime (7.6). We have therefore found a second, more general definition for the lifetime of an elementary excitation.

The real part of (7.8) means an energy (or frequency) shift compared with the result of the harmonic approximation. We have still to add contributions from H_4 to this term, since the number of particles can remain unchanged in a four-phonon process. Thus here a first-order calculation provides contributions. The two relevant processes are also illustrated in Fig. 7.2. As first-order processes they have no energy denominator and therefore no imaginary parts either. They make no contributions to the lifetime. Altogether the real frequency shift following from H_3 and H_4 is

$$\begin{aligned}\Delta(\boldsymbol{q}j) = {} & \frac{\hbar}{8N\omega_{\boldsymbol{q}j}} \sum_{\boldsymbol{q}_1 j_1} \frac{|\Phi(-\boldsymbol{q}j, \boldsymbol{q}j, \boldsymbol{q}_1 j_1, -\boldsymbol{q}_1 j_1)|^2}{\omega_{\boldsymbol{q}_1 j_1}} (2n_1 + 1) \\ & + \frac{\hbar}{16N\omega_{\boldsymbol{q}j}} \sum_{\boldsymbol{q}_1 j_1 \boldsymbol{q}_2 j_2} \frac{|\Phi(-\boldsymbol{q}j, \boldsymbol{q}_1 j_1, \boldsymbol{q}_2 j_2)|^2}{\omega_{\boldsymbol{q}_1 j_1}\omega_{\boldsymbol{q}_2 j_2}} \delta_{\boldsymbol{q}_1 + \boldsymbol{q}_2 - \boldsymbol{q}, \boldsymbol{K}_m} \\ & \times \left\{ (n_1 + n_2 + 1)\left[P\left(\frac{1}{\omega - \omega_1 - \omega_2}\right) - P\left(\frac{1}{\omega + \omega_1 + \omega_2}\right)\right] \right. \\ & \left. + (n_1 - n_2)\left[P\left(\frac{1}{\omega + \omega_1 - \omega_2}\right) - P\left(\frac{1}{\omega - \omega_1 + \omega_2}\right)\right]\right\}. \end{aligned} \tag{7.9}$$

This frequency shift must be taken into account, for example, in the determination of phonon frequencies by resonance experiments (neutron scattering, etc.). The finite lifetime shows itself in a broadening of the lines.

7.3 The Anharmonic Contributions to the Free Energy, Thermal Expansion

Thermal expansion, i.e., a temperature dependence of the lattice constant, is not contained in the harmonic approximation of Section 3.3. The equilibrium positions of the lattice ions are determined by minimizing the potential energy and are regarded as temperature independent. The lattice constant does not explicitly appear in the parameters associated with this approximation. Consider the phonon frequencies $\omega_{\boldsymbol{q}j}$ for example. For the linear chain they are given by (3.112). The lattice constant appears there in the combination qa. q itself is equal to $2\pi/aN$ times an integer. The lattice constant thus drops out again.

Along with the absence of a thermal expansion, the harmonic approximation leads to identical adiabatic and isothermal elastic constants (independent of pressure and temperature), and to a temperature independence of the specific heat above the Debye temperature. All these results no longer apply strictly when we take account of lattice anharmonicity.

Yet the lattice constant is implicitly contained in the harmonic approximation since the mean potential Φ_0 depends on the equilibrium positions of the ions and since the force constants (Sec. 3.3.5) are also defined for these equilibrium positions. When we consider lattice anharmonicity by including higher terms from the expansion of the potential energy, it is appropriate, even for the contributions from the harmonic approximation, to interpret the equilibrium positions of the ions as free parameters (*quasi-harmonic approximation*). One then determines the actual lattice constants by minimizing the free energy. The parameters thus obtained are temperature dependent and deviate from the values found from the minimum potential energy.

To describe thermal expansion, and all other thermal and caloric data affected by lattice anharmonicity, we must start from the *free energy*. We use, from Section 2.1.4,

$$F = -k_B T \ln Z, \qquad Z = \sum_n \langle n| \exp(-H/k_B T)|n\rangle . \tag{7.10}$$

The Hamiltonian is the sum of the Hamiltonians for the quasi-harmonic approximation H_0, and the anharmonic contributions H_3 and H_4. We neglect higher terms in H. We regard the sum $H_3 + H_4$ as a small perturbation, and expand the free energy in powers of this perturbation. In this it is convenient to write H in the form $H_0 + \delta(H_3 + H_4)$, to expand in powers of δ, and finally to put $\delta = 1$. Thus we find

$$
\begin{aligned}
F &= -k_B T \ln Z(\delta) \\
&= -k_B T \ln Z(0) - k_B T \frac{\partial}{\partial \delta} \ln Z \bigg|_{\delta=0} \delta - \frac{k_B T}{2} \frac{\partial^2}{\partial \delta^2} \ln Z \bigg|_{\delta=0} \delta^2 - \cdots \\
&= -k_B T \ln Z(0) - k_B T \frac{Z'(0)}{Z(0)} \delta - \frac{k_B T}{2} \left\{ \frac{Z''(0)}{Z(0)} - \left[\frac{Z'(0)}{Z(0)} \right]^2 \right\} \delta^2 - \cdots .
\end{aligned}
\tag{7.11}
$$

We obtain $Z(0)$ and its derivatives by expanding the matrix element $M_{nm} = \langle n | \exp(-H/k_B T) | m \rangle$ in powers of δ

$$
\begin{aligned}
M_{nm} &= \exp(-E_m/k_B T)\delta_{nm} + \delta N_{nm} \left[\frac{\exp(-E_n/k_B T)}{E_n - E_m} + \frac{\exp(-E_m/k_B T)}{E_m - E_n} \right] \\
&\quad + \delta^2 \sum_p N_{np} N_{pm} \left[\frac{\exp(-E_n/k_B T)}{(E_n - E_m)(E_n - E_p)} + \text{cyclic in } n, m, p \right].
\end{aligned}
\tag{7.12}
$$

E_n are the eigenvalues of H_0 $[\langle n | \exp(-H_0/k_B T) | n \rangle = \exp(-E_n/k_B T)]$, and the N_{nm} are matrix elements $\langle n | H_3 + H_4 | m \rangle$. Thus we have

$$
\begin{aligned}
Z &= Z(0) + \delta Z'(0) + \frac{\delta^2}{2} Z''(0) = \sum_n M_{nn} \\
&= \sum_n \exp(-E_n/k_B T) + \delta \sum_n \left[-\frac{N_{nn}}{k_B T} \exp(-E_n/k_B T) \right] + \frac{\delta^2}{2} \sum_{np} 2 N_{np} N_{pn} \\
&\quad \times \frac{\exp(-E_p/k_B T) - \exp(-E_n/k_B T) - \frac{E_n - E_p}{k_B T} \exp(-E_n/k_B T)}{(E_n - E_p)^2} + \cdots .
\end{aligned}
\tag{7.13}
$$

From this $Z(0)$, $Z'(0)$, $Z''(0)$ can be determined immediately. $Z''(0)$ can be simplified by exchanging the summation indices in the summation terms. Furthermore we can again use the fact that in N_{nn} only contributions from H_4, and in the product $N_{np} N_{pn}$ only the contributions from H_3, need be included in first order of approximation. Then in $Z'(0)$ only an H_4-term appears, in $Z''(0)$ only an H_3-term, and $[Z'(0)]^2$ can be neglected compared with $Z''(0)$. Taking all this together, one obtains for the free energy the expression

$$
\begin{aligned}
F &= -k_B T \ln \sum_n \exp(-E_n/k_B T) \\
&\quad + \frac{\sum_n \left(\langle n | H_4 | n \rangle + \sum_{m(\neq n)} \frac{|\langle m | H_3 | n \rangle|^2}{E_n - E_m} \right) \exp(-E_n/k_B T)}{\sum_n \exp(-E_n/k_B T)} .
\end{aligned}
\tag{7.14}
$$

The first term on the right is the free energy from the quasi-harmonic approximation. Inserting (3.133) for E_n yields

$$F_0 = k_B T \sum_{qj} \ln \left[2 \sinh \frac{\hbar\omega_j(\boldsymbol{q})}{2k_B T} \right]. \tag{7.15}$$

We can rewrite the anharmonic contribution to give

$$F = F_0 + \overline{\langle n|H_4|n\rangle + \sum_{m(\neq n)} \frac{|\langle m|H_3|n\rangle|^2}{E_n - E_m}}. \tag{7.16}$$

The averaging bar means the thermal average used earlier [(3.134)]: $\bar{A} = \sum_n A \exp(-E_n/k_B T)/\sum_n \exp(-E_n/k_B T)$.

From the free energy we can calculate all thermal and caloric quantities—the internal energy, entropy; the anharmonic contributions to specific heat; the thermal and caloric equation of state; etc. By equating to zero the derivatives of the free energy with respect to lattice constant, the constant can be found as a function of temperature. One notes that to do this (7.16) must be extended by the now lattice-constant-dependent term Φ_0.

The first step in determining all these quantities is the evaluation of the matrix element in (7.16) and the subsequent thermal averaging. The calculation becomes rather complicated, without however raising any fundamental problems. In the literature usually only the linear chain is worked out explicitly. On this topic we refer the reader above all to the treatments by *Leibfried* and *Ludwig* mentioned in Section 7.1.

7.4 The Thermal Conductivity of the Lattice

In this section we discuss the role of the phonon-phonon interaction in heat conduction by insulators. In a gas of noninteracting phonons, a local increase in temperature will spread out with the velocity of elastic waves. The locally delivered thermal energy will be distributed throughout the crystal by phonons.

The fact that the thermal current density is proportional to the temperature gradient means a thermal resistance which stems from phonon interaction processes. If we exclude an interaction with an electron system (insulator), several scattering mechanisms still remain, for example scattering of phonons at imperfections and at the surface of the crystal, and of course phonon-phonon interactions.

It seems at first impossible that a flow of heat can be attenuated by phonon-phonon interaction. Let us consider a heat current which has a fixed quasi-momentum (total wave vector) $\boldsymbol{Q} = \sum_{qj} n_{qj}\boldsymbol{q}$. In a phonon-phonon interaction process, energy and wave vector are conserved. The total quasi-momentum therefore remains constant.

This statement is only true when we restrict ourselves—as so far—to normal processes. The law of conservation of quasi-momentum in interaction processes is only valid of course modulo $\boldsymbol{K}_m$: The sum of the $\boldsymbol{q}$-vectors can be equal to a translation vector of the reciprocal lattice. If $\boldsymbol{K}_m \neq 0$, we speak of Umklapp or *U*-processes (Sec. 4.1.2). By restriction to normal processes ($\boldsymbol{K}_m = 0$) there is no thermal resistance in a perfect infinite crystal.

Figure 7.3 shows the meaning of an Umklapp process for an interaction process in which two phonons $\boldsymbol{q}'$ and $\boldsymbol{q}''$ are absorbed and a phonon $\boldsymbol{q}$ is emitted. The branches to which the phonons belong is for the moment of secondary importance. Depending on the direction of $\boldsymbol{q}''$, the vector $\boldsymbol{q} = \boldsymbol{q}' + \boldsymbol{q}''$ lies in the Brillouin zone (*N*-process) or outside it (*U*-process). The vector $\boldsymbol{q}$ reduced by a lattice vector $-\boldsymbol{K}_m$, points in a *U*-process in the opposite direction to $\boldsymbol{q}'$ and $\boldsymbol{q}''$. The total quasi-momentum $\boldsymbol{Q}$ is diminished in this process by $\boldsymbol{K}_m$.

It is not important here that the choice of the Brillouin zone in $\boldsymbol{q}$-space is not unique. By displacing the Brillouin zone, certain *U*-processes do admittedly become *N*-processes, but likewise certain *N*-processes become *U*-processes. Only the total loss of quasi-momentum is significant, and one can show that however the Brillouin zone is chosen the *U*-processes deliver the same contribution in each case.

In calculating the transport properties of a phonon system in Section 4.2.2, we set up a Boltzmann equation of the form

$$\dot{\boldsymbol{r}}\cdot\nabla_{\boldsymbol{r}} g = \left.\frac{\partial g}{\partial t}\right|_{\text{coll}}. \tag{7.17}$$

Here $g = g_j(\boldsymbol{r}, \boldsymbol{q}, t)$ is a phonon distribution function similar to the distribution function $f_n(\boldsymbol{r}, \boldsymbol{k}, t)$ introduced for the electrons. Here too we must therefore construct a wave packet from states $\boldsymbol{q}j$, where the extent of the wave packet in

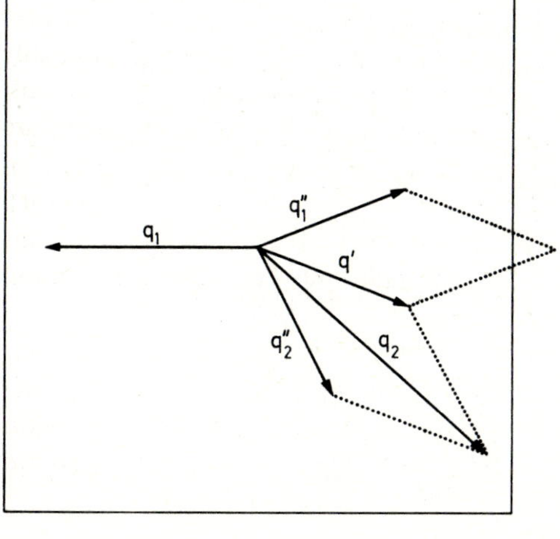

Fig. 7.3. Absorption of two phonons with emission of a third phonon. A normal process and an Umklapp process are illustrated.

real space and in $\boldsymbol{q}$-space is related through the uncertainty principle. All other $\boldsymbol{r}$- and $\boldsymbol{q}$-dependent parameters in the theory are required to be practically unchanged over the extent of the packet. In the following, let $\boldsymbol{r}$ and $\boldsymbol{q}$ indicate the centre of gravity of the wave packet.

We consider first the collision term. The number of phonons in a volume element $d\tau_{\boldsymbol{q}} d\tau_{\boldsymbol{r}}$ changes due to four "scattering processes": a) a phonon $\boldsymbol{q}j$ is absorbed, two phonons $\boldsymbol{q}'j'$ and $\boldsymbol{q}''j''$ are emitted; b) $\boldsymbol{q}j$ is emitted, $\boldsymbol{q}'j'$ and $\boldsymbol{q}''j''$ are absorbed; c) $\boldsymbol{q}j$ and $-\boldsymbol{q}'j'$ are absorbed, $\boldsymbol{q}''j''$ is emitted, d) $\boldsymbol{q}j$ and $-\boldsymbol{q}'j'$ are emitted, $\boldsymbol{q}''j''$ is absorbed. The processes a) and c) reduce, and b) and d) increase the number of phonons in $d\tau_{\boldsymbol{q}} d\tau_{\boldsymbol{r}}$.

The transition probability has the form (4.10) with the operator H_3 from (7.4). In summing over the possible processes, one must pay attention to the fact that the factor 1/3! in H_3 drops out, since 3! terms from H_3 each contribute to the same process. The following expression remains for the total transition probability:

$$\frac{\pi\hbar}{4N} \sum_{\substack{\boldsymbol{q}'\boldsymbol{q}'' \\ j'j''}} \frac{|\Phi(-\boldsymbol{q}j, \boldsymbol{q}'j', \boldsymbol{q}''j'')|^2}{\omega_{\boldsymbol{q}j}\omega_{\boldsymbol{q}'j'}\omega_{\boldsymbol{q}''j''}} \delta_{\boldsymbol{q}'+\boldsymbol{q}''-\boldsymbol{q},\boldsymbol{K}_m} \{[n''(n+1)(n'+1) - nn'(n''+1)]$$
$$\times\, \delta(\omega - \omega' - \omega'') + \frac{1}{2}[n'n''(n+1) - n(n'+1)(n''+1)]$$
$$\times\, \delta(\omega - \omega' - \omega'')\}\,. \tag{7.18}$$

The factor 1/2 in the last term arises because the processes a) and b) are counted twice in the double summation over $\boldsymbol{q}'$ and $\boldsymbol{q}''$.

The n in (7.18) are the phonon occupation numbers of the individual processes. To arrive at the collision term we have to convert from the n to the thermally averaged $\bar{n}$, and from these to the distribution function g. We divide the distribution function into its equilibrium value g_0 (Bose distribution) and the deviation δg. In the transport term $\dot{\boldsymbol{r}} \cdot \nabla_{\boldsymbol{r}} g$ of the Boltzmann equation (7.17), we can neglect δg and obtain, since $\partial g_0/\partial T = g_0(g_0 + 1)\hbar\omega_j(\boldsymbol{q})/k_{\mathrm{B}}T^2$ and $\dot{\boldsymbol{r}} = \nabla_{\boldsymbol{q}}\omega_j(\boldsymbol{q})$, the form given further below in (7.19).

The collision term vanishes in equilibrium. This follows from (7.18) by replacing the n by g_0, and taking account of energy conservation. In non-equilibrium, similar to (4.47), we can put $\delta g = g_0(g_0 + 1)\delta\gamma$ and replace the summation over $\boldsymbol{q}'$ by an integration over $g(\boldsymbol{q}')d\tau_{\boldsymbol{q}'}$. We then obtain for the Boltzmann equation the final form

$$g_0(g_0 + 1)\frac{\hbar\omega}{k_{\mathrm{B}}T^2}\nabla_{\boldsymbol{q}}\omega \cdot \nabla T = \frac{\hbar}{32\pi^2 N}\int d\tau_{\boldsymbol{q}'} \sum_{j'j''} \frac{|\Phi(-j\boldsymbol{q}, j'\boldsymbol{q}', j''\boldsymbol{q}'')|^2}{\omega\omega'\omega''}$$
$$\times \Big[g_0 g_0'(g_0'' + 1)(\delta\gamma'' - \delta\gamma - \delta\gamma')$$
$$+ \frac{1}{2} g_0(g_0' + 1)(g_0'' + 1)(\delta\gamma' + \delta\gamma'' - \delta\gamma) \Big]\,. \tag{7.19}$$

Here the q'' are fixed by the condition $q = q' + q'' + K_m$. Since furthermore q'' must lie in the Brillouin zone, K_m is also known uniquely for given q and q'.

With the same argument as that which followed (4.43), one can show that N-processes are not sufficient to restore the phonon distribution to equilibrium after a disturbance. If one puts $\delta\gamma \propto c \cdot q$ with q-independent vector c, then the right side of (7.19) becomes zero since $q = q' + q''$ for N-processes only. The current-carrying state described by this expression is not attenuated by the phonon-phonon interaction.

It is difficult to solve the Boltzmann equation (7.19). One can exploit the variational method here too. We refer to the literature, e.g., to *Ziman* [35].

We restrict ourselves here to two statements. At low temperatures Umklapp processes are very improbable. The probability of a process in which a phonon q decays into two others q' and q'' is proportional to $g_0 \approx \exp(-\hbar\omega_q/k_B T)$. For a U-process to occur we must have $q + q' + q'' \geq K_m$; each individual q however must lie in the Brillouin zone. Let us take a Debye model for simplicity

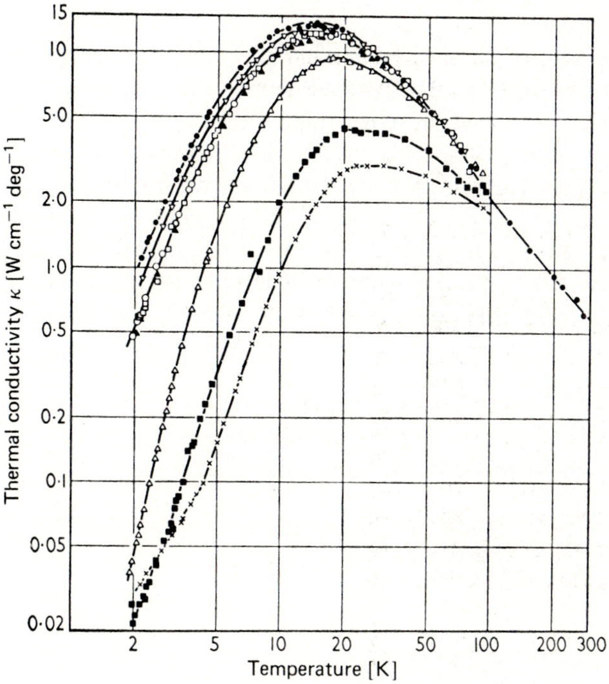

Fig. 7.4. Thermal conduction in germanium. Only in the high-temperature branch common to all curves is the thermal conductivity limited by phonon-phonon interaction. The thermal resistivity at low temperature is significantly influenced by the introduction of defects in different concentrations; the thermal conductivity decreases [from J. A. Carruthers et al.: Proc. Roy. Soc. A **238**, 502 (1957)].

($\omega = sq$, radius of the Brillouin zone = radius of the Debye sphere = $q_D = \omega_D/s$). Furthermore, let all phonons belong to the same branch: $s = s' = s''$. Energy conservation gives $\boldsymbol{q} = \boldsymbol{q}' + \boldsymbol{q}''$. Since $\boldsymbol{q}' + \boldsymbol{q}''$ ought to lead out of the Brillouin zone, we must have $\boldsymbol{q}' + \boldsymbol{q}'' = \boldsymbol{q} \geq \boldsymbol{K}_m/2$. Umklapp processes therefore set in only when $q = \boldsymbol{K}_m/2$ or $\hbar\omega_{\boldsymbol{q}} = k_B\theta_D/2$. At low temperatures one therefore finds that thermal resistance due to U-processes sets in proportional to $\exp(-\theta_D/2T)$. The important temperature-dependent factor in the thermal conductivity thus has the form $\exp(\theta_D/2T)$. Here we must introduce a correction: One can readily show that energy and $\boldsymbol{q}$-conservation can only be satisfied if the spectrum has two branches (longitudinal and transverse), $\boldsymbol{q}$ belonging to the upper, and at least $\boldsymbol{q}'$ or $\boldsymbol{q}''$ to the lower. The condition $s = s' = s''$ is not therefore satisfied. This leads merely to a factor in the exponent which differs little from 1/2.

At *high temperatures* many phonons are excited, and in (7.19) $g_0 + 1$ can be replaced by g_0. The g_0 themselves are given by $k_BT/\hbar\omega_{\boldsymbol{q}}$. Hence the left side of (7.19) becomes proportional to ∇T. The temperature dependence of the right side is $T^3\delta\gamma$ or $T\delta g$. The perturbation of the distribution function and hence the thermal current density also becomes proportional to $(1/T)\nabla T$; the thermal conductivity becomes proportional to $1/T$.

All this leads to a decrease of the thermal conductivity which is exponential for $T \ll \theta_D$, and proportional to T^{-1} for $T \gg \theta_D$, if phonon-phonon interaction is the dominant scattering mechanism. In pure samples there will be additional scattering at the surfaces; this will be effective at low temperatures where the "mean free paths" of the phonons are long. In samples with lattice imperfections or free charge carriers, the interaction with these must be included. This leads to a thermal conductivity which rises at low temperatures, goes through a maximum, and falls again. The phonon-phonon scattering discussed here only dominates above this maximum. All of this is illustrated in Fig. 7.4.

8. Local Description of Solid-State Properties

8.1 Localized and Extended States

In describing theoretically the properties of a solid, one can adopt two points of view. *First* one can regard the solid as composed of individual atoms with given properties. One then considers phenomena which occur in the solid to be *local processes* taking place at individual atoms, but which are influenced by the fact that the atoms are embedded in a lattice. Local excitations in the solid can propagate through the solid by the interaction between these atoms. *Secondly* one can take the arrangement of atoms into a lattice of given structure to be the most important feature of the solid. Then the interpretation of solid-state phenomena is based on the *collective properties* of the solid.

The second viewpoint corresponds to the concept of elementary excitations. The vibrational state of the lattice is seen, not as a sum of individual local oscillations of the atoms in the lattice, but as a superposition of harmonic lattice waves with associated energy quanta (phonons). Spin waves, and the magnons associated with them, are introduced to describe the spin orientation of the lattice ions. Localized excitation of an electron-hole pair is interpreted in terms of the collective excitation "exciton". A feature common to all these excitations is the fact that a fixed value of the wave vector is associated with each state in their energy spectrum, while the location of the state is undefined. The states extend throughout the entire solid. In this sense they are *delocalized.*

We find the same delocalized description in the quasi-particles of the band model. The translation invariance of the crystal lattice leads to plane waves modulated with the lattice periodicity as solutions of the one-electron Schrödinger equation. The probability of finding a Bloch electron at a certain position within a Wigner-Seitz cell is the same as for all equivalent positions in the other cells. This delocalization of the band model states is expressed by the designation *extended states.*

In the band model description, the collective properties of the lattice are included in the properties of the crystal electron. Further interactions in the lattice are described by electron transitions from one delocalized state to another, with change of electron energy and wave vector. In contrast to this, in a localized description the electron transfers from one localized state at one atom into another localized state, the transfer involving an interaction with all the charged particles in its vicinity.

Although the description of solid-state phenomena in terms of delocalized excitations and their interaction often possesses significant advantages, there are, however, cases where it is more appropriate, or even essential, to introduce *localized states*. These are the cases we shall be concerned with in this and the following chapters.

In Section 3.2.2 we saw that the exciton concept could be introduced either via delocalized Bloch functions or via localized Wannier functions. The two limiting cases of the Wannier and the Frenkel exciton described states which emphasized, in the one case, the free movement of the exciton and, in the other, its localization at a lattice ion. The nature of the solid to be investigated determines which description is the more suitable.

Nonlocal and local forms of description are equivalent in the case of a full energy band (e.g., the valence band of an insulator). In the nonlocal description this band does not contribute to conductivity since, in a fully occupied band, the contributions from each two electrons of identical energy but opposite wave vector exactly cancel. In a local description there can be no transitions from one atom to the next since the corresponding states of the neighbouring atoms are also occupied. One would choose the band model description, for example, if one wants to examine the optical excitation of an electron out of a fully occupied band by the absorption of a photon of given energy and wave vector. The local description is often preferred when questions are being asked about chemical bonding. We shall explore this latter area in the sections which follow. There we shall establish where the concepts of the local description can promote a qualitative (and also partly quantitative) understanding of the different types of chemical bond in solids.

In a perfect, infinite lattice a local description becomes important when the band model approximations (neglect of the explicit electron-electron and electron-phonon interactions in defining one-electron states) break down. We shall look at this question in Section 8.3.

In the chapters following we turn to *distorted lattices*. We start in Section 9.1 with *local imperfections*, which include impurity atoms built into an otherwise strictly periodic lattice and local disorder like vacancies and interstitial atoms. Such imperfections have truly localized states, in which electrons can be trapped. The introduction of such states is not a question of suitability of one or other type of description. They are present at the same time as the elementary excitations, and can interact with them. This interaction leads to a finite lifetime for an electron in a localized state, and simultaneously to a finite lifetime of the elementary excitations.

In addition to states localized at point imperfections, one finds localized states at (one-dimensional) *dislocations* and (two-dimensional) *surfaces*. Section 9.2 is devoted to the surface states.

Finally in Chapter 10 we examine disordered lattices. Because of the lack of long-range order, i.e., of strict periodicity, we shall have to revise the concepts developed for the band model. In the energy spectrum of the electrons both regions with localized states and with delocalized states appear.

8.2 The Chemical Bond

8.2.1 Introduction

The chemical bond in a solid is concerned with the mutual interaction of valence electrons of all the atoms in the lattice. The factors determining the nature of the bond are the electron configuration of the free atoms (number of electrons outside closed shells, symmetry of the wave functions of the occupied states) and the atomic environment of an atom in the crystal lattice (type, number, and arrangement of neighbouring atoms). There are two main types of bond. If the number of nearest neighbours is equal to the number of valence electrons an atom possesses, the electrons can be assigned in pairs to individual bonds between nearest neighbours. The bonds can then be described by pairs of *localized electrons*. If the atom does not have enough valence electrons, then a valence electron interacts with the electrons of several neighbouring atoms. The bond is *delocalized*.

We begin in Section 8.2.2 with the consideration of the localized single bond. This case can be treated similarly to the chemical bond of a diatomic molecule. An extension to molecules with many atoms in Section 8.2.3 leads to short-range order in the vicinity of an atom, and hence to solids. In this connection we shall find criteria when a local description of the bond should be selected, and when a nonlocal description is a better approximation. We turn to the localized bond in Section 8.2.4. It embraces the two limiting types of chemical bond, *covalent* and *ionic*, as well as their mixed forms, found in semiconductors and insulators. In Section 8.2.5 we present a recently developed theory for the covalent bond.

Metals possess a delocalized bond. We shall discuss it in Section 8.2.6. In addition to these three principal types of chemical bond—metallic, covalent and ionic—there are other weaker types of bond (van der Waals bond between neutral molecules or rare gas atoms, hydrogen bond between H-atoms in molecular crystals). We shall not investigate these.

Literature on the topics treated in this chapter mostly starts with a consideration of the chemical bond in molecules. We refer particularly to the books by *Coulson* [89] and *Pauling* [90]. Two books by *Phillips* [91] are relevant to the covalent bond theory which we outline in Section 8.2.5.

In the following sections we shall often make use of the properties of a free atom. We remind ourselves therefore of the most significant results from the quantum-mechanical treatment of the free atom.

The Schrödinger equation for an atom with n electrons is

$$H\Psi \equiv \left(-\frac{\hbar^2}{2m}\sum_{i=1}^{n}\nabla_i^2 - \sum_{i=1}^{n}\frac{Ze^2}{4\pi\varepsilon_0 r_i} + \sum_{ij}{}' \frac{e^2}{8\pi\varepsilon_0 r_{ij}}\right)\Psi = E\Psi\,. \tag{8.1}$$

Similar to Section 1.3, one goes over from the n-electron problem to a one-electron approximation, parts of the electron-electron interaction being also carried over. One arrives at a one-electron Schrödinger equation with the form

$$\left[-\frac{\hbar^2}{2m}\nabla^2 + V(\boldsymbol{r})\right]\psi = E\psi\,, \tag{8.2}$$

in which the atomic potential $V(\boldsymbol{r})$ contains the screening of the remaining electrons and can be calculated self-consistently. The solutions with negative energy are the eigenfunctions $\psi^{\mathrm{at}}_{nlm}(\boldsymbol{r})$. They are characterized by the quantum numbers n ($n = 1, 2, 3\ldots$), l ($l = 0, 1, 2\ldots, n-1$), and m ($|m| \leqslant l$). Fig. 8.1 shows a few examples. Eigenfunctions with $l = 0, 1, 2, \ldots$ are also known

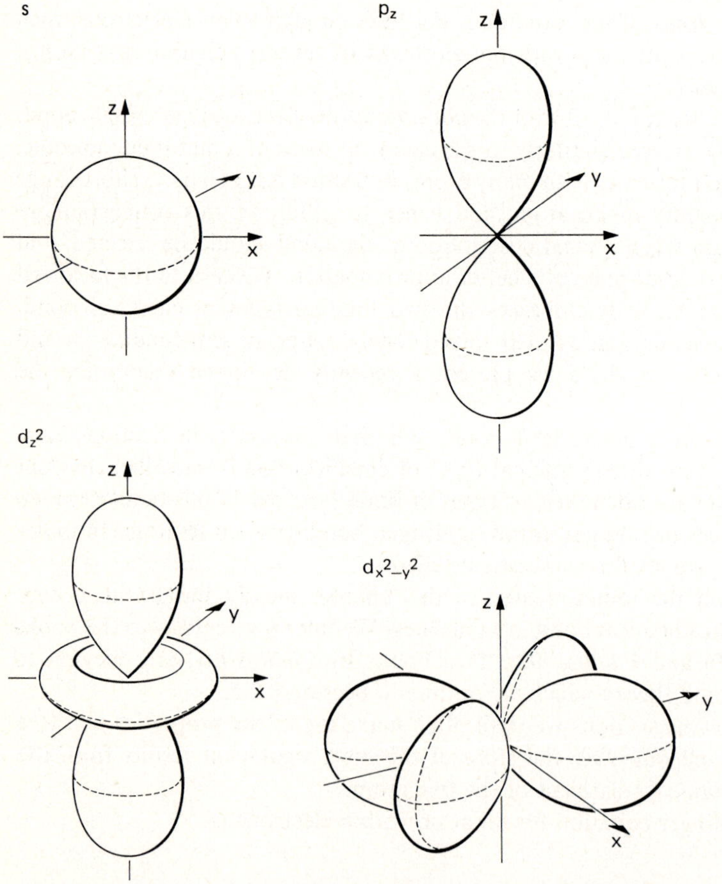

Fig. 8.1. Absolute value of the angular component of the atomic orbitals with $l = 0, 1, 2$. The s-function is isotropic, the three p-functions p_x, p_y, p_z have the x, y, z-axis, respectively, as preferred directions. Of the five d-functions only d_{z^2} and d_{xy} are given. $d_{x^2-y^2}$, d_{yz}, and d_{zx} are only differently oriented.

as s-, p-, d-functions. To allow further discrimination, indices are added to indicate the angular dependence of the respective function, e.g., $d_{xz} \propto xz \propto \cos\theta \sin\theta \cos\psi$.

In the theory of the chemical bond these eigenfunctions are also referred to as *atomic orbitals*.

The energy levels given by the one-electron approximation (8.2) are occupied by the electrons in accordance with the Pauli principle. This determines the *electron configuration* of the atom. The electrons group into shells (quantum number n). Completely filled shells have no net spin and no net angular momentum. The electrons in shells which are not filled—characterized by the symmetries of the atomic orbitals associated with them, and by their spin—are the ones which determine the nature of the chemical bond.

8.2.2 The Localized Single Bond

We examine the following simplified model: Two atomic orbitals ψ_A^{at} and ψ_B^{at} of two neighbouring atoms A and B significantly overlap. In the free atom, each of these orbitals is occupied by one electron. The two electrons interact with the nuclei A and B, as well as with each other. We take all further interactions into account in the atomic potentials V_A and V_B. We thus assume that the two electrons are uniquely associated with the individual bond A-B considered. The problem is equivalent to a diatomic molecule with two valence electrons. The Hamiltonian is

$$H = -\frac{\hbar^2}{2m}\nabla_1^2 - \frac{\hbar^2}{2m}\nabla_2^1 + V_{A1} + V_{A2} + V_{B1} + V_{B2} + V_{12}. \tag{8.3}$$

The meaning of the terms is self-evident. In what follows we are interested almost entirely in the ground state of the system defined by (8.3), and only a little in its excited states. The usual way of solving for the eigenfunctions of the Schrödinger equation $H\psi = E\psi$ proceeds by choosing a trial function for the wave function which contains free parameters, calculating the expectation value of energy, and determining the free parameters in such a way that E is an extremum. It is obvious that a well-chosen trial function greatly eases the solution.

Two methods have proved to be particularly successful: The method of *molecular orbitals* (MO method) reduces the Schrödinger equation to a one-electron equation by neglecting electron-electron interaction. Eq. (8.3) then separates into a sum of two one-electron operators. The ansatz $\psi(\boldsymbol{r}_1, \boldsymbol{r}_2) = \psi_1(\boldsymbol{r}_1)\psi_2(\boldsymbol{r}_2)$ leads to two one-electron Schrödinger equations

$$H_i\psi_i \equiv \left(\frac{\hbar^2}{2m}\nabla_i^2 + V_{Ai} + V_{Bi}\right)\psi_i = E_i\psi_i, \qquad i = 1, 2. \tag{8.4}$$

As trial functions $\psi_i(\boldsymbol{r}_i)$ we choose atomic orbitals, which we combine linearly (linear combination of atomic orbitals, LCAO method)

$$\psi_i = N_i[\psi_A^{at}(i) + \lambda_i \psi_B^{at}(i)]. \tag{8.5}$$

Here N_i is an normalization facor, λ_i a parameter which has to be determined by variation. The function $\psi(\boldsymbol{r}_1, \boldsymbol{r}_2) = \psi(1, 2)$ then becomes

$$\begin{aligned}\psi(1, 2) &= N[\psi_A^{at}(1) + \lambda_1 \psi_B^{at}(1)][\psi_A^{at}(2) + \lambda_2 \psi_B^{at}(2)] \\ &= N[\psi_A^{at}(1)\psi_A^{at}(2) + \lambda_1\lambda_2 \psi_B^{at}(1)\psi_B^{at}(2) \\ &\qquad + \lambda_1 \psi_B^{at}(1)\psi_A^{at}(2) + \lambda_2 \psi_A^{at}(1)\psi_B^{at}(2)].\end{aligned} \tag{8.6}$$

To satisfy the Pauli principle $\psi(1, 2)$ has to be symmetrized such that, when the coordinates are exchanged, it transforms into itself or changes its sign. To make the total wave function antisymmetric we have to add a combination of spin functions which, in the first case, are antisymmetric and, in the second case, symmetric.

In the *valence bond* method (VB method) one considers both electrons at the same time. By distributing both electrons over the possible atomic orbitals, one assigns each electron to a particular nucleus. Each of the distributions possible defines a trial wave function, which is inserted with one free parameter into the total function. In our example this means three possibilities: a) one electron at A, one at B, b) both electrons at A, c) both electrons at B. The spatial component of the wave function for possibility a) is

$$\psi(AB) = \psi_A^{at}(1)\psi_B^{at}(2) \pm \psi_B^{at}(1)\psi_A^{at}(2). \tag{8.7}$$

The "ionic states" b) and c) are represented by

$$\psi(AA) = \psi_A^{at}(1)\psi_A^{at}(2); \quad \psi(BB) = \psi_B^{at}(1)\psi_B^{at}(2), \tag{8.8}$$

so that the trial function for the spatial component of the wave function in the VB method has the form

$$\psi(1, 2) = \bar{N}[\psi(AB) + \bar{\lambda}_1 \psi(AA) + \bar{\lambda}_2 \psi(BB)]. \tag{8.9}$$

We now look more closely at both methods and then compare them.

MO Method

We start with the simplest case: identical atoms A and B. ψ_A^{at} and ψ_B^{at} are then the same orbitals centred around nucleus A or B. By introducing the abbreviations

$$S = \int \psi_A^{at} \psi_B^{at} d\tau \tag{8.10}$$

$$C = \int \psi_A^{at} H_i \psi_A^{at} d\tau_i = \int \psi_B^{at} H_i \psi_B^{at} d\tau_i \tag{8.11}$$

$$A = \int \psi_B^{at} H_i \psi_A^{at} d\tau_i = \int \psi_A^{at} H_i \psi_B^{at} d\tau_i \tag{8.12}$$

the normalized wave function (8.5) for the one-electron problem becomes

$$\psi = \frac{\psi_A + \lambda \psi_B}{\sqrt{1 + \lambda^2 + 2\lambda S}} \tag{8.13}$$

and the expectation value of energy is

$$E = \int \psi H_i \psi d\tau_i = \frac{(1 + \lambda^2)C + 2\lambda A}{1 + \lambda^2 + 2\lambda S} \,. \tag{8.14}$$

We determine the coefficients λ such that E is minimized. From $dE/d\lambda = 0$ it follows that $\lambda = \pm 1$ and

$$\psi_\pm = \frac{1}{\sqrt{2(1 \pm S)}} (\psi_A^{at} \pm \psi_B^{at}), \qquad E_\pm = \frac{C \pm A}{1 \pm S} \,. \tag{8.15}$$

ψ_+ and ψ_- are called *molecular orbitals*. The probability of finding the electron at a particular position follows from (8.15)

$$|\psi_\pm|^2 = \frac{1}{2(1 \pm S)} (|\psi_A^{at}|^2 + |\psi_B^{at}|^2 \pm 2\psi_A^{at}\psi_B^{at}) \,. \tag{8.16}$$

This function (along the line A-B) is given in Fig. 8.2 for the case of an electron in the field of two protons (H_2^+-molecule ion). The positive sign is associated with an increase, the negative with a decrease in the probability of finding the electron between the two nuclei. In all the cases of interest to us, the value of the integral A is negative. The lower eigenvalue then is E_+. The ground state E_+ is consequently a *bonding state* with an "electron bond" between both nuclei; the excited state E_- is an *antibonding state* with reduced probability of the electron being between the nuclei.

In accordance with (8.15) (symmetric spatial component of the wave function), the two electrons occupying the bonding state E_+ have opposite spin. In the ground state the bond is therefore maintained by a localized *spin-saturated electron pair*. In the antibonding state the spins are parallel.

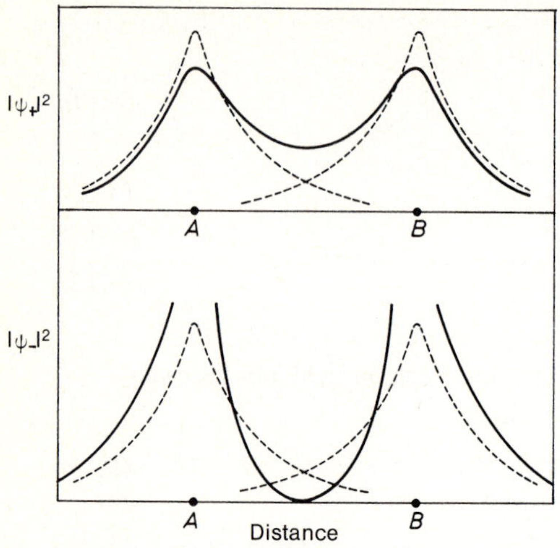

Fig. 8.2. Probability of finding an electron along the line joining the nuclei A and B in a H_2-molecule, according to (8.16) using hydrogen 1s-orbitals. Upper curve: bonding state, lower curve: antibonding state. The absolute squares of the free atomic orbitals are shown dashed.

When the atoms A and B are different, the parameter λ becomes $\neq \pm 1$. The bond is then asymmetric; the probability that an electron will be near to one nucleus is then greater than the probability that it will be near to the other. λ is therefore often called the *polarity* of the bond.

By combining other atomic orbitals of the atoms A and B, one can correspondingly form molecular orbitals for *excited states* of the bond considered. Symmetry considerations show that not every combination of a ψ_A^{at} with a ψ_B^{at} leads to nonvanishing integrals S and A. Similar to the one-electron approximation of the band model, we can make use of group theory to classify the possible eigenfunctions and hence the possible eigenvalues: σ-bonds are molecular orbitals with rotation symmetry about the line joining the nuclei. If one makes this axis of symmetry the z-axis, the following combinations of atomic orbitals lead to σ-bonds: $s - s$, $s - p_z$, $p_z - p_z$, $p_z - d_z$, etc.. π-bonds are molecular orbitals with a nodal plane containing the line between the nuclei ($p_x - p_x$, $p_y - p_y \ldots$). The molecular orbitals of δ-bonds possess two nodal planes ($d_{xy} - d_{xy} \ldots$). Hence the molecular orbitals are classified by the atomic orbitals from which they stem, by their symmetry properties, and by the distinction "bonding/antibonding". Accordingly the orbitals in Fig. 8.2 have the designation $1s\sigma^b$ and $1s\sigma^*$ (index b = bonding, index * = antibonding).

This method consequently provides a scheme of energy levels which can be populated by the electrons involved in the bond. This is important for diatomic molecules, for which the MO method was developed, less so however for the localized single bond in a solid considered here.

VB Method

We look first of all once again at the simpler case of a symmetric bond between two identical atoms. In (8.9) we have then $\bar{\lambda}_1 = \bar{\lambda}_2$. We cannot determine the contribution from the ionic states without a quantitative formulation of the problem. We therefore omit these terms for the moment. The trial function $\psi(\text{AB})$ from (8.7) then remains. Similar to (8.10–12) we define

$$S = \int \psi_A^{at}(1)\psi_A^{at}(2)\psi_B^{at}(1)\psi_B^{at}(2)d\tau_1 d\tau_2 \tag{8.17}$$

$$C = \int \left(\frac{e^2}{4\pi\varepsilon_0 R} + \frac{e^2}{4\pi\varepsilon_0 r_{12}} - V_{A2} - V_{B1}\right)|\psi_A^{at}(1)|^2|\psi_B^{at}(2)|^2 d\tau_1 d\tau_2 \tag{8.18}$$

$$A = \int \left(\frac{e^2}{4\pi\varepsilon_0 R} + \frac{e^2}{4\pi\varepsilon_0 r_{12}} - V_{A2} - V_{B1}\right)\psi_A^{at}(1)\psi_A^{at}(2)\psi_B^{at}(1)\psi_B^{at}(2)d\tau_1 d\tau_2 \,. \tag{8.19}$$

A has the meaning of an exchange integral, as one can readily see from a comparison with the integral in (1.25).

It follows that

$$E_\pm = E_A + E_B + \frac{C \pm A}{1 \pm S}\,, \tag{8.20}$$

where $E_A = E_B$ are the energies for $R \to \infty$ (no overlapping of the atomic orbitals). The signs correspond to those of (8.7). The exchange integral is again negative. Hence E_+ is the ground state. One can readily demonstate that E_+ is a bonding state with increased probability of an electron being between the nuclei. Correspondingly E_- is an antibonding state.

Since in the ground state $\psi(\text{AB})$ is symmetric in the electron coordinates, we have to introduce an antisymmetric combination of spin functions. The bonding electron-pair is thus spin-saturated, in agreement with the result from the MO method. In the antibonding state, the spin component of the wave function must be symmetric. Since there are three possible ways to realize this [$\alpha(1)\alpha(2)$, $\beta(1)\beta(2)$, $\alpha(1)\beta(2) + \beta(1)\alpha(2)$] the antibonding state is a triplet state, the bonding state, on the other hand, is a singlet state.

Hence we find the same result from both methods, namely that the two identical atoms are bonded symmetrically by a spin-saturated electron pair. This type of bond is called *covalent*. In covalent bonds the factor $\bar{\lambda}$ in (8.9) determines the addition of ionic states. $\bar{\lambda}$ is often called the *ionicity* of the bond. The asymmetry of the bond, for different atoms A and B, which is given in the MO method by the polarity λ, is determined in the VB method by the ratio $\bar{\lambda}_1/\bar{\lambda}_2$ [(8.9)].

It is useful to compare the two trial functions (8.6) and (8.7) for the covalent bond. Eq. (8.6) (with $\lambda = 1$) is a linear combination of the four possibilities (1A)(2B), (1B)(2A), (1A)(2A), and (1B)(2B) with identical weighting. On the other hand, the two ionic states (1A)(2A) and (1B)(2B) are missing in (8.7). They are added in the next step, the transition from (8.7) to (8.9). A corresponding extension is also possible in the MO method. One combines in the trial function the wave functions of the bonding and antibonding states

$$\psi = [\psi_A^{at}(1) + \psi_B^{at}(1)][\psi_A^{at}(2) + \psi_B^{at}(2)] + \alpha[\psi_A^{at}(1) - \psi_B^{at}(1)][\psi_A^{at}(2) - \psi_B^{at}(2)]\,. \tag{8.21}$$

Eqs. (8.21) and (8.9) are identical if one chooses $\bar{\lambda}_1 = \bar{\lambda}_2 = (1 + \alpha)/(1 - \alpha)$.

8.2.3 Localized and Delocalized Bonds

In the last section we treated the localized single bond between two nearest neighbours in a crystal lattice in an analogous way to a diatomic molecule. We can make the transition to the crystal lattice by going over to multiatomic molecules, and attaching more and more atoms until we have constructed a lattice.

In this case too we can start out from either the MO or VB approximation. However before we discuss this, we extend the number of trial functions by inclusion of the so-called hybrid functions. There was a physical basis for our choice of atomic orbitals as trial functions so far. There is no fundamental reason however why any other function of the electron coordinates should not have been used. For example, in (8.6–8), one possibility instead of individual atomic orbitals would be to use linear combinations of atomic orbitals of an atom, with coefficients selected at will. There was no inducement to do this in the case of the diatomic molecule. If, however, we examine the bonds to all the nearest neighbours of an atom simultaneously, then the spatial symmetry is another feature to be taken into account. It is then convenient in the trial functions to use combinations of atomic orbitals which are matched to the symmetry with which nearest neighbours are arranged around a particular atom. The best known example is the sp^3-hybrid function of the C-atom in the diamond lattice. From the $2s$-orbital and the three $2p$-orbitals the following four linear combinations are formed:

$$\begin{aligned} \sigma_1 &= \frac{1}{2}(s + p_x + p_y + p_z) \qquad & \sigma_2 &= \frac{1}{2}(s + p_x - p_y - p_z) \\ \sigma_3 &= \frac{1}{2}(s - p_x + p_y - p_z) \qquad & \sigma_4 &= \frac{1}{2}(s - p_x - p_y + p_z)\,. \end{aligned} \tag{8.22}$$

Each of these four functions has a preferred direction to one of the four corners of a tetrahedron, respectively, (Fig. 8.3). *Together* they therefore have the symmetry properties required by the arrangement of the nearest neighbours in the diamond lattice. In addition the sp^3-functions of a pair of nearest neighbours overlap much more strongly than any other combination of atomic orbitals. These properties suggest that hybrid wave functions (also called valence states) should be used as well as atomic orbitals in the trial functions. Some

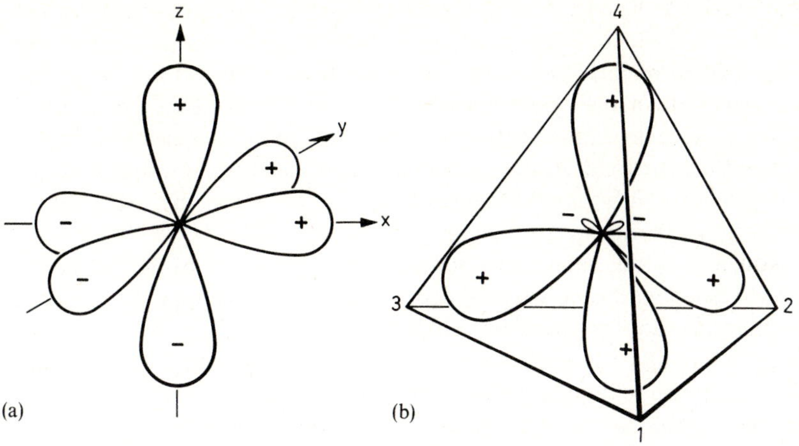

Fig. 8.3. If following (8.22) one combines one s-function with the three p-functions (a), one obtains four sp^3-hybrids (b), which extend preferentially in the directions of the four tetrahedral axes *1–4*.

important hybrid functions and their symmetries are: sp^2 (trigonal plane), sp^3 (tetrahedral), d^2sp^3 (octahedral), d^4sp^3 (dodecahedral).

We consider now the *MO method*. The transition from the description of an atom to the description of a diatomic molecule lay in the addition of a second (screened) nucleus in the interaction term of the one-electron Schrödinger equation. A logical approach would be to consider corresponding one-electron equations for the motion of an electron in the field of N nuclei—in the limit where we are concerned with a solid, the number N would embrace all the atomic nuclei in the crystal. This is precisely the one-electron problem of the *band model*. The electron observed moves in the field due to all the ions (extended to include parts of the electron-electron interaction). The Bloch function is then nothing other than a molecular orbital of the "crystal-molecule". In this formulation of the MO method the bond is delocalized. Each molecular orbital extends throughout the entire crystal. The splitting of an atomic orbital into a bonding and an antibonding molecular orbital for a diatomic molecule then corresponds to the splitting into a band with N states. Each band (as long as there is no hybridization) can be related to atomic orbitals of the atoms in the lattice (see Fig. 2.23).

This description—important as it is in the one-electron approximation and the treatment of the properties of solids in earlier chapters—is of no further help on questions of chemical bonding. The fact that many properties of a crystal can be described by the assumption of localized single bonds suggests that in the MO approximation too we should respectively combine atomic orbitals (or hybrid functions) of two nearest neighbours into molecular orbitals, thus uniquely associating the electrons they contain with a particular bond.

We can then divide the electrons of a lattice atom into four groups: a) electrons in atomic orbitals which have practically no overlap with orbitals of neighbouring atoms (electrons in closed shells, etc.); electrons such as these only make their presence felt in their contribution to the atomic potential; b) electrons in bonding molecular orbitals between nearest neighbours; c) electrons in antibonding orbitals between nearest neighbours; d) electrons in atomic orbitals which do overlap with neighbouring orbits, but for which the exchange and overlap integrals disappear on symmetry grounds.

We now look at the same problem from the point of view of the *VB method.* In the diatomic molecule we assigned a spin-saturated electron pair to the bond. In a multiatomic configuration (we shall not make the transition to a crystal until later), we assign all the electrons outside the closed shells (i.e., the valence electrons) in pairs to bonds between two nearest neighbours. We can also include ionic states in this, i.e., associate electron pairs with a single atom. This assignment of electron pairs to bonds and to individual atoms is called a *valence structure.* It corresponds to the notation commonly used in chemistry to describe covalent bonding in molecules, e.g., H – H or H: H for the H_2-molecule, N ≡ N or :N: : :N: for the N_2-molecule, etc. By an ansatz made up of a product of factors which describe the individual electron pairs, we can then set up a trial wave function for this valence structure. In practically all cases, various valence structures will be possible, so the grouping of the valence electrons to localized pairs will not be unique. We then have to construct a trial function for the variational method which is a linear combination of the wave functions of the possible valence structures. Determination of the coefficients by minimizing the energy of the ground state then shows whether one of the valence structures is dominant, or whether many structures contribute equally to the bond. In the first case one calls the bond *localized,* in the second *delocalized.* The designation "localized bond" is used here in a somewhat different sense to that in which it was used in the MO method.

For the crystal lattice the same difference applies between localized and delocalized bond. Here it is certainly not possible to employ quantitative calculations of this kind for the delocalized bond. The number of valence structures possible is too large. We can, however, immediately point to a fundamental difference between the two types of bond. Localized bonds will be preferred when the number of valence electrons belonging to an atom is equal to the number of nearest neighbours. The assignment of electrons to bonds in pairs then is unique. The electrons are bound at fixed locations in the lattice and cannot follow weak external forces. The solids concerned are *insulators* or semiconductors. An example is diamond, in which the four valence electrons of the C-atoms are localized in sp^3-hybrids.

One finds *delocalized bonds* when the number of valence electrons is much less than the number of nearest neighbours (e.g., Na with one valence electron per atom, but 12 nearest neighbours). The delocalization then induces an easy movability of the electrons, so that such solids are metals.

In addition to the *metallic bond* there are other important types of bond—the *covalent bond* and the *ionic bond.* The two latter are the limiting cases of the localized bond already mentioned in Section 8.2.2. We shall look more closely at these types in the next sections.

We terminate this section with a note of caution. To describe the chemical bond we have introduced concepts like localization and delocalization, atomic orbital and molecular orbital, ionicity and polarity, bonding and antibonding states. We shall encounter further concepts in the next section (resonance, electronegativity, etc.). All these concepts are aids which stem from the fact that we have used atomic orbitals (i.e., eigenfunctions of the *free* atom) for the ansatz which we propose to use in a variational method for the approximate solution of the Schrödinger equation for our problem. One should be clear about the facts that neither atomic orbitals nor valence states exist in a molecule or a solid, and that the valence structures from which the trial functions are formed for the nonlocalized bond are not realized individually and can therefore not be measured.

If one recognizes this danger of overvaluing these aids, they can be of considerable assistance in qualitative estimates, in systematic comparisons of the properties of solids, and in classifying different types of bond.

8.2.4 Solids with Localized Bonds: Insulators and Semiconductors

We now take up again the discussion in Section 8.2.2. For localized single bonds, in which electrons belonging to adjacent atoms are uniquely associated in pairs, we can distinguish two limiting cases.

1) The bond is *covalent.* The bond formed by the electron pair is symmetric. In the lattices of elements this is the case when the number of valence electrons of an atom equals the number of nearest neighbours (example: diamond).
2) The bond is *ionic.* The electron pair is taken up in the outermost shell of one of the two atoms in the bond. The bonding partners consequently become ions with opposite charge. The bond results from electrostatic attraction. This bond is possible when the rearrangement of the valence electrons leads to ions with closed shells (example: NaCl).

Between these two extremes there can occur all forms of mixed bond. We return to this point further below. First we look at the cohesive energy in the two limiting cases.

One can calculate the cohesive energy in *ionic crystals* by the following classical model. By including the bonding electron pair in the electron shell of one of the partners, we lose the unique association of the pair with one bond. If we look for example at a binary crystalline compound AB in which the bonding electron pairs reside with the A atoms, the lattice consists of negatively charged A-ions and positively charged B-ions. The bonding energy of an ion has contributions from the Coulomb attraction of nearest neighbours, the repulsion of next-nearest neighbours, etc. The full Coulomb interaction energy can be written as a series with terms of the form $z_j e_i e_j / 4\pi\varepsilon_0 r_{ij}$, where r_{ij} is the separation between the ith ion (charge e_i) and the "j-next" neighbour, and z_j and e_j represent their number and charge. A restriction to Coulomb interaction with nearest neighbours only is not possible. The series can, however, always be summed in the form $-Ae^2/4\pi\varepsilon_0 R$ (R = separation between nearest neighbours). A is called the Madelung constant (E. Madelung 1909). For binary lattices it lies mostly between 1.5 and 2.0; for ternary and higher lattices it can take on much larger values.

In addition to the Coulomb interaction of the ions, we have to introduce a short-range repulsive interaction, caused by the repulsive forces which arise from a penetration of the electron shells of neighbouring ions. One usually takes for this an exponential law $E_{ij} = a \exp(-r_{ij}/b)$ or a power law $E_{ij} = ar_{ij}^{-n}$, each with two free parameters. If we remain with the example of a binary lattice having N positively and N negatively charged ions, the cohesive energy is (since each interaction may only be taken once in the sum over all lattice ions)

$$E = NE_i = N[za \exp(-R/b) - Ae^2/4\pi\varepsilon_0 R]. \tag{8.23}$$

Here z is again the number of nearest neighbours of a lattice ion. One determines a from the fact that the binding energy is a minimum for the equilibrium separation R: $dE/dr = 0$ for $r = R$. Then

$$E = -\frac{NAe^2}{4\pi\varepsilon_0 R}\left(1 - \frac{b}{R}\right). \tag{8.24}$$

The parameter b can be related to the experimentally known compressibility. This is proportional to the second derivative of E with respect to volume, hence also with respect to R: $\kappa = V(d^2E/dV^2)$ or, since $V = 2NR^3$, $dE/dR = 0$ after a little calculation: $\kappa = (d^2E/dR^2)(18NR)^{-1}$.

More difficult to estimate is the cohesive energy of solids with covalent bonds. Because of the unique association of each pair of electrons in a valence structure, we have to add the bonding energies of all pairs. For an electron pair belonging to atoms A and B, it is given by (8.20).

This equation can readily be simplified if the overlap of the orbitals of the two electrons is so small that one can neglect the integral S [(8.17)] compared with 1. It then follows for the electron pair that

$$E_{\pm}(\mathrm{A-B}) = E_{\mathrm{A}} + E_{\mathrm{B}} + C_{\mathrm{AB}} \pm A_{\mathrm{AB}}. \tag{8.25}$$

A summation over all bonds leads to a cohesive energy of the crystal having the form

$$E = E_0 + C + \sum_{\text{pairs}} A_{ij}, \tag{8.26}$$

i.e., a sum over the total contribution of all lattice atoms E_0, of all Coulomb energies (the total ion-ion interaction of the ion cores has also to be included here) and of the exchange energies of all bonding electron pairs. This relation can be improved by taking account of the interaction of all electrons, in so far as it is not included in the contributions from (8.25). This adds to C the Coulomb interaction of each electron with all the electrons contained in other bonds. One has further to add the exchange energy of these "unpaired" electrons. Since the spins of the unpaired electrons are not correlated, one has to average over the possibility of antiparallel spin [positive sign in (8.25)] and the three possibilities of parallel spin (negative sign). This yields a term $-(1/2) \sum A_{ij}$ (summation over all unpaired electrons) to be added to (8.26).

Equation (8.26) shows which interactions contribute to the energy of the ground state of a covalent-bonded solid. It is not possible to calculate the energy quantitatively since the exchange integrals cannot be determined accurately.

One has therefore to resort to empirical methods to determine the energy of a single covalent bond. Here we are helped by the observation that a covalent bond between two atoms A and B does not seem to depend much on the surroundings, or the solid, or the molecule in which it appears. This follows, for example, from the possibility of associating with each atom A a *covalent radius* in such a way that the *bonding length* of a covalent bond A-B (i.e., the separation of the atoms A and B in the bond) is equal to the sum of these covalent radii: $R = r_A + r_B$. The energies of covalent bonds A-A in a solid or a multiatomic molecule can then be estimated from the cohesive energy, i.e., from the dissociation energy of diatomic A_2-molecules.

We turn now to the mixed forms lying between the two limiting cases. There are three possible ways to describe them.

1) Starting from the covalent (symmetric) bond, one regards the mixed form as a *polarization of the bond* towards one of the two lattice atoms.
2) Starting from the ionic bond (full incorporation of the electron pair in the electron shell of one of the partners) one regards the bond as a *polarization of the electron shell* towards the other partner.
3) One regards the covalent and the ionic bond as two limiting structures which *resonate* with one another.

The description in terms of a *resonance* between limiting structures, introduced by *Pauling*, is linked most closely to the approximation methods discussed above. There too limiting structures were considered, each represented respectively by a trial function. These trial functions are superimposed with undefined coefficients. The free coefficients are then determined by minimizing the expectation value of energy. One finds from this that the individual limiting structures "contribute" to the bond in a given proportion. In the trial function (8.9), for example, the three terms represent three resonating limiting structures structures A-B, A^-B^+ and A^+B^-. For the delocalized bond treated in the previous section the valence structures resonate with one another. In the case of interest here the two limiting structures of the localized bond resonate. A *mixed covalent/ionic bond* is built.

We describe such a bond by

$$\psi = a\psi_{\text{cov}} + b\psi_{\text{ion}} . \tag{8.27}$$

Here ψ_{cov} is the wave function for the covalent limit (8.7) and ψ_{ion} is one of the two functions (8.8). The ratio b/a determines the ionic component of the bond. It is often also referred to as the *ionicity* of the bond. There is a possibility here of confusion with the ionicity (symmetric addition of ionic states to the covalent bond) defined in Section 8.2.2.

The more trial functions one superimposes (with free parameters) in the variational procedure, the better does the approximation become, and hence the lower the total energy. In terms of the resonance description this means: The binding energy of a resonant state is always greater than the binding energy of one of the limiting states.

This increase in binding energy denoted *resonance strengthening* is defined empirically for predominantly covalent-bonded materials by the following statement: Let the energy of a bond $E(\text{A} - \text{B})$ lie below the binding energy $E_{\text{cov}}(\text{A} - \text{B})$ by an ionic *resonance* energy Δ. $E_{\text{cov}}(\text{A} - \text{B})$ certainly lies between the energies $E_{\text{cov}}(\text{A} - \text{A})$ and $E_{\text{cov}}(\text{B} - \text{B})$, which can be found from the arguments presented above. Usually the arithmetic mean (sometimes the geometric mean) of the two energies is chosen ad hoc for $E_{\text{cov}}(\text{A} - \text{B})$. Hence let

$$E(\text{A} - \text{B}) = \frac{1}{2}[E(\text{A} - \text{A}) + E(\text{B} - \text{B})] - \Delta . \tag{8.28}$$

Empirical estimates of Δ show that for a large number of bonds Δ can be represented by

$$\Delta \sim (x_{\text{A}} - x_{\text{B}})^2 , \tag{8.29}$$

where x_{A} and x_{B} are numbers which can be assigned to every element (*electronegativities*). The difference in the electronegativities determines the departure of the bond from the covalent limit, and hence the ionic component of the bond.

The concept of *effective ion charges* e^* offers another possible way to define the ionicity. As an example we look at a bond in the semiconductor GaAs, which crystallizes in a zinc blende lattice. This lattice differs from the diamond lattice only in the occupation of lattice positions by two different types of atoms. From each Ga atom there are four bonds to As atoms, and vice versa. Ga and As appear in the lattice as Ga^{3+}- and As^{5+}-ions. This positive charge is compensated by the eight valence electrons of each Ga–As pair. For purely covalent bonding, from each bond half of the charge of the electron pair is ascribed to each of the two neighbours. With four bonds this means that the

Ga atoms have a single negative charge, and the As atoms a single positive charge: $Ga^{1-}As^{1+}$. In the limit of ionic bonding the three valence electrons from each Ga atom transfer to the As atoms: $Ga^{3+}As^{3-}$. In going over to ionic bonding, the effective charge e^* of the Ga atom increases from -1 to $+3$.

This description has the advantage that it points out a special case—the *neutral bond.* In our example it is defined by $e^* = 0$. The bonds are polarized towards the As atoms just enough that the regions associated with the Ga- and the As-atoms—approximately defined by the covalent radii of the Ga- and As-atoms—each have zero charge.

The determination of the ionicity of the chemical bond is important for a systematic comparison of the properties of insulators and semiconductors. We must, however, take a brief look at the obstacles presented to a quantitative determination of the ionicity. The concept is not even unique in its definition. The b/a-ratio [(8.27)] is a quantity from a theoretical model, which can scarcely be correlated with other parameter , although attempts have been made to derive electronegativity differences and effective ion charges from b/a-values. The problem lies in the fact that the x_A can only be given very approximately, and differences in the x_A often have large bounds of error, which, for example, makes it impossible to set up a sequence of increasing ionicity within a group of related solids. Consequently we shall consider a further definition of the ionicity of a bond in the following section.

The determination of the effective charge e^* by different methods also presents problems. Effective charges can be found from electron density distributions. One can make measurements of reflection in the far infrared to determine the difference between the static and high-frequency dielectric constants, and obtain from this an effective ion charge (cf. Sec. 6.3.2). Here, however, we immediately run into the difficulty that these measurements mean interference with the charge distribution, and that we must therefore distinguish between a *static* effective ion charge (the quantity of interest to us here) and a *dynamic* effective ion charge. They can be very different.

The connection between bonding and crystal structure is an important problem. Such a connection must certainly exist. Goldschmidt made early use of the *ion radii* to advance predictions about the occurrence of zinc blende, wurtzite, NaCl, and CsCl lattices in predominantly ionic-bonded solids. According to these predictions lattices with a ratio between the radii of cation and anion $r_C/r_A < 0.22$ are unstable; for $0.22 < r_C/r_A < 0.41$ the zinc blende and wurtzite lattice would be preferred; for $0.41 < r_C/r_A < 0.72$ the NaCl lattice should dominate, while above 0.72 the CsCl lattice takes over. These rules are often well satisfied for ionic lattices, but poorly reproduce the relationships in many semiconductors. We return to this question in the next section.

The ionicity is also important in other questions of crystal formation. Thus the possibility of *mixed crystal formation* is not only dependent on whether the lattice constant is approximately the same in the two phases. These must also be a correspondence in the ionic component of the bond, in order to avoid strong distortions of the lattice. Thus, for example, there is no continuous series of mixed crystals between the covalent-bonded Ge and GaAs, while such a series does exist between GaAs and GaP.

We close these considerations by inquiring about the relationship between chemical bond and semiconducting properties, which we have so far described via the band model. The most important parameter is the width E_G of the energy gap, i.e., the energy needed to lift an electron from the valence band into the conduction band. In the chemical bond picture this is the energy required to break an electron free of a localized bond. This makes it possible to explain qualitatively trends in the E_G-values of different semiconductors. Thus the reduction in bonding strength in the homologous sequence diamond–Si–Ge–gray tin is paralleled by the reduction in the value of the energy gap. Through Fig. 2. 33 (Sec. 2.2.12), we see, however, that there are numerous problems in the way of a quantitative comparison. The upper edge of the valence band $E_V(\boldsymbol{k}_V)$ and the lower edge of the conduction band $E_C(\boldsymbol{k}_C)$ can lie at different $\boldsymbol{k}$-vectors. Thus in Ge E_C lies at point L in the Brillouin zone, in Si it lies along the Δ-axes. E_G therefore means energy differences between quite different subbands in the two semiconductors. For a systematic comparison one should, however, only look at energy differences between definite subbands at definite $\boldsymbol{k}$-vectors (e.g., the energies of direct optical transitions at Γ). There is also a second difficulty: In the localized chemical bond model the electrons involved in a bond occupy atomic orbitals or hybrid states (in Ge, for example, all four valence electrons occupy sp^3-orbitals). The valence band in the band model, on the other hand, often consists of subbands which contain states with very different symmetry. Within a subband the character of the symmetry changes with $\boldsymbol{k}$. The wave functions can have "s-character" at Γ, as $\boldsymbol{k}$ increases the addition of "p-character" can increase, etc.

Figures 8.4–8.6 are instructive examples of how seldom in localized bonding a subband can be associated with an atomic orbital or a hybrid orbital of the lattice atoms. Fig. 8.4 shows the arrangement of atoms in the layer lattice of GaSe. The two 4s- and the one 4p-electron per Ga-atom and the two 4s- and four 4p-electrons per Se-atom are available for bonding. For every four Ga- and Se-atoms in the elementary cell there are thus 36 valence electrons per cell. Fig. 8.5 shows the band model calculated by the pseudopotential method. The valence band reduces into five groups of subbands. The electron distribution of all these electrons has been calculated for each of these five groups. Fig. 8.6(a-f) presents the result. The eight 4s-electrons of Ga lie in a band below the structure shown. The lowest valence bands (group I) contain the 4s-electrons of Se. The electron density is consequently radially symmetric about the Se ions. The contributions to group II stem mainly (but not exclusively!) from Ga-p_z-orbitals, which arrange themselves into a bonding Ga-Ga-molecular orbital. The third group contains the associated antibonding Ga-Ga-molecular orbitals with a significant blend of Ga–Se bonds. The fourth group, as well as Ga–Ga bonds, contains p_z-orbitals from the Se-ions which contribute to

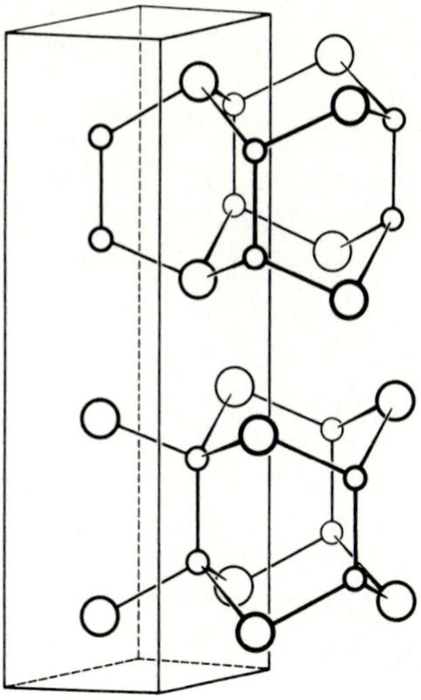

Fig. 8.4. Arrangement of atoms in the layer lattice of GaSe.

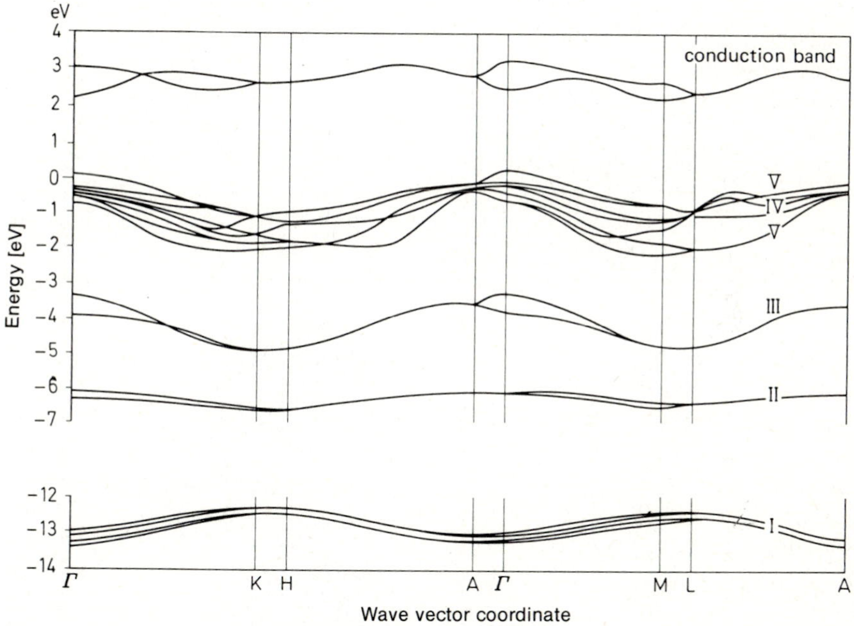

Fig. 8.5. Band structure of GaSe along the most important axes of symmetry in the Brillouin zone of the hexagonal lattice (Fig. 2.22d). The five groups of subbands in the valence band, and the lowest group of conduction bands are plotted [from M. Schlüter: Nuov. Cim. **13B**, 313 (1973)].

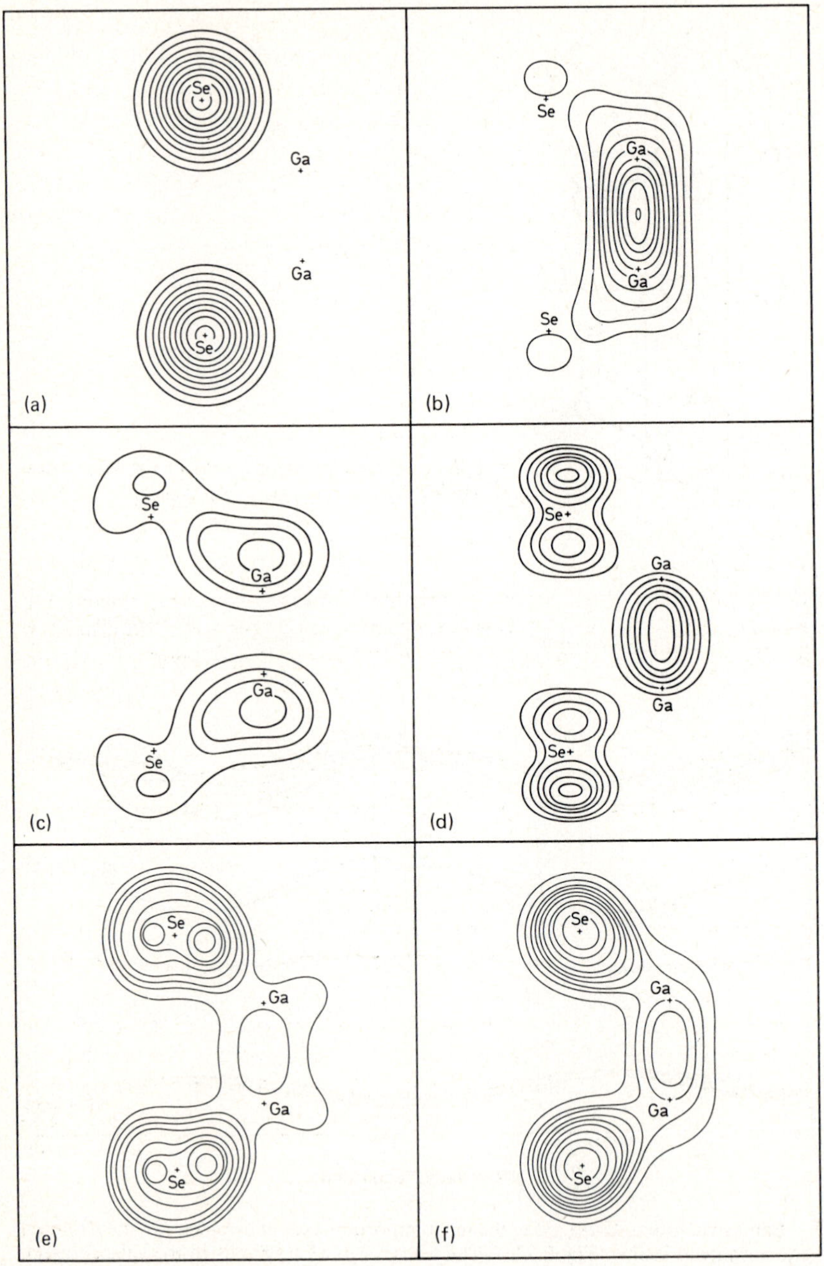

Fig. 8.6a-f. Electron distribution in the five groups of subbands in the valence band of GaSe (a)-(e) and distribution of all valence electrons (f) [from M. Schlüter: Nuov. Cim. **13B**, 313 (1973)].

the bond between the layers of the lattice. The fifth group mainly contains bonds formed from the p_x- and p_y-orbitals of Se. If one adds all these density distributions together to produce a total electron distribution over all valence subbands [(Fig. 8.6(f)], one clearly sees the formation of electron bonds between lattice atoms, the characteristic feature of the covalent bond; however, we also see the concentration around the lattice ions which reveals the admixture of ionic states. In this overall bond, covalent and ionic components can be defined, and effective ion charges e^* can be calculated. The connection between bonding and the band model, and also the fundamental points of difference, are evident in this example.

8.2.5 The Dielectric Theory of the Covalent Bond

We shall treat the chemical bond problem once again, but from a different standpoint. We regard the lattice as a framework of positively charged ions, and inquire how the valence electrons screen these charges.

In the case of a metal, we can view the valence electrons as a free electron gas. We have already considered the screening of a perturbing potential $V_a(\boldsymbol{r}, t)$ by a gas of free electrons in Section 3.1.6. We found there that the screened potential $V(\boldsymbol{r}, t)$ is determined from $V_a(\boldsymbol{r}, t)$ by division by a wave-vector- and frequency-dependent dielectric constant $\varepsilon(\boldsymbol{q}, \omega)$. We also arrived at the Lindhard equation (3.58) for $\varepsilon(\boldsymbol{q}, \omega)$. Following (3.65), the static potential of a lattice ion $V_i(\boldsymbol{r}) = -Ze^2/4\pi\varepsilon_0 r$, in the limit of small $\boldsymbol{q}$, is screened exponentially according to $\varepsilon(\boldsymbol{q}) = 1 + \lambda^2/q^2$. Thus

$$V(r) = -\frac{Ze^2}{4\pi\varepsilon_0 r}\mathrm{e}^{-\lambda r}, \qquad V(\boldsymbol{q}) = -\frac{Ze^2}{\varepsilon_0 V_g}\frac{1}{q^2 + \lambda^2}, \qquad \lambda^2 = \frac{3e^2 n}{2\varepsilon_0 E_F}. \tag{8.30}$$

The most important result of the screening is that the Fourier transform of the potential $V(q)$ is no longer singular in the limit $q \to 0$ [the singularities of $V_i(\boldsymbol{q})$ and $\varepsilon(\boldsymbol{q})$ mutually cancel].

The relationships are different in an insulator (semiconductor), where the electron gas fully occupies the valence band. The occupied states are separated from the unoccupied by an energy gap. We shall show in the following that in this case $\varepsilon(\boldsymbol{q})$ remains finite for $\boldsymbol{q} \to 0$. This results in an incomplete screening of the ion.

We can obtain the $\boldsymbol{q}$-dependent dielectric constant from the equation

$$\varepsilon(\boldsymbol{q}) = 1 - \frac{e^2}{\varepsilon_0 q^2}\sum_{\boldsymbol{k}\boldsymbol{K}_m} |\langle\boldsymbol{k}|\exp(\mathrm{i}\boldsymbol{q}\cdot\boldsymbol{r})|\boldsymbol{k}+\boldsymbol{q}+\boldsymbol{K}_m\rangle|^2 \frac{f_0(\boldsymbol{k}+\boldsymbol{q}+\boldsymbol{K}_m) - f_0(\boldsymbol{k})}{E(\boldsymbol{k}+\boldsymbol{q}+\boldsymbol{K}_m) - E(\boldsymbol{k})}. \tag{8.31}$$

f_0 is the Fermi distribution, $\boldsymbol{K}_m$ a lattice vector in the reciprocal lattice. The $|\boldsymbol{k}\rangle$ are the wave functions of the band states considered. Eq. (8.31) is an extension of the Lindhard equation (3.58). If one inserts plane waves into the matrix element one obtains exactly (3.58) for the limit $\omega=0$ considered here. We shall not derive (8.31). See in this connection *Ziman* [23], where another derivation of (8.35) below is also given.

To link up most closely with the free electron case, we shall use the so-called *Penn model* to describe the semiconductor. We remind ourselves of the results from the nearly free electron model in Section 2.2.6. The energy of a free electron in the vicinity of the surface of the Brillouin zone follows from a secular determinant of the form [cf. (2.101)]

$$\begin{vmatrix} \dfrac{\hbar^2 k^2}{2m} - E(\boldsymbol{k}) & V(\boldsymbol{K}_m) \\ V(-\boldsymbol{K}_m) & \dfrac{\hbar^2}{2m}(\boldsymbol{k} - \boldsymbol{K}_m)^2 - E(\boldsymbol{k}) \end{vmatrix} = 0\,. \tag{8.32}$$

On the surface [defined by the condition $k^2=(\boldsymbol{k}-\boldsymbol{K}_m)^2$], E jumps by an amount $E_{\mathrm{G}}=2|V(\boldsymbol{K}_m)|$.

This result was derived in Section 2.2.6 for the one-dimensional case. In the Penn model a three-dimensional isotropic case is considered: one regards a sphere of radius k_{F} as the "Brillouin zone" and assumes that the determinantal equation (8.32) is valid for each direction of $\boldsymbol{k}$. One then replaces $\boldsymbol{K}_m$ by $2k_{\mathrm{F}}(\boldsymbol{k}/k)$. The energy becomes a function of k alone. By solving the determinant, one obtains

$$\begin{aligned} E_{\pm} &= \frac{1}{2}\left\{\frac{\hbar^2}{2m}(k^2 + k'^2) \pm \left[\frac{\hbar^4}{4m^2}(k^2 - k'^2) + E_{\mathrm{G}}^2\right]^{1/2}\right\}, \\ \boldsymbol{k}' &= \boldsymbol{k}\left(1 - \frac{2k_{\mathrm{F}}}{k}\right). \end{aligned} \tag{8.33}$$

This "band structure" is illustrated in Fig. 8.7. The associated wave functions following (2.97) are made up of two terms (plane waves with wave vectors $\boldsymbol{k}$

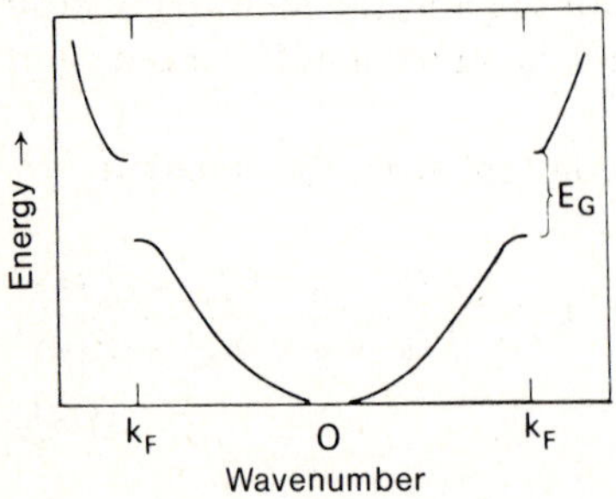

Fig. 8.7. Band structure for the isotropic Penn model, from (8.33).

and $\boldsymbol{k}' = \boldsymbol{k} - \boldsymbol{K}_m$)

$$\psi_\pm = \frac{1}{\sqrt{1+\alpha_\pm^2}}[\exp(\mathrm{i}\boldsymbol{k}\cdot\boldsymbol{r}) + \alpha_\pm \exp(\mathrm{i}\boldsymbol{k}'\cdot\boldsymbol{r})], \qquad \alpha_\pm = \frac{E_\mathrm{G}/2}{E_\pm - \dfrac{\hbar^2 k'^2}{2m}}. \tag{8.34}$$

One can use (8.33) and (8.34) to evaluate (8.31). As an approximate solution one finds in the limit $q \to 0$

$$\varepsilon(0) = 1 + \left(\frac{\hbar\omega_\mathrm{p}}{E_\mathrm{G}}\right)^2 A; \qquad A = 1 - \frac{E_\mathrm{G}}{4E_\mathrm{F}} + \frac{1}{3}\left(\frac{E_\mathrm{G}}{4E_\mathrm{F}}\right)^2 \approx 1. \tag{8.35}$$

ω_p is the plasma frequency previously introduced: $\omega_\mathrm{p} = (ne^2/\varepsilon_0 m)^{1/2}$. For the Fourier transform of the potential, and hence for the potential itself, follows

$$\lim_{q\to 0} V(q) = -\frac{Ze^2}{V_\mathrm{g}\varepsilon_0\varepsilon(0)q^2} \quad \text{or} \quad V(r) = -\frac{Ze^2}{4\pi\varepsilon_0\varepsilon(0)r} \quad \text{for large } r. \tag{8.36}$$

This is the potential of an ion having an effective charge $Ze/\varepsilon(0)$. If one constructs a crystal with ions screened in this way, additional negative charges have to be introduced to screen the residual charge remaining and hence to maintain the neutrality of the crystal as a whole. The dielectric theory of the covalent bond postulates that these screening charges are contained *in the localized* bonds. Along with screened ions one therefore introduces *bonding charges* which one usually takes to be concentrated at the midpoint of the line joining two nearest neighbours. If the number of nearest neighbours is z, each electron bond has the bonding charge $2Ze/z\varepsilon(0)$.

There are numerous possible applications of this model. We restrict ourselves here to that which is most important to bonding: the possibility of a new definition for the ionicity of a localized bond. See the references given at the end of this section for other applications of the model.

In order to be able to discuss the ionicity of the bond at all, we must first extend the Penn model to solids with two different ions in the Wigner-Seitz cell. The Penn model is a three-dimensional extension of a one-dimensional result. We begin, therefore, with a one-dimensional potential ansatz. Let a "Wigner-Seitz cell" of length l contain two ions at a separation 2τ

$$V(x) = V_1(|x-\tau|) + V_2(|x+\tau|). \tag{8.37}$$

The Fourier transform of the potential is then

$$V(q) = \frac{1}{2}[V_1(q) + V_2(q)]\cos q\tau + \frac{\mathrm{i}}{2}[V_1(q) - V_2(q)]\sin q\tau \tag{8.38}$$

with

$$V_{1,2}(q) = \frac{2}{l}\int V_{1,2}(x)\exp(-iqx)dx\,.$$

$V(q)$ is complex. The real part is the mean value of the potential. The imaginary part is a measure of the difference between the two potentials.

We carry this over to the three-dimensional Penn model. There the absolute value of $V(\boldsymbol{K}_m)$ is identified as half the band separation E_G, following (8.32). Since V is now complex, we correspondingly put

$$2|V(\boldsymbol{K}_m)| = E_G = |E_c + iC| = \sqrt{E_c^2 + C^2}\,. \tag{8.39}$$

The band gap thus consists of a "covalent" component E_c and a component due to the dissimilarity of the two atomic potentials. The latter can be used as a measure of the *ionic component* of the bond.

To apply this concept we have to provide a connection between the idealized Penn model and the band model of the semiconductor whose bond we are considering. The first difficulty this presents is that a band in the Brillouin zone of a solid contains as many states (each able to contain two electrons with opposite spin) as the crystal has Wigner-Seitz cells. We must, however, be able to accommodate all the valence electrons in the band E_+ of the Penn model. We go over, therefore, to the extended zone scheme and take (with $2n$ valence electrons per Wigner-Seitz cell) the first n Brillouin zones together. This combined zone is referred to as the *Jones zone*. For diamond structure the Jones zone contains the first four Brillouin zones. It is shown in Fig. 8.8. The tetrahedral bonded semiconductors with zinc blende and wurtzite structure have the same Brillouin zone and consequently the same Jones zones. The

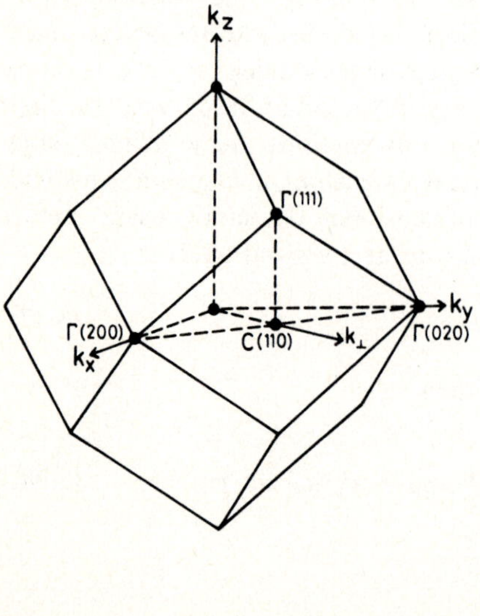

Fig. 8.8. The Jones zone for the diamond structure.

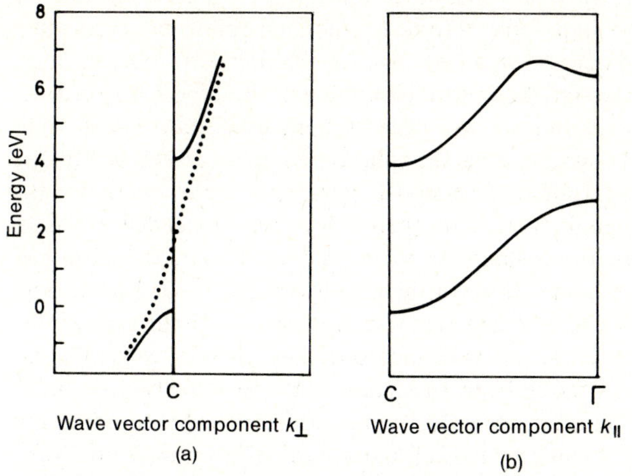

Fig. 8.9 a and b. Energy bands of silicon in the vicinity of point C on the surface of the Jones zone (Fig. 8.8) (a) perpendicular to the surface, (b) along the surface (from *Heine* and *Weaire* [101.24]).

Jones zone for these semiconductors has an approximately spherical shape. Near the surface the band structure (unfolded on the Jones zone) resembles that for free electrons. The band separation at various points on the surface is also approximately the same. Fig. 8.9 shows an example. Thus we can identify the parameter E_G in the Penn model as the *mean band separation* of the actual band structure on the surface of the Jones zone.

When we do this, we first of all find that (8.35) gives the values of the static dielectric constant for the elemental semiconductors diamond, Si, Ge, α-Sn quite accurately. All the other tetrahedral-bonded semiconductors can be derived from the group IV semiconductors by replacing half the lattice atoms with $(4-n)$-valent atoms and the other half with $(4+n)$-valent atoms (III–V, II–VI, I–VII compounds). In view of the dissimilarity of the nearest neighbours these compounds have a mixed covalent/ionic bond. Experimental results indicate that the band separation E_G with such a bond is larger than that of the isoelectronic element. Following (8.39), one can regard the square of the mean band separation as the sum of the squares of the band separation E_c of the covalent-bonded isoelectronic element and the parameter C. We thus have as a quantitative definition for the *ionicity of a bond*

$$f_i = \frac{C^2}{E_c^2 + C^2}; \quad 0 \le f_i \le 1 . \tag{8.40}$$

One knows E_c from the corresponding elemental semiconductor. E_G follows from (8.35). Hence C and f_i can be found for each of the tetrahedral-bonded semiconductors.

The ionicity defined by (8.40) allows a much more precise distinction to be drawn between the individual tetrahedral-bonded semiconductors than do, for example, the ionicities obtained via the Pauling electronegativities, which often only roughly assign the same ionicity to a group of semiconductors.

The theory presented in this section permits both quantitative and semi-empirical statements to be made about the cohesive energies, crystal structures, ionization energies, and about band-model parameters particular to the tetrahedral-bonded semiconductors. The theory has been extended to other classes of semiconductors and insulators, but we cannot pursue this here and refer the reader to [91]. As an example of the results obtained, we show in Fig. 8.10 a diagram in which the covalent band separation is plotted against the parameter C for all semiconductors with diamond, zinc blende, wurtzite, and NaCl structure. One can deduce from the figure that the ionicity $f_i = 0.785$ draws a sharp dividing line between the semiconductors with coordination number 4 (diamond, zinc blende, wurtzite) and coordination number 6 (NaCl structure).

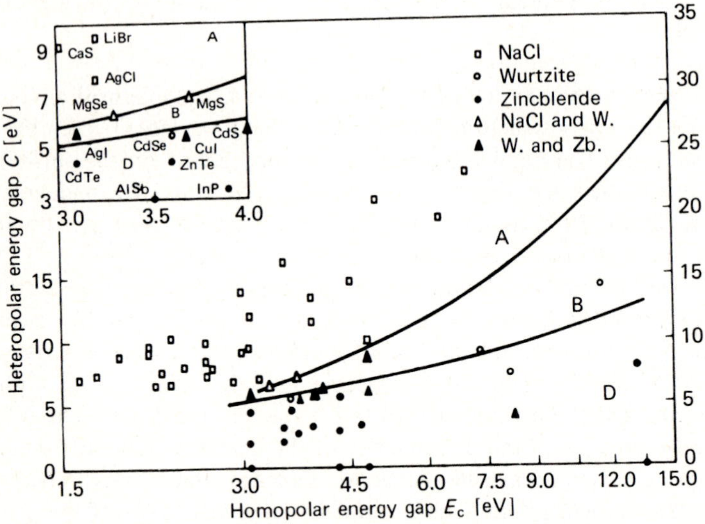

Fig. 8.10. Dependence of the covalent band gap on the parameter C of (8.40), for all semiconductors with diamond, zinc blende, wurtzite, and NaCl structure. Particular values of the parameter $f_i = C^2/(E_c^2 + C^2)$ (solid-line curves) separate the regions in which the different structures exist (from *Phillips* [91]).

8.2.6 Solids with Delocalized Bonds: Metals

The metals are characterized by two particular features. The coordination number of the lattice (the number of nearest neighbours) is greater than the number of valence electrons a lattice atom possesses. Homogeneous alloys exist, that is, metallic phases are formed from different components whose atoms are statistically distributed on the lattice sites. Both aspects point to a delocalized bond.

A characteristic property of the delocalized bond is the movability of the valence electrons about the crystal. The lattice ions are embedded in the valence electron gas, which holds the lattice together. Since directional chemical forces are absent, the lattices of most metals are close-packed structures (hexagonal or cubic). Depending on the electron configuration of the lattice atoms, on the appearance of nonclosed *d*-shells, and on the size ratio of the different atoms in the lattice, more complicated structures can also occur. We do not want to consider these aspects.

The question of the *ground state* of a metal raises problems completely different from those we treated in the last section. It is much more closely linked with the electron theory we presented in the earlier chapters of the book. The bonding energy of a metal (cohesive energy) is defined as the energy required to break the metal down into neutral atoms.

As an example we shall consider a monovalent metal. When the atoms are brought together to form a cohesive crystal lattice, each atom contributes a single valence electron (apart from its closed shells). The collection of valence electrons forms an electron gas. The cohesive energy can be divided into two parts: the energy of each electron as a particle in an interacting electron gas, and the correction needed to account for the fact that the electron gas is embedded in a crystal lattice.

We have already calculated the first part for the jellium model (interacting electron gas in a homogeneously distributed positive charge background) in the Hartree-Fock approximation in Section 8.1.3. We found in (3.11) for the energy of a Hartree-Fock electron the expression

$$E(k) = \frac{\hbar^2 k^2}{2m} - \frac{e^2 k_F}{8\pi^2 \varepsilon_0}\left(2 + \frac{k_F^2 - k^2}{k k_F} \ln\left|\frac{k + k_F}{k - k_F}\right|\right). \tag{8.41}$$

The mean energy per electron follows by integration over the Fermi sphere and division by the number of electrons. The value of the first term in (8.41) is already known from (2.36): $\bar{E} = (3/5)E_F$.

In the literature the electron energy is often given as a function of the mean electron separation in the electron gas r_0 [defined by $(4\pi/3)r_0^3 = 1/n$, n = electron concentration], or by the dimensionless quantity $r_s = r_0/a_0$ (a_0 = Bohr radius). In most metals r_s lies between 2 and 6. Large r_s means low electron concentration, and vice versa.

Following (8.41), the mean energy of a Hartree-Fock electron (of effective mass m^*) as a function of r_s is

$$\bar{E}_{HF} = \left(\frac{2.21}{r_s^2}\left(\frac{m}{m^*}\right) - \frac{0.916}{r_s}\right) \text{ryd}. \tag{8.42}$$

The second term is the contribution from the exchange interaction taken into account in the Hartree-Fock approximation.

We consider the effect of the crystal in two steps. First we replace the homogeneous positive background of the jellium model by a point lattice of ions. In a monovalent metal each ion approximately occupies the volume of a sphere of radius r_0. Thus Na crystallizes in a body-centred cubic lattice whose Wigner-Seitz cell (Fig. 2.12c) can be closely approximated by a sphere. In (3.10) the Coulomb interaction of the electrons was exactly compensated by the positive background. If we omit the background we have to add a term to (8.42), which represents the interaction energy of all the electrons (taken to be homogeneously distributed throughout the crystal). Expressed in terms of r_s this yields a contribution per electron of $+1.2/r_s$ ryd.

The next step is to add the Coulomb energy of the lattice ions and the interaction energy of the electrons with the lattice ions. The contribution of an electron to this is the energy of its ground state in the crystal. The following method of calculation has proved successful for monovalent metals (Wigner-Seitz approximation): One places a Wigner-Seitz cell around each lattice ion, and approximates it with a sphere. One assumes that in the ground state of the electron gas there is just one electron in each cell. This electron moves in the field of the associated ion; the interactions with the other ions and the electrons outside the cell just compensate. The only difference from a free atom is then the containment of the atom within a sphere of radius r_0. This alters the boundary conditions for the radial component of the wave function. Its derivative with respect to r must now vanish at $r = r_0$ instead of at infinity. The energy of the ground state of the electron is thus changed relative to its ionization energy in the free atom. This change of energy has to be added to the contribution the electron makes to the binding energy.

We have now examined all the contributions which can be taken account of within the Hartree-Fock approximation by a first-order perturbation calculation. As we saw in Section 3.1.3, however, the electron-electron interaction is only incompletely accounted for in this approximation. Correlations between the electrons are contained in the exchange term for electrons of parallel spin. The assumption of one single electron per cell in the Wigner-Seitz approximation also implies a correlation of course. There are, however, many ways to compute the electron-electron interaction more exactly. The contributions, which are added to the energy derived in the Hartree-Fock approximation, are referred to as *correlation energy*.

One might as a next step, still within the Hartree-Fock approximation, consider adding higher order contributions from the perturbation calculation. Such contributions are represented in the graphs in Fig. 3.2 This approach, as we saw in Section 3.1.2, does not lead any further, since the second-order contribution of Fig. 3.2c diverges logarithmically. The divergence stems from the long-range Coulomb forces involved in this approximation.

In Section 3.1.4, by introducing a screened Coulomb interaction and by replacing the contributions thereby removed with plasmon contributions, we arrived at an adequate method for describing the interacting electron gas. In this formulation the Hamiltonian is given by (3.23). By neglecting the electron-plasmon interaction terms, (3.23) becomes

$$H = \sum_i \frac{p_i^2}{2m} + \frac{e^2}{2V_g\varepsilon_0} \sum_{ij}' \sum_{k>k_c} \frac{\exp[ik\cdot(r_i - r_j)]}{k^2} + \frac{1}{2} \sum_{k<k_c}' \left(P_k^* P_k + \omega_p^2 Q_k^* Q_k - \frac{ne^2}{\varepsilon_0 k^2} \right). \tag{8.43}$$

The first line describes the kinetic energy and interaction energy of screened electrons; the second describes the energy of the plasmons less the long-range component of the self-energy of the electron gas.

The correlation energy has two parts.

1) The difference between the energy resulting from the second term in (8.43) and the corresponding term in the Hartree-Fock approximation. Since the difference lies in the screening, i.e., the restriction of the $\boldsymbol{k}$-summation to $k > k_c$, this difference is just the sum of the missing terms

$$E_1 = \frac{e^2}{2V_g\varepsilon_0} \sum_{ij}' \sum_{k<k_c}' \frac{\exp[ik\cdot(r_i - r_j)]}{k^2}. \tag{8.44}$$

2) No plasmons are excited in the ground state. The plasmon contribution to the second line in (8.43) is thus the plasmon zero-point energy $\hbar\omega_p/2$

$$E_2 = \sum_{k<k_c}' \left(\frac{\hbar\omega_p}{2} - \frac{ne^2}{2k^2\varepsilon_0} \right). \tag{8.45}$$

Equation (8.44) can be evaluated by transforming the sums to integrals. The integration is somewhat complicated, so we shall not carry it out here. The result (with the abbreviation $\beta = k_c/k_F$) is

$$E_1 = \frac{0.916}{r_s} \left(\frac{4}{3}\beta - \frac{\beta^2}{2} + \frac{\beta^4}{48} \right) \text{ryd}. \tag{8.46}$$

Correspondingly we find from (8.45)

$$E_2 = \left(\frac{0.866\beta^3}{r_s^{3/2}} - \frac{1.222\beta}{r_s} \right) \text{ryd}. \tag{8.47}$$

These expressions still contain the unknown cut-off parameter k_c. It is determined such that the correlation energy is minimized. One finds $\beta = 0.353\, r_s^{1/2}$. Eqs. (8.46) and (8.47) together then lead to the correlation energy

$$E_c = -(0.019 - 0.0003 r_s)\,\text{ryd}. \tag{8.48}$$

This value can be improved when one includes the interaction terms from (3.23) which were neglected in (8.43).

There are other methods of calculating the correlation energy, methods connected with the general problem of describing an interacting many-body system. We cannot pursue this problem further here. We refer the reader to the presentation in *Pines* [27, 101.1], and to the treatment of the electron gas in [36–42]. For the lattice contribution, particularly the Wigner-Seitz approximation, see the article by *Wigner* and *Seitz* in [101.1] and the references given there. For the connection between crystal structure and cohesion using the theory of pseudopotentials, and hence for a more recent development of this area which also links up with the results of the last section, we refer the reader to the review article by *Heine* and *Weaire* in [101.24].

8.3 Local Versus Nonlocal Description in Unperturbed Lattices

8.3.1 Introduction

The one-electron approximation of the band model corresponds to the MO approximation in chemical bond theory. The Bloch functions extend over the entire lattice; every electron is delocalized. We have already encountered the far-reaching applications for this model in earlier chapters.

In this chapter we pose the question: Under what conditions is it reasonable to adopt a local description when we are dealing with a homogeneous solid of infinite extent, or alternatively, when is it more appropriate to use a description similar to the VB method instead of a description in terms of the band model?

Three important assumptions were made in the band theory.

1) Strict periodicity of the lattice.
2) Replacement of the actual electron-electron interaction with the mean interaction (1.32) of the Hartree-Fock approximation.
3) Neglect of lattice vibrations, with the assumption that these could be taken account of later by perturbation calculations (weak electron-phonon coupling).

The restriction to *periodic lattices* is needed to be able to introduce a $\boldsymbol{k}$-vector at all, and from it the delocalized states of the band model. The band model loses validity when we consider alloys or amorphous phases. We shall explore in Chapter 10 to what extent we are then still able to apply band model concepts. There we shall also discuss a localization of the electrons associated with perturbation of the periodicity.

The introduction of a *mean electron-electron interaction* is needed to allow us, in the one-electron approximation, to regard the possible states of the electron observed as being totally independent of the occupation of other states by electrons. The behaviour of a Bloch electron is determined solely by the periodic potential in which it moves. This implies the neglect of *correlations* between valence electrons in a crystal. In the following section we shall investigate how far it is possible to include correlation into the band model. We shall find that, in the so-called Hubbard model, we can follow the transition from the nonlocal description of the valence electrons by the band model to a local description.

Experimental indications of failures of the band model are most apparent in solids whose band structures contain narrow *d*-bands. We are familiar with such bands through the transition metals (cf., e.g., Fig. 2.28). There are no anomalies there, however, since the *d*-bands are overlapped by an *s*-band. In many *transition metal compounds* the relative positions of the bands shift in such a way that the *d*-bands separate from lower and higher bands and cover the energy range in which the Fermi energy lies. According to the band model all these compounds should be metals (as long as their *d*-bands are not completely filled). In practice, however, one finds both metals and insulators among them, with differences in conductivity of the order of 10^{20}. A few of them change from insulators to metals with increase in temperature. We shall look at such (and other) *metal-insulator transitions* in Section 8.3.3.

We used the assumption of *weak electron-phonon coupling* to define the concept of a polaron in Section 4.1. The only way in which this quasi-particle differs significantly from a Bloch electron is in its effective mass m^{**} [(4.31)]. The inclusion of parts of the electron-phonon coupling in the band model description fails if the coupling becomes too strong [the coupling parameter α defined in (4.31) becomes too large]. This leads to the limiting case of the *small polaron*, which we did not consider in Section 4.1. We shall discuss it in Section 8.3.5. The transport theory presented in Section 4.1 starts from the Boltzmann equation. We shall see that when the one-electron approximation fails, the Boltzmann equation is no longer valid (Sec. 8.3.4). In this connection we introduce a more general expression for the conductivity, the *Kubo formula.*

The chapter ends with a section on the transport mechanism in polar solids. There we shall see that in addition to band conduction, electron or polaron *hopping* from one localized state to another can take place.

8.3.2 Correlations, the Hubbard Model

We first look at two examples where correlation between electrons clearly plays a vital role.

Consider a monovalent metal. Each lattice atom carries one valence electron (*s*-electron) with it. The valence band of the metal is then half filled. For simplicity we neglect the possibility of band overlap. Occupied and empty states are adjacent in the *s*-band. The valence electrons are delocalized and move freely about the crystal. As far as the band model is concerned, this is all we need to know to explain the metallic properties of the crystal.

We shall now increase the lattice constant, while still maintaining the crystal structure, i.e., the relative arrangement of the lattice ions. The consequence is a reduction in the width of the s-band. In we increase the lattice constant to such an extent that there is practically no interaction between the lattice atoms, the band reduces to the discrete s-level of the isolated atom. We discussed this previously in Section 2.2.11 (Fig. 2.23a). However in the limit of isolated atoms, each atom is certainly neutral, i.e., it has one of the valence electrons *localized* in its vicinity. Metallic conduction is no longer possible, although according to the band model approximation we still have a half-filled band. The band model approximation thus breaks down for narrow bands. The localization of the electrons on the lattice atoms means a correlation between the electrons. For narrow bands such correlations have to be considered.

We find a similar occurrence in the interacting electron gas (jellium model). As the electron concentration decreases (increasing mean electron separation r_s) the kinetic energy contribution in (8.42) (first term on the right) decreases relative to the potential energy contribution (second term on the right). When the second term is much more significant, one finds that in the state of lowest energy the electrons are organized in a crystal-like array, i.e., the electrons arrange themselves in such a way as to be as far away from each other as possible. This also means a localization of the electrons through correlation effects (Wigner crystallization).

Both experiments here are conceptual; they reveal what influence correlation can have on the statements derived from the one-electron approximation. For a quantitative formulation we now outline an approach which, by taking correlation effects into account in the band model, leads to a localized description. We shall base our development of the approach on the first example here, and we shall therefore examine a (narrow) s-band which is just half filled with N electrons from the N lattice sites.

The states of the band are described by the energies $E(\boldsymbol{k})$ and the wave functions $\psi(\boldsymbol{k}, \boldsymbol{r})$. We omit the band index. To take account of correlation in an approximate manner in the Hamiltonian, we remove from $E(\boldsymbol{k})$ the mean electron-electron interaction contained in the band model description, and replace it with the full electron-electron interaction.

For the first step we can proceed very similarly to the formalism developed in Section 3.2.2. for the exciton problem. From (3.70) the difference $E(\boldsymbol{k}) - W(\boldsymbol{r})$ is given in Bloch representation by

$$E_1 = W(\boldsymbol{k}) - \sum_{\boldsymbol{\kappa}} (2\langle \boldsymbol{k}, \boldsymbol{\kappa}|g|\boldsymbol{k}, \boldsymbol{\kappa}\rangle - \langle \boldsymbol{k}, \boldsymbol{\kappa}|g|\boldsymbol{\kappa}, \boldsymbol{k}\rangle)\nu_{\boldsymbol{\kappa}\sigma}\,. \tag{8.49}$$

Here $g = e^2/4\pi\varepsilon_0|\boldsymbol{r} - \boldsymbol{r}'|$, $W(\boldsymbol{k})$ is the one-electron energy of the Hartree-Fock approximation in the Bloch representation; the summation runs over all states of the band, the occupation number $\nu_{\boldsymbol{\kappa}\sigma}$ guaranteeing that only occupied states are counted. The Hamiltonian associated with (8.49) can be put into the occupation number representation and has the form $H_1 = \sum_{\boldsymbol{k}\sigma} E_1 c^+_{\boldsymbol{k}\sigma} c_{\boldsymbol{k}\sigma}$.

We add to H_1 the electron-electron interaction. The overall Hamiltonian becomes

$$H = \sum_{k\sigma} E_1 c_{k\sigma}^+ c_{k\sigma} + \frac{1}{2} \sum_{\substack{k_1 k_2 \\ k_1' k_2' \\ \sigma_1 \sigma_2}} \langle k_1, k_2 | g | k_1', k_2' \rangle c_{k_1\sigma_1}^+ c_{k_2\sigma_2}^+ c_{k_2'\sigma_2} c_{k_1'\sigma_1} . \tag{8.50}$$

Our aim is to describe correlations in narrow bands. Since in the process we are approaching a local description, it seems appropriate to adopt the Wannier representation (used in Sec. 3.2.2) instead of the Bloch representation. We use (2.104) and (3.71) to define the Wannier functions $a(\boldsymbol{R}_m, \boldsymbol{r})$, and introduce creation and annihilation operators for the Wannier representation. We further define

$$\begin{aligned} T_{ij} &= \frac{1}{N} \sum_{k} W(\boldsymbol{k}) \exp [i\boldsymbol{k} \cdot (\boldsymbol{R}_i - \boldsymbol{R}_j)] \\ v_{ij} &= \frac{1}{N} \sum_{k\sigma} v_{k\sigma} \exp [i\boldsymbol{k} \cdot (\boldsymbol{R}_i - \boldsymbol{R}_j)] . \end{aligned} \tag{8.51}$$

We then find for the Hamiltonian in the Wannier representation

$$\begin{aligned} H = &\sum_{ik\sigma} [T_{ik} - \sum_{jl} (2\langle ij|g|kl\rangle - \langle ij|g|lk\rangle v_{jl})] c_{i\sigma}^+ c_{k\sigma} \\ &+ \frac{1}{2} \sum_{ijkl} \sum_{\sigma\sigma'} \langle ij|g|kl\rangle c_{i\sigma}^+ c_{j\sigma'}^+ c_{l\sigma'} c_{k\sigma} , \end{aligned} \tag{8.52}$$

where now the matrix elements are to be formed using Wannier functions. We interpret these matrix elements as describing interactions between electrons associated with different lattice ions (at locations $\boldsymbol{R}_i$). The interaction between electrons at the same ion will of course play the major role. As the first step in considering correlation we consequently only take matrix elements in which $i = j = k = l$. We put $\langle ii|g|ii\rangle = U$ and thereby obtain the simplified Hamiltonian

$$H = \sum_{ik\sigma} T_{ik} c_{i\sigma}^+ c_{k\sigma} - U \sum_{i\sigma} v_{ii} c_{i\sigma}^+ c_{i\sigma} + \frac{U}{2} \sum_{i\sigma\sigma'} c_{i\sigma}^+ c_{i\sigma'}^+ c_{i\sigma'} c_{i\sigma} . \tag{8.53}$$

Now $v_{ii} = (1/N) \sum_{k\sigma} v_{k\sigma}$ is a number, just as is the sum $\sum_{i\sigma} c_{i\sigma}^+ c_{i\sigma} = \sum_{i\sigma} n_{i\sigma}$, which is left in the second term on the right. The second term on the right thus yields a constant contribution to the energy, and can be omitted in the following discussion which is concerned only with energy differences. The first term on the right in (8.53) describes transitions of an electron from the kth to the ith lattice position. Here terms where $\boldsymbol{R}_i$ and $\boldsymbol{R}_k$ are adjacent lattice positions will dominate.

The last term on the right in (8.53) becomes $(U/2)\sum_{i\sigma\sigma'} n_{i\sigma}n_{i\sigma'} = U\sum_i n_{i\sigma}n_{i,-\sigma}$. We therefore obtain the operator

$$H = \sum_{ik\sigma} T_{ik}c_{i\sigma}^{+}c_{k\sigma} + U\sum_{i} n_{i\sigma}n_{i,-\sigma}\,. \tag{8.54}$$

This Hamiltonian was first discussed by Hubbard in connection with correlations in narrow band (*Hubbard Hamiltonian*). It contains three parameters: $T_0 = T_{ii}$, $T_1 = T_{ik}$ ($\boldsymbol{R}_i$, $\boldsymbol{R}_k$ = nearest neighbours), and U.

The meaning of these parameters can best be shown if, instead of Wannier functions, we use atomic orbitals. We can then link up directly with the results of Section 2.2.7 and, for the energy $W(\boldsymbol{k})$ in (8.51), use (2.109)

$$W(\boldsymbol{k}) = E^{\mathrm{at}} + C + \sum_{\substack{n\\ \mathrm{n.n.}}} \exp(\mathrm{i}\boldsymbol{k}\cdot\boldsymbol{R}_n)A(\boldsymbol{R}_n)\,. \tag{8.55}$$

For a given nearest-neighbour configuration (8.55) can be summed. For a Bravais lattice with a centre of inversion, each pair of nearest neighbours in opposite directions contributes $2A(a)\cos(k_i a)$. Since the cosine can take values between -1 and $+1$, the band described by (8.55) has a width $2A(a)z$, with z = number of nearest neighbours.

With (8.55) we find

$$T_{ij} = \frac{1}{N}\sum_{k}\,[E^{\mathrm{at}} + C + \sum_{\substack{l\\ \mathrm{n.n.}}} A(\boldsymbol{R}_l)\exp(\mathrm{i}\boldsymbol{k}\cdot\boldsymbol{R}_l)]\exp[\mathrm{i}\boldsymbol{k}\cdot(\boldsymbol{R}_i - \boldsymbol{R}_j)]\,. \tag{8.56}$$

The summations are of the type $(1/N)\sum_{\boldsymbol{k}}\exp(\mathrm{i}\boldsymbol{k}\cdot\boldsymbol{R}) = \delta_{\boldsymbol{R},0}$. Hence

$$T_{ij} = (E^{\mathrm{at}} + C)\delta_{ij} + \sum_{\substack{l\\ \mathrm{n.n.}}} A(\boldsymbol{R}_l)\delta_{i+l,j}\,. \tag{8.57}$$

$T_0 = T_{ii}$ is thus the mean energy of the band; $T_1 = T_{ik}$ (i,k nearest neighbours) is equal to half the width of the band.

The meaning of the parameter U follows from a consideration of the limit of infinite lattice constant. Clearly then $T_1 = 0$ and the Hamiltonian becomes diagonal. The energy becomes

$$E = \sum_{i}\,[T_0(n_{i\sigma} + n_{i,-\sigma}) + Un_{i\sigma}n_{i,-\sigma}] = N_1T_0 + N_2(2T_0 + U)\,, \tag{8.58}$$

where N_1 is the number of lattice positions each occupied by one electron, and N_2 the number of sites occupied by two electrons. T_0 is therefore the energy needed to bind an electron on an isolated atom. $T_0 + U$ is the energy needed to attach a second electron, of opposite spin. Hence U is the Coulomb interaction energy of two electrons located at the same atom.

In the ground state the N electrons available have the energy T_0, i.e., one electron is accommodated at each atom ($N_1 = N$, $N_2 = 0$). In this limit we therefore find a strict localization of the electrons. The Hubbard approximation thus leads from the band model to a local description.

We now look at the ground state of a system with finite lattice constant in which each lattice atom possesses one electron. The spin direction is taken to change from neighbour to neighbour (antiferromagnetic ground state). If we introduce into this system a further electron with a given spin, it can be accommodated at one of the $N/2$ atoms which already has an electron with opposite spin. The Pauli principle forbids its placement with any of the othe $N/2$ atoms. The energy of this electron amounts to $T_0 + U$ at isolated atoms [(8.58)]. Due to the interaction between all $N/2$ states which can accept the electron, this energy splits up into a band centred around $T_0 + U$. The same arguments lead to the splitting up of the energy T_0 into a corresponding band. As long as the widths of the bands are smaller than the separation $(T_0 + U) - T_0 = U$, there will be a gap between the two bands. At a critical amount of splitting (determined by the lattice constant) the gap will disappear. The transition from the localized description to the band model then takes place.

This result can be derived quantitatively from the Hamiltonian (8.54). The calculations however are too lengthy to be reproduced here. We refer instead to the review article by *Adler* [101.21] and Hubbard's original work cited there. In the example considered here (half-filled s-band, antiferromagnetic ground state) one finds, by assuming a simple density of states, the behaviour shown in Figs. 8.11 and 8.12. Depending on the ratio Δ/U (width of band in the band model without correlations/effective Coulomb repulsion of an electron pair at an ion), the s-band splits into two separate bands or only alters its density

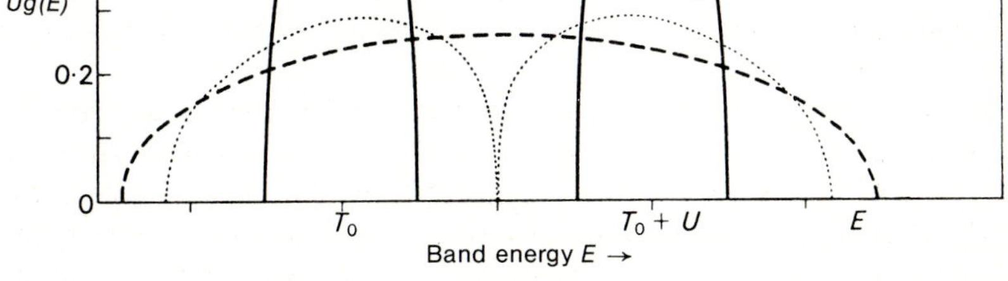

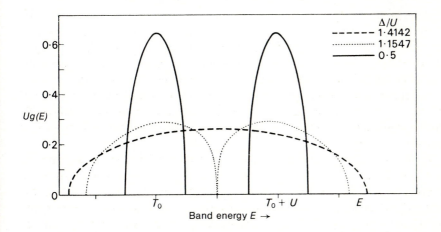

Fig. 8.11. Band splitting by electron-electron interaction in the Hubbard model. $g(E)$—density of states, Δ—bandwidth in the band model (no correlations), U—effective Coulomb repulsion of two electrons on the same ion [from J. Hubbard: Proc. Roy. Soc. **A281**, 401 (1965)].

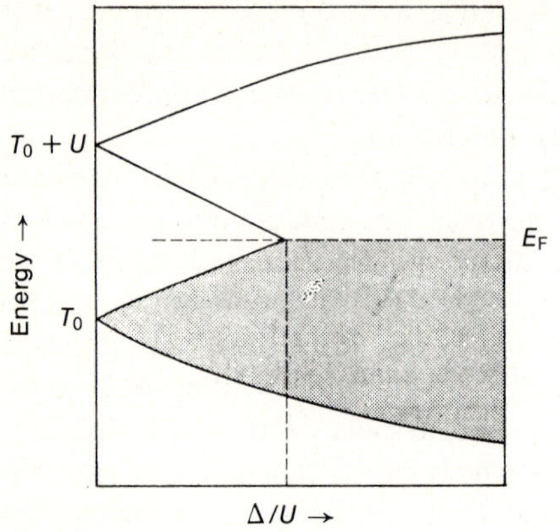

Fig. 8.12. Transition from localized states to delocalized states in a (half-filled) energy band in the Hubbard model (cf. Fig. 8.11). T_0—energy needed to attach an electron to a free ion, $T_0 + U$—energy needed to attach a second electron to the same ion.

of states. In the ground state the lower half of the states is occupied. We have thus in the above been able to show qualitatively the transition from metallic to insulating behaviour in a solid.

For a quantitative discussion of the Hubbard approximation, the possibilities for its extension, and its points of weakness, we again refer to the article by *Adler* in [101.21], additionally to *Doniach* [111a.18], *Mott* and *Zinamon* [111c.21], and contributions in [125].

8.3.3 Metal-Insulator Transitions

We mentioned in Section 8.3.1 that the large group of transition metal compounds provides numerous examples of solids which cannot be adequately described by the band model. Among this group one finds both metals and insulators.

Furthermore the group also provides examples of another important phenomenon: with increasing temperature a jump in the conductivity is observed in certain compounds which can amount to many orders of magnitude. A possible explanation is that with increase of temperature the lattice constant crosses the threshold value at which the localized electrons become delocalized, i.e., the two Hubbard bands in Fig. 8.12 merge. An argument advanced by *Mott*, which we now develop, shows that such a transition can lead to a sharp increase in conductivity.

If, in the antiferromagnetic ground state we considered in the last section, we move an electron from its lattice ion to another, this means in the Hubbard model the creation of a free electron in the upper band and a free hole in the lower (filled) band. The interaction between the two particles is transmitted by the potential $V = -e^2/4\pi\varepsilon\varepsilon_0 r$, where ε is the dielectric constant of the lattice. This potential is long range. Electron and hole form a bound state (exciton). In spite of the excitation of an electron the crystal remains an insulator. (One should note that this argument goes beyond the Hubbard model, which only includes interactions within a lattice ion). If we excite many electrons, the electron gas will screen the electron-hole interaction according to (3.66). The potential is

$$V = -\frac{e^2}{4\pi\varepsilon\varepsilon_0 r}\,\mathrm{e}^{-\lambda r} \quad \text{with} \quad \lambda^2 = \frac{3e^2 n}{2E_F\varepsilon\varepsilon_0} = \frac{3e^2 nm}{\hbar^2(3\pi^2 n)^{2/3}\varepsilon\varepsilon_0}. \tag{8.59}$$

The screening weakens the binding energy of the electron-hole pair. At a critical electron concentration, the screening constant λ becomes so large that electron and hole no longer occupy a bound state. This happens when the reciprocal screening constant is smaller than the electron-hole separation (Bohr radius a_0). From (8.59) it follows that

$$a_0\lambda > 1, \quad a_0 = \frac{4\pi\hbar^2\varepsilon\varepsilon_0}{me^2} \quad \text{and hence} \quad n^{1/3}a_0 > \frac{1}{4}\left(\frac{\pi}{3}\right)^{1/3} \approx \frac{1}{4}. \tag{8.60}$$

If we use the reduced electron-electron separation introduced in Section 8.2.6, the condition becomes $r_s < 2.5$.

This estimate is in good agreement with the criterion for Wigner crystallization. The mean energy of a Hartree-Fock electron in the jellium model is from (8.42) the sum of a kinetic and a potential component. Depending on which dominates, the electrons in the electron gas are localized or not. The limit is given by (8.42) as $r_s < 2.4$, if one neglects the difference between m and m^*. This value is little changed when the correlation energy is also added to E_{HF}.

This mechanism allows an abrupt rise in conductivity at the metal-insulator transition even though the Hubbard model predicts a continuous, if fast, increase in the concentration of free carriers. A sharp transition between metallic and insulating character like this is known as a *Mott transition.*

Metal-insulator transitions do not arise only in transition metal compounds. In glasses, in amorphous semiconductors, in the phenomenon of impurity band conduction, in dilute solutions, etc., one also finds abrupt leaps in conductivity as a particular parameter is varied (temperature, defect concentration, etc). Not all of these are Mott transitions. In many cases it is not yet clear what the responsible mechanism is.

The following possibilities need to be examined.

Overlapping Bands (Wilson Transition)

According to the band model the valence band of divalent elements is filled, as long as it does not overlap with a higher-lying band. A solid with this band structure is an insulator. On the other hand, if the valence band overlaps with a higher band, unoccupied states lie directly above the highest occupied states, and the solid behaves like a metal. The maximum of the valence band then lies at an energy higher than the minimum of the higher band. Both extrema however lie at different values of $\boldsymbol{k}$ in the Brillouin zone (Fig. 8.13). A change

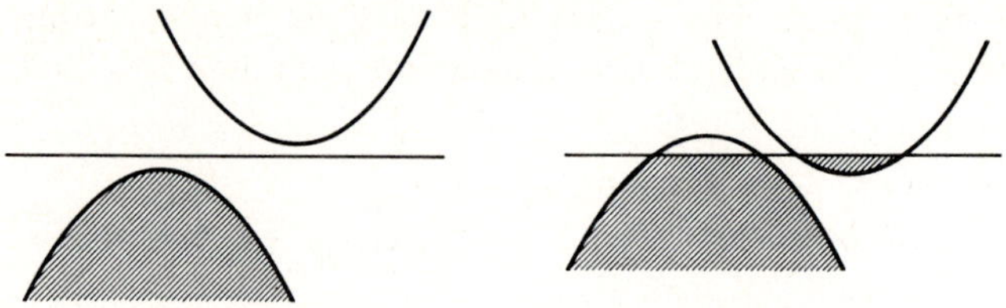

Fig. 8.13. If the Fermi level lies between two bands which do not overlap, the solid is an insulator (semiconductor). If change of temperature or pressure causes the two bands to overlap, occupied and unoccupied states will be adjacent in both bands. The solid exhibits metallic characteristics (Wilson transition).

of lattice constant (by change of temperature or pressure) can effect a displacement of the two band edges relative to each other, and hence cause a metal-insulator transition. A condensation of free charge carriers to excitons is also possible here. Instead of the argument based on absence of screening at low concentrations, another possible explanation can be used: The energy needed to excite an exciton is the energy of the band gap E_G minus the binding energy of the electron-hole pair E_e. If we allow the two bands to approach, before the transition semiconductor-semimetal at $E_G = 0$, we will reach the case where $E_G - E_e$ becomes negative. This leads to a spontaneous condensation of excitons.

Halperin and *Rice* [101.21] discuss the formation of a new phase with superstructure as an intermediate state possible in semimetals with overlapping bands (*excitonic insulator*).

Change in Structure

The appearance of an energy gap within a band when a critical temperature threshold is crossed can also be caused by a change in structure (Fig. 8.14). In the manner shown in the figure, a small displacement of neighbouring ions can produce a structure with double the lattice constant (and a corresponding increase in the number of basis atoms). The increase in size of the Wigner-Seitz cell corresponds to a reduction in the size of the Brillouin zone, and hence a

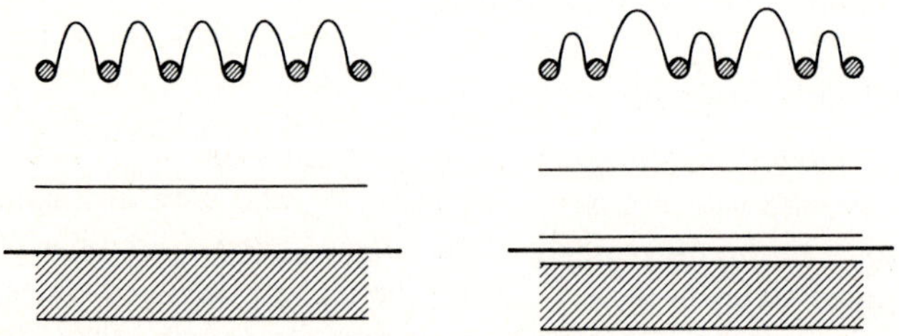

Fig. 8.14. Metal-semiconductor transition by change of structure. If the lattice constant is doubled by a small shift of the regularly arranged lattice atoms, then a band can split (a half-filled band in the case considered), and a metal can turn into an insulator.

possible splitting of previously connected bands. Examples of this are discussed by *Adler* [101.21].

Slater used a similar argument to interpret the insulating properties of several transition metal oxides. The antiferromagnetic spin organization in the ground state of a metal with half-filled valence band means a doubling of the lattice constant, at least for the exchange-part of the lattice potential. From this once again the possibility arises of a splitting of the band between the occupied and the unoccupied states.

Perturbation of the Lattice Periodicity

In disordered lattices the states near the band edges become localized (see Sec. 10.1.2). By changing the degree of disorder, delocalized states near the Fermi energy become localized and a metal-insulator transition can occur (*Anderson transition*). Since in this chapter we are only concerned with strictly periodic lattices, we leave the discussion of this localization possibility until Chapter 10.

In this section we have restricted ourselves to an enumeration of possible mechanisms for metal-insulator transitions. We cannot discuss here the question of which possibility is responsible for a given experimental observation. On this we refer the reader above all to *Mott* [93]. Further literature on the contents of this section is to be found in [101.21, 111a.33].

8.3.4 Limits of the Boltzmann Equation, the Kubo and Kubo-Greenwood Formulae

Having exceeded the limits of the band model, the validity of the Boltzmann equation as the basis for the description of transport effects is also limited.

The Boltzmann equation is based on the concept that the electrons in a solid move freely under the influence of external fields, and that these movements are interrupted by interaction processes with the lattice (phonon emission and absorption). Electrons considered in this way are represented as wave packets, using Bloch functions. The centre of gravity of the wave packet ($\boldsymbol{k}_0$, $\boldsymbol{r}_0$) defines the wave vector and location of the electron. In this description the extension of the wave packet in $\boldsymbol{k}$-space must be small compared with the dimension of the Brillouin zone ($\propto 1$/lattice constant). From the uncertainty principle the extension of the wave packet in real space must then stretch over several lattice constants. The use of Bloch functions is thus essential since only then does the rapidly changing potential of the lattice ions drop out of the explicit description. Furthermore all other space-dependent quantities (external fields, inhomogeneities of the lattice, etc.) are then only permitted to vary slowly along the length of the wave packet.

The Boltzmann equation is further founded on the assumption that an electron is accelerated as a free particle of mass m^* between two "collisions". In doing so it absorbs energy from the external field and gives it up to the lattice at the collision. The model fails when the energy absorbed from an external field becomes comparable with the width of the energy band. This must be taken into consideration with small bands. The collisions are regarded as instantaneous processes which interrupt free flight: the interaction time must be small compared with the time between two interactions. That means weak electron-phonon coupling.

By using the relaxation time approximation, the relaxation time τ leads to a *mean free path* $l = \tau v_{\mathrm{th}}$ (v_{th} = thermal velocity $\sqrt{8k_{\mathrm{B}}T/\pi m^*}$). Defining the electron mobility by $\mu_n = e\tau/m^*$, we find

$$\mu_n \approx 7l(\text{Å})\sqrt{\frac{m}{m^*}}\sqrt{\frac{300\ \mathrm{K}}{T}}\,\frac{\mathrm{cm}^2}{\mathrm{Vs}}\,. \tag{8.61}$$

With experimental mobilities of the order $1 - 10\ \mathrm{cm^2V^{-1}s^{-1}}$, the mean free path becomes comparable with the lattice constant. The Boltzmann equation then is no longer suited to the description of transport properties.

It becomes clear from this argument that, for small bands, for lattice inhomogeneities, and for strong electron-phonon coupling, we need other methods for the treatment of transport phenomena.

In this section we develop a more general formulation of conductivity theory. We treat the case of strong electron-phonon interaction in the following section. The question of lattice inhomogeneities will be discussed in the subsequent chapters.

We seek the most generalized expression for the electrical conductivity of a solid. We start with the statistical operator ρ introduced in Section 2.1.4 and used to calculate expectation values $\langle f\rangle$ from the associated operators f. We adopt (2.23) and (2.24) from Section 2.1.4, and (3.49) from Section 3.1.6.

$$\begin{aligned}
&\langle f\rangle = \mathrm{Trace}\,\{f\rho\}\\
&\rho = Z^{-1}\exp(-H/k_{\mathrm{B}}T)\,, \qquad Z = \mathrm{Trace}\,\{\exp(-H/k_{\mathrm{B}}T)\}\\
&\mathrm{i}\hbar\dot{\rho} = [H, \rho]\,.
\end{aligned} \tag{8.62}$$

We apply these equations to the determination of the expectation value of current. We start similar to Section 3.1.6, and divide the Hamiltonian for the system into the operator H_0 for the zero-field case and the perturbation operator δH associated with the external field. For simplicity we take the latter to be constant in space. Let the field be applied at $t = -\infty$ and grow adiabatically to its value at $t = 0$. We therefore put

$$H = H_0 + \delta H = H_0 + \lim_{\alpha\to 0}\,[e\boldsymbol{E}\cdot\boldsymbol{r}\exp(-\mathrm{i}\omega t + \alpha t)]; \quad \rho = \rho_0 + \delta\rho. \tag{8.63}$$

As in (3.48), we linearize the equations of motion for ρ

$$i\hbar\delta\dot{\rho} = [H_0, \delta\rho] + [\delta H, \rho_0]\,. \tag{8.64}$$

We obtain $\delta\rho$ by defining a $\Delta\rho$ by transferring to the interaction picture

$$\delta\rho = \exp\left(-\frac{\mathrm{i}}{\hbar}H_0 t\right)\Delta\rho\,\exp\left(\frac{\mathrm{i}}{\hbar}H_0 t\right). \tag{8.65}$$

If we continue to neglect quadratic perturbation terms it then follows that

$$\begin{aligned} i\hbar\Delta\dot{\rho} &= \exp\left(\frac{\mathrm{i}}{\hbar}H_0 t\right)[\delta H, \rho_0]\exp\left(-\frac{\mathrm{i}}{\hbar}H_0 t\right) \\ &= \lim_{\alpha\to 0}\left(\exp(-\mathrm{i}\omega t + \alpha t)\exp\left(\frac{\mathrm{i}}{\hbar}H_0 t\right)[e\boldsymbol{r}, \rho_0]\exp\left(-\frac{\mathrm{i}}{\hbar}H_0 t\right)\cdot\boldsymbol{E}\right). \end{aligned} \tag{8.66}$$

$\Delta\rho$ and $\delta\rho$ take the same values at $t = 0$. Furthermore at $t = -\infty$, both are zero. Consequently one obtains from (8.66) by integration

$$\begin{aligned} \delta\rho(t = 0) = \lim_{\alpha\to 0}\Bigg(\frac{1}{\mathrm{i}\hbar}\int_{-\infty}^{0} & dt\,\exp(-\mathrm{i}\omega t + \alpha t) \\ & \times \exp\left(\frac{\mathrm{i}}{\hbar}H_0 t\right)[e\boldsymbol{r}, \rho_0]\exp\left(-\frac{\mathrm{i}}{\hbar}H_0 t\right)\cdot\boldsymbol{E}\Bigg). \end{aligned} \tag{8.67}$$

Inserting (8.67) into the expectation value for the current density $\boldsymbol{j}$ yields

$$\begin{aligned} \langle\boldsymbol{j}\rangle = \mathrm{Trace}\,\{\boldsymbol{j}\delta\rho\} = \frac{1}{\mathrm{i}\hbar}\lim_{\alpha\to 0}\int_{-\infty}^{0} & dt\,\exp(-\mathrm{i}\omega t + \alpha t) \\ & \times \mathrm{Trace}\left\{\boldsymbol{j}\exp\left(\frac{\mathrm{i}}{\hbar}H_0 t\right)[e\boldsymbol{r}\cdot\boldsymbol{E}, \rho_0]\exp\left(-\frac{\mathrm{i}}{\hbar}H_0 t\right)\right\}. \end{aligned} \tag{8.68}$$

From (8.68) it follows that a component of the conductivity tensor is given by the *Kubo formula*

$$\sigma_{\mu\nu} = \lim_{\alpha\to 0}\int_{-\infty}^{0} dt\,\exp(-\mathrm{i}\omega t + \alpha t)K_{\mu\nu} \tag{8.69}$$

with

$$K_{\mu\nu} = \frac{1}{\mathrm{i}\hbar}\mathrm{Trace}\left\{j_\mu\exp\left(\frac{\mathrm{i}}{\hbar}H_0 t\right)[er_\nu, \rho_0]\exp\left(-\frac{\mathrm{i}}{\hbar}H_0 t\right)\right\}. \tag{8.70}$$

Equation (8.70) is often given in another form. For this one rearranges it using (8.62)

$$[r_\nu, \rho_0] = \rho_0(\rho_0^{-1} r_\nu \rho_0 - r_\nu) = \rho_0(\exp(H_0/k_B T) r_\nu \exp(-H_0/k_B T) - r_\nu)$$

$$= \rho_0 \int_0^{1/k_B T} d\lambda \frac{d}{d\lambda}(\exp(\lambda H_0) r_\nu \exp(-\lambda H_0))$$

$$= \rho_0 \int_0^{1/k_B T} d\lambda \exp(\lambda H_0)[H_0, r_\nu] \exp(-\lambda H_0). \tag{8.71}$$

The commutator $[H_0, r_\nu]$ can be replaced according to the quantum mechanical equation of motion by $-i\hbar\dot{r}_\nu$. If we further note that $-e\dot{r}_\nu = j_\nu$, then

$$[er_\nu, \rho_0] = i\hbar\rho_0 \int_0^{1/k_B T} d\lambda \exp(\lambda H_0) j_\nu \exp(-\lambda H_0) \tag{8.72}$$

and hence

$$K_{\mu\nu} = \int_0^{1/k_B T} d\lambda \, \mathrm{Trace}\left\{\rho_0 j_\mu \exp\left(\frac{i}{\hbar} H_0(t - i\hbar\lambda)\right) j_\nu \exp\left(-\frac{i}{\hbar} H_0(t - i\hbar\lambda)\right)\right\}. \tag{8.73}$$

The final step is to replace the current operators by time-dependent operators from the Heisenberg picture: $\boldsymbol{j}(t) = \exp[(i/\hbar)Ht]\boldsymbol{j}\exp[-(i/\hbar)Ht]$. Writing $\mathrm{Trace}\{\rho_0 f\} = \langle f \rangle$ we find

$$K_{\mu\nu} = \int_0^{1/k_B T} d\lambda \langle j_\mu(0) j_\nu(t - i\hbar\lambda)\rangle. \tag{8.74}$$

Eq. (8.69) together with (8.74) is the customary form of the Kubo formula.

The Kubo formula is of considerable importance as the starting point for the calculation of the electrical conductivity. In view of its complexity, we shall not however use it in what follows. Greenwood has developed a simplified form, which is of interest for the definition of the localization or delocalization of one-electron states.

For its derivation we start with the one-electron Hamiltonian $H = H_0 + H'$. Let H_0 describe the free electron and H' the perturbation due to a constant electric field. Let the energy and electron wave function then be E and ψ_E. As in the derivation of the Kubo formula, we use the statistical operator ρ to calculate the conductivity. By constructing the Trace it follows from (8.6.2) that

$$\langle \boldsymbol{j} \rangle = \mathrm{Trace}\{\rho \boldsymbol{j}_{op}\}, \tag{8.75}$$

where the current operator is given by

$$\boldsymbol{j}_{\mathrm{op}} = \mathrm{Re}\left[-\frac{e}{V_{\mathrm{g}}}\frac{\hbar}{im}\nabla\right]. \tag{8.76}$$

By using $\boldsymbol{E} = -\dot{\boldsymbol{A}}$, and assuming a time-dependence of the vector potential and the electric field of the form $\exp(-i\omega t + \alpha t)$ (in the limit $\alpha \to 0$), we put for H'

$$H' = \frac{e}{m}\boldsymbol{A}\cdot\boldsymbol{p} = \frac{\hbar e}{im}\boldsymbol{A}\cdot\nabla = -\frac{\hbar e}{m\omega}\boldsymbol{E}\cdot\nabla . \tag{8.77}$$

For the perturbation component of the statistical operator $\rho = \rho_0 + \delta\rho$ we find similar to (3.53)

$$\langle\psi_{E'}|\delta\rho|\psi_E\rangle = \frac{f_0(E') - f_0(E)}{E' - E - \hbar\omega - \mathrm{i}\hbar\alpha}\langle\psi_{E'}|H'|\psi_E\rangle . \tag{8.78}$$

Equation (8.75) can be written explicitly in the form

$$\langle\boldsymbol{j}\rangle = V_{\mathrm{g}}^2\iint dEdE'g(E)g(E')\langle\psi_{E'}|\delta\rho|\psi_E\rangle_{\mathrm{av}}\langle\psi_E|\boldsymbol{j}_{\mathrm{op}}|\psi_{E'}\rangle_{\mathrm{av}} . \tag{8.79}$$

Here $g(E)$ is the density of states of the electron states E. The index "av" (not explicitly written in the subsequent equations) means that the matrix element is to be averaged over all states in the intervals dE and dE'. Inserting (8.76–78) into (8.75), one obtains

$$\sigma_{ij}(\omega) = \mathrm{Re}\left\{\lim_{\alpha\to 0}\frac{\hbar^2 e^2 V_{\mathrm{g}}}{im^2\omega}\iint dEdE'g(E)g(E')\right.$$
$$\left.\times\left\langle\psi_{E'}\left|\frac{\partial}{\partial x_i}\right|\psi_E\right\rangle\left\langle\psi_E\left|\frac{\partial}{\partial x_j}\right|\psi_{E'}\right\rangle\frac{f_0(E') - f_0(E)}{E' - E - \hbar\omega - \mathrm{i}\hbar\alpha}\right\} \tag{8.80}$$

for the ij-component of the conductivity tensor. Using the transformation

$$\mathrm{Re}\left\{\lim_{\alpha\to 0}\frac{1}{\mathrm{i}}\frac{1}{E' - E - \hbar\omega - \mathrm{i}\hbar\alpha}\right\} = \pi\delta(E' - E - \hbar\omega) \tag{8.81}$$

it follows that

$$\sigma_{ij}(\omega) = \frac{e^2\hbar^2\pi V_{\mathrm{g}}}{m^2\omega}\int g(E)g(E + \hbar\omega)\left\langle\psi_{E+\hbar\omega}\left|\frac{\partial}{\partial x_i}\right|\psi_E\right\rangle$$
$$\times\left\langle\psi_E\left|\frac{\partial}{\partial x_j}\right|\psi_{E+\hbar\omega}\right\rangle[f_0(E + \hbar\omega) - f_0(E)]dE . \tag{8.82}$$

We find the direct-current conductivity by going to the limit $\omega \to 0$. The difference in the Fermi distributions for E and E', divided by $\hbar\omega$, becomes $\partial f_0/\partial E$. For simplification we shall assume in what follows that the conductivity is isotropic ($\sigma_{ij} = \sigma\delta_{ij}$). The product of the two matrix elements can be replaced by the negative of the absolute square of the one matrix element. In this way one obtains the usual form for the *Kubo-Greenwood formula*

$$\sigma = -\int \sigma_E(0)\frac{\partial f_0}{\partial E}dE \tag{8.83}$$

with

$$\sigma_E(0) = \frac{e^2\hbar^2\pi V_g}{m^2}[g(E)]^2\left|\left\langle\psi_E\left|\frac{\partial}{\partial x}\right|\psi_E\right\rangle\right|^2. \tag{8.84}$$

The Kubo-Greenwood formula can be clearly interpreted. We know that only electrons from a region within a few k_BT of the Fermi energy contribute to conduction. It is just this region which is specified by the positive factor $-\partial f_0/\partial E$ in the integrand. For $T = 0$, $-\partial f_0/\partial E$ is a δ-function at $E = E_F$. The remaining factor $\sigma_E(0)$ is the contribution to conductivity from electrons in states in the region (E, dE).

Equation (8.83) has a form which is particularly useful when we are looking at the contribution to conductivity from states in the vicinity of the Fermi energy. When the states involved in the conductivity are several k_BT above the Fermi energy, an alternative form is more advantageous. For $E - E_F \gg k_BT$, $f_0 \approx \exp[(E_F - E)/k_BT]$ and hence $-\partial f_0/\partial E \approx f_0/k_BT$.
Thus

$$\sigma = \frac{1}{k_BT}\int \sigma_E(0)f_0 dE \equiv \int e\mu(E)n(E)dE \tag{8.85}$$

with

$$n(E) = g(E)f_0(E), \qquad \mu(E) = \frac{\sigma_E(0)}{k_BTeg(E)}. \tag{8.86}$$

According to this the conductivity is made up of contributions from individual states, whose localization or delocalization is given by the magnitude of the "mobility" $\mu(E)$. We return to this in Chapter 10.

8.3.5 The Small Polaron

We now turn to a discussion of a strong electron-phonon interaction. Some of the aspects of this problem were already discussed in Section 4.1.3. In the approximation used there we were able to introduce a new quasi-particle, the polaron. Its major properties, compared with a Bloch electron, were a) a reduction in its energy by an amount $\alpha\hbar\omega_L$ in its self-induced potential well, and b) a change in its effective mass from m^* to $m^{**} = m^*/(1 - \alpha/6)$. Here $\hbar\omega_L$ is the energy of the longitudinal phonons which form the phonon cloud of the polaron. α is the coupling constant for the electron-phonon interaction. The results in Section 4.1.3 were limited to the range $\alpha \ll 1$.

An important parameter which was not discussed in Section 4.1.3 is the size of the polaron. The self-induced charge distribution of the polaron can be calculated in a similar way to the result (4.33). One obtains a distribution which decreases exponentially with increasing distance, the characteristic length being $r_0 = (\hbar/2m^*\omega_L)^{1/2}$. r_0 can be interpreted as the "radius" of the polaron. For solids having coupling constants $\alpha \leqslant 1$, the radius takes on values between 10 and 100 lattice constants.

The meaning of r_0 also follows from a simple estimate. Consider a polaron at rest. By absorption and emission of virtual phonons its uncertainty in energy is $\Delta E = \hbar\omega_L$. Since $E = \hbar^2k^2/2m$, this is therefore associated with an uncertainty in wave number $\Delta k = (2m^*\omega_L/\hbar)^{1/2}$. This leads to an uncertainty in position of $\Delta r = 1/\Delta k = r_0$.

If the coupling constant exceeds unity, the approximation in Section 4.1.3 breaks down. Admittedly one can still take higher terms in the perturbation series into account, but it is better to rearrange the Hamiltonian with a canonical transformation. This was done by Lee, Low and Pines, and improved on subsequently by other authors (cf. Problem 4.2). The most important results one finds is that the effective polaron mass is $m^{**} = m^*(1 + \alpha/6)$ and that the polaron radius diminishes according to $1/\alpha$.

This approximation fails when the polaron radius is reduced to the order of the lattice constant. The basis of all approximations was of course the description of the lattice as a continuum, which is only justified if the polaron radius is sufficiently large.

Polarons which only extend over regions of the order of a lattice constant are called *small polarons*, in contrast to the large polarons treated so far.

The validity of the large polaron model is limited not only by the order of magnitude of the coupling constant but also by the fact that the uncertainty in energy of the polaron ΔE must be small in relation to the energy of the electron. For free electrons, this is the Fermi energy. In an energy band it is at most the width of the band. Thus when we consider narrow energy bands in solids, a correction to the polaron model may also be necessary from this point of view.

The direction of the correction is easy to see. Because of its large effective mass m^{**}, the large polaron is less mobile than an electron without a polarization cloud. With increasing electron-phonon coupling m^{**} increases and at the same time so does the self-induced potential well which the electron creates. The limit is an immobile *localized* electron. Although we started out with a lattice which was translation invariant, we find here the possibility of localized states.

We can readily estimate the binding energy of the small polaron in its potential well. For a rigid lattice the potential of an electron is $-e/4\pi\varepsilon(0)\varepsilon_0$. Considering lattice deformation it becomes $-e/4\pi\varepsilon(\infty)\varepsilon_0$. The difference between these two expressions is the self-induced potential well. We give the polaron a finite radius r_p (which is not related to the radius of the large polaron defined above). We assume that within $r = r_p$ the potential is constant, and that outside r_p it is given by $-(e/4\pi\varepsilon_0 r)\{[1/\varepsilon(\infty)] - [1/\varepsilon(0)]\}$ (Fig. 8.15). The depth of the well is then $-(e/4\pi\varepsilon_0 r_p)\{[1/\varepsilon(\infty)] - [1/\varepsilon(0)]\}$. The kinetic energy of the electron restricted to a sphere of radius r_p is added to this energy. From the uncertainty principle this is $\hbar^2/2m^* r^2$. Finally we must add the energy required to polarize the surroundings. It is equal to half the (positive) potential energy, so that this contribution only yields a numerical factor. Altogether we find

$$E(r_p) = \frac{\hbar^2}{2m^* r_p^2} - \frac{e^2}{8\pi\varepsilon_0 r_p}\left[\frac{1}{\varepsilon(\infty)} - \frac{1}{\varepsilon(0)}\right]. \tag{8.87}$$

One determines the still unknown radius r_p by minimizing (8.87). Finally

$$r_p^{-1} = \frac{m^* e^2}{8\pi\varepsilon_0 \hbar^2}\left[\frac{1}{\varepsilon(\infty)} - \frac{1}{\varepsilon(0)}\right] \tag{8.88}$$

and

$$E = -\frac{e^2}{16\pi\varepsilon_0 r_p}\left[\frac{1}{\varepsilon(\infty)} - \frac{1}{\varepsilon(0)}\right] = -\frac{1}{4}\alpha^2\hbar\omega_L. \tag{8.89}$$

Thus the binding energy is proportional to α^2.

More important than this estimate is the question of the polaron states. The starting point for this is neither the effective-mass approximation, nor the continuum approximation. The most suitable approach is to transform the electron-phonon interaction by a canonical transformation, as we did in

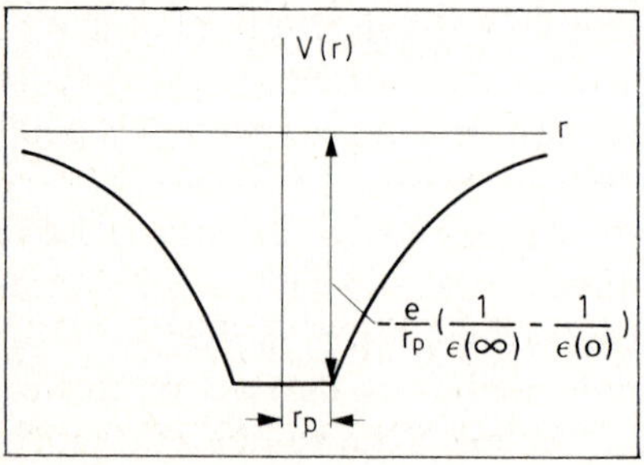

Fig. 8.15. Potential well of the small polaron.

generalized form in Section 3.1.5. We treat this problem in a wider context in the next section.

Literature for this section is: *Appel* [101.21], *Mott* and *Davis* [94], and several articles in [118].

8.3.6 Hopping Conductivity in Polar Solids

In the last section we saw that for strong electron-phonon coupling, polarization can lead to a localization of an electron in its self-induced potential well (small polaron). The movement of a small polaron takes place (except at very low temperature) by *hopping processes* from one lattice atom to another. In this it is of secondary importance whether the lattice is ordered or not. We look at such hopping processes in this section.

Before we turn to a quantitative description, let us treat the problem qualitatively. We describe the lattice distortion in the vicinity of a lattice atom by a configuration coordinate q (Fig. 8.16). (See also Sec. 9.1.9 for the configuration coordinate concept). The deformation energy depends quadratically on q: $E_{def} = Aq^2$. We put the energy of the electron in its potential well proportional to q: $E_{el} = -Bq$. Thus the total polaron energy is $E = Aq^2 - Bq$. It has a minimum at $q_0 = B/2A$, so that we can also write $E = -Aq_0^2 + A(q - q_0)^2$. Aq_0^2 is the binding energy of the small polaron [(8.89)]. We now consider two lattice atoms, which we describe (along with their surroundings) by configuration coordinates q_1, q_2. Let there be a small polaron localized on one of the atoms. An electron transition from 1 to 2, and hence the formation of the small polaron on atom 2, takes place when the electron energies are the same at both sites: $Bq_1 = Bq_2$. The energy needed to construct the state $q_1 = q_2$ is $A(q_1 - q_0)^2 + Aq_2^2 = A(q_1 - q_0)^2 + Aq_1^2$. It has a minimum at $q_1 = q_0/2$. At least an energy $W_H = Aq_0^2/2$ (that is half the binding energy of the polaron) must be delivered by lattice vibrations to make a change of place possible. The hopping transition probability w has consequently a temperature

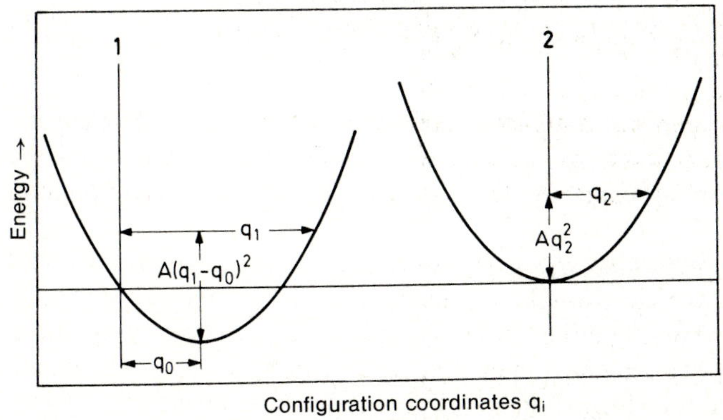

Fig. 8.16. Description of hopping processes in polar solids by the configuration coordinate model.

dependence proportional to $\exp(-W_H/k_BT)$. w is equal to the product of this exponential and a characteristic frequency, of the order of the phonon frequency. This transition probability can be used to estimate a diffusion constant and, from this, a hopping mobility. Maximum small polaron hopping mobilities of the order of 0.1 $cm^2V^{-1}s^{-1}$ follow.

We shall come back to similar considerations in Section 10.2 in the context of hopping mechanisms in disordered lattices. Here we want to refine this coarse estimate with a quantitative treatment of the movement of localized electrons in polar and nonpolar lattices, taking account of the electron-phonon interaction. We link up with some work by *Schnakenberg*,[1] to which we refer for further details.

The Hamiltonian consists of an electron term, a phonon term, and the electron-phonon interaction. We write for the electron term

$$H_{el} = \sum_n E_n c_n^+ c_n + \sum_{mn}' V_{mn} c_m^+ c_n \,. \tag{8.90}$$

The E_n here can be equal, or assumed to be distributed over a given range of energy. The c_n^+, c_n are creation and annihilation operators for an electron localized at lattice position $\boldsymbol{R}_n$. V_{nm} are integrals which relate the transitions between states at $\boldsymbol{R}_n$ and $\boldsymbol{R}_m$. Neglecting the zero-point energy, we write for the phonon part, from (3.132)

$$H_{ph} = \sum_{j=1}^{2} \sum_{\boldsymbol{q}} \hbar\omega_{\boldsymbol{q}j} a_{\boldsymbol{q}j}^+ a_{\boldsymbol{q}j} \,. \tag{8.91}$$

Here the index $j = 1$ combines the acoustic branches, while $j = 2$ combines the optical. The $a_{\boldsymbol{q}j}^+$ and $a_{\boldsymbol{q}j}$ are creation and annihilation operators for phonons of wave number $\boldsymbol{q}$ in the jth branch.

For the electron-phonon interaction finally we have

$$H_{el-ph} = -\sum_{j=1}^{2} \sum_{n,\boldsymbol{q}} \sqrt{u_{\boldsymbol{q}j}}\,\omega_{\boldsymbol{q}j} [\exp(i\boldsymbol{q}\cdot\boldsymbol{R}_n) a_{\boldsymbol{q}j} + \exp(-i\boldsymbol{q}\cdot\boldsymbol{R}_n) a_{\boldsymbol{q}j}^+] c_n^+ c_n \,. \tag{8.92}$$

This equation can be derived from (4.4) by forming each of the matrix-elements in (4.5) with localized functions. In addition to the phonon frequencies the $u_{\boldsymbol{q}j}$ also contain coupling constants for acoustic and optical electron-phonon interaction.

The complete Hamiltonian $H_{el} + H_{ph} + H_{el-ph}$ contains two perturbation components. One is the electron-phonon interaction (8.92), and the second is the term in (8.90) which describes processes involving a change of position. Since the electron-phonon interaction in polar solids can be strong, we eliminate H_{el-ph} by a canonical transformation and thus transfer from an electron system

[1] J. Schnakenberg: phys. stat. sol. **28**, 623 (1968).

with interactions to a polaron system, in which only a part of the interactions still has to be considered explicitly.

We take for the new Hamiltonian $\bar{H} = \exp(-s)H\exp(s)$ with

$$s = \sum_{jnq} \sqrt{u_{qj}}[\exp(-i\boldsymbol{q}\cdot\boldsymbol{R}_n)a_{qj}^+ - \exp(i\boldsymbol{q}\cdot\boldsymbol{R}_n)a_{qj}]c_n^+c_n. \tag{8.93}$$

For this it follows that

$$\bar{H} = \sum_n \eta_n c_n^+ c_n + \sum_{jq} \hbar\omega_{qj} a_{qj}^+ a_{qj} + \sum_{mn} V_{mn} B_{mn} c_m^+ c_n + \bar{H}_{pp}, \tag{8.94}$$

with

$$\begin{aligned} \eta_n &= E_n - \sum_{jq} u_{qj}\hbar\omega_{qj}, \\ B_{mn} &= \exp(\sum_{jq} \sqrt{n_{qj}}\{[\exp(-i\boldsymbol{q}\cdot\boldsymbol{R}_m) - \exp(-i\boldsymbol{q}\cdot\boldsymbol{R}_n)]a_{qj}^+ - \text{c.c.}\}. \end{aligned} \tag{8.95}$$

The first term in (8.94) describes the unperturbed polaron system with renormalized energies η_n; the second describes the phonon system. The third term describes polaron transitions from lattice point $\boldsymbol{R}_m$ to lattice point $\boldsymbol{R}_n$. The last term represents an effective polaron interaction, which we shall neglect.

In what follows we are interested most of all in the third term, which we rewrite as follows

$$\begin{aligned} \sum_{mn} V_{mn} B_{mn} c_m^+ c_n &= \sum_{mn} \tilde{V}_{mn} c_m^+ c_n + \sum_{mn} V_{mn}(B_{mn} - \bar{B}_{mn}) c_m^+ c_n \\ \tilde{V}_{mn} &= \bar{B}_{mn} V_{mn}. \end{aligned} \tag{8.96}$$

Here $\bar{B}_{mn}$ is the expectation value of the operator B_{mn} for the unperturbed system ($V_{mn} = 0$). We find for $\bar{B}_{mn}$

$$\bar{B}_{mn} = \exp\left[-\sum_{jq} u_{qj} \sin^2 \frac{\boldsymbol{q}\cdot(\boldsymbol{R}_m - \boldsymbol{R}_n)}{2} \coth \frac{\hbar\omega_{qj}}{2k_B T}\right]. \tag{8.97}$$

The two terms in (8.96) describe polaron transitions between different points in the lattice. The first term describes processes, in which the phonon system remains unchanged, the second term processes in which each individual step is associated with the absorption or emission of phonons. This means the possibility of *band conduction* and *thermally activated hopping*. The integral V_{mn} decides the extent of band conduction. We can see here that the polaron band is narrower than the electron band, and that its width diminishes with increasing temperature.

We shall only discuss the hopping part of equation (8.96) in the following. It can be calculated using the Kubo formula (8.69). In addition to the Hamiltonian H, one needs the current operator $\boldsymbol{j}$ for the calculation. From the definition $\boldsymbol{j} = e\dot{\boldsymbol{r}} = (e/i\hbar)[\boldsymbol{r}, H]$ with $\boldsymbol{r} = \sum_n \boldsymbol{R}_n c_n^+ c_n$, the latter follows as

$$\boldsymbol{j} = \frac{e}{i\hbar} \sum_{mn} V_{mn}(\boldsymbol{R}_m - \boldsymbol{R}_n) c_m^+ c_n \, . \tag{8.98}$$

If one subjects it to the same canonical transformation as H, only a factor B_{mn} is added to (8.98). $\boldsymbol{j}$ can be reduced to a band term and a hopping term, where the hopping term contains the factor $(B_{mn} - \bar{B}_{mn})$.

Insertion of (8.98) and the relevant parts of (8.94) and (8.96) into the Kubo formula then leads to the hopping conductivity. We shall not explicitly give the laborious calculation or the final result; instead we refer the reader to the work of *Schnakenberg* mentioned. The important results are the following:

Hopping conductivity is represented as the sum of contributions from all individual hops. From it the hopping probability can be derived in nonpolar solids (restriction to the index $j = 1$) and polar solids (including $j = 2$) as special cases. Two temperature regions can be clearly distinguished in polar solids, depending on whether the polar energy is greater or less than the energy of an optical phonon. From this, one finds regions with different temperature dependence in the conductivity.

We continue the discussion of hopping conductivity in solids in Section 10.2.2

9. Localized States

9.1 Point Imperfections

9.1.1 Introduction

The periodicity of a lattice can be disturbed over large regions, at surfaces or inner boundaries, along dislocation lines, or at individual lattice sites. Correspondingly one makes the distinction between three-dimensional, two-dimensional, one-dimensional, and zero-dimensional lattice imperfections. In this chapter we shall be concerned with the zero-dimensional *point imperfections* (or *lattice defects*) and their influence on the properties of a solid.

The most important point imperfections are chemical *impurities*, *vacant lattice sites*, and *interstitial atoms*. The feature they have in common is the ability to bind and to release electrons. They thus create *localized states* in the lattice.

In this chapter we consider these localized states from the following points of view:

1) Can we continue to use the concept of elementary excitations in disturbed lattices? We shall see that for sufficiently low defect concentrations the properties of the elementary excitations are only slightly modified. Elementary excitations and imperfections can then be considered together.
2) What interactions arise between elementary excitations and defects?
3) How are the physical properties of a solid (optical or transport properties for example) affected by lattice imperfections?

We start in Section 9.1.2 with an examination of the effect of point imperfections on the band model. It will emerge that the delocalized band states and the localized states associated with defects can be treated together in the same energy diagram.

In this discussion, the defect will be interpreted as a disturbance of the host lattice. One can pose the opposite question: How far does the host lattice modify the properties of a free atom? *Crystal field theory* has something to say about this, as we shall see in Section 9.1.3.

Just as the band model describes the properties of the quasi-particle "crystal electron", the phonon dispersion spectrum describes the properties of the lattice vibrations. In Section 9.1.4, we shall find that individual states of the otherwise delocalized phonon states become localized at defects, and split off from the continuous spectrum of a branch (*localized lattice vibrations*).

We then consider the electrons again. Starting with the possibility of uniting the electronic defect states with the band-model Bloch states in an approximate way, we ask in Section 9.1.5 how the statistics of Sections 2.1.4 and 2.2.10 should now be extended. A change in meaning of the concepts introduced there will make it possible to go over to a description in which not only the electrons in band states and in localized levels, but also the defects themselves are subjected to equilibrium conditions. This leads to reaction kinetics, which we need in Section 9.1.6 for the description of disorder equilibria. In Section 9.1.7 we treat the important subject of defect kinetics.

The last four sections of this chapter are concerned with the effect of defects on the optical and transport properties of solids. In addition to *recombination centres* and *traps*, we discuss luminescence, the broadening of absorption and emission spectra by electron-phonon interaction during optical transitions, and the influence of bound excitons on emission spectra. Finally we look briefly at the part played by defects in transport phenomena. Here we shall be interested most in the scattering of electrons at defects (impurity scattering) as a mechanism taking place concurrently with electron-phonon interaction.

9.1.2 Description Within the Framework of the Band Model

Band theory assumes a strictly periodic lattice structure. This periodicity is perturbed by defects. In this section we consider a single defect in an otherwise periodic potential.

Instead of the Schrödinger equation (2.72)

$$H_0\psi_n(\boldsymbol{k}, \boldsymbol{r}) \equiv \left[-\frac{\hbar^2}{2m}\nabla^2 + V(\boldsymbol{r})\right]\psi_n(\boldsymbol{k}, \boldsymbol{r}) = E_n(\boldsymbol{k})\psi_n(\boldsymbol{k}, \boldsymbol{r}) \tag{9.1}$$

[$V(\boldsymbol{r})$: periodic lattice potential, $\psi_n(\boldsymbol{k}, \boldsymbol{r})$: Bloch function], we now have the equation

$$[H_0 + U(\boldsymbol{r})]\psi = E\psi\,. \tag{9.2}$$

Here $U(\boldsymbol{r})$ is the additional potential introduced by the defect. For an impurity atom at a lattice position, for example, it is the potential of the impurity less the potential of the lattice atom which has been substituted. $U(\boldsymbol{r})$ may, however, also have to take account of a local lattice distortion caused by the impurity atom.

$U(\boldsymbol{r})$ can be negative or positive. Electrons can thus be bound to the defect or repelled by it. We consequently anticipate that (9.2) has solutions which are localized about the defect, and whose energy levels lie below or above the states of the energy band considered.

An examination of (9.2) for simple models reveals this behaviour. We reproduce one result in Fig. 9.1. The general result is that the presence of the defect in an otherwise periodic potential leads to a splitting off of one state from the band considered. If $U(\boldsymbol{r})$ is positive, the uppermost state splits off; if $U(\boldsymbol{r})$ is

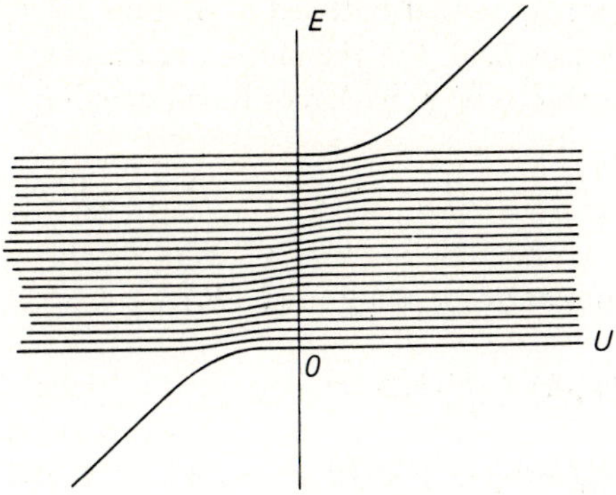

Fig. 9.1. Energy band for a simple three-dimensional potential model (periodic potential with an isolated defect) as a function of the deviation U of the defect potential from the potential at an undisturbed lattice site [cf. (9.2)] [from *Koster* and *Slater* (*Slater* [106/XIX])]. Depending on the sign of U, the state of highest or lowest energy splits out from the quasi-continuous band. The split-off state is spatially localized to the vicinity of the defect. The remaining delocalized band states are only displaced slightly in energy. This result justifies the retention of the band model and its extension with defect levels to describe crystals distorted by a small concentration of defects.

negative the lowest splits off. There are only minor displacements of the states within the band. While the wave function in the band remains approximately a delocalized Bloch function, the wave function associated with the split-off state is localized. We shall not carry out the calculations which lead to Fig. 9.1. The models to be discussed below show the same properties and are more valid for the problem of a defect in a crystal. We treat a simple example of a similar case in Section 9.1.4 (influence of a defect on the vibrational spectrum).

It is difficult to discuss (9.2) in general because of the unknown form of the function $U(\boldsymbol{r})$. We obtain the simplest form if we take the defect to be a donor atom in an elemental semiconductor, i.e., we take a $n + 1$-valent atom in a covalent-bonded lattice of n-valent atoms. The donor atom then introduces one electron and one additional positive charge to the nucleus. This electron is not required for the covalent bonds to the nearest neighbours. $U(\boldsymbol{r})$ in this case is the potential of the additional positive charge, in which the additional electron moves.

For large separations of the electron from the positive charge, the crystal lattice screens the Coulomb potential just like a homogeneous medium of dielectric constant ε, in accordance with (8.36). If we use this continuum approximation, and thus put in (9.2) the potential

$$U(r) = -\frac{e^2}{4\pi\varepsilon\varepsilon_0 r} \tag{9.3}$$

we have reduced the perturbing potential to that of an H-atom in a medium of dielectric constant ε.

We proceed to solve (9.2) in a way similar to that outlined in Section 2.2.9 for the motion of a wave packet in an electric field. For the motion of the electron in the field $U(r)$ of the defect, we construct a wave packet of Bloch functions

$$\psi = \sum_{n,\boldsymbol{k}} c_n(\boldsymbol{k})\psi_n(\boldsymbol{k}, \boldsymbol{r}) . \tag{9.4}$$

We put this function into (9.2) and obtain, corresponding to (2.146)

$$\begin{aligned}\sum_{n,\boldsymbol{k}} c_n(\boldsymbol{k})(H_0 + U)\psi_n(\boldsymbol{k}, \boldsymbol{r}) &= \sum_{n,\boldsymbol{k}} c_n(\boldsymbol{k})(E_n(\boldsymbol{k}) + U)\psi_n(\boldsymbol{k}, \boldsymbol{r}) \\ &= \sum_{n,\boldsymbol{k}} c_n(\boldsymbol{k})(E_n(-\mathrm{i}\nabla) + U)\psi_n(\boldsymbol{k}, \boldsymbol{r}) = E\psi . \end{aligned} \tag{9.5}$$

Further rearrangement only becomes possible when we restrict the number of Bloch functions used to construct the wave packet.

As a first restriction we use only conduction band functions. This we can do if the energy with which the electron is bound to the defect is small compared to the energy with which the valence electrons are bound to the lattice (the energy gap E_G). In the other case, Bloch functions from the valence band have also to be included. Defects to which these restrictions can be applied are called *shallow impurities*. The condition is fulfilled for most donors. Defects with binding energies comparable with E_G act as traps and recombination centres (Sec. 9.1.8). The binding energies for the most important donors in Ge and Si are less than 1% of the band gap.

Thus in (9.5) we can omit the summation over n and hence the index itself. We find

$$[E(-\mathrm{i}\nabla) + U]\psi = E\psi \qquad \text{with} \quad \psi = \sum_{\boldsymbol{k}} c(\boldsymbol{k})\psi(\boldsymbol{k}, \boldsymbol{r}) . \tag{9.6}$$

The second restriction follows from (9.3). Replacement of the lattice with a homogeneous medium of given dielectric constant contains the assumption that the orbit of the electron bound to the defect traverses many lattice cells. The extent of the wave packet in space is thus large compared with the lattice constant. Consequently its extent in $\boldsymbol{k}$-space is small compared with the dimensions of the Brillouin zone. Thus only $\boldsymbol{k}$-vectors from a narrow region around the band minimum contribute to (9.4). If we look first at the case of a simple, isotropic parabolic minimum at $\boldsymbol{k} = 0$, the summation in (9.4) only runs over small values of $\boldsymbol{k}$. Since the lattice periodic part u in the Bloch function $\psi(\boldsymbol{k}, \boldsymbol{r}) = u(\boldsymbol{k}, \boldsymbol{r})\cdot\exp(\mathrm{i}\boldsymbol{k}\cdot\boldsymbol{r})$ only changes slowly with $\boldsymbol{k}$, we can replace $u(\boldsymbol{k}, \boldsymbol{r})$ by $u(0, \boldsymbol{r})$. We thus obtain the wave packet

$$\psi = \sum_{\boldsymbol{k}} c(\boldsymbol{k})u(0, \boldsymbol{r}) \exp(\mathrm{i}\boldsymbol{k}\cdot\boldsymbol{r}) = \Big[\sum_{\boldsymbol{k}} c(\boldsymbol{k}) \exp(\mathrm{i}\boldsymbol{k}\cdot\boldsymbol{r})\Big]\psi(0, \boldsymbol{r}) \equiv F(\boldsymbol{r})\psi(0, \boldsymbol{r}) . \tag{9.7}$$

If one puts this into (9.6) then with (2.143) and

$$E(-\mathrm{i}\nabla)F(\boldsymbol{r})\psi(0,\boldsymbol{r}) = \sum_m E_m F(\boldsymbol{r}+\boldsymbol{R}_m)\psi(0,\boldsymbol{r}+\boldsymbol{R}_m)$$
$$= \psi(0,\boldsymbol{r})\sum_m E_m \exp(\boldsymbol{R}_m\cdot\nabla)F(\boldsymbol{r}) = \psi(0,\boldsymbol{r})E_n(-\mathrm{i}\nabla)F(\boldsymbol{r}) \tag{9.8}$$

it follows that

$$[E(-\mathrm{i}\nabla) + U]F(\boldsymbol{r}) = EF(\boldsymbol{r}). \tag{9.9}$$

Equation (9.9) differs from (9.6) in that the rapidly changing function ψ has been replaced by the slowly changing "envelope-function" $F(\boldsymbol{r})$. One can then expand the operator $E(-\mathrm{i}\nabla)$ and terminate the expansion after the quadratic term

$$E(-\mathrm{i}\nabla) = E_C - \frac{\hbar^2}{2m^*}\nabla^2$$
$$\text{(with } E_C = \text{lower edge of the conduction band)}. \tag{9.10}$$

This leads here to

$$\left(-\frac{\hbar^2}{2m^*}\nabla^2 - \frac{e^2}{4\pi\varepsilon\varepsilon_0 r}\right)F(\boldsymbol{r}) = (E - E_C)F(\boldsymbol{r}), \tag{9.11}$$

where we have inserted the explicit form (9.3) for $U(r)$.

Equation (9.11) is the Schrödinger equation for an electron with effective mass m^* in the field of a positive charge in a medium of dielectric constant ε. We know the solution to this equation from the H-atom. The eigenvalues are

$$E_l = E_C - \frac{1}{l^2}\frac{e^4 m^*}{32\pi^2\varepsilon^2\varepsilon_0^2\hbar^2}, \qquad l = 1, 2, \ldots. \tag{9.12}$$

The envelope function for the ground state is

$$F(r) = \frac{1}{\sqrt{\pi a_0^{*3}}}\exp\left(-\frac{r}{a_0^*}\right), \qquad a_0^* = \frac{4\pi\varepsilon_0\hbar^2}{me^2}\frac{m}{m^*}\varepsilon. \tag{9.13}$$

The eigenfunctions for the excited states can readily be found in a similar way.

We note first of all that the bound states of the electron form a spectrum, similar to that of the H-atom, which lies below the lower edge of the conduction band. The Bohr radius for the orbit of the ground state is increased relative to that of the free H-atom (0.53 Å) by a factor $\varepsilon(m/m^*)$. For Si and Ge this leads to values between 20 and 50 Å. Further below we shall see that the binding energy for such orbits is reproduced fairly well by this approximation and that for the even larger orbits of the excited states there is almost exact agreement.

We obtain the extent of the wave packet in $\boldsymbol{k}$-space by evaluating the coefficients $c(\boldsymbol{k})$ of the expansion in (9.7) for known $F(\boldsymbol{r})$. We find for the coefficients

$$c(\boldsymbol{k}) = \frac{8\pi^{1/2}}{V_g^{1/2} a_0^{*5/2}} (k^2 + a_0^{*-2})^{-2} . \tag{9.14}$$

The wave packet is limited in $\boldsymbol{k}$-space to a region having a radius of the order of $1/a_0^*$. In Si and Ge this is only a few percent of the mean radius of the Brillouin zone.

The results (9.12) and (9.14) suggest that in the $E(\boldsymbol{k})$-diagram of the band model we should represent *donor levels* as discrete states below the minimum of the conduction band. The width of the level [given by (9.14)] can also be shown on the diagram, as a measure of the concentration of the wave packet (9.7) in $\boldsymbol{k}$-space. This is shown in Fig. 9.2 for the case discussed so far of a donor atom in a semiconductor with isotropic, parabolic conduction band minimum.

We have used Si and Ge as our examples of the order of magnitude to be expected for the binding energies and Bohr radii, although these semiconductors show marked deviations from the isotropic model. We can readily find the corrections needed when we note that these corrections concern assumptions we made in the context of the effective mass approximation. In Si and Ge we find, instead of an isotropic, parabolic minimum, a number of equivalent minima centred around the $\boldsymbol{k}$-vectors of a star in the Brillouin zone. The energy surfaces

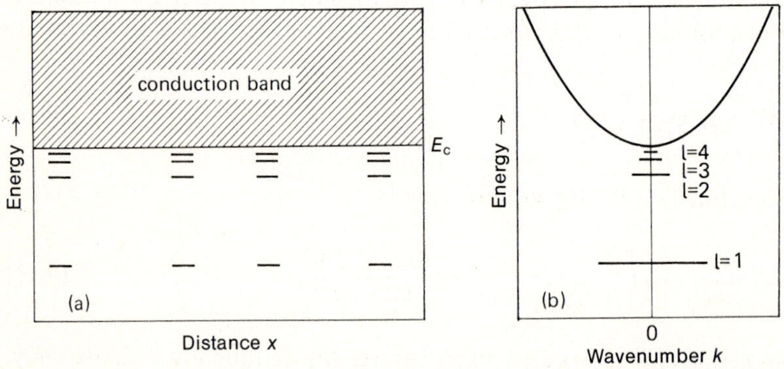

Fig. 9.2a and b. Impurity levels in the band model. (a) In an energy-space diagram, the localized impurity levels lie below the delocalized conduction band states. (b) In an E-k diagram, the localized impurity levels lie below the minimum of the conduction band (here assumed isotropic and parabolic). The width of the impurity levels gives the extent of the state in $\boldsymbol{k}$-space, according to (9.14).

around these minima are rotational ellipsoids, which are defined by two effective masses

$$E = E_C + \frac{\hbar^2}{2}\left(\frac{k_i^2}{m_l} + \frac{k_j^2 + k_k^2}{m_t}\right). \tag{9.15}$$

Here m_l is the longitudinal and m_t the transverse mass (reciprocal of the curvature parallel and perpendicular to the major axis of the ellipsoids). k_i, k_j, k_k are the components of $\boldsymbol{k}$ in a suitably oriented set of cartesian coordinates, with centre at the respective $\boldsymbol{k}$-vector of the minimum $\boldsymbol{k}_C$. We handle this anomaly by setting up an effective mass equation (9.11) for each extremum, and write the kinetic energy operator in the form

$$-\frac{\hbar^2}{2m_l}\frac{\partial^2}{\partial x_i^2} - \frac{\hbar^2}{2m_t}\left(\frac{\partial^2}{\partial x_j^2} + \frac{\partial^2}{\partial x_k^2}\right). \tag{9.16}$$

The elliptical shape of the energy surfaces leads to a preferred direction. Eigenvalues of the Schrödinger equation (9.11) which were degenerate in the isotropic case then split. The split levels are classified by a further quantum number m, which corresponds to the magnetic quantum number in the free atom case. Deviations from the Coulomb spectrum thus appear.

In Fig. 9.3 we compare results from a theory improved in this way with

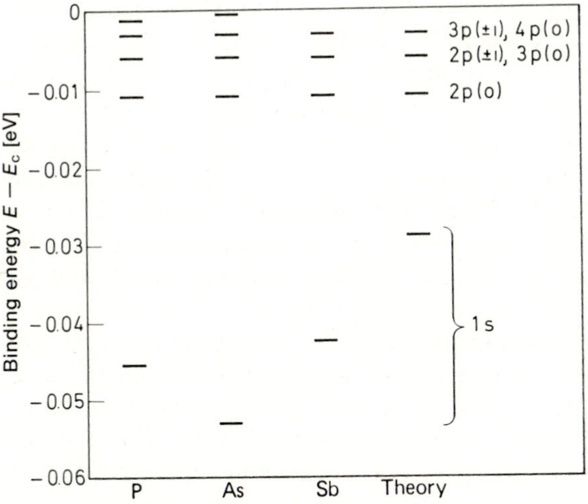

Fig. 9.3. Donor levels in silicon. Experimental values for P-, As-, and Sb-impurities. The energies of the excited states ($l = 1$, $m = -1, 0, +1$) agree well with each other and with theoretical predictions. The approximation used in this section fails as far as the ground state is concerned (from *Kohn* [101.5]).

donor spectra in Si. One can see an almost quantitative agreement between theoretical predictions and experiment for the excited states. The agreement is not so good for the ground state. Here the approximation (9.3) for the potential $U(r)$ is not adequate. It is apparent that the discrepancies stem from the failure of this approximation since the energies of the excited states are practically independent of the nature of the donor, while the energy of the ground state is different for P, As, and Sb.

So far we have restricted our attention to *shallow donors.* A corresponding effective mass equation (9.11) can be set up for *shallow acceptors* (replacement of electrons in the conduction band by holes in the valence band). $F(\boldsymbol{r})$ is then the envelope function of a wave packet constructed out of states from the valence band. The Coulomb spectrum (9.12) consists of a series of levels which lie above the valence band.

Here too we have to apply corrections to the simple theory when—as in all cubic semiconductors—several valence bands are degenerate at their upper edge. This leads to a new problem: In many semiconductors, nonequivalent band extrema lie close in energy. In the cubic semiconductors, for example, below the maximum of the valence band lies the maximum of a further subband, likewise at $\boldsymbol{k} = 0$. At only a small energy separation above the lowest conduction band minimum at $\boldsymbol{k} = 0$ in GaAs there are a number of equivalent minima lying along the Δ-axes. Defect levels can also occur near these extrema. As long as such states lie within the valence or conduction band they are degenerate with the delocalized Bloch states. An electron, therefore, only remains bound for a short time. Or put another way: the lifetime in such *resonance states* is finite. Associated with this is a broadening of the energy levels of these states. Fig. 9.4 shows examples. It also shows a discrete level to which states of different extrema collectively contribute. For further details see *Bassani* et al [111c.37, 113b, c]. We also refer the reader to this and to an article by *Kohn* [101.5] for further information on the theory of shallow impurity states.

Defects with larger electron binding energies lead to levels which lie deeper in the energy gap. Electrons in these centres are strongly localized. An effective mass approximation is impossible. States from throughout the entire Brillouin zone and from several bands are needed to construct a wave packet. Defects such as these are called *deep defects* (see *Queisser* [103.XI] and *Pantelides* [103.XV]).

There is a further obstacle to setting up a theory for deep impurities. Impurity atoms—whether at lattice sites or interstitial—do not fit in well to the host lattice. A displacement of the equilibrium positions of neighbouring atoms arises when they are inserted. Hence $U(\boldsymbol{r})$ consists not only from contributions from the impurity atom itself, but also from the lattice atoms which surround it.

We restrict ourselves to the example of the insertion of nitrogen atoms into the diamond lattice. The N-atoms are built in substitutionally. The lattice configuration which results is unstable. The four neighbouring lattice atoms are

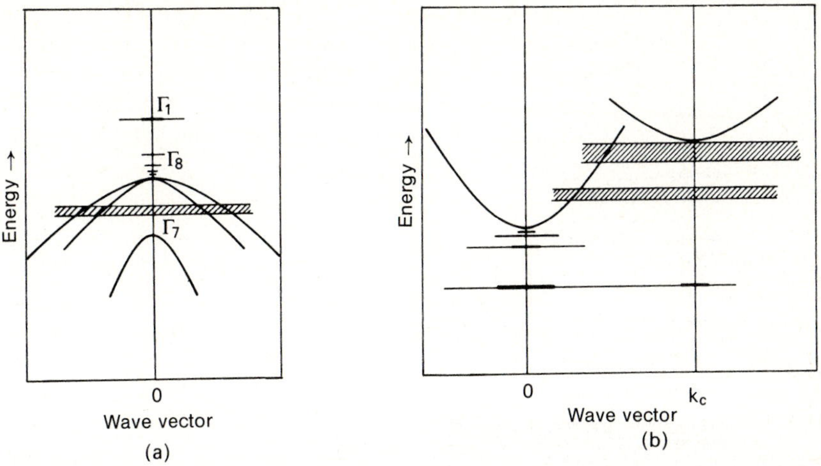

Fig. 9.4a and b. Impurity levels in the energy gap and resonances in the bands for (a) a degenerate valence band, and (b) two energetically close conduction bands with minima at different points in the Brillouin zone. The resonances in the bands are unstable. This is indicated by a "lifetime-broadening" of the levels (from *Bassani* [113b]).

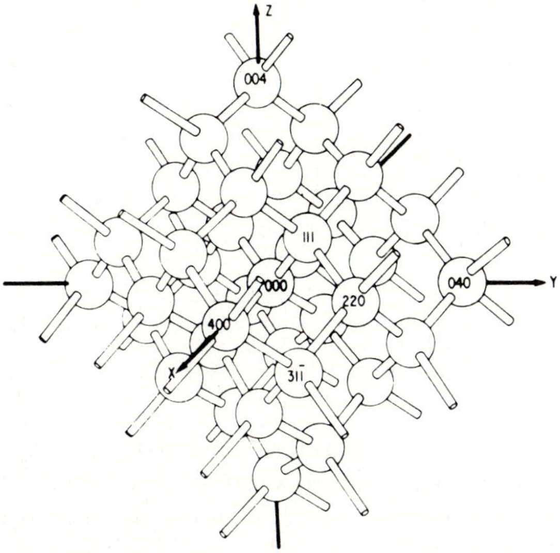

Fig. 9.5. "Molecule" made up of 35 C-atoms, as a model lattice for the approximate computation of the levels of deep impurities.

displaced trigonally. The new equilibrium state possesses lower total energy (Jahn-Teller effect, cf. following section).

We show the results of a model calculation in Figs. 9.5 and 9.6. The energy levels of a "molecule" with 35 C-atoms were calculated numerically. Fig. 9.5 shows the arrangement of the atom in the molecular cluster. Fig. 9.6a presents the associated "band structure". It is already in fairly good agreement with the

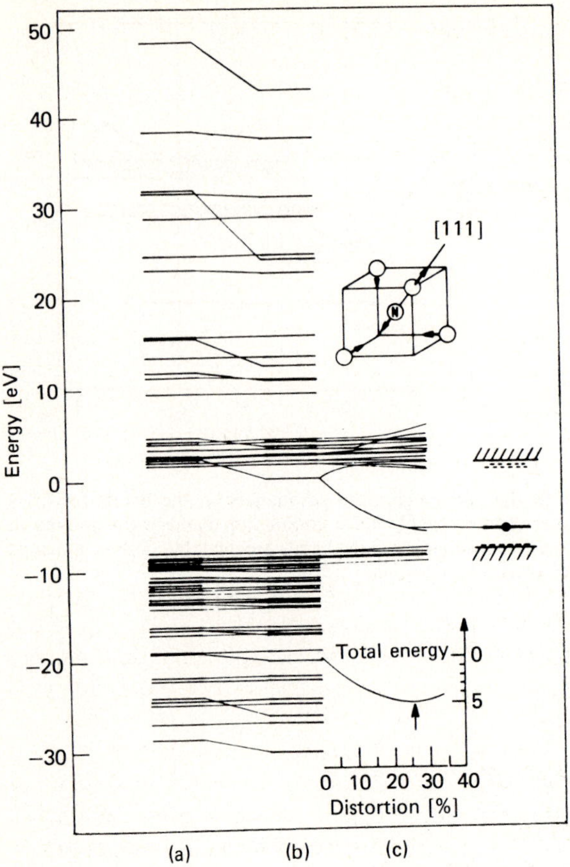

Fig. 9.6. (a) Energy levels of the "C-molecule" of Fig. 9.5. (b) The same spectrum after replacing the central C-atom with a N-atom. (c) Consideration of the Jahn-Teller shift. Diagram below right: Total energy as a function of the Jahn-Teller distortion (from G. D. Watkins, R. P. Messmer: Proc. X. Int. Conf. Semiconductor Physics, Cambridge, Mass. 1970).

known data on the width of valence band, conduction band, and energy gap for diamond. The cluster can be viewed as an approximation to the diamond crystal. Fig. 9.6b shows the scheme of levels for the same cluster, but where the central atom has been substituted by an N-atom. A level splits off (with localized wave function) from the conduction band. Fig. 9.6c illustrates how the energy of this level depends on the trigonal distortion of the vicinity of the impurity atom. The total energy has a maximum at a distortion of about 26%. At this value the defect level lies 2.2 eV above the valence band. The experimental value is 1.5 eV. One sees the need to consider the lattice distortion in the vicinity of a deep defect in order to get a reasonable value for the binding energy. The appearance of further shallow levels just below the conduction band

and just above the valence band is connected with the Jahn Teller shift of the deep level.

This example lays bare the difficulties which face a theory of deep defects. The potential of the one electron equation (9.2) can no longer be treated as a small perturbation. The relationships only become simpler in the other limit where $U(\mathbf{r})$ is not a small perturbation of the periodic potential, but instead $V(\mathbf{r})$ is a small perturbation of the defect potential. We now examine this case.

9.1.3 Crystal Field Theory

In the last section we inquired about the influence that the potential of an isolated defect would have on the one-electron states of the band model. In this, $U(\mathbf{r})$ was only a small perturbation in the Hamiltonian of (9.2). We now want to ask the reverse question: To what extent is the energy scheme of a free atom influenced by its incorporation into a lattice with given symmetry?

The Hamiltonian of the free atom is invariant to all rotations and reflections in space which leave the position of the nucleus unchanged. The group of the Hamiltonian is the (infinite) three-dimensional rotation group. The degeneracies of the energy levels of the free atom are given by the irreducible representations of this group. If the atom is substituted at a lattice site, the point group of the lattice determines the symmetry-induced degeneracies of the atomic energy levels. The most important effect we have to consider is thus the *splitting of atomic levels in the crystal field.*

Crystal field theory (also called ligand field theory in molecular physics) is concerned with examining the effect of the electric field of symmetrically arranged neighbouring atoms on a particular lattice atom (substitutional defect or atom of the host lattice). All other interactions with the neighbouring atoms are neglected. Thus we do not encompass with this theory the valence electrons handled in the last section, but electrons in deeper-lying partially filled shells. Transition metal ions with incomplete *f*-shells are examples.

Equation (9.2) does not give us an adequate point from which to start. We are not interested in the movement of a single electron in the field of the ion and its surroundings. We rather have to examine all the electrons of the atom, at least the electrons in the incomplete shells. For this purpose we first write down explicitly all the interaction terms in the Hamiltonian. For even when the crystal field is taken to be only a small perturbation, the order of magnitude of the contributions from individual terms in the Hamiltonian must first be estimated to facilitate the choice of an adequate approach to the solution of the Schrödinger equation.

The full Hamiltonian is

$$H = H_{\mathrm{kin}} + H_{\mathrm{el-nucl}} + H_{\mathrm{el-el}} + H_{\mathrm{SO}} + H_{\mathrm{CF}} + H_{\mathrm{M}} + H_{\mathrm{ex}} \tag{9.17}$$

where H_{kin} is the kinetic energy of all electrons of the atom considered $\sum_i p_i^2/2m$, $H_{\mathrm{el-nucl}}$ is the interaction of all electrons with the nucleus $-\sum_i Z_k e^2/4\pi\varepsilon_0 r_{ik}$,

H_{el-el} is the electron-electron interaction in the atom: $(e^2/8\pi\varepsilon_0)\sum'_{ij} 1/r_{ik}$, H_{SO} is the spin-orbit interaction $\sum_{i<j} r_{ij}\boldsymbol{l}_i \cdot \boldsymbol{s}_j$, H_{CF} is the influence of the crystal field, H_M is the magnetic interaction between electrons and nucleus, and H_{ex} is the interaction with external (electric or magnetic) fields.

We do not consider H_M and H_{ex} further here. H_M is smaller than all other contributions, except in special cases. H_{kin} and $H_{el-nucl}$ represent the principal contributions. We combine them in the zeroth approximation into an operator H_0. H_{el-el}, H_{SO}, and H_{CF} are perturbations of H_0. The approach which suggests itself is to take first only one of the three operators along with H_0, then to use perturbation techniques to extend the solutions further by accommodating the contribution from a second operator, etc.

For the free atom we assume that the splitting of levels, degenerate in the zeroth approximation, due to electron-electron interaction is larger than that due to spin-orbit interaction. For the atom incorporated in the lattice we can then distinguish three cases.

1) *Strong crystal fields*: $H_{CF} > H_{el-el} > H_{SO}$.
2) *Medium crystal fields*: $H_{el-el} > H_{CF} > H_{SO}$.
3) *Weak crystal fields*: $H_{el-el} > H_{SO} > H_{CF}$.

The three cases decide the point at which the crystal field has to be considered in the perturbation calculation.

The incorporation of an impurity atom into a host lattice also disturbes the symmetry of the surroundings. The crystal field need not therefore be identical with the field that a lattice atom "sees". If the lattice distortion is symmetric, i.e., if the substitution only changes the distances to the nearest neighbours, the point group of the host lattice continues to determine the crystal field splitting. Asymmetric lattice distortions change the symmetry in the vicinity of the atom considered. Such distortions have often to be taken into account. The reason can be seen from the following: First we fix the equilibrium positions of neighbouring atoms and substitute a lattice atom by a impurity atom. The atomic levels split up in the crystal field. Next we allow neighbouring atoms to be displaced, while retaining lattice symmetry. This results in a shift in the energy levels, but no further splitting. If a reduction in the energy of the ground state is associated with the displacements, a symmetrical lattice distortion will lead to a new equilibrium state. As a further step we permit displacements which reduce the symmetry in the vicinity of the foreign atom. If the ground state is not degenerate, such displacements will be associated with an increase in energy. If it is degenerate, the reduction in symmetry will lead to a splitting up of the ground state. For weak splitting the centre of gravity of the levels is unchanged. One of the split levels must therefore lead to a lower ground state energy. This is a special case of the general *Jahn-Teller theorem*, according to which each arrangement of atoms with a degenerate ground state of the electron system is unstable, except for pure spin degeneracy or a linear arrangement of the atoms. We met an example of this in the last section. For a general presentation on the Jahn-Teller effect see *Sturge* [101.20].

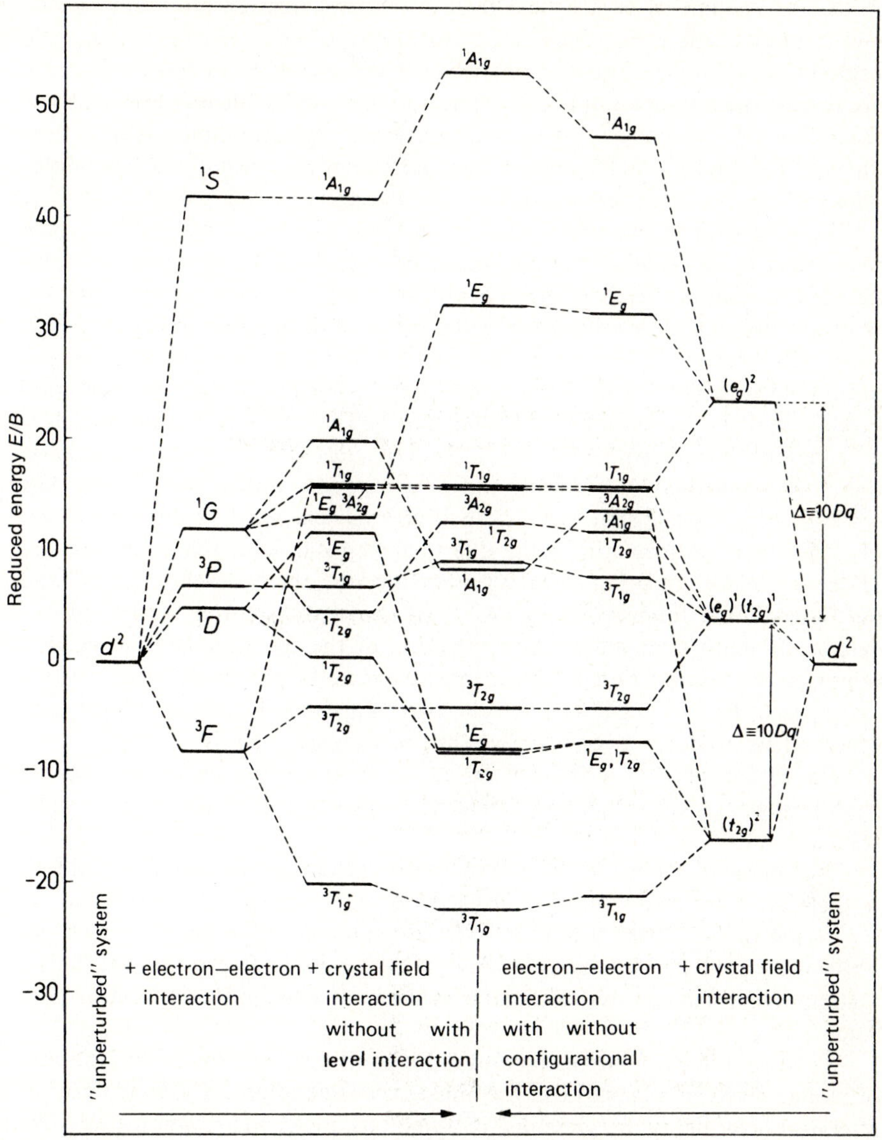

Fig. 9.7. Steps in the approximate calculation of the splitting of a d^2-level in a cubic lattice according to crystal field theory. Ordinate: Energy, abscissa: from left: starting from the weak crystal field approximation, from right: starting from the strong crystal field approximation [from H. L. Schläfer, G. Gliemann: Einführung in die Ligandenfeldtheorie (Akademische Verlagsanstalt, Frankfurt 1967)].

As an example of crystal field theory, we shall treat an atom having two d-electrons outside closed shells in a host lattice of cubic symmetry (O_h). We neglect spin-orbit coupling and restrict ourselves therefore to cases a) and b). We discuss the results using Fig. 9.7. The notation used to identify levels there is the following: A_1 and A_2 are one-dimensional representations, E is a two-dimensional, and T_1 and T_2 are three-dimensional representations. The multiplicity $2S + 1$ is given above left. The index g refers to even parity.

We start with the case of a *weak crystal field.* As a first step we have then to consider the splitting of a doubly occupied d-level in the free atom. The splitting of the resulting levels in the crystal field is taken account of subsequently. In the atom each d-electron has the ten states with quantum numbers $l = 2$, $m = -2, -1, 0, 1, 2$, $s = \pm 1/2$. The electrons obeying the Pauli principle can occupy 45 different states. These states are classified according to their total orbital angular momentum L and total spin S. There is one ^1S-state with $L = S = 0$, nine ^3P-states with $L = S = 1$, five ^1D-states with $L = 2$, $S = 0$, twenty-one ^3F-states with $L = 3$, $S = 1$, and nine ^1G-states with $L = 4$, $S = 0$. The d^2-level (Fig. 9.7 left) splits up into the five levels classified by different values of L. Further splitting takes place due to the crystal field. While the ^1S- and ^3P-levels remain singlet, the ^1D-level splits into two, the ^3F-level into three, and the ^1G-level into four levels. Fig. 9.7 (left) shows the splitting separately calculated for both interactions. Consideration of the so-called level-interaction improves the results: Levels with identical symmetry properties (here the two $^1A_{1g}$-levels, the two ^1E_g-levels, and the two $^3T_{1g}$- and $^1T_{2g}$-levels) "repel each other". This produces a considerable shift of the energy levels.

We do not want to look in detail at the *strong crystal field* case. Fig. 9.7 (right) shows the individual approximation steps which lead finally to the same result.

The energy levels in Fig. 9.7 and the steps in the approximations which lead to them are calculated with three given parameters: the splitting Δ between the $(t_{2g})^2$- and $(e_g)^1(t_{2g})^1$-levels and two "Racah-parameters" B and C, which describe the separations between the S-, P- . . . levels in the free atom. One finds a quantitative discussion of this example in the book cited in the caption for Fig. 9.7.

We have limited ourselves here to a qualitative explanation of the splitting of atomic levels in a crystal using symmetry considerations. The fundamentals of crystal field theory are presented in numerous books and review articles. We refer the reader especially to articles by *Bassani* in [113b], *Herzfeld* and *Meijer* in [101.12], and *McClure* in [101.9]. The article by *McClure* pays particular attention to the optical spectra of individual ions in crystals.

9.1.4 Localized Lattice Vibrations

Just like the electron states in the band model, the states in the vibrational spectrum of the lattice are modified by isolated defects. The most important results from Section 9.1.2 can be adopted: slight influence of defects on the states in the branches of the phonon spectrum; appearance of localized states between the acoustic and optical branches and above the optical branches; and possibility of resonance states within the branches.

We look first at the appearance of localized states with the simple example of a linear, infinite chain of identical spheres (mass M) with uniform separation a connected by springs with force constant f (cf. Fig. 3.6a). We therefore link up with the discussion in Section 3.3.2. For the unperturbed chain we found the frequency spectrum in (3.112) to be

$$\omega(q) = 2\sqrt{\frac{f}{M}}\left|\sin\frac{qa}{2}\right| = \omega_0\left|\sin\frac{qa}{2}\right| . \tag{9.18}$$

We have here introduced the upper frequency limit $\omega_0 = 2\sqrt{f/M}$.

We now assume that the sphere $n = 0$ has a slightly different mass $M_0 = M(1 - \varepsilon)$. ε can be positive or negative. Instead of (3.109), we then have the equations of motion

$$\begin{aligned} M_0\ddot{s}_0 &= f(s_1 + s_{-1} - 2s_0)\,, \\ M\ddot{s}_n &= f(s_{n+1} + s_{n-1} - 2s_n) \quad n \neq 0\,. \end{aligned} \tag{9.19}$$

Using the time dependence $s_n \propto \exp(-i\omega t)$, we can put it into the form

$$\begin{aligned} s_1 + s_{-1} + \left[\frac{4\omega^2}{\omega_0^2}(1-\varepsilon) - 2\right]s_0 &= 0\,, \\ s_{n+1} + s_{n-1} + \left(\frac{4\omega^2}{\omega_0^2} - 2\right)s_n &= 0 \quad n \neq 0\,. \end{aligned} \tag{9.20}$$

For the n-dependence of the displacements we choose the ansatz

$$s_n = A\lambda^n + B\lambda^{-n}\,. \tag{9.21}$$

Inserting this into (9.20) gives the following secular equation for λ:

$$\lambda^2 + 1 + \left(\frac{4\omega^2}{\omega_0^2} - 2\right)\lambda = 0\,. \tag{9.22}$$

Only solutions with $\omega < \omega_0$ are feasible for the undisturbed chain. λ then becomes complex. If one writes λ in the form $\exp(i\alpha)$, one finds that the solutions of the equation of motion are plane waves with frequencies ω, related through (9.18) with $\alpha = qa$.

For the perturbed chain we likewise restrict ourselves first to the region $\omega < \omega_0$. We can always construct a general solution out of two standing waves, one symmetric ($s_n = s_{-n}$) and the other antisymmetric ($s_n = -s_{-n}$). For the second case the perturbing atom at $n = 0$ is at rest. This part is not altered by the perturbation We only look therefore for solutions with $s_n = s_{-n}$. For these we can write

$$s_n = A \exp(iqa|n|) + B \exp(-iqa|n|). \tag{9.23}$$

The first equation (9.20) then becomes

$$\frac{A \exp(iqa) + B \exp(-iqa)}{A + B} + \left[\frac{2\omega^2}{\omega_0^2}(1 - \varepsilon) - 1\right] = 0. \tag{9.24}$$

One can readily show that for $\varepsilon = 0$ (unperturbed chain), $A = B$. For the perturbed chain we put

$$A = \frac{C}{2} \exp(-i\delta), \qquad B = \frac{C}{2} \exp(i\delta). \tag{9.25}$$

From (9.23) and (9.24) the following equations result:

$$s_n = C \cos(|n|qa - \delta) \quad \text{and} \quad \tan\delta = \varepsilon \tan\frac{qa}{2}. \tag{9.26}$$

This means only a slight change compared with the solutions for the unperturbed chain.

What is more important is that a solution is now possible for $\omega > \omega_0$. λ is then real and negative! One of the two real solutions of (9.22) is less than unity, the other (reciprocal) greater than unity. Both yield the same result when one exchanges A and B. We can thus restrict ourselves to one solution and choose it to be the (negative) λ with $|\lambda| < 1$.

We must now distinguish between the two cases $n > 0$ and $n < 0$ in (9.21). In each case one of the two terms diverges for $|n| \to \infty$. The associated amplitude must then vanish. We put

$$s_n = A\lambda^n \quad \text{for} \quad n > 0, \qquad s_n = B\lambda^{-n} \quad \text{for} \quad n < 0 \quad (|\lambda| < 1). \tag{9.27}$$

If we eliminate the displacement s_0 from the two equations (9.20) with $n = 1$ and $n = -1$, we are left with a relation between the displacements s_2, s_1, s_{-1}, s_{-2}. If we use (9.27) in this relation, we find that the amplitudes A and B of (9.27) must be equal. If we then insert (9.27) into the first equation (9.20), we find

$$2\lambda + \frac{4\omega^2}{\omega_0^2}(1 - \varepsilon) - 2 = 0. \tag{9.28}$$

This is a relation between λ, ω, and ε. ω can be eliminated using (9.22). Then

$$(\lambda - 1)\left[2 - \frac{\lambda - 1}{\lambda}(1 - \varepsilon)\right] = 0, \tag{9.29}$$

or since $\lambda \neq 1$

$$\lambda = \frac{\varepsilon - 1}{\varepsilon + 1}. \tag{9.30}$$

Inserting this solution into (9.27) and (9.22) gives

$$s_n = s_0(-1)^n\left(\frac{1 - \varepsilon}{1 + \varepsilon}\right)^{|n|} \quad \text{and} \quad \omega^2 = \frac{\omega_0^2}{1 - \varepsilon}. \tag{9.31}$$

This solution describes an oscillation in which neighbouring spheres vibrate in opposite direction (Fig. 9.8). The displacements diminish with increasing $|n|$. The vibration is localized!

We must pay attention here to our assumption that λ is negative and $|\lambda| < 1$. From (9.30) this means that a solution only exists in the region $\omega > \omega_0$, if

(a)

(b)

(c)

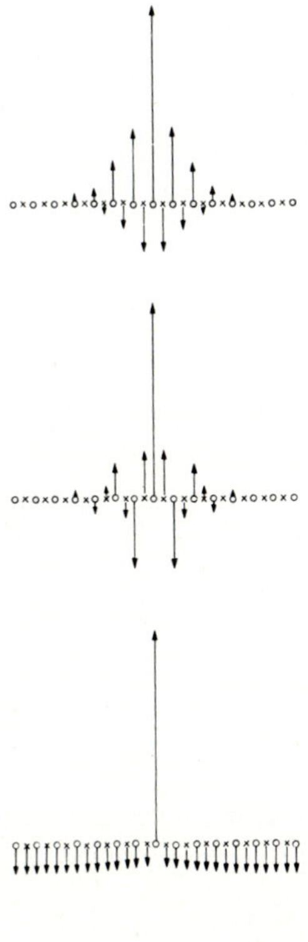

Fig. 9.8. Localized lattice vibrations: (a) for identical atoms in the chain—localized vibrations above the frequency limit ω_0, for a chain with a basis—localized vibrations above the optical branch; (b) localized vibrations between the acoustic and optical branch (gap modes); (c) resonance in the acoustic branch (from *Sievers* [124]).

ε is positive and less then unity. Localized vibration states are only possible when the mass M_0 of the perturbing atom is less than the mass M of a lattice atom.

Even if this one-dimensional example is not applicable in all respects to the three-dimensional case, it does show the most important properties of localized lattice vibrations. The uppermost level $\omega = \omega_0$ of the unperturbed spectrum is split by the perturbation, and moves with increase in $\varepsilon = (M - M_0)/M$ to higher frequencies; the vibration which is a delocalized plane wave in the unperturbed spectrum becomes localized.

The same result can be obtained by a different route, one which can readily be generalized to apply to the three-dimensional case. Assuming $s_n \propto \exp(-i\omega t)$, one writes (9.19) as

$$-\omega^2 M s_n - f(s_{n+1} + s_{n-1} - 2s_n) = -M\varepsilon\omega^2 s_0 \delta_{n0} \equiv F_n . \tag{9.32}$$

This equation can be read as an equation of motion for an unperturbed chain with an external force $F_n = -M\varepsilon\omega^2 s_0 \delta_{n0}$ acting on the nth sphere. By Fourier transformation

$$s_n = \frac{1}{\sqrt{N}} \sum_q s(q) \exp(-iqan) , \qquad F_n = \frac{1}{\sqrt{N}} \sum_q F(q) \exp(-iqan) \tag{9.33}$$

(N—number of spheres in the chain, periodic boundary conditions), one finds the equation

$$-\omega^2 M s(q) + \left(4f \sin^2 \frac{qa}{2}\right) s(q) = F(q) \tag{9.34}$$

hence

$$s(q) = \frac{F(q)/M}{\omega_q^2 - \omega^2} ; \qquad \omega_q^2 \equiv \frac{4f}{M} \sin^2 \frac{qa}{2} . \tag{9.35}$$

Multiplying by $N^{-1/2} \exp(-iqam)$ and summing over q yields

$$s_m = \frac{1}{MN} \sum_{qn} \frac{F_n \exp[iqa(n - m)]}{\omega_q^2 - \omega^2} . \tag{9.36}$$

Inserting F_n from (9.32), it follows for $m = 0$ that

$$\frac{\omega^2}{N} \sum_q \frac{1}{\omega^2 - \omega_q^2} = \frac{1}{\varepsilon} . \tag{9.37}$$

This equation has the same structure as (5.14). It can be solved graphically. This is done in Fig. 9.9. One sees again the same result as obtained above:

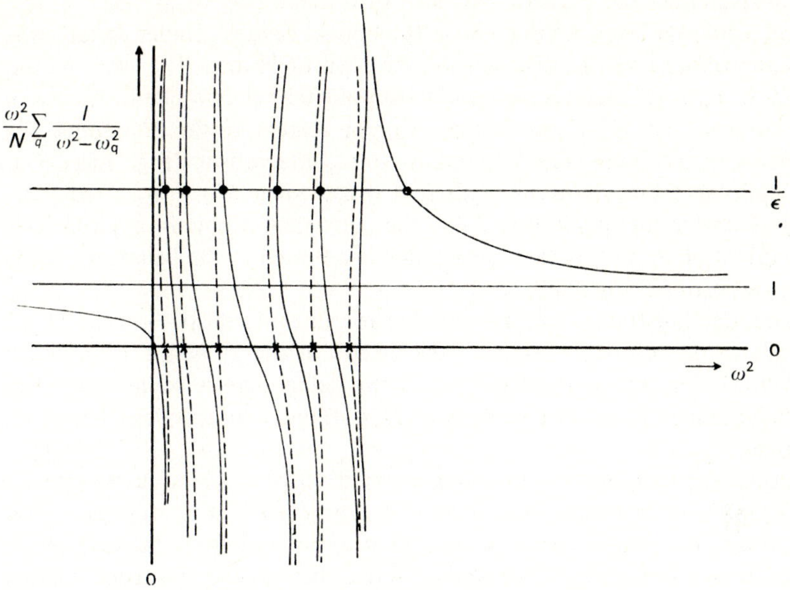

Fig. 9.9. Graphical solution of (9.37) (from *Pryce* [133.II]).

For $\omega < \omega_0$ one finds only weak displacements compared with the case $\varepsilon = 0$; for $\omega > \omega_0$ a solution splits off if $M_0 < M$.

Equation (9.37) can be extended to the three-dimensional case. For cubic lattices one has only to replace the scalar k by a vector $\boldsymbol{k}$, to extend the summation to cover all branches of the phonon spectrum, and to divide by the number of branches.

The one-dimensional example examined so far is limited to a chain with identical spheres. Only an acoustic branch then appears, extending over a frequency range from zero to the frequency limit ω_0. It can thus be seen that defects with positive ε (smaller mass M_0) alone give rise to localized lattice vibrations. As illustrated in Fig. 9.1, discrete levels split off from the bottom or top of a band, depending on whether the potential (the mass) of the defect is larger or smaller than the lattice potential (larger or smaller than the mass M). Since the acoustic branch here starts at zero frequency, it is only possible for a localized vibration to split off "upwards". This is caused by the defects with $M_0 < M$.

According to these considerations in lattices with a basis, which have both acoustic and optical branches in their vibration spectrum, localized vibrations must not only occur above the highest frequency, but also in the energy gap between acoustic and optical branches (gap modes). A one-dimensional model with diatomic basis (Figs. 3.6b and 3.7) already shows this. We shall not perform the calculations again here. From Fig. 9.8 the result is the following: if the mass

M_0 is smaller than the mass of the atom substituted (M_1 or M_2, $M_1 > M_2$), localized vibration levels appear above the optical branch. Under certain conditions a localized level can also split off the acoustic branch. The level splitting off from the optical branch is the uppermost optical level associated with $q = 0$. In the latter all the M_1-atoms vibrate in phase relative to the M_2-atoms. The corresponding localized vibration retains this characteristic (Fig. 9.8a). Gap modes split off from vibrations with $q = \pm\pi/a$ when $M_1 > M_0 > M_2$. According to the results of Section 3.3.2, the equivalent atoms in adjacent cells vibrate in antiphase in this case. Again this is reflected in the behaviour of the localized vibrations (Fig. 9.8b).

Figure 9.8c illustrates a resonance vibration in the acoustic branch. This is analogous to the resonance states of the defects in energy bands, discussed in Section 9.1.2. The strong displacement of the defect here is connected with a small displacement of numerous neighbouring atoms (long wavelength acoustic vibrations).

Localized lattice vibrations are often infrared active and can thus be detected in the absorption spectrum of a crystal. Energy conservation applies for the optical transition. Wave vector conservation is not necessary because of the distorted lattice periodicity. One finds discrete lines in the spectrum. Isotope defects are examples of cases to which the above model (extended to three dimensions) can be applied—these are lattice atoms which differ from the other lattice atoms only in their mass, and not in their chemical nature. In the case of impurity atoms, along with a change of mass, at least a change of the force constants f connecting them to the neighbouring atoms in the host lattice is also involved.

The absorption lines of localized vibrations are important in the analysis of the defects and defect complexes present in a crystal. As an example we look at Si atoms as defects in GaAs. A tetravalent Si-atom can occupy a (trivalent) Ga-site as readily as a (pentavalent) As-site. Along with Si_{Ga} and Si_{As} it is also possible to have defect pairs on adjacent sites (Si_{Ga}–Si_{As}). Si_{Ga} and Si_{As} defects act as donors and acceptors, and thus deliver free charge carriers to the lattice. If the absorption lines of interest are not to be swamped by simultaneous free carrier absorption, electrically active Si defects have to be compensated by the introduction of other defects. In the example in Fig. 9.10, the desired effect is achieved by the diffusion of ^{7}Li or by electron bombardment. The spectrum correspondingly shows numerous discrete lines which can be associated with agglomerates of the possible lattice defects.

Localized elementary excitations other than the localized phonons treated here are possible (localized magnons, plasmons, excitons). We particularly refer the reader to the conference report [124]. For further information on localized phonons, see the contributions by *Maradudin* in [101.18/19, 102.3], by *Spitzer* in [103.XI], and by *Price* in [133.II]. There are also many contributions in [117, 121]. We shall be examining localized excitons in Section 9.1.11.

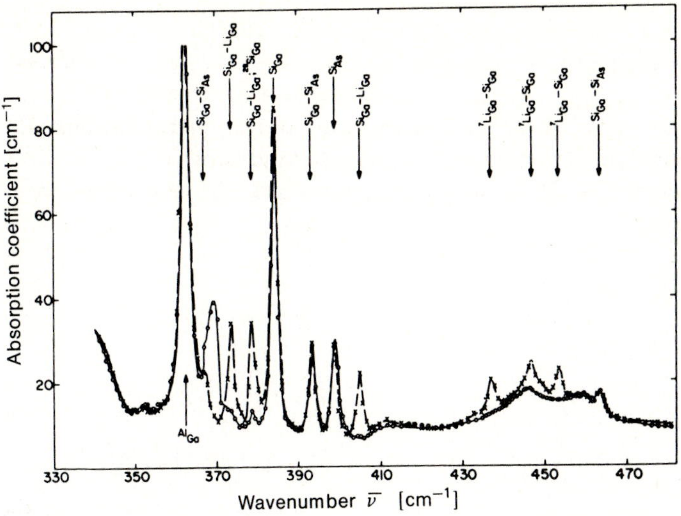

Fig. 9.10. Absorption spectrum of Si-doped GaAs at low temperature (—— compensation of the Si-impurities with Li, ---- compensation by introducing lattice defects by ion bombardment) (from *Spitzer* [103.11]).

9.1.5 Defect Statistics, Reaction Kinetics

If one describes defects by localized levels in the band model as in Fig. 9.2, one has to include these levels in the equilibrium statistics. From Sections 2.1.4 and 2.2.10 the concentration of electrons in an energy interval dE at an energy E is given by the Fermi distribution (2.30) and the density of states (2.152). Both equations were derived for electrons in an energy band $E_n(\boldsymbol{k})$ only.

Defect levels appear in the energy gap between conduction and valence band of a semiconductor. If the concentration of one sort of defect is n_d and the energy of the defect level E_d, the contribution of the defects to the density of states is

$$g_\mathrm{d}(E)dE = n_\mathrm{d}\delta(E - E_\mathrm{d})dE\,. \tag{9.38}$$

The Fermi distribution (2.30) needs a small correction for defects. If, for example, we examine a shallow impurity with a hydrogen-like spectrum, the outermost electron of such a defect can be inserted in two different ways according to its two possible spin directions. The occupied state of the defect is thus degenerate, the unoccupied nondegenerate. Conversely the spin of an electron bound to a defect can be determined by a second electron in an equivalent state in the defect. Then the occupied state is nondegenerate and the unoccupied degenerate.

If g_b and g_fr are the degrees of degeneracy of the occupied (bound) and free states, respectively, instead of (2.30), we find for the occupation probability, with $g = g_b/g_{fr}$

$$f(E) = (1 + g \exp[(E(\boldsymbol{k}) - \mu)/k_B T])^{-1}. \tag{9.39}$$

We thus obtain the following for the concentration of free charge carriers and of carriers in donors and acceptors in a doped semiconductor:

a) Electron concentration in the conduction band and hole concentration in the valence band

$$n = \int_{E_C}^{\infty} \frac{g(E)dE}{1 + \exp\left(\dfrac{E - \mu}{k_B T}\right)},$$

$$p = \int_{-\infty}^{E_V} \left[1 - \frac{1}{1 + \exp\left(\dfrac{E - \mu}{k_B T}\right)}\right] g(E)dE = \int_{-\infty}^{E_V} \frac{g(E)dE}{1 + \exp\left(\dfrac{\mu - E}{k_B T}\right)} \tag{9.40}$$

where $g(E)$ is given by (2.152).

b) Concentration of occupied (neutral) and unoccupied (positively charged) donors

$$n_{D^\times} = n_D\left[1 + g \exp\left(\frac{E_D - \mu}{k_B T}\right)\right]^{-1}; \quad n_{D^+} = n_D\left[1 + \frac{1}{g}\exp\left(\frac{\mu - E_D}{k_B T}\right)\right]^{-1}. \tag{9.41}$$

c) Concentration of neutral and negatively charged acceptors

$$n_{A^\times} = n_A\left[1 + g \exp\left(\frac{\mu - E_A}{k_B T}\right)\right]^{-1}; \quad n_{A^-} = n_A\left[1 + \frac{1}{g}\exp\left(\frac{E_A - \mu}{k_B T}\right)\right]^{-1}. \tag{9.42}$$

All these concentrations contain the chemical potential of electrons μ as a free parameter. In neutral semiconductors it is determined by the requirement that, in each volume element, the sum of electrons and negatively charged defects is on the average equal to the sum of holes and positively charged defects (*condition of neutrality*)

$$n + n_{A^-} = p + n_{D^+}. \tag{9.43}$$

In (9.40–42) we have transferred from a description of the distribution of electrons over all energy states available, to a description in which we have combined groups of electrons and groups of unoccupied states into "collectives".

We are thus now interested in the *total* number of electrons in the conduction band, of holes in the valence band, of electrons in the donors, etc. The concentration of the particles in such a collective can be determined for a given band structure (density of states), temperature, and defect concentration.

All this assumes that the energy states available to the electrons are known and that their concentration and energy do not depend on the electron distribution. This is not so when we include lattice disorder (vacancies and interstitials) which is itself subject to equilibrium conditions. At a given temperature not only are electrons distributed over given band and defect levels in accordance with Fermi statistics, but the total concentration of vacancies and interstitials is itself a function of temperature. The concentration of impurity atoms can also be a function of temperature, for example when the solid is in contact with a gaseous phase containing atoms which readily diffuse in.

To cope with such problems, we go to a different method of description (*reaction kinetics*). To do this we have to reformulate the concept of a "collective". We have in fact already deviated formally in (9.41) from the model of collectives of band states, between which electrons can transfer. Instead of electrons in the valence band, we introduced the holes in the valence band (describing them still by the chemical potential of the electrons). We now go a stage further. Instead of combining different groups of electrons into collectives, we combine all the particles in the solid which can "*react*" with each other into collectives. Along with the electrons (in the conduction band) and the holes (in the valence band), we regard the charged and uncharged defects likewise as special collectives. In this description, for example, the relaxation of an electron from conduction band into a donor level means the reaction of a free electron with a positively charged donor atom with formation of a neutral donor atom. We introduce the concentration n_p for the particles of the pth collective and associate a chemical potential μ_p with it. The conditions of equilibrium then require that the free energy F possess an extremal value

$$\delta F = \sum_p \mu_p \delta n_p = 0\,. \tag{9.44}$$

We consider a special "reaction"

$$\nu_1 P + \nu_2 Q \leftrightarrows (-\nu_3) R \quad (\nu_1 P + \nu_2 Q + \nu_3 R \rightleftarrows 0)\,, \tag{9.45}$$

where the $|\nu_i|$ give the number of particles of the collectives P, Q, or R participating in the reaction, with the sign distinguishing between particles vanishing and appearing in the reaction. The equilibrium condition is then

$$\nu_1 \mu_P + \nu_2 \mu_Q + \nu_3 \mu_R = 0\,. \tag{9.46}$$

In equilibrium, the sum of the chemical potentials multiplied by the numbers of particles involved in the reaction is equal to zero. For an explicit representation of this equilibrium condition, one usually divides the chemical potentials into concentration-dependent and -independent parts

$$\mu_p = E_p + k_B T \ln \frac{n_p}{n_p^0}, \tag{9.47}$$

where n_p^0 is an initially unknown reference concentration. This division is, of course, only possible as long as $n_p \ll n_p^0$, and thus limited as we shall see later to nondegenerate semiconductors with not too high a concentration of defects.

One chooses for the E_p the energies needed to take a particle from the deepest collective state to infinity, and to reduce it there into its independent constituents. The division (9.47) yields, when one inserts it into the generalized equation (9.46) ($\sum_p \mu_p \nu_p = 0$)

$$\prod_p \left(\frac{n_p}{n_p^0}\right)^{\nu_p} = \exp\left(-\frac{1}{k_B T}\sum_p \nu_p E_p\right). \tag{9.48}$$

This is the *law of mass action*, which describes the equilibrium between the different collectives. The mass action constant (apart from a factor formed from the n_p^0) is an exponential whose exponent contains the energy exchanged in the reaction.

We shall look at two examples and in the process compare the results from band model statistics with those from reaction kinetics. We restrict ourselves, in accordance with (9.47), to small electron and hole concentrations (nondegenerate semiconductors). Eq. (9.40) then simplifies to

$$n = n_0 \exp\left(\frac{\mu - E_C}{k_B T}\right), \quad p = p_0 \exp\left(\frac{E_V - \mu}{k_B T}\right), \quad \left.\begin{matrix} n_0 \\ p_0 \end{matrix}\right\} = 2\left(\frac{m_{n,p} k_B T}{2\pi\hbar^2}\right)^{3/2}. \tag{9.49}$$

If n or p exceeds the concentrations n_0, p_0, the electron- or hole-gas becomes degenerate.

1) Dissociation of a Donor: $D^+ + \ominus \rightleftarrows D^\times$

The law of mass action on the one hand gives

$$\frac{n n_{D^+}}{n_{D^\times}} = \frac{n^0 n_{D^+}^0}{n_{D^\times}^0} \exp\left[-(E_- + E_{D^+} - E_{D^\times})/k_B T\right]. \tag{9.50}$$

On the other hand, it follows from (9.49) and (9.41) by using the condition of neutrality $n = n_{D^+}$ and by eliminating the chemical potential, that

$$\frac{n^2}{n_D - n} = \frac{n_0}{g} \exp\left(-\frac{E_C - E_D}{k_B T}\right). \tag{9.51}$$

One equation leads to the other. The left sides agree because of the condition of neutrality. $E_{D^+} - E_{D^\times}$ is the energy needed to change a charged donor into an uncharged, i.e., it equals the energy needed to bring the electron from infinity to the donor. But this is just the energy $-E_D$ in the band model. E_- is equal to the energy of the lower edge of the conduction band E_C. The exponents then agree. Finally the coefficients on the right and the left are equal, if one takes the electron reference concentration to be n_0 and the coefficient g as the ratio of the reference concentrations $n_{D^\times}/n_{D^+}$.

2) Creation and Recombination of an Electron-Hole Pair: $\ominus + \oplus \rightleftarrows 0$
The law of mass action gives

$$np = n^0 p^0 \exp[-(E_- + E_+)/k_B T]. \tag{9.52}$$

E_- is the energy needed to take an electron from the edge of the conduction band to infinity, i.e., it equals E_C. E_+ is the energy needed to take a hole from the edge of the valence band to infinity, or to bring an electron from infinity to the band edge: $E_+ = -E_V$. If further one again takes the concentrations n_0 and p_0 as the reference concentrations, (9.52) becomes

$$np = n_0 p_0 \exp[-(E_C - E_V)/k_B T] = n_0 p_0 \exp\left(-\frac{E_G}{kT}\right). \tag{9.53}$$

This is the well-known relation between electron and hole concentrations in nondegenerate semiconductors. Within the framework of band model statistics, it follows by eliminating μ from (9.49).

The advantages and disadvantages of both methods can be seen in these examples. Reaction kinetics is free of the assumption of fixed defect levels and concentrations. On the other hand, band model statistics allow the reference concentrations to be calculated explicitly. Furthermore it is not restricted to low electron concentrations.

We shall also meet examples of the application of reaction kinetics in the next section.

9.1.6 Disorder Equilibria

The two types of lattice disorder "vacancies" and "interstitials" differ fundamentally from impurity atoms in that the latter, in the ideal case, can be completely removed from the crystal. Lattice disorder on the other hand is unavoidable. At a given temperature a distinct equilibrium is set up between these types of imperfections.

The formation of this type of lattice disorder—for example, the removal of a lattice atom from its lattice site and its insertion at some interstitial position—requires energy. As the energy of the crystal increases, so does the entropy.

The free energy, which decides the equilibrium condition for given volume and temperature, changes likewise.

First we determine the equilibrium concentrations for the simplest case. Consider a monatomic lattice; let temperature and volume be constant; let vacancies and interstitials be so far away from each other that we can take them to be independent. The crystal may have N lattice sites and N' locations where interstitial atoms can be accommodated.

At the so-called *Frenkel disorder*, lattice atoms are accommodated interstitially. Let the number of vacancy/interstitial pairs be n. The interstitials can be arranged in $N'!/(N'-n)!n!$ different ways, and the vacancies in $N!/(N-n)!n!$ different ways. This yields an increase in entropy

$$S_n = k_B \ln\left[\frac{N!}{(N-n)!n!}\frac{N'!}{(N'-n)!n!}\right]. \tag{9.54}$$

The energy to create the vacancy/interstitial pairs is $U_n = nW_F$, where W_F is the energy necessary to remove a lattice atom to a sufficiently distant inderstitial site.

Equilibrium is found by varying the free energy with respect to particle number. We find

$$\left.\frac{\partial F}{\partial n}\right|_T = \left.\frac{\partial U_n}{\partial n}\right|_T - T\left.\frac{\partial S_n}{\partial n}\right|_T = 0. \tag{9.55}$$

Inserting the values for U_n and S_n, and applying Stirling's theorem $\ln n! = n \ln n - n$ for large n, leads to an equation defining the equilibrium concentrations. It follows that

$$n = \sqrt{NN'} \exp(-W_F/2k_BT). \tag{9.56}$$

One can also interpret (9.56) as the result of the law of mass action applied to the reaction

occupied lattice site + unoccupied interstitial site

$\rightleftarrows$ vacancy + interstitial atom.

With the reaction energy W_F we find

$$\frac{(N-n)(N'-n)}{n^2} = \exp(W_F/k_BT), \tag{9.57}$$

and from this, (9.56) follows for $N, N' \gg n$.

At a *Schottky disorder*, only vacancies are created. The lattice atoms removed are brought to the surface. In the logarithm in (9.54), only the quotient with $N!$ occurs. If the energy needed to form a vacant site is W_S, we then find

$$n = N \exp(-W_S/k_B T). \tag{9.58}$$

In binary lattices AB, for example in ionic crystals with oppositely charged sublattices, several types of imperfections are possible: Frenkel disorder in the A-lattice; Frenkel disorder in the B-lattice; formation of pairs of vacancies V_A, V_B; formation of interstitial pairs I_A, I_B; exchange of two atoms between the two sublattices (A_B, B_A). The formation of pairs in each case is needed to preserve neutrality. Of the types of imperfections mentioned, only Frenkel disorder and vacancy pair-formation are important.

Let W'_S be the energy needed to form a pair of vacancies. Instead of (9.58), we have

$$n = N \exp(-W'_S/2k_B T). \tag{9.59}$$

This equation too can readily be explained as a consequence of the law of mass action.

We now extend our considerations of disorder equilibrium to the case where, as well as vacancies and interstitials, the crystal contains electrons, holes, and impurities, whose concentrations are in each case subject to equilibrium conditions. We further assume that the crystal is a binary compound of anions (A) and cations (C), and is in contact with a gas phase of A_2 molecules. A-atoms from the gas phase can be accommodated into the anion sublattice with simultaneous formation of vacancies V_C in the cation sublattice. Equilibrium between crystal and gas phase is determined by the reaction

$$\frac{1}{2}A_2^{(g)} \rightleftarrows V_C^{\times} + AC. \tag{9.60}$$

The concentration of cation vacancies is then given by the associated law of mass action

$$[V_C^{\times}] = K_A^{(g)} p_{A_2}^{(g)1/2}. \tag{9.61}$$

Here $p_{A_2}^{(g)}$ is the partial pressure of molecules A_2 in the gas phase and $K_A^{(g)}$ is the associated mass action constant. We have expressed the concentrations here, and in the formulae which follow, within square brackets.

Various reactions can take place in the crystal. First, cations and anions on interstitial sites (C_i, A_i), and cation- and anion-vacancies (V_C, V_A), can accept and release charge carriers. The reactions and their mass-action laws are

$$C_i^\times \rightleftarrows C_i^+ + \ominus \qquad \frac{[C_i^+]n}{[C_i^\times]} = K_{C_i}, \tag{9.62}$$

$$V_C^\times \rightleftarrows V_C^- + \oplus \qquad \frac{[V_C^-]p}{[V_C^\times]} = K_{V_C}, \tag{9.63}$$

$$A_i^\times \rightleftarrows A_i^- + \oplus \qquad \frac{[A_i^-]p}{[A_i^\times]} = K_{A_i}, \tag{9.64}$$

$$V_A^\times \rightleftarrows V_A^+ + \ominus \qquad \frac{[V_A^+]n}{[V_A^\times]} = K_{V_A}. \tag{9.65}$$

Further possible reactions and mass-action laws are: For Frenkel and anti-Frenkel disorder (vacancy/interstitial pairs in the cation or anion sublattice)

$$C_i^+ + V_C^- \rightleftarrows 0 \qquad [C_i^+][V_C^-] = K_C \tag{9.66}$$

$$A_i^- + V_A^+ \rightleftarrows 0 \qquad [A_i^-][V_A^+] = K_A. \tag{9.67}$$

For Schottky and anti-Schottky disorder (vacancy pairs or interstitial pairs)

$$V_C^- + V_A^- \rightleftarrows -CA \qquad [V_C^-][V_A^+] = K_V \tag{9.68}$$

$$C_i^+ + A_i^- \rightleftarrows +CA \qquad [C_i^+][A_i^-] = K_i. \tag{9.69}$$

Finally, between electrons and holes we have

$$\ominus + \oplus \rightleftarrows 0 \qquad np = K_{np}. \tag{9.70}$$

In the reaction equations, "0" means the undisturbed lattice. Thus (9.66) describes the creation of a vacancy/interstitial pair from the unperturbed cation sublattice and its destruction etc.

All these laws are characterized by mass-action constants. The K_α are functions of temperature and of the reaction energies (cf. the previous section).

So far we have only looked at disorder in a system without impurities. As an example for a system with additional impurities, we consider donors in the cation sublattice. The following equations apply:

$$D_C^+ + \ominus \rightleftarrows D_C^\times \qquad \frac{[D_C^+]n}{[D_C^\times]} = K_D. \tag{9.71}$$

Equations (9.62–71) are not sufficient to determine the equilibrium between the different defect types and the electrons and holes. In addition to the equation for equilibrium with an external phase (9.61), we must include the *neutrality condition*, and the requirement that the sum of neutral and charged impurities is equal to the total impurity concentration

$$n + [V_C^-] + [A_i^-] = p + [V_A^+] + [C_i^+] + [D_C^+] \quad (9.72)$$

$$[D_C^\times] + [D_C^+] = [D_C^{tot}]\,. \quad (9.73)$$

With these we now have all equations needed to determine the equilibrium concentrations.

Let us consider the application of such a system of equations by means of an example (Fig. 9.11). We examine a crystal with Schottky disorder. We thus have to take account of electrons, holes, neutral and negatively charged cation vacancies, and neutral and positively charged anion vacancies. To determine the six concentrations, we make use of (9.63) and (9.65) (dissociation of neutral vacancies), (9.70) (creation and recombination of electron-hole pairs), (9.68) (interaction between vacancies in both sublattices), (9.72) (condition of neutrality), and (9.61) (dependence on the partial pressure of A_2).

Figure 9.11a shows the resulting concentrations in schematic form. For small A_2 partial pressure, the anion vacancies dominate. In the main they are dissociated and have thus given up their electron. With increasing partial pressure

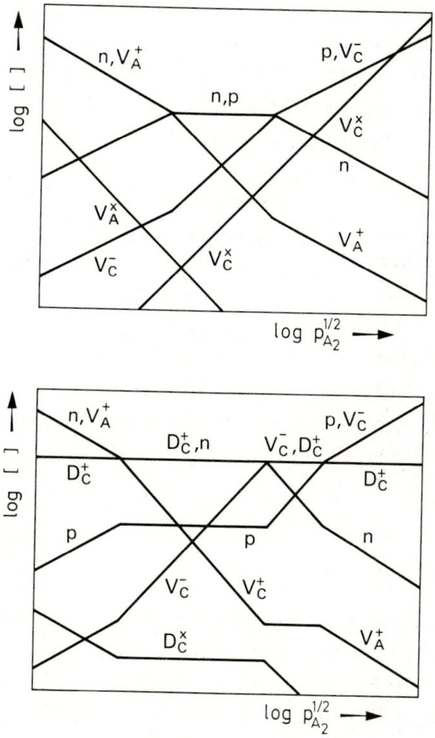

Fig. 9.11. Equilibrium concentrations of defects and free carriers as a function of the vapour pressure of a gas phase with which the solid can exchange anions. See the text for the model and for discussion.

the concentration of cation vacancies increases. These too can be taken to be largely dissociated. If the concentration of positive holes is greater than that of the positively charged anion vacancies, we can define a second zone, in which $n = p$. If the concentration of negatively charged cation vacancies exceeds the electron concentration, a third zone follows with $p = V_C^-$, approximately.

If we include donors (neutral and positively charged) as additional imperfections, we have also to take (9.71) and (9.73). We then find the concentrations illustrated in Fig. 9.11b. Here we can distinguish four zones. In the first $n = V_A^+$ approximately, and both concentrations are small compared with all others. In the second n equals the number of positively charged donors, in the third D_C^+ and V_C^- are approximately the same, and in the fourth $p = V_C^-$.

Figures 9.11a, b can look quite different for another choice of mass action constants. One (or more) of the zones can disappear, one or more of the concentrations shown can remain so small that it cannot be detected. This is why only electron or only hole conduction can be detected in some semiconductors and why lattice disorder is important in some semiconductors yet apparently absent in others. For further details see, e.g., *Kröger* and *Vink* in [101.3].

9.1.7 Diffusion and Ionic Conduction

The *kinetics* of imperfections is important in diffusion and ionic conduction. Defects take up equilibrium positions in a crystal. Vacancies are missing lattice atoms. Interstitial atoms are situated in the potential wells between lattice sites. In their motion through the crystal, defects jump from one equilibrium position to a neighbouring one. In what follows we look first at the transition probability for an individual process. The parameters which determine diffusion and ionic conduction then follow therefrom.

We take as our model a crystal with cubic arrangement of equilibrium positions for interstitials. The following considerations apply equally to vacancies. Let the crystal extend in the x-direction over a distance $L = Na$ (a = separation between two wells in the x-direction). We inquire about the number of interstitials which pass in a time t through a plane $x = x'$ between two planes x_0 and x_1. If on the average there are m interstitials in the wells of each plane, this number is equal to mtw where w is the probability of an individual jump.

The Hamiltonian function for an interstitial atom and the lattice ions surrounding it is

$$H = \frac{p^2}{2M} + T(P) + \Phi(\boldsymbol{r}, Q)\,. \tag{9.74}$$

Here $\boldsymbol{r}$, $\boldsymbol{p}$, and M are position, momentum, and mass of the interstitial atom. Q and P represent all the coordinates and momenta of the lattice ions, T is the kinetic energy of the lattice ions, and Φ the potential energy which depends on all the particles.

During a time dt the plane $x = x'$ will be crossed from left to right by all those interstitial atoms which have a positive velocity component v_x and which are within a distance $dx = v_x\, dt$ to the left of x'. The number of such atoms is

$$ndx \frac{\int \exp(-H/k_B T) dp_y dp_z dy dz dP dQ}{\int \exp(-H/k_B T) d\tau_p d\tau_r dP dQ}, \tag{9.75}$$

where n is the total number of interstitials ($n = Nm$). By integrating over all positive velocities, and using (9.74), we find

$$mwdt = dt \int_0^\infty dv_x n v_x \frac{\int \exp(-H/k_B T) dp_y dp_z dy dz dP dQ}{\int \exp(-H/k_B T) d\tau_p d\tau_r dP dQ}$$

$$= ndt \frac{\int \exp(-\Phi/k_B T) dy dz dQ}{\int \exp(-\Phi/k_B T) d\tau_r dQ} \frac{\int_0^\infty \frac{p_x}{M} \exp(-p_x^2/2Mk_B T) dp_x}{\int_{-\infty}^{+\infty} \exp(-p_x^2/2Mk_B T) dp_x}. \tag{9.76}$$

We now define a mean potential energy $V(x)$ by

$$\exp[-V(x)/k_B T] = \int \exp[-\Phi(x \ldots)/k_B T] dy dz dQ \tag{9.77}$$

and hence obtain for the probability w

$$w = \frac{\exp[-V(x')/k_B T]}{\frac{1}{N}\int \exp[-V(x)/k_B T] dx} \frac{\int_0^\infty \frac{p_x}{M} \exp(-p_x^2/2Mk_B T) dp_x}{\int_{-\infty}^{+\infty} \exp(-p_x^2/2Mk_B T) dp_x}. \tag{9.78}$$

The function $V(x)$ is periodic in the well-separation a. The integral over x in (9.78) is thus N times the integral over the distance from x_0 to x_1. If the height of the potential barrier between x_0 and x_1 is large compared with $k_B T$, we can replace $V(x)$ in the integral by the first two terms of a Taylor expansion $V(x) = V(x_0) + (K/2)(x - x_0)^2$. The integration is then easily performed. The second quotient in (9.78) is the mean velocity $\bar{v}$, which becomes $(k_B T/2\pi M)^{1/2}$, when the integral is evaluated. Taking it all together, we find

$$w = \frac{1}{2\pi}\sqrt{\frac{K}{M}} \exp(-\Delta V/k_B T); \qquad \Delta V = V(x') - V(x_0). \tag{9.79}$$

This equation can be simplified by recognizing that the factor in front of the exponential is the oscillation frequency in the x-direction of the interstitial in the potential well $V(x) = V(x_0) + (K/2)(x - x_0)^2$. Denoting this frequency ν, we finally have for the transition probability

$$w = \nu \exp(-\Delta V/k_B T). \tag{9.80}$$

In equilibrium, the same number of transitions take place in each direction. Under the action of an *electric field* the interstitials move (if charged) in the direction of the field. One can obtain the mobility if one assumes that the electric field only changes the height of the potential barrier ΔV. If the interstitial has a single positive charge, one has to subtract the amount $eEa/2$ from ΔV in the direction of the field, and add the corresponding amount in the opposite direction. The difference between the transitions probabilites in the field direction and opposite to it, multiplied by the distance jumped (a), is the mean velocity of an interstitial. The mobility follows by dividing this by the electric field E. Hence

$$\mu = \frac{(w_{\rightarrow} - w_{\leftarrow})}{E} a = \frac{\nu a}{E} \exp\left(-\frac{\Delta V}{k_B T}\right)\left[\exp\left(\frac{eaE}{2k_B T}\right) - \exp\left(-\frac{eaE}{2k_B T}\right)\right]$$
$$\approx \frac{ea^2 w}{k_B T}. \tag{9.81}$$

The mobility of the lattice defects and their concentrations following from mass-action laws determine the *ionic conduction* in an electric field. The charge and mass transport associated with ionic conduction is due either to interstitials, or to the movement of vacancies. In the latter case lattice atoms successively make individual jumps to fill vacant lattice sites, thereby displacing the vacancies in the opposite direction.

Diffusion in solids is also mainly associated with the movement of lattice defects. Individual atoms can exchange positions with nearest neighbours, but diffusion in homogeneous solids is mostly due to the movement of vacancies or interstitials. We do not want to present here the considerations leading to the diffusion law. It follows analogously to Brownian motion (statistical succession of individual jumps, random walk). As long as there are no correlation effects, i.e., dependence of individual jumps on previous jumps or on jumps of other defects, these considerations lead to the well-known statement that the diffusion current flows in the direction of the negative concentration gradient. The constant of proportionality (diffusion coefficient) is linearly related to the mobility.

Correlation effects are of particular importance when we consider the diffusion of impurity atoms. If the impurity atoms lattice sites, a diffusion step is only possible when there is a vacant site nearby. Since impurity atom and vacancy change places in the jump, the diffusions of both defects are closely correlated.

There are a great many comprehensive reports on the thermodynamics and kinetics of lattice defects in ionic conductors and metals. We refer, among others, to the articles by *Lidiard* in [106.XX, 113b] and particularly for diffusion in metals to the reports by *Lazarus* in [101.10] and *Peterson* in [101.22].

9.1.8 Recombination Processes at Imperfections

We now consider how imperfections influence optical processes in solids. In the treatment of the electron-photon interaction in Chapter 6, we restricted ourselves to absorption processes, in which photon absorption leads to an electron making a transition from valence to conduction band in a semiconductor. We did not consider the reverse process of an electron falling back into the valence band.

In the usual semiconductor model, the absorption process means the creation of an electron-hole pair. Correspondingly the reverse process is known as *recombination* of an electron-hole pair. In band-to-band recombination, the photon is emitted again (*radiative recombination*). However *nonradiative recombination* is also possible, i.e., transitions where the recombination energy is transferred to another electron (*Auger recombination*) or where the energy is given to the lattice by a multi-phonon process.

The conservation laws for energy and momentum must be obeyed in each recombination process. Along with the transition probability, these laws determine the "lifetime" of an excited electron-hole pair. The transition probability can be increased significantly when recombination occurs as a two-step process via an imperfection (*recombination centre*). The electron is first captured by the defect and then delivered to the valence band. In contrast to the donors and acceptors considered so far, recombination centres must therefore have sufficiently large effective cross sections for interaction processes with both the conduction and valence bands.

Recombination processes are important for two reasons: The excited state is marked by the existence of freely moving charge carriers. It is therefore connected with the appearance of *photoconduction*. Recombination determines the extent and the time behaviour of the photoconductivity in a solid. Numerous semiconductor phenomena are characterized by the lifetime of excited electron-hole pairs and their recombination.

The second important area, *luminescence*, concerns the radiative recombination processes. If photon emission takes place immediately after absorption, i.e., within times corresponding to the lifetime of an excited state, one speaks of *fluorescence*. If the recombination is delayed (by mechanisms we have yet to discuss) one speaks of *phosphorescence*.

In luminescence, i.e., light emission by recombination of excited states in a crystal, we distinguish two limiting cases.

1) Excitation of the crystal causes electrons from the valence band to be transferred into the conduction band. Recombination occurs in several stages

involving defect levels. Some of the stages involve radiation, others do not. This is the theme of this section.

2) Excitation and relaxation into the ground state occur *within* a defect. We shall deal with this later.

We start with a general discussion of the kinetics of recombination processes. In this we initially neglect the question of whether a transition takes place with or without radiation.

When defects are not involved, the only possible recombination processes are electron transitions from the conduction band into the valence band (Fig. 9.12a). Let the concentrations of electrons and holes be n and p, respectively. The recombination law is then

$$\frac{dn}{dt} = \frac{dp}{dt} = G - r(np - n_{eq}p_{eq}) . \tag{9.82}$$

G is the number of electron-hole pairs generated in unit time by external influences. The second term on the right describes the recombination, which is proportional to the concentrations of electrons and holes. The third accounts for the fact that electron-hole pairs will be generated in any event by thermal excitation—their number is independent of the concentrations n and p—and that in equilibrium ($G = 0, n = n_{eq}, p = p_{eq}$), the thermal generation of electron-hole pairs exactly balances the recombination.

For small deviations from equilibrium concentrations ($n = n_{eq} + \delta n$, $p = p_{eq} + \delta n$, $\delta n \ll n_{eq}, p_{eq}$), (9.82) becomes

$$\frac{d}{dt}\delta n = G - r(n_{eq} + p_{eq})\delta n \equiv G - \frac{\delta n}{\tau} . \tag{9.83}$$

This defines the lifetime τ of an electron-hole pair. In the steady state ($dn/dt = 0$) it is related to the excess concentration δn by $\delta n = G\tau$ and, when external excitation is removed ($G = 0$), it leads to an exponential decay law $\delta n \sim \exp(-t/\tau)$.

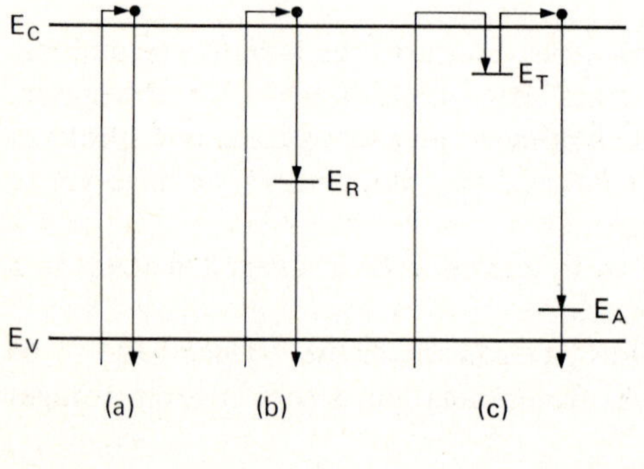

Fig. 9.12a-c. Electron transitions between valence and conduction bands: (a) band-band transitions, (b) involvement of recombination centres, (c) transitions into traps and activators.

If recombination takes place via *recombination centres*, i.e., as a two-step process involving a defect (Fig. 9.12b), electron and hole recombinations are decoupled

$$\frac{dn}{dt} = G - U_n, \qquad \frac{dp}{dt} = G - U_p, \tag{9.84}$$

where the U_n and U_p are the recombination rates of electrons and holes, respectively. We leave the calculation of the lifetimes of electrons and holes for this case to the reader (see Problem 9.5).

The mechanism of recombination via a recombination centre is not sufficient to describe luminescence in crystalline phosphors. Two types of defects have always to be considered, the *activators* and the *traps*. Activators are recombination centres (usually with levels relatively near to a band) which make a radiative transition possible. Traps capture free charge carriers for a while and thereby delay recombination. A typical example is illustrated in Fig. 9.12c. An electron is raised from valence into conduction band. There it is captured by a trap and thermally freed after a time. It then makes a radiative transition into an activator and from there finally returns radiationless to the valence band.

This model calls for a large number of balance equations of the type (9.84). Separate equations have to be set up for the change of carrier concentrations in both bands and in each defect level with time. The different concentrations are related through the condition of neutrality. For the levels shown in Fig. 9.13 the first balance equation, for example, is

$$\frac{dn}{dt} = G - \tau np + \alpha(nn_{T^\times} - n_1 n_{T^-}) + \beta(nn_{A^+} - n_2 n_{A^-}), \tag{9.85}$$

where the $n_{T^\times}$ and n_{T^-} are, respectively, the concentrations of the neutral and negatively charged traps and the n_{A^+} and $n_{A^\times}$ the corresponding concentrations of the activators. Corresponding equations have to be set up for the change with time of n_{T^-}, n_{A^+}, and p.

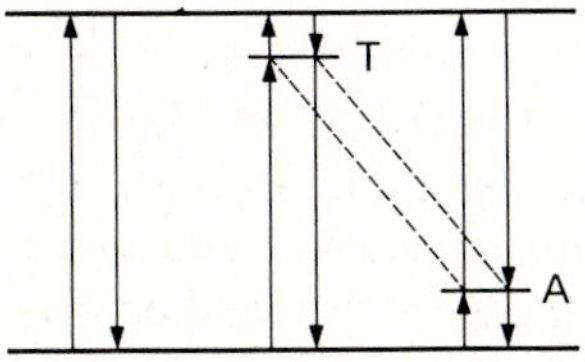

Fig. 9.13. Level scheme for the model of (9.85).

Such systems of equations help to answer many important questions. The solutions for the electron and hole concentrations yield the *photoconductivity*. If external fields or concentration gradients are present, then the left side of (9.85) has to be extended by the divergence of the electron and hole current. These systems of equations also allow us to calculate the transition rates of individual subprocesses. If the transition conduction-band → activator shown in Fig. 9.12 is radiative, for example, but the competing transitions are nonradiative, then the efficiency and the decay time of the luminescence can be calculated.

The systems of equations can usually be simplified when one knows the effective cross sections of the competing processes. Individual terms in the balance equations can dominate. The luminescence efficiency can often be described by the two limiting cases

$$\frac{dn}{dt} = G' - L \qquad \text{with} \quad L = \alpha n^2 \quad \text{or} \quad = \alpha n \tag{9.86}$$

(bimolecular or monomolecular processes). The first case arises when charge carriers fall back into the state they were excited from, since then the number of charge carriers is the same as the number of final states attainable. The other case arises when the number of final states is so large that the recombination rate is only determined by the number of charge carriers recombining. In both cases one finds, for the variation with time of the electron concentration n and intensity of luminescence L,

$$\begin{aligned} &n = n_0(1 + n_0\alpha t)^{-1}, \qquad L = L_0(1 + \sqrt{\alpha L_0}\, t)^{-2} \\ &\qquad\qquad\qquad\qquad\qquad \text{for bimolecular processes} \\ &n = n_0 \exp(-\alpha t), \qquad L = L_0 \exp(-\alpha t) \\ &\qquad\qquad\qquad\qquad\qquad \text{for monomolecular processes.} \end{aligned} \tag{9.87}$$

The model illustrated in Fig. 9.13 is only one of many possible cases. The activator can have donor character; the roles of electrons and holes can be interchanged; the radiative transition can take place between two levels on adjacent defects (donor-acceptor pair), etc. If, in addition, one irradiates with light of another frequency, one of the levels needed for the radiative transition can become occupied or the carriers captured by the traps can be set free. All these possibilities lead to special sets of equations like (9.87) and complicate any general discussion of the kinetics of recombination processes.

We shall not go into a discussion of the transition probabilities for radiative and nonradiative transitions here. For the *radiative transitions*, the theory of direct transitions developed in Section 6.2.2 is largely applicable. In contrast to the band-to-band transition case treated there, here band-defect and defect-defect transitions are also possible. However, this alters few of the fundamentals of the theoretical model. For the line shapes, we also have to take electron-phonon and electron-photon interaction into account. This interaction is dominant in the *nonradiative transitions*. Two processes are important: multi-phonon processes and cascade processes, i.e., a sequence of one-phonon processes. As mentioned above, the electron-electron interaction can also be important (Auger recombination). The latter possibility, in which the recombination energy and the associated momentum are passed to another electron, demands particular attention when large amounts of energy are liberated by recombination (some tens of eV and more).

9.1.9 Optical Transitions at Imperfections, Configuration Coordinates

We now look to the second limiting case: excitation and relaxation of an electron take place within an imperfection, their interaction with the crystal lattice being restricted to the immediate vicinity.

The electronic charge distribution of the defect atom is changed when it is excited from the ground into an excited state. The bonding with the nearest neighbours in the lattice will therefore be influenced. The equilibrium configuration of the neighbouring ions becomes unstable; the ions take up new equilibrium positions. Part of the excitation energy is thus transferred to the lattice. This process takes a long time compared with the electronic transition. One can assume that the latter occurs while the lattice configuration is still unchanged (Franck-Condon principle).

The electronic transition back to the ground state takes place under the new lattice configuration. The ground state is not therefore reached immediately. The electronic transition is instead followed by a rearrangement of the ions into the original configuration. Since energy has then been given up twice to the lattice, the electron recombination energy is less than the excitation energy. Compared with the absorption spectrum, the emission spectrum is displaced to longer wavelengths (Stokes shift).

The concept of *configuration coordinates* is often employed to describe this state of affairs. If one formally combines the changes in all lattice coordinates between both lattice configurations into one configuration coordinate, the potential energy of the system "defect electron + surrounding lattice" can be represented as in Fig. 9.14. In the ground state and the excited state the potential energy is a quadratic function of the configuration coordinate, with a minimum at the respective equilibrium configuration. The transitions mentioned above can readily be discussed with the aid of such a diagram.

To gain a better understanding of the configuration coordinate model, we outline in the following a line of thought which leads to Fig. 9.14.

We start with the Hamiltonian for the system electron + lattice

$$H = T_e + T_i + V(r, R). \tag{9.88}$$

T_e and T_i are the kinetic energy operators for the electron and the lattice ions. $V(r, R)$ is the interaction energy, where the electron coordinates are combined in r and the lattice ion coordinates are combined in R.

To decouple the two systems, we introduce a product ansatz for the wave function: $\Psi = \psi(r, R_0)\varphi(R)$. Here ψ is meant to be the solution of the Schrödinger equation for the ground state of the electron when the lattice

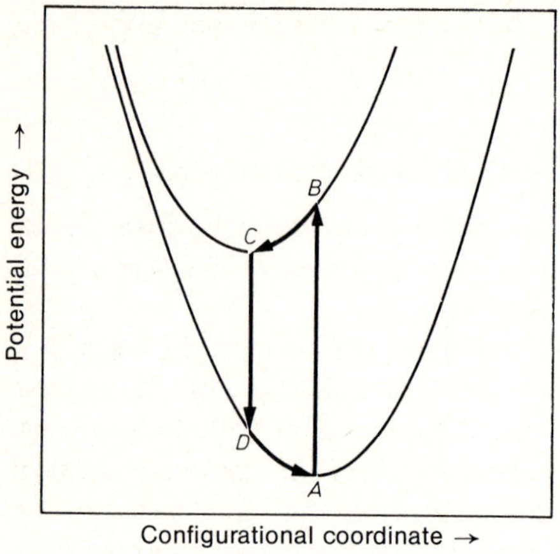

Fig. 9.14. The configuration coordinate model for describing optical transitions at defects: AB—excitation of the defect, BC—relaxation of the lattice (setting up of a new equilibrium configuration), CD—recombination of the electron ($E_{CD} < E_{AB}$), DA—return of the lattice to its original configuration.

configuration is held at R_0 (the equilibrium position of the lattice ions)

$$H_e\psi = [T_e + V(r, R_0)]\psi = E(R_0)\psi \,. \tag{9.89}$$

The remaining part of the Hamiltonian describes the lattice vibrations. If we express the deviations of the ions $R - R_0$ by (real) normal coordinates, we can write for (9.88)

$$H = H_e + \frac{1}{2}\sum_k (P_k^2 + \omega_k^2 Q_k^2) \,. \tag{9.90}$$

We can regard the part

$$E(R) = E(R_0) + \frac{1}{2}\sum_k \omega_k^2 Q_k^2 \tag{9.91}$$

as the potential energy of the system in the ground state. Eq. (9.91) describes the lower curve in Fig. 9.14, where E is plotted as function of one of the Q_k.

We now examine the excited state of the defect. Let the electron energy be $\bar{E}(R)$. It differs from $E(R)$, the energy of the ground state for configuration R, by an amount $\Delta E(R)$. We then put

$$\begin{aligned}\bar{E}(R) &= E(R) + \Delta E(R)\\ &= E(R_0) + \frac{1}{2}\sum_k \omega_k^2 Q_k^2 + \Delta E(R_0) + \sum_k e_k Q_k + \sum_{kk'} e_{kk'} Q_k Q_{k'} + \cdots,\end{aligned} \tag{9.92}$$

where the e_k, $e_{kk'}$ are derivatives of ΔE with respect to the Q_k. By the transformation

$$Q_k = Q'_k + \sum_{k'(\neq k)} \frac{e_{kk'}}{\omega_{k'}^2 - \omega_k^2} Q'_{k'} \tag{9.93}$$

it follows that

$$\bar{E}(R) = E(R_0) + \Delta E(R_0) + \frac{1}{2}\sum_k \bar{\omega}_k^2 Q_k'^2 + \sum_k e_k Q'_k \tag{9.94}$$

with $\bar{\omega}_k^2 = \omega_k^2 + e_{kk}$. $\bar{E}(R)$ has a minimum when $\partial \bar{E}/\partial Q'_k = \bar{\omega}_k^2 Q'_k + e_k = 0$. This defines a new equilibrium value $Q'_{k0} = -e_k/\bar{\omega}_k^2$. If we displace the Q'_k by this value $\bar{Q}_k = Q'_k - Q'_{k0}$, it follows that

$$\bar{E}(R) = \bar{E}(R_0) - \frac{1}{2}\sum_k \bar{\omega}_k^2 Q_{k0}'^2 + \frac{1}{2}\sum_k \bar{\omega}_k^2 \bar{Q}_k^2 = \bar{E}(\bar{R}) + \frac{1}{2}\sum_k \bar{\omega}_k^2 \bar{Q}_k^2 \,. \tag{9.95}$$

$\bar{R}$ is the equilibrium configuration in the excited state. This equation gives—as a function of the $\bar{Q}_k$—the upper curve in Fig. 9.14. The e_k give the magnitude of the shift of both parabolae, the e_{kk} their different curvature.

To be able to apply the concept in a reasonable way, only one oscillation (described by one pair of normal coordinates Q_k, P_k) may be excited in the electronic transition.

In this qualitative derivation we have made a number of approximations which we should really justify. Among others we have neglected the possibility that localized vibrations are excited. For an application of these conditions in the next section we state: In the electronic ground state and in the excited state the lattice vibrations are described by *different* normal coordinates. The lattice component φ of the product ansatz for the wave function can be written as a product of oscillator eigenfunctions. In the ground state these are functions of the Q_k, in the excited state of the $\bar{Q}_k$.

In the next section we shall return to the configuration coordinate model. For further details of the application of this concept to luminescence see, for example, *Dexter* [101.6], *Klick* and *Schulman* [101.5]; for an instructive one-dimensional model and its use to carry out quantitatively the derivation outlined above see also *Markham* [102.8] and the literature cited there.

9.1.10 Electron-Phonon Interaction at Imperfections

Electronic transitions at imperfections are often associated with the creation (or absorption) of phonons. Such processes are important in determining the structure of absorption spectra. Band-band transitions always lead to continuous absorption spectra since any number of transition energies are possible above a threshold value. Defect transitions on the other hand give sharp absorption lines as long as no phonons are involved. Electron-phonon coupling can lead to broadening of lines and to absorption and emission bands with characteristic structure.

The defect contribution to optical absorption is usually described by the effective cross section of the defect for the absorption process. For a sufficiently low defect concentration, these processes are independent, and the defect contribution to the absorption is simply the product of defect concentration and effective cross section. Analogous to the direct band-band transitions treated in Section 6.2.2, we set the effective cross section proportional to the square of the transition matrix element and to a delta-function which guarantees energy conservation [cf. (6.49) and (6.55)]

$$\sigma(\omega) \propto |\langle j'| \exp(-\mathrm{i}\boldsymbol{\kappa}\cdot\boldsymbol{r})\boldsymbol{e}\cdot\nabla|j\rangle|^2 \delta(E_{j'} - E_j - \hbar\omega) . \tag{9.96}$$

The transition here is between two states j, j' of the defect. $\boldsymbol{\kappa}$ is the photon wave vector, $\hbar\omega$ its energy. If phonons are involved in the transition, we have to extend (9.96). Instead of the wave functions $|j\rangle$ and $|j'\rangle$ of the electron states, we have to use wave functions $|j, n\rangle$ and $|j', n'\rangle$, where the n, n' describe the state of the phonon field before and after the transition. In accordance with the last section, these functions are products of the wave functions $|j\rangle$ or $|j'\rangle$ and products of oscillator eigenfunctions, which depend on the normal coordinates Q, $\bar{Q}$ of both states. To simplify matters, we assume that the $e_{kk'}$ in (9.93) are zero. The Q and $\bar{Q}$ then only differ in the separation coordinate of the two lattice configurations: $\bar{Q} = Q - Q_0$.

The wave functions then become

$$|j, n\rangle = |j\rangle \prod_k \chi_{n_k}(Q_k) , \qquad |j', n'\rangle = |j'\rangle \prod_k \chi_{n_k'}(Q_k - Q_{k0}) . \tag{9.97}$$

In (9.96) we have to sum over all phonon final states, and to average over all possible initial states given by the n, taking account of their respective statistical weights. Thus instead of (9.96) we start out from

$$\sigma(\omega) \propto \operatorname*{Av}_{n} \sum_{n'} |\langle j', n'| \exp(-\mathrm{i}\boldsymbol{\kappa}\cdot\boldsymbol{r})\boldsymbol{e}\cdot\nabla|j, n\rangle|^2 \delta(E_{j'n'} - E_{jn} - \hbar\omega) . \tag{9.98}$$

Here E_{jn} is the energy of the system electron + phonons.

The matrix element in (9.98) can readily be reduced to the corresponding matrix element in (9.96). Since the operator in the matrix element only operates on the electron coordinates, we have

$$\begin{aligned} &\langle j'n'| \exp(-\mathrm{i}\boldsymbol{\kappa}\cdot\boldsymbol{r})\boldsymbol{e}\cdot\nabla|j, n\rangle \\ &\quad = \langle j'| \exp(-\mathrm{i}\boldsymbol{\kappa}\cdot\boldsymbol{r})\boldsymbol{e}\cdot\nabla|j\rangle \prod_{k=1}^{3N} \left[\int dQ_k \chi_{n_k'}(Q_k - Q_{k0})\chi_{n_k}(Q_k) \right] \\ &\quad \equiv \langle j'| \exp(-\mathrm{i}\boldsymbol{\kappa}\cdot\boldsymbol{r})\boldsymbol{e}\cdot\nabla|j\rangle\langle n'|n\rangle . \end{aligned} \tag{9.99}$$

This allows us to put (9.98) into the form

$$\sigma(\omega) \propto |\langle j'| \exp(-i\boldsymbol{\kappa}\cdot\boldsymbol{r})\boldsymbol{e}\cdot\nabla|j\rangle|^2 G(\omega) \tag{9.100}$$

where $G(\omega)$ is the following function:

$$G(\omega) = \underset{n}{\text{Av}} \sum_{n'} |\langle n'|n\rangle|^2 \delta(E_{j'n'} - E_{jn} - \hbar\omega)\,. \tag{9.101}$$

It decides the shape of the absorption line or band.

Thus the problem has been reduced to the determination of the factor $|\langle n'|n\rangle|^2$. We only give the most important results here. For a derivation, see, for example *Pryce* [121] and *Chiarotti* [113b]. One has to distinguish between the approximations of weak and strong electron-phonon coupling. Of importance here is the magnitude of the factor $S = \sum_k (n_k + \frac{1}{2})Q_{k0}^2$ appearing in the $|\langle n'|n\rangle|^2$. If S is small compared with unity, the zero- and one-phonon processes dominate; if S is large compared to unity, multi-phonon processes are important.

In the weak coupling approximation, $|\langle n'|n\rangle|^2 = \exp(-S)$ and hence the contribution of the zero-phonon transition to $G(\omega)$ is

$$G_0(\omega) = \exp(-S)\delta(E_{j'} - E_j - \hbar\omega)\,. \tag{9.102}$$

This is a sharp spectral line.

One-phonon processes give

$$G_1(\omega) = \exp(-S) \sum_k \frac{1}{2} \begin{Bmatrix} \bar{n}_k + 1 \\ \bar{n}_k \end{Bmatrix} Q_{k0}^2 \delta(E_{j'} - E_j - \hbar\omega \pm \hbar\omega_{\text{ph}}) \tag{9.103}$$

for phonon emission or absorption. The $\bar{n}_k$ are the thermal averages of the phonon distribution in the initial state.

Contributions from multi-phonon processes can be neglected. If one inserts a Bose distribution for the $\bar{n}_k$, one finds a spectrum of the form shown in Fig. 9.15a. For $T = 0$, the spectrum only contains tails for photon energies above the zero-phonon line, since no phonons are excited which can be absorbed.

For strong coupling one finds

$$G(\omega) = \exp(-S) \sum_{r=0}^{\infty} \frac{S^r}{r!} B_r(\omega); \qquad \int d\omega B_r(\omega) = 1\,. \tag{9.104}$$

The $B_r(\omega)$ are very complicated expressions which we shall not give here. Each term in the series describes processes involving r phonons.

The shape of the absorption band at $T = 0$ (no phonon absorption) is shown in Fig. 9.15b. The contributions from the r-phonon processes combine to give a

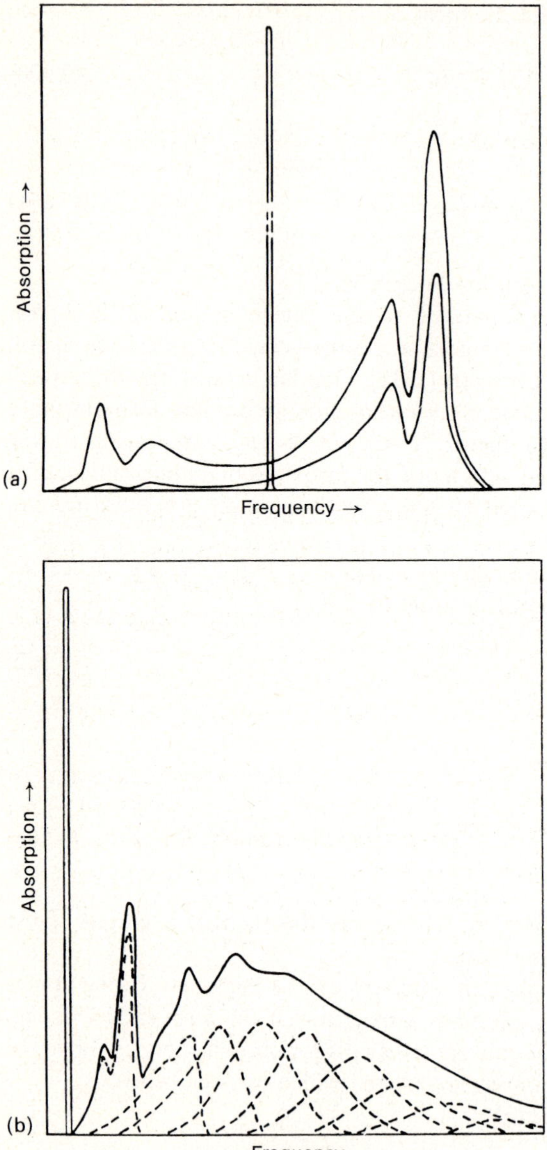

Fig. 9.15a and b. Contribution to the absorption (a) from zero- and one-phonon processes at two different temperatures [according to (9.102) and (9.103)] and (b) from multi-phonon processes at $T = 0$ [according to (9.104)], schematic (from *Pryce* [121]).

spectrum with a complicated form, which is separated from the absorption line of the zero-phonon process. At $T \neq 0$ a side-band also appears at lower frequencies. Depending on the magnitude of S, maximum absorption is at the zero-phonon line or in the multi-phonon bands.

In this section we have concentrated exclusively on the coupling of normal phonons to the electronic transition. However, localized phonons and resonances can also be coupled to it. The structure of absorption and emission bands is not only influenced by whether the electron-phonon coupling is weak or strong, but also by whether the lattice distortion in the vicinity of the defect is weak or strong. In the latter case localized phonons and resonances play an important part.

9.1.11 Bound Excitons

Recombination centres allow the transition of an electron from the conduction band of a semiconductor into the valence band to take place in two steps. First the electron is bound to the centre, then it transfers into the valence band. In the usual semiconductor model, this transition is described as the recombination of an electron-hole pair at the defect: The electron is captured by the centre; subsequently a hole recombines with the bound electron (or vice versa: a hole is captured and then recombines with an electron).

This process can be preceded by an intermediate state, in which the electron and hole are both bound at a recombination centre. This state can be interpreted as a *bound exciton.*

It is easy to appreciate the existence of bound excitons. We know from the hydrogen atom that it is possible to add an electron, thus forming an H^--ion (stable in its ground state). Neutral hydrogen-like defects can store charge carriers in a similar way. A donor can capture a second electron, an acceptor a second hole. Let us stay for a moment with the donor example. The storage of the electron on the neutral donor atom involves a short-range interaction. The complex thus produced can form a bound state along with a hole: $\oplus \,\overline{\underline{\;-\;}}\, +$. This arrangement has similarities with an H_2-molecule. Apart from its presence in a crystal lattice the only important difference between this complex and the H_2-molecule is the smaller effective mass of the hole. The binding energy depends strongly on the ratio of the electron and hole masses m_n and m_p. Nevertheless there are bound states for every ratio of the masses. Fig. 9.16 shows the binding energy for different possibilities of binding electrons and holes at a donor atom.

An exciton can also attach itself to an ionized donor D^+: $\oplus - +$. This corresponds to the H_2-molecule ion. Bound states are, however, only possible for a mass ratio above $m_n/m_p = 1.4$. The energy needed to localize the hole is too large when mass ratios are smaller than this.

Excitons are not only bound to donors and acceptors. *Isoelectronic defects* are also important. These include impurity atoms which substitute a lattice ion and have an identical electron configuration. A much studied example is

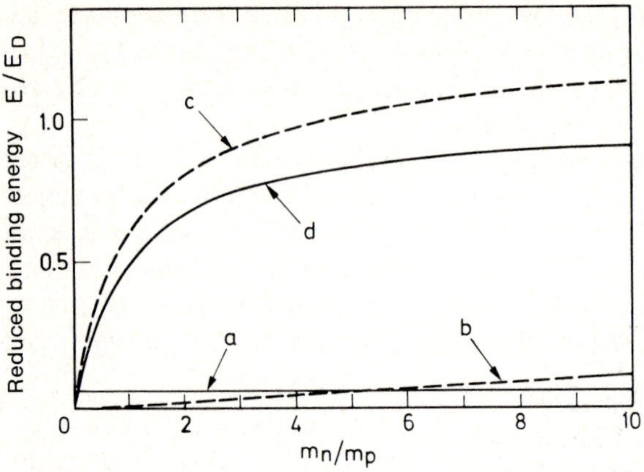

Fig. 9.16. Binding energy of a) an electron, b) a hole, c) an electron and a hole to a neutral donor; d) an exciton. Abscissa: Ratio of the effective masses of electron and hole. Ordinate: Binding energy in units of the binding energy of the donor electron (from J. J. Hopfield: Proc. Int. Conf. Semiconductor Physics, Paris 1964).

nitrogen on phosphorus sites in GaP (GaP: N). In the neutral state such isoelectronic defects have levels deep in the energy gap which can be occupied by electrons or holes. They are therefore typical recombination centres. The recombination of a bound exciton, i.e., the last step in electron-hole recombination, takes place radiatively. The luminescence spectra of such crystals contain lines, which give information about the states of the bound exciton. Fig. 9.17 shows an example. The two lines A and B are zero-phonon transitions in which

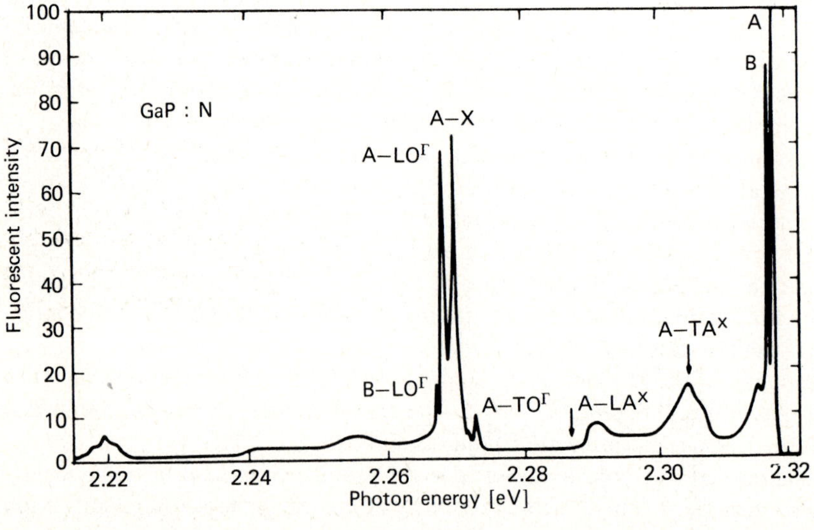

Fig. 9.17. Exciton lines in the luminescence spectrum of N-doped GaP (from *Czaja* [103.XI]).

bound excitons from a state with $J = 1$ and $J = 2$ recombine into the ground state $J = 0$. In addition to these transitions one can see side-bands, where phonons with energies corresponding to critical points in the vibration spectrum are additionally emitted.

For the theory of bound excitons and the characteristics of isoelectronic defects, see, among others, *Czaja* [103.XI], *Schröder* [103.XIII] and numerous contributions in [124].

9.1.12 Imperfections as Scattering Centres, the Kondo Effect

The scattering of electrons at defects contributes to the collision term in the Boltzmann equation and hence restricts the mobility of the free carriers similarly to electron-phonon coupling. The most important mechanism is scattering at charged defects.

We look at the simple case of scattering of a free electron of effective mass m^* at a simple positively or negatively charged defect in a medium with dielectric constant ε. This model can be applied to hydrogen-like (shallow) defects whose concentration is so small that the scattering processes at various defects are independent of each other. We do not yet want to tie ourselves to a semiconductor (nondegenerate electron gas) or a metal (degenerate electron gas). The interaction potential is a screened Coulomb potential of the form $V(r) = \pm(e^2/4\pi\varepsilon\varepsilon_0 r)\cdot\exp(-\lambda r)$. For metals, the screening constant is, from (3.66), $\lambda = (3e^2 n/2\varepsilon\varepsilon_0 E_F)^{1/2}$. For semiconductors we use the Debye length $\lambda = (\varepsilon\varepsilon_0 k_B T/e^2 n)^{1/2}$.

With this ansatz for $V(\boldsymbol{r})$ and using plane waves $|\boldsymbol{k}\rangle$, we calculate the transition matrix element and obtain [cf. (3.5)]

$$\langle \boldsymbol{k}'|V|\boldsymbol{k}\rangle = \pm\frac{e^2}{\varepsilon\varepsilon_0 V_g}\frac{1}{\lambda^2 + (\boldsymbol{k}' - \boldsymbol{k})^2}\,. \tag{9.105}$$

The scattering process is elastic ($k = k'$), so that with θ as the angle between $\boldsymbol{k}$ and $\boldsymbol{k}'$, we find $(\boldsymbol{k} - \boldsymbol{k}')^2 = 2k^2(1 - \cos\theta)$. We use the matrix element (9.105) to calculate the transition probability (4.10). We insert this into the collision term of the Boltzmann equation. For this we can use (4.56), which is valid for elastic scattering processes, and which yields the reciprocal relaxation time directly. If we carry out the integration in angle in (4.56), we find for the final result

$$\frac{1}{\tau(E)} = a\left[\ln(1 + b) - \frac{b}{1 + b}\right]. \tag{9.106}$$

The parameters a and b have different forms for semiconductors and metals, because of the different screening lengths. When we go from scattering at a single defect to scattering at $N = nV_g$ independent defects, we obtain

$$\text{for semiconductors:}\quad a = \frac{e^4 n}{16\pi\sqrt{2m^*}\varepsilon^2\varepsilon_0^2 E^{3/2}}, \quad b = \frac{8m^*\varepsilon\varepsilon_0 k_B T}{\hbar^2 e^2 n} E, \tag{9.107}$$

$$\text{for metals:}\quad a = \frac{\pi e^4 n\hbar k_F}{2m^* E_F^2}, \quad b = \frac{2\pi\hbar^2 k_F}{me^2}. \tag{9.108}$$

We see that the relaxation time does not depend on whether the scattering potential is attractive or repulsive. We obtain the temperature dependence of the electrical resistivity of metals from (4.127), and that of semiconductors from (4.128). Since from (9.108) $\tau(E)$ for metals is temperature independent and no other temperature-dependent factors arise in (4.127) [$\sigma = (e^2/m^*)n\tau$], we find that scattering at defects leads to an electrical resistivity which is temperature independent. This is the *residual resistivity* in metals (*Mathiessen's rule*) mentioned in connection with Fig. 4.7.

Many factors in (4.128) contribute to the temperature dependence of the resistivity in semiconductors. If one takes them all together, one finds for the mobility a $T^{3/2}$-law instead of the $T^{-3/2}$-law for electron-LA phonon interaction. If the two scattering mechanisms compete, at high temperatures one finds that the electron-phonon interaction is dominant, whereas at low temperatures the defect scattering (impurity scattering) dominates. The mobility is limited in either case by the dominant mechanism. When we examine the temperature dependence of the conductivity in semiconductors we should note that over most of the temperature range the temperature dependence of the electron or hole concentration is the major factor.

Just as electrical conductivity is limited by defect scattering, thermal conductivity is also limited at low temperatures by defect scattering. We gave an example in Fig. 7.4. We do not want to become involved in the theory of this. Instead we want to turn to a scattering process in which not only the charge but also the spin of the scattered electron plays a part.

We consider a defect with a local magnetic moment, for example a d-electron of a transition metal atom in a nonmagnetic lattice. Then, along with the Coulomb interaction of the scattered electron with the localized d-electron, the exchange interaction between the two particles can play a role. We shall see that this interaction makes a contribution to the electrical resistivity, which decreases with increasing temperature. Together with the constant residual resistivity and the resistivity of electron-phonon interaction (which increases as T^5), this leads to a resistivity minimum in metals at low temperature (*Kondo effect*).

In the following we want to find out which temperature-dependent contributions to the reciprocal relaxation time stem from the exchange interaction between free and localized electrons. We describe the free electron by a Bloch function $\psi(\boldsymbol{k}, \boldsymbol{r})$, the localized electron by an atomic function of the scattering centre situated at $\boldsymbol{R}_n$: $\Phi(\boldsymbol{r} - \boldsymbol{R}_n)$. We take electron-electron interaction into account according to Section 1.3, and use the occupation number representation. The interaction term of the Hamiltonian is then

$$H' = \sum_{kk'\sigma\sigma'n} \left(\langle k'n|V|kn\rangle c^+_{k'\sigma}c_{k\sigma}c^+_{n\sigma'}c_{n\sigma'} - \langle k'n|V|nk\rangle c^+_{k'\sigma}c_{k\sigma'}c^+_{n\sigma'}c_{n\sigma}\right). \quad (9.109)$$

The $c_k^{(+)}$ and $c_n^{(+)}$ are creation and annihilation operators for the free and the localized electron, respectively. The index σ gives the spin direction. The matrix elements are

$$\langle k'n|V|kn\rangle = \int \psi^*(k', r_1)\Phi^*(r_2 - R_n)V\psi(k, r_1)\Phi(r_2 - R_n)d\tau_1 d\tau_2 \quad (9.110)$$

and

$$\langle k'n|V|nk\rangle = \int \psi^*(k', r_1)\Phi^*(r_2 - R_n)V\Phi(r_1 - R_n)\psi(k, r_2)d\tau_1 d\tau_2 . \quad (9.111)$$

The first term in (9.109) means an interaction of both electrons with spin conservation, the second an interaction with exchange of spin (including the case $\sigma = \sigma'$). By carrying out the spin summation one obtains

$$\begin{aligned} H' = &\sum_{kk'n} (c^+_{k'\uparrow}c_{k\uparrow} + c^+_{k'\downarrow}c_{k\downarrow})(c^+_{n\uparrow}c_{n\uparrow} + c^+_{n\downarrow}c_{n\downarrow})\left(\langle k'n|V|kn\rangle - \frac{1}{2}\langle k'n|V|nk\rangle\right) \\ &- \sum_{kk'n}\left[(c^+_{k'\uparrow}c_{k\uparrow} - c^+_{k'\downarrow}c_{k\downarrow})\left(\frac{1}{2}c^+_{n\uparrow}c_{n\uparrow} - \frac{1}{2}c^+_{n\downarrow}c_{n\downarrow}\right)\right. \\ &\left. + c^+_{k'\downarrow}c_{k\uparrow}c^+_{n\uparrow}c_{n\downarrow} + c^+_{k'\uparrow}c_{k\uparrow}c^+_{n\downarrow}c_{n\uparrow}\right]\langle k'n|V|nk\rangle . \end{aligned} \quad (9.112)$$

According to (3.199, 200) the combinations of the c_n are precisely the spin operators S_z, S_+, and S_-. Hence we can write for the second (exchange) term in (9.112)

$$H'_{\text{ex}} = -\sum_{kk'n} \langle k'n|V|nk\rangle[(c^+_{k'\uparrow}c_{k\uparrow} - c^+_{k'\downarrow}c_{k\downarrow})S_{nz} + c^+_{k'\downarrow}c_{k\uparrow}S_{n+} + c^+_{k'\uparrow}c_{k\downarrow}S_{n-}] . \quad (9.113)$$

This operator describes four processes by which an electron can transfer from state $\boldsymbol{k}$ to state $\boldsymbol{k}'$. In the initial state the electron has one of the two possible spin directions, and in the final state it has either the same or the opposite spin direction. In a first-order perturbation calculation, these processes only contribute terms which are independent of temperature. Of more importance are processes of second order, which proceed through an intermediate state. We look in particular at processes from state $\boldsymbol{k}\uparrow$ into $\boldsymbol{k}'\uparrow$. There are then four possibilities.

a) The electron makes a transition into an unoccupied state $\boldsymbol{k}''$, and from there transfers to $\boldsymbol{k}'$. In $\boldsymbol{k}''$ it can retain its spin direction, or it may be reversed.
b) An electron first transfers from an occupied state $\boldsymbol{k}''$ into $\boldsymbol{k}'$, the electron $\boldsymbol{k}$ then jumps into the hole $\boldsymbol{k}''$. Here, too, two spin directions are possible for $\boldsymbol{k}''$.

We simplify (9.113) for further calculation by taking the exchange matrix element (9.111) to be constant and negative. We call it $-J$. For the process $\boldsymbol{k}\uparrow$ into $\boldsymbol{k}'\uparrow$, for fixed n, the second-order contribution to the Hamiltonian which then follows from possibility a) (spin-reversal in the intermediate state) is

$$H'^{(2)}_{\mathrm{ex,a}} = \sum_{\boldsymbol{k}''} J^2 \frac{c^+_{\boldsymbol{k}'\uparrow} c_{\boldsymbol{k}''\downarrow} c^+_{\boldsymbol{k}''\downarrow} c_{\boldsymbol{k}\uparrow} S_{n-} S_{n+}}{E(\boldsymbol{k}) - E(\boldsymbol{k}'')} \tag{9.114a}$$

and from possibility b) (also with spin-reversal) it is

$$H'^{(2)}_{\mathrm{ex,b}} = \sum_{\boldsymbol{k}''} J^2 \frac{c^+_{\boldsymbol{k}''\downarrow} c_{\boldsymbol{k}\uparrow} c^+_{\boldsymbol{k}'\uparrow} c_{\boldsymbol{k}''\downarrow} S_{n+} S_{n-}}{E(\boldsymbol{k}) - [E(\boldsymbol{k}) + E(\boldsymbol{k}') - E(\boldsymbol{k}'')]} . \tag{9.114b}$$

The corresponding expressions without spin-reversal can be obtained by reversing the spin-arrows in the $c^{(+)}_{\boldsymbol{k}''}$, and replacing the S_- and S_+ by S_z. Before adding these contributions we note that, apart from the sign, the denominator in (9.114b) is equal to the denominator in (9.114a), since $E(\boldsymbol{k}) = E(\boldsymbol{k}')$. We further rearrange the c-operators in the two equations into $c^+_{\boldsymbol{k}'\uparrow} c_{\boldsymbol{k}\uparrow} c_{\boldsymbol{k}''\downarrow} c^+_{\boldsymbol{k}''\downarrow}$ and $-c^+_{\boldsymbol{k}'\uparrow} c_{\boldsymbol{k}\uparrow} c_{\boldsymbol{k}''\downarrow} c^+_{\boldsymbol{k}''\downarrow}$, and take thermal averages for the products $c^+_{\boldsymbol{k}''} c_{\boldsymbol{k}''}$ [occupation probability $f(\boldsymbol{k}'')$ and nonoccupation probability $1 - f(\boldsymbol{k}'')$]. Finally we use the commutation relations $S_+S_- = S_-S_+ + 2S_z$. The sum of (9.114a) and (9.114b) is then found to be

$$H'^{(2)}_{\mathrm{ex}} = \sum_{\boldsymbol{k}''} J^2 \frac{c^+_{\boldsymbol{k}'\uparrow} c_{\boldsymbol{k}\uparrow}}{E(\boldsymbol{k}) - E(\boldsymbol{k}'')} [S_{n-} S_{n+} + 2S_z f(\boldsymbol{k}'')] . \tag{9.115}$$

To this we have to add the first-order contribution from (9.113) (factor at $c^+_{\boldsymbol{k}'\uparrow} c_{\boldsymbol{k}\uparrow}$), and the second-order contribution without spin reversal in the intermediate state. Using the relation $\frac{1}{2}(S_+S_- + S_-S_+) + S_z^2 = S^2$, which comes from the definition of the S_+ and S_-, this leads to

$$H'_{\mathrm{ex}} = \left\{ J S_{nz} + \sum_{\boldsymbol{k}''} J^2 \frac{S_n^2 + S_{nz}[2f(\boldsymbol{k}'') - 1]}{E(\boldsymbol{k}) - E(\boldsymbol{k}'')} \right\} c^+_{\boldsymbol{k}'\uparrow} c_{\boldsymbol{k}\uparrow} . \tag{9.116}$$

The temperature dependence of the sum in braces can only come from the factor $f(\boldsymbol{k}'')$, the Fermi distribution. The square of this sum enters the transition probability. As a linear term in $f(\boldsymbol{k}'')$ it contains the contribution

$$W(\boldsymbol{k}\uparrow, \boldsymbol{k}'\uparrow) \propto J^2 S_{nz}^2 \sum_{\boldsymbol{k}''} \frac{2f(\boldsymbol{k}'') - 1}{E(\boldsymbol{k}) - E(\boldsymbol{k}'')} . \tag{9.117}$$

In the vicinity of E_F the numerator of the terms in the sum in (9.117) goes from $+1$ to -1, the denominator has a pole in this region. Replacing the sum by an integral over the energy $E(\mathbf{k}'')$, one finds for (9.117)

$$W(\mathbf{k}\uparrow, \mathbf{k}'\uparrow) \propto -\ln\left|\frac{\delta E}{E_F}\right|. \qquad (9.118)$$

δE is the separation of the energy $E(\mathbf{k}) = E(\mathbf{k}')$ from the Fermi energy. It is of the order of $k_B T$. Thus the result follows: The transition energy due to exchange interaction between two electrons, and hence the contribution this mechanism makes to the electrical resistivity, increases logarithmically with decreasing temperature.

Thus we can explain the resistivity minimum observed. However the many approximations in the derivation outlined above do not allow a quantitative comparison between theory and experiment. This approximation has to be improved, if only because (9.118) leads to a logarithmic divergence of the resistivity at $T = 0$. For a review of the Kondo effect see, for example, the articles by *Kondo* and by *Heeger* in [101.23].

9.2 Localized States and Elementary Excitations at Surfaces

9.2.1 Introduction

The defects examined in Section 9.1 were (zero-dimensional) point imperfections of the ideal lattice. Dislocations form one-dimensional imperfections, while grain boundaries and crystal *surfaces* form two-dimensional imperfections.

Localized states can occur at such imperfections, too. There is, however, an important difference compared with the zero-dimensional defects. A surface represents a two-dimensional periodic arrangement of atoms. Consequently one can define elementary excitations which are *localized* to a narrow region in the direction normal to the surface, but *extended* parallel to the surface. Quasi-particles and collective excitations occur. The quasi-particles are electrons localized in *surface states*. We shall examine these states in Section 9.2.2. Corresponding to the collective excitations of the lattice atoms, we can introduce *surface phonons* and *surface polaritons* as collective excitations of the surface layer and *surface plasmons* as collective excitations of the electron gas near the surface, etc. We discuss such possibilities in Section 9.2.3.

The periodic structure of a surface is either the same as the structure inside the solid (ideal surface), or it is a superstructure created by the rearrangement of the surface atoms. If the surface is covered with an adsorbed layer, this too can have a structure which deviates from the crystalline arrangement within the solid. If the adsorption layer is incomplete, or if the surface is distorted locally, local surface states can appear—they correspond to the states treated in Section 9.1. We shall not discuss these states here, nor do we want to go into a discussion of elementary excitations associated with dislocations, which can be defined in a wholly analogous way.

9.2.2 Electronic Surface States

To explore the influence of a surface on the energy spectrum of the electrons we examine a simplified model of an idealized surface. Let the periodic potential of a crystal stretch over the half-space $z < 0$ of a cartesian coordinate system. In the half-space $z > 0$ (vacuum) let the potential be constant, V_0. The surface thus represents an abrupt transition between the strongly periodic lattice and the vacuum.

We make the following further assumptions:

a) We regard the lattice potential as a weak perturbation, we thus use the approximation for nearly free electrons (Sec. 2.2.6).
b) We reduce the problem to a one-dimensional model: periodic potential $V(z) = V(z + na)$ for $z < 0$ (a = lattice constant), $V(z) = V_0$ for $z > 0$.

Initially we solve the Schrödinger equation for the one-dimensional problem

$$\left[-\frac{\hbar^2}{2m}\frac{d^2}{dz^2} + V(z)\right]\psi(z) = E\psi(z) \tag{9.119}$$

separately for $z > 0$ and $z < 0$, and then match the solutions at $z = 0$ (continuous values and derivatives).

Only solutions of the Schrödinger equation which decrease with increasing z are physically relevant for the vacuum

$$\psi = a \exp\left[-\sqrt{\frac{2m}{\hbar^2}(V_0 - E)}z\right]. \tag{9.120}$$

As solutions for the periodic chain, we adopt the starting equations of Section 2.2.6. The band structure is given by Fig. 2.16. The most important details are shown again in Fig. 9.18a. The first Brillouin zone extends from $k = -\pi/a$ to $+\pi/a$. The $E(k)$-parabola for free electrons is distorted near the surface of the Brillouin zone; bands separated by an energy gap occur.

From (2.97) we have in the vicinity of $k = +\pi/a$

$$\psi(k, z) = \alpha \exp(ikz) + \beta \exp\left[i\left(k - \frac{2\pi}{a}\right)z\right]. \tag{9.121}$$

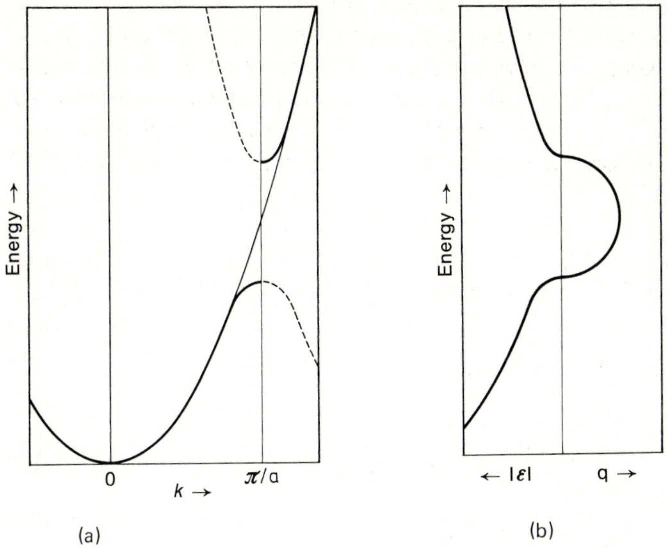

Fig. 9.18. (a) Section of the band structure for a one-dimensional periodic potential of lattice constant a, in the nearly free electron approximation. (b) In the energy gap between the two bands, solutions with imaginary k can appear.

α and β can be found from (2.102)

$$\begin{aligned}&\left[\frac{\hbar^2}{2m}k^2 - E(k)\right]\alpha + V\left(\frac{\pi}{a}\right)\beta = 0\\ &V^*\left(\frac{\pi}{a}\right)\alpha + \left[\frac{\hbar^2}{2m}\left(k - \frac{2\pi}{a}\right)^2 - E(k)\right]\beta = 0\,.\end{aligned} \tag{9.122}$$

With $k = \pi/a + \varepsilon$ and $\gamma = (\hbar^2\pi/ma|V|)\varepsilon$, it follows that

$$\psi = b\left[\exp\left(\mathrm{i}\frac{\pi}{a}z\right) + \frac{|V|}{V}(-\gamma \pm \sqrt{\gamma^2 + 1})\exp\left(-\mathrm{i}\frac{\pi}{a}z\right)\right]\exp(\mathrm{i}\varepsilon z) \tag{9.123}$$

and

$$E = \frac{\hbar^2}{2m}\left(\frac{\pi}{a} + \varepsilon\right)^2 \pm |V|(-\gamma \pm \sqrt{\gamma^2 + 1})\,. \tag{9.124}$$

For real ε, (9.124) gives the bands shown in Fig. 9.18a. The wave functions (9.120) and (9.123) can be connected together for each E. In the half-space $z < 0$, one needs for this the two solutions $\psi(k, z)$ and $\psi(-k, z)$ which, when linearly combined, connect with the vacuum solutions. The energy bands of the infinite lattice then—except for minor corrections—remain unchanged.

Along with the bands, we now find *solutions localized to the surface*. Since (9.123) is only a solution in the half-space $z < 0$, the parameter ε can also become *imaginary*. For $\varepsilon = -iq$, with real positive q, solutions which decrease exponentially into the crystal appear. With $\gamma = \mathrm{i}\sin(2\delta) = -\mathrm{i}(\hbar^2\pi/ma|V|)q$ we find from (9.123)

$$\psi = c\left\{\exp\left[\mathrm{i}\left(\frac{\pi}{a}z \pm \delta\right)\right] \pm \frac{|V|}{V}\exp\left[-\mathrm{i}\left(\frac{\pi}{a}z \pm \delta\right)\right]\right\}\exp(qz)\,. \tag{9.125}$$

The energy associated with this solution is

$$E = \frac{\hbar^2}{2m}\left[\left(\frac{\pi}{a}\right)^2 - q^2\right] \pm |V|\left[1 - \left(\frac{\hbar^2\pi q}{ma|V|}\right)^2\right]^{1/2}. \tag{9.126}$$

It is real for $0 \leq q \leq q_{\max} = ma|V|/\hbar^2\pi$.

For $q = 0$, one obtains the solutions

$$\psi = c\left[\exp\left(\mathrm{i}\frac{\pi}{a}z\right) \pm \frac{|V|}{V}\exp\left(-\mathrm{i}\frac{\pi}{a}z\right)\right] \propto \begin{cases} \cos\frac{\pi}{a}z & \text{for} \quad V > 0 \\ \sin\frac{\pi}{a}z & \text{for} \quad V < 0\,, \end{cases} \tag{9.127}$$

which correspond to the energies of the two band edges. Depending on the sign of the Fourier coefficient V of the potential, the lower eigenvalue is associated with the sin function and the higher with the cos function, or vice versa.

To allow us to connect the solutions in the inner and outer space, two free parameters are available—the ratio of the coefficients a and c from (9.120) and (9.125), and the energy E. Both are determined by the condition of continuity of the wave function and its derivative at $z = 0$.

We thus obtain the result: While the solutions of the Schrödinger equation (9.119) with real k correspond to the usual band solutions, solutions are possible for imaginary k, which decrease with increasing distance from the surface. The associated energy values, according to (9.126), lie in the energy gap between the bands (Fig. 9.18b). *One* of these solutions can be connected to the solution for the outer space. It represents a state in which the electron is localized to a narrow region at the surface. This is the surface state sought.

A deeper discussion shows that a match can only be obtained for $V > 0$. Thus in this model, surface states *can* occur, but they need not do so in every case.

In the one-dimensional model a surface state has a discrete level in the energy gap. Extending the model to three dimensions, we can regard the results as characterizing the component of $\boldsymbol{k}$ perpendicular to the surface. For each fixed value of the components of $\boldsymbol{k}$ parallel to the surface, we can expect a different position of the energy level of the surface state. Instead of individual levels we thus find *energy bands* for the surface states as well. Since the energy gap in which each surface level must lie is different for each value of $\boldsymbol{k}$, the density of states of the surface states can overlap with the density of states of the bands of the interior (Fig. 9.19).

The one-dimensional model discussed here, and its qualitative extension to three dimensions, is still unrealistic in several respects. A surface is not an abrupt transition from an unperturbed periodic potential to an outer space. In spite of this, states of the type discussed are possible for many surfaces.

Corrections are needed above all to account for three characteristics of a real surface.

1) In the uppermost atomic layer of the lattice, the lattice forces are only one-sided. This leads at least to a deformation of the periodic potential which changes the lattice constant near the surface.
2) The free valence bonds at the surface can connect differently from the way they connect in the interior of the crystal. This gives rise to a superstructure, i.e., a change of symmetry in the surface layer.
3) The surface can be covered with an ordered adsorption layer.

All these corrections change the positions of the surface bands and hence alter their density of states. They do not, however, remove the possibility that surface states will appear.

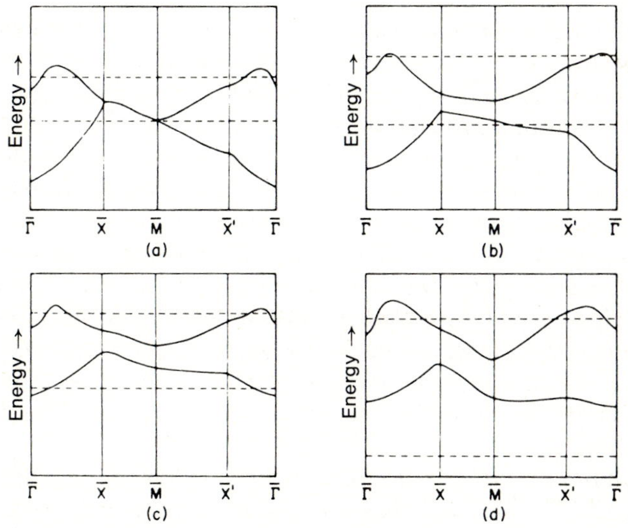

Fig. 9.19a-d. Surface bands for a 110-surface of (a) Ge, (b) GaAs, (c) InP, (d) ZnSe. The dotted lines show the upper edge of the valence band and the lower edge of the conduction band of these semiconductors (from *Jones* [128]).

In addition to these bands of delocalized states—i.e., states extended over the entire surface—we find localized states as discrete levels associated with local distortions of the surface (individual adsorbed atoms, incomplete adsorption layers, steps in the surface layer, etc.). Such localized states are observed when the conditions for matching (9.120) and (9.121) at the surface cannot be satisfied, i.e., when surface bands are absent.

Surface states of this kind can most readily be detected in semiconductors. In metals they are hidden by the high density of states in the bands. In insulators they are often hidden by the large number of localized volume defect states. If the energy gap of a semiconductor contains a large number of surface states, a surface charge will form due to the exchange of electrons between surface and volume states. This will be positive or negative, depending on the direction of the electron shift needed to set up equilibrium. The localized surface charge will expel any carriers of similar charge from a region just below the surface, or attract carriers of opposite charge in this region. The *space charge layers* which thus appear beneath semiconductor surfaces play a decisive role in many typical semiconductor phenomena.

In this section we have only illustrated basic principles of the origin of surface states. For more realistic models and more detailed discussions see above all the articles by *Davison* and *Levine* in [101.25] and by *Forstmann* in [113b], also the volumes [126–128].

9.2.3 Surface-Phonons, -Polaritons, and -Plasmons

Of the various collective excitations which can be localized at a surface, we look first at the surface-phonons and surface-polaritons.

In the vibration spectrum of a solid one finds excitations of acoustic and optical type localized at surfaces. If we restrict ourselves to the long wavelength limit, corresponding to the elastic vibrations of the continuum (acoustic branch), we find elastic surface waves, which propagate along the surface in a layer with the thickness of a wavelength. These are the so-called *Rayleigh waves*. Along with optical vibrations of the continuum, solids with a basis can have corresponding localized excitations. It is these that we now want to discuss further. We link up with the discussion in Section 3.3.8, where we examined the long wavelength limit for an ionic crystal with diatomic basis. In the infinite medium we found two types of propagating waves, the longitudinal waves (irrotational component of the lattice vibrations, frequency limit ω_L) and the transverse waves (divergence-free component, frequency limit $\omega_T < \omega_L$). Associated with them are the LO- and TO-phonons.

Other types of vibrations are possible in a finite medium. Let us consider the boundary between a solid with dielectric constant $\varepsilon(\omega)$ and a vacuum ($\varepsilon = 1$). Let the solid again extend over the half-space $z < 0$, the vacuum over $z > 0$. If one looks for solutions with time dependence $\exp(-i\omega t)$, $\mathbf{w}$ can be eliminated from the two equations (3.186) and, since $\boldsymbol{D} = \varepsilon_0 \boldsymbol{E} + \boldsymbol{P} = \varepsilon_0 \varepsilon(\omega) \boldsymbol{E}$, one finds

$$\varepsilon(\omega) = \varepsilon(\infty) + \frac{\varepsilon(0) - \varepsilon(\infty)}{1 - (\omega/\omega_{\mathrm{T}})^2} . \tag{9.128}$$

A wave localized at the surface, which propagates along the surface (e.g., in the x-direction) and falls off exponentially in the $+z$- and $-z$-directions will be described by

$$\boldsymbol{E} = -\nabla\varphi, \qquad \varphi = \varphi_0 \exp\left[\mathrm{i}(k_x x - \omega t)\right] \exp\left(-\kappa|z|\right) . \tag{9.129}$$

When one puts $\kappa = k_z$, (9.129) describes an irrotational *and* divergence-free solution.

We still have to connect the two exponentially decaying parts of the surface wave through the condition of continuity of $\boldsymbol{D}$ at the surface. This provides a condition for $\varepsilon(\infty)$ and determines the frequency ω. One finds $\varepsilon = -1$, and hence from (9.128)

$$\omega = \omega_{\mathrm{T}} \sqrt{\frac{\varepsilon(0) + 1}{\varepsilon(\infty) + 1}} \qquad (\omega_{\mathrm{T}} < \omega < \omega_{\mathrm{L}}) . \tag{9.130}$$

Thus we have shown that—in addition to the either divergence-free or irrotational solutions of (3.186)—divergence-free *and* irrotational solutions are still possible in a bounded medium and correspond to surface excitations. For a more detailed discussion we extend the model by including all Maxwell's equations (transition from surface phonons to surface polaritons, cf. Sec. 6.1.3) and by choosing another geometry. Instead of a medium bounded by a surface, we examine a plate of thickness $2a$. The surfaces lie in the x-y-plane at $z = \pm a$. We describe the medium in the continuum approximation (long wavelength limit) by the frequency-dependent dielectric constant (9.128).

The Maxwell equations we seek to solve are

$$\nabla \cdot (\varepsilon\varepsilon_0 \boldsymbol{E}) = 0 , \qquad \nabla \cdot \boldsymbol{H} = 0 , \qquad \nabla \times \boldsymbol{E} = -\mu_0 \dot{\boldsymbol{H}} , \qquad \nabla \times \boldsymbol{H} = \varepsilon\varepsilon_0 \dot{\boldsymbol{E}} . \tag{9.131}$$

We look for solutions of the type

$$\boldsymbol{E} = \boldsymbol{E}(z) \exp\left[\mathrm{i}(k_x x - \omega t)\right] , \qquad \boldsymbol{H} = \boldsymbol{H}(z) \exp\left[\mathrm{i}(k_x x - \omega t)\right] . \tag{9.132}$$

Putting this into (9.131) we find, after rearrangement,

$$\frac{\partial^2 E_x}{\partial z^2} = \alpha^2 E_x , \qquad \frac{\partial E_z}{\partial z} = -\mathrm{i} k_x E_x \tag{9.133}$$

where $\alpha^2 = k_x^2 - \varepsilon(\omega/c)^2$. From this the solutions in the medium are

$$E_x(z) = \exp(\alpha z) \mp \exp(-\alpha z)\,, \qquad E_z(z) = -\mathrm{i}\frac{k_x}{\alpha}[\exp(\alpha z) \pm \exp(-\alpha z)] \tag{9.134}$$

and in the outer space $x > a$, we find the exponentially decaying solutions

$$E_x(z) = A\exp(-\alpha_0 z)\,, \qquad E_z(z) = \mathrm{i}\frac{k_x}{\alpha_0}A\exp(-\alpha_0 z)$$
$$\left[\alpha_0 = \left(k_x^2 - \frac{\omega^2}{c^2}\right)^{1/2}\right], \tag{9.135}$$

and correspondingly for $x < -a$.

The continuity requirements at the surface $x = a$ for E_z and $D_x = \varepsilon_0\varepsilon(\omega)E_x$ lead to the condition

$$\varepsilon = \frac{\alpha}{\alpha_0}\left[\frac{\exp(-\alpha a) \mp \exp(\alpha a)}{\exp(-\alpha a) \pm \exp(\alpha a)}\right]. \tag{9.136}$$

For large thicknesses of the plate (9.136) becomes: $\varepsilon = -|\alpha/\alpha_0|$, from which we obtain

$$\omega = k_x c\sqrt{\frac{\varepsilon(\omega) + 1}{\varepsilon(\omega)}} \tag{9.137}$$

by inserting the expressions for α and α_0. When k_x is large one finds from this $\varepsilon(\omega) \to -1$, and hence (9.130).

For smaller thickness of the plate, volume oscillations appear which are determined by the boundary conditions at the two surfaces. The surfaces then govern the entire oscillation spectrum of the plate. Only for large thicknesses does one obtain the usual volume oscillations in an unbounded medium and *additionally* localized surface oscillations.

The solutions (9.134–136) however contain even more information. First we see that solutions which decay exponentially away from the boundary are only possible for $k_x > \omega/c$. For $k_x < \omega/c$, α_0 becomes imaginary. Eq. (9.135) describes waves which propagate outwards.

Secondly (9.136) leads to a restriction of the range of ω possible. For thick plates we find $\varepsilon = -|\alpha/\alpha_0|$, i.e., a negative value of ε. From (9.128) this can be satisfied only when $\omega_\mathrm{T} < \omega < \omega_\mathrm{L}$.

A complete discussion of all Maxwell's equations then builds up the following picture (Fig. 9.20): Each mode of oscillation of the type (9.132) for the vacuum-plate-vacuum system can be constructed by superimposing (in both

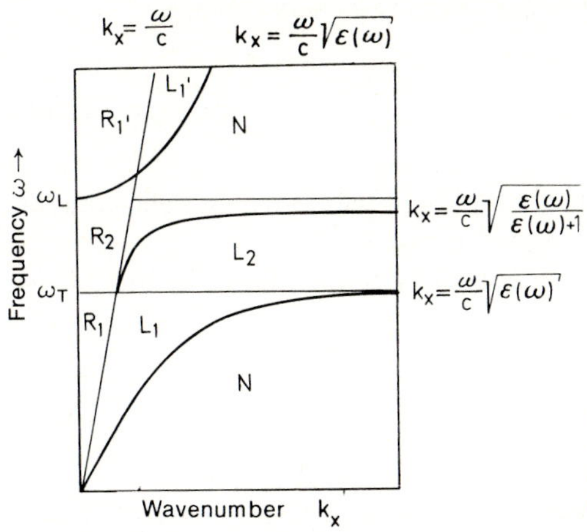

Fig. 9.20. Dispersion spectrum of surface phonons. See the text for a discussion of the individual regions (from *Ibach* [103.XI]).

z-directions) undamped propagating and damped standing waves in the medium, and outwardly decaying or propagating waves in the vacuum. The ω-k_x diagram can be divided into regions in which only particular types of waves are possible. The region indicated by L_2 in Fig. 9.20 only allows localized surface waves. The regions R_1, R_2, $R_{1'}$, contain the waves which propagate into the vacuum, connected with waves propagating in the medium (R_1, $R_{1'}$) or waves which decay into the interior (R_2). Within L_1 and $L_{1'}$ we find corresponding waves, damped outside and undamped inside. No solutions are possible in the region N.

In the above we have only touched on one aspect of surface oscillations. It is clear that for small crystal dimensions (crystallites, powder, thin layers) the vibrational spectrum is almost completely determined by the surfaces. This can be seen in the infrared absorption. One finds totally different forms of oscillations for layers, cylinders, spheres, and other geometric shapes.

The surface vibrations in larger samples, however, also deserve our attention. Surface phonons can be excited by slow electrons. They too can be seen in the optical absorption, if the surface geometry is suitable. This is not so, of course, for the planar surface just treated above, since from Fig. 9.20 the solutions for surface polaritons (surface phonons) lie to the right of the line $k_x = \omega/c$. Excitation by photons is not possible—in contrast to volume polaritons where the momentum transfer is due to partial reflection at the surface.

We refer the reader to reports by *Ruppin* and *Englman* in [111.XXXIII] and *Otto* [103.XIV] for further results on surface-phonons and -polaritons.

We find analogous types of oscillations in the electron gas of a solid. Along with the plasma oscillations of the gas (plasmons), localized oscillations (*surface plasmons*) are also possible.

For the dispersion relations of surface plasmons we can adopt (9.137) immediately, since this formula was derived directly from the Maxwell equations (without reference to lattice vibrations). We gave the dielectric constant for a free electron gas in (3.62), and that for an electron gas embedded in a solid in

(6.105). We take (6.105) but omit the damping ω_0 introduced there

$$\varepsilon(\omega) = \varepsilon_L\left(1 - \frac{\omega_p^2}{\omega^2}\right) \tag{9.138}$$

ε_L is the dielectric constant of the medium without the electron gas contribution, and ω_p is the plasma frequency $\omega_p = (me^2/m^*\varepsilon_0\varepsilon_L)^{1/2}$.

At large k_x (9.138) approaches a frequency limit determined by $\varepsilon(\omega) = -1$. From (9.138) we find

$$\omega = \omega_p\left(1 + \frac{1}{\varepsilon_L}\right)^{-1/2}. \tag{9.139}$$

As an example, Fig. 9.21 compares experimental measurements of surface plasmon dispersion curves in InSb with theory. For planar unperturbed surfaces, the surface plasmons are just as little affected by light excitation as the surface polaritons. A line grating of lattice constant d had consequently to be engraved on the surface in order to obtain the measurements presented in Fig. 9.21. One obtains components of the **k**-vector tangential to the surface with values $k_x = (\omega/c)\sin\alpha + 2\pi m/d$, where α is the angle of incidence and m takes up all integer values. Surface plasmons (and polaritons) can then be excited. On this see the reference given in the caption of Fig. 9.21.

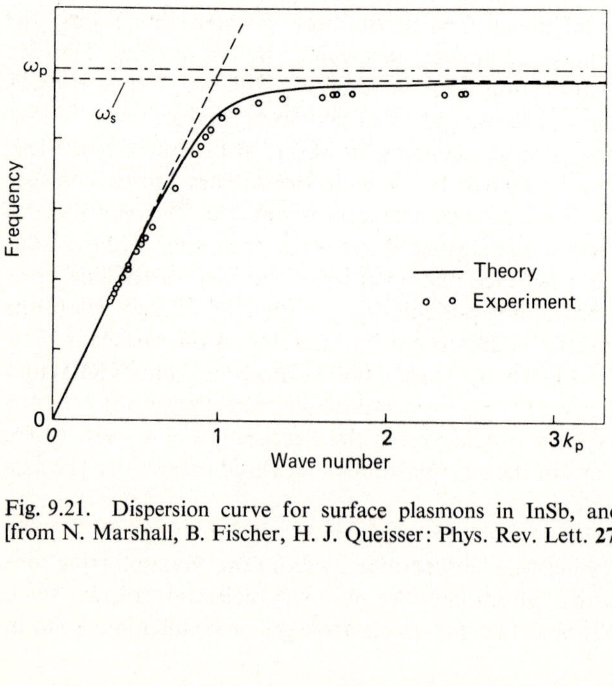

Fig. 9.21. Dispersion curve for surface plasmons in InSb, and comparison with theory [from N. Marshall, B. Fischer, H. J. Queisser: Phys. Rev. Lett. **27**, 95 (1971)].

10. Disorder

10.1 Localized States in Disordered Lattices

10.1.1 Introduction

The ideal, infinite crystal is characterized by the regular arrangement of the atoms in its lattice. Two aspects of this order are important.

Short-range order. By this we mean the regular arrangement of lattice atoms in the immediate vicinity of the particular atom considered. It determines the crystal field in which the atom is embedded.

Long-range order. By this we mean the strict periodicity and hence the translation invariance of the crystal lattice. Long-range order connects the regions of short-range order in such a way that atoms at equivalent lattice sites have the same surroundings in the same orientation.

Long-range order is very important for the theory of the band model. In Section 2.2.5 the translation invariance of the lattice allowed us to define delocalized states with given wave vector $\boldsymbol{k}$ and to represent these states in a Brillouin zone in $\boldsymbol{k}$-space. Long-range order was also vital to the definition of elementary excitations with distinct momentum and hence to the formulation of interactions in solids.

It must be noted, however, that strict long-range order scarcely influences the physical properties of a solid. The unperturbed, infinite lattice is an idealization which can be very important as the *zeroth approximation* in the calculation of solid-state properties. In reality a crystal is always perturbed, be it by the finite extent of the crystal, or by the elementary excitations as *dynamic perturbations* and by the point imperfections (treated in the last chapter) as *static perturbations.* So far we have regarded these perturbations in the first approximation as small, and we have treated their effect on the solutions of the zeroth approximation by perturbation theory. They lead to lifetimes for the elementary excitations in the stationary states of the zeroth approximation, and hence also to characteristic lengths, as for example, the mean free path of a Bloch electron between two interactions with the phonon system. Such lengths provide the standard by which long-range order can be meaningfully defined.

We shall refer to a lattice as *ordered* in the following if it is possible to explain its characteristics by starting with an infinite lattice with ideal long-range order as the zeroth approximation and to include the dynamic and static perturbations by perturbation theory. We shall call an arrangement of atoms *disordered* when this approximation is not meaningful.

Another restriction is necessary here. The question of whether, and in what approximation, we have to take account of disorder in a solid also depends on which of its characteristics we are interested in. A microcrystalline metal behaves electrically and optically like a single crystal; its mechanical properties, however, depend strongly on the size of the crystallites. In many respects, alloys do not behave differently from solids with an ideal lattice. One could give many further examples.

Even a disordered (according to our definition above) array of atoms possesses features which are typical of order. A wholly disordered, random lattice, as is used, for example, as an approximation in the theory of fluids, is usually not at all suited to a description of disordered solids. To understand this better, we consider Fig. 10.1. The two-dimensional cubic point lattice of Fig. 10.1a has three important signs of order. All lattice atoms are equal, the neighbours of a particular lattice atom are arranged in a geometrically fixed short-range order, and the coordination number, i.e., the number of nearest neighbours, is the same for all atoms.

The first sign of order is absent in Fig. 10.1b: two sorts of atoms are statistically distributed over the available lattice sites. This type of disorder is often called *compositional*. It occurs in alloys. We can imagine making the transition from Fig. 10.1a to 10.1b by successively inserting substitutional defects into the ideal, ordered lattice. For low defect concentrations the lattice in Fig. 10.1a is a good zeroth approximation. With increasing insertion of the second sort of atoms this approximation becomes increasingly poor.

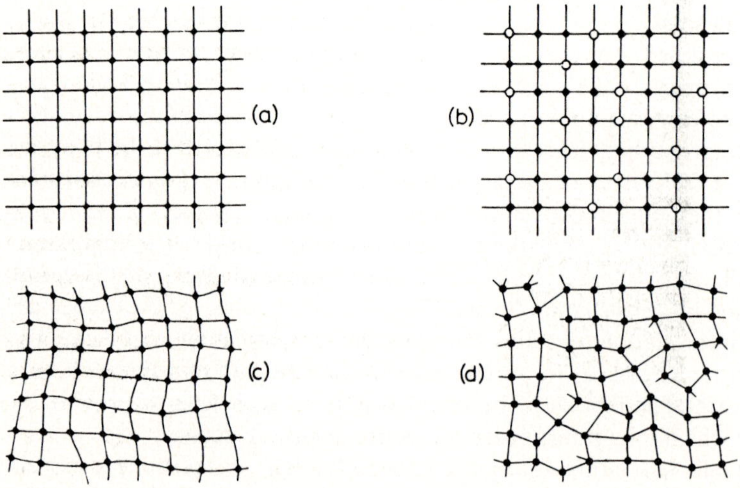

Fig. 10.1a-d. Possible types of lattice disorder: (a) ordered lattice, (b) mixed-crystal formation by statistical distribution of two sorts of atoms over the lattice sites: compositional disorder, (c) positional disorder by distortion of the lattice, (d) topological disorder with simultaneous formation of dangling bonds.

Figure 10.1c illustrates another type of disorder: *positional* disorder. Here all the lattice atoms are the same as in the ordered lattice, the geometric arrangement of the nearest neighbours, however, is statistically disturbed. A degree of positional disorder which can still be treated by perturbation theory is the statistical displacement of each lattice atom by its thermal vibration. Positional disorder is the characteristic of amorphous phases of a solid.

A further sign of disorder can be added to positional disorder, namely *topological* disorder (Fig. 10.1d). Here the topology of the lattice is perturbed. In addition to the four-atom rings which are characteristic for the cubic net, Fig. 10.1d shows the appearance of rings with five and six atoms. If the coordination number is maintained despite topological disorder, the bonds between nearest neighbours remain intact. If it is not maintained, individual valences remain unsaturated (*dangling bonds*). These form additional localized defects, similar to vacancies in an ordered lattice. The appearance of dangling bonds is often, but not necessarily, connected with topological disorder in amorphous semiconductors. The so-called Polk model is a model of a disordered lattice with tetrahedral short-range order, in which the coordination number and the separations between nearest neighbours are identical with those of the diamond lattice, and the distortion needed in the bonding angle remains less than 20°.

Further types of disorder are often defined in the literature, for example, *magnetic disorder*, which only involves the spin system in an otherwise ordered crystal. The limiting case of a weak perturbation here is described by the magnons.

We can define the random insertion of impurities into a heavily doped semiconductor as another type of disorder. An ordered crystal lattice will then have a disordered lattice of greater lattice constant superimposed on it. The two sublattices contribute in different ways to the physical properties of the semiconductor.

The number of different types of disorder clearly makes it impossible to have a *single* all-embracing theory for the disordered solid—to correspond to the theory for crystalline solids. The features which arise from the residual order, for example, play a decisive part in determining the properties of a given solid. If we are to deal with disorder in this final chapter, we must pose two main questions.

First we must ask which *general properties* are common to all disordered solids, i.e., how disordered solids are different in general from ordered solids. We shall find that along with the delocalized states—as found in the band model—*localized states* have an important, indeed even a decisive, role. We shall, therefore, have to first define the concept of localization more precisely.

Secondly we ask which results from the theory presented in the earlier chapters we can carry over to disordered solids. It will certainly not be the full band model theory with all its concepts, such as Bloch states, Brillouin zones, etc. For without translational invariance $\boldsymbol{k}$ no longer is a good quantum number.

If it is at all possible to define one-electron states in disordered solids—and there is no reason to doubt that this is possible to a good approximation—the concept of the density of (one-electron) states will also certainly apply. We have, therefore, to develop methods of determining the density of states. For this we shall look at different models. It will emerge that models intended to describe the common properties of all disordered solids are too unrealistic to allow quantitative statements to be made about particular substances.

Finally we ask how transport via localized states differs from transport via delocalized states. We have already prepared some of the fundamentals required for this in Section 8.3.6.

With the formulation of these questions, we have defined the content of this chapter. However, it will only be possible, within the context of this book, to present a qualitative discussion of important aspects. With the loss of the translation invariance of the lattice, more extensive mathematical aids are required. We cannot develop these here—particularly since they were only needed for this final chapter. Instead we shall concern ourselves with setting up important features required in formulating the theory needed to make the transition from order to disorder, and we shall give the reader references to allow him to explore this area in greater depth. We shall put the emphasis on the topic of amorphous semiconductors, where the difference between the crystalline and the noncrystalline version of a solid can be seen particularly clearly.

The most important references on disordered solids are books by *LeComber* and *Mort* [131], *Mitra* [114f], *Mott* [93], *Mott* and *Davis* [94], and *Tauc* [99] and the conference reports [116, 119–132] and further review articles by *Adler* in [110.2] and by *March* and *Stoddard* [111c.31]. Additional references are given in the following sections.

10.1.2 Localized States

We have already encountered the concept of localized states in Chapter 8, in connection with the local description of unperturbed lattices, and in Chapter 9, in connection with localized states at point imperfections. We did not, however, define the concept there at all precisely. This we now try to do in connection with the localization of one-electron states, which can occur with disorder.

In principle every state is localized if its associated wave function vanishes sufficiently rapidly at infinity, i.e., if the integral $\int \psi^*\psi \, d\tau$ is finite. As we shall see below, this definition does not take us very far. We start with a one-dimensional example, which will quickly show some important aspects of the problem.

We assume a succession of one-dimensional potential barriers, for which for simplicity we choose delta-functions

$$V(x) = \sum_i \frac{\hbar^2}{2m} \Omega_i \delta(x - x_i) . \tag{10.1}$$

At first we make no assumptions about the distribution of the x_i. With the potential chosen, the Schrödinger equation is

$$[-\nabla^2 + \sum_i \Omega_i \delta(x - x_i) - k^2]\psi(x) = 0 \qquad \text{with} \quad k = \sqrt{\frac{2mE}{\hbar^2}}. \tag{10.2}$$

We first look at a single δ-function at $x = 0$. For $x < 0$, the solution to the Schrödinger equation is $\psi_-(x) = A \exp(ikx) + B \exp(-ikx)$. For $x > 0$ we put $\psi_+(x) = C \exp(ikx) + D \exp(-ikx)$. The boundary condition at $x = 0$ requires a continuous connection between the two solutions $[\psi_-(0) = \psi_+(0)]$ and a step in their first derivative by an amount proportional to the strength of the δ-potential $[\psi'_+(0) = \psi'_-(0) + \Omega_i\psi(0)]$.

We now ask under what conditions the absolute value of ψ_+, for any $x = a$, becomes equal to the absolute value of $\psi_-(0)$: $\psi_+(a) = \exp(iKa)\psi_-(0)$. A short calculation shows that the condition $\cos(Ka) = \cos(ka) + (\Omega_i/k)\sin(ka)$ must apply. The two parameters K and a can then be found for all k (i.e., for each energy E) satisfying this condition.

This result immediately shows when an *extended state* can appear for a chain of δ-potentials. The absolute value of the wave function only becomes periodic with "lattice constant" a, when all potentials Ω_i are equal and have the same separation a, and when the energy E lies within the "bands" given by the condition stated above. This is the result of the well-known *Kronig-Penney model.* Each sequence of potential barriers, with statistically distributed separations $x_{i-1} - x_i$ or statistically varying potential Ω_i, leads to wave functions which either diverge or approach zero as x increases. Some of the latter also approach zero as $x \to -\infty$. These solutions represent *localized states.* All other solutions diverge and are not, therefore, physically meaningful. Hence we can state: All physically allowed (nondiverging) wave functions for the one-dimensional chain with statistically distributed potential barriers represent *localized states.* We have only indicated the proof of this statement here, and for a further discussion we refer to *Economou* et al. in [99] and to the literature cited there. In the example discussed here, the extended state has to have strict *phase coherence.* A knowledge of the phase of the wave function at one point allows it to be given at all other points.

This result can *not* simply be carried over to the three-dimensional case. Even the definition of an extended state here is not as clear as in the one-dimensional case. We shall explain this using a classical example which is often discussed in this connection, namely the motion of a particle of given energy in a random potential (Fig. 10.2).

Since there can be no tunneling in the classical problem, the particle finds allowed and forbidden regions. The two-dimensional example in Fig. 10.2 can be described in terms of the motion of a particle moving on a water surface of height E in the valleys of the potential mountains. At the lowest water level,

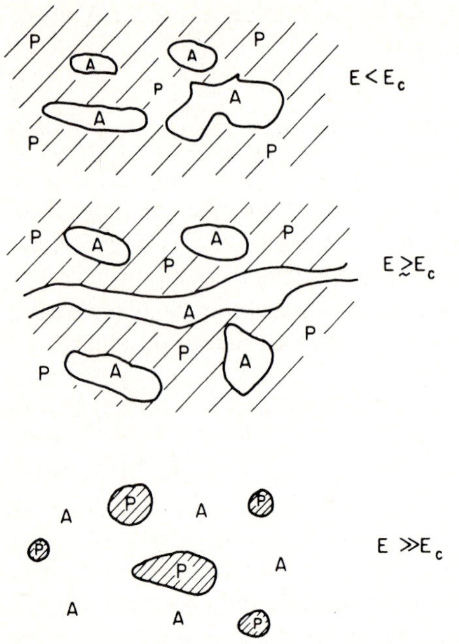

Fig. 10.2. The motion of a classical particle in a random potential. The "allowed" regions A grow with increasing energy. Isolated lakes link up via channels. The first channels which extend to infinity appear above the energy E_c (from M. H. Cohen: Proc. Int. Conf. Semiconductor Physics, Cambridge, Mass. 1970).

only lakes exist; the particle is *localized.* As the water level rises channels form which connect the lakes. The particle becomes delocalized when an ocean has formed, i.e., when the channels extend to infinity. The mountain tops penetrating the ocean surface remain as scattering centres for the extended states of the particle. Only when these too have been submerged can the particle move anywhere on the (two-dimensional) ocean.

This example—the quantum mechanical extension of which we shall discuss later—shows first that we must define the extended state concept anew. In the crystalline case the probability of an electron being in a given extended Bloch state was the same for all equivalent lattice points in the infinite crystal. Now we must also add to the extended states those states, whose wave function does admittedly extend to infinity, but for which $|\psi|^2$ can fluctuate markedly in space.

We can also formulate this result in a different way. Let us start with an ideal extended state of a crystalline lattice. The conductivity σ_E defined by means of the Kubo-Greenwood formula (8.83–85) and the mobility $\mu(E)$ are then infinite: An electron in this state can diffuse unscattered to infinity. Disturbances of the ideal lattice, the phonons, point impurities, or a small degree of disorder restrict the *mean free path* of the electron. Or put another way: As a result of the perturbations, the phase coherence of the wave function is limited to a finite coherence length. As the degree of disorder grows, the mean free path and the coherence length reduce. In spite of this the state is still extended; the wave function still extends to infinity. Further increase in disorder can then additionally lead to localized states, i.e., states restricted to finite regions. The extent of these can be described by a suitably defined *localization length*.

A further difficulty in distinguishing between localized and extended states lies in the fact that a physically meaningful definition has to include the length of the solid under consideration. A localized state whose localization length is indeed finite but is nevertheless large compared with the length of the sample will thus appear in an experiment to be extended. The ratio of localization length or phase coherence length to samp'e length therefore determines the character of a one-electron state of a given noncrystalline solid.

Summarizing, we can for the moment only say that in the disordered solid—in contrast to the crystalline infinite medium—localized states *can* occur.

A model first used by Anderson is important in obtaining a quantitative definition of localization. He considers a three-dimensional point lattice occupied by "atoms", each of which has just one single state E_n. If all E_n are equal, an energy band of width B results. For the discussion of the states in a disordered lattice, he maintains the positions of the atoms in the point lattice, but takes the E_n to be statistically distributed over a range of width W (Fig. 10.3).

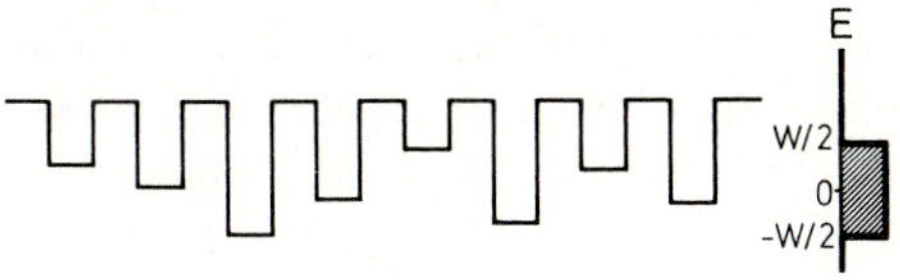

Fig. 10.3. Anderson model: Potential wells of different depths at the lattice sites of a three-dimensional point lattice.

The Hamiltonian can then be written (cf. Sec. 8.3.2)

$$H = \sum_n E_n c_n^+ c_n + \sum_{mn} V_{mn} c_m^+ c_n \tag{10.3}$$

with creation and annihilation operators in the Wannier representation. To simplify matters, in the second term only transitions between nearest neighbours are admitted and for them $V_{mn} = V = \text{const}$ is assumed. Starting from an initial state in which an electron is located at a given lattice point, one can inquire about the probability of finding the electron at this point again as $t \to \infty$. Diffusion of the electron in the lattice can, of course, occur since transitions are made possible by the second term in the Hamiltonian. If the initial position belongs to a localized state, diffusion is restricted to a finite volume. The probability of return for $t \to \infty$ is then nonzero. If, however, the electron can diffuse to infinity, the probability of return is zero. Anderson was able to show that the magnitude of the ratio W/B decides between these two alternatives. In particular, for the state $E = 0$ (mean value of the E_n distribution in the disordered lattice, middle of the band in the ordered lattice) the probability of return is zero if W/B falls below a fixed value of the order of 5. The state is then delocalized (extended). For larger ratios W/B, i.e., when the width of the spread of energy levels significantly exceeds the band width B, the state $E = 0$ is localized.

This definition of localization does not allow us to distinguish in a real case between localized and extended states. It can, however, help us to understand the increasing localization of band states as the transition is made from an ordered to a disordered lattice.

We can follow this transition qualitatively, without having to go into the detailed calculations carried out by many authors following the original work of Anderson. In Section 9.1.2 (Fig. 9.1) we saw that a single imperfection leads to the splitting-off (and simultaneous localization) of a state from the band edge. With increasing number of imperfections the number of localized states outside the band increases. The defect levels combine into a band (*impurity band*) which can overlap with the band of the delocalized states if the defect concentration is sufficiently high. We can imagine that the same phenomenon occurs with increasing disorder of a lattice. The states at the edges of an energy band become localized first, and simultaneously shift into the energy gap. The band thus acquires tails with localized states at its top and bottom edges. Fig. 10.4 shows the result of a calculation by *Economou* and *Cohen*. With growing disorder, the limits E_c and $E_{c'}$ approach one another from either side and meet in the middle of the band. When they meet, all band states are localized. This occurs just when the condition given by Anderson for the ratio W/B is fulfilled.

Finally we take a closer look at the boundary between localized and extended states of a band. According to the classical model of Fig. 10.2, this boundary is diffuse. Along with extended channels, isolated lakes occur. This need no longer be so if tunneling through P regions is allowed. The lakes are then connected with the channels. A distinct energy E_c separates localized and extended states. One can also appreciate the existence of a sharp boundary from the Kubo-Greenwood formula. If one approaches from the extended state side, the mean free path (coherence length) becomes smaller. However, since the mean free path cannot become smaller than a lattice constant, a minimum conductivity σ_{E_c} remains. Below E_c the states are localized. The conductivity σ_E (at $T = 0!$) then is zero. At E_c an abrupt step occurs in σ_E, and hence in the mobility defined by

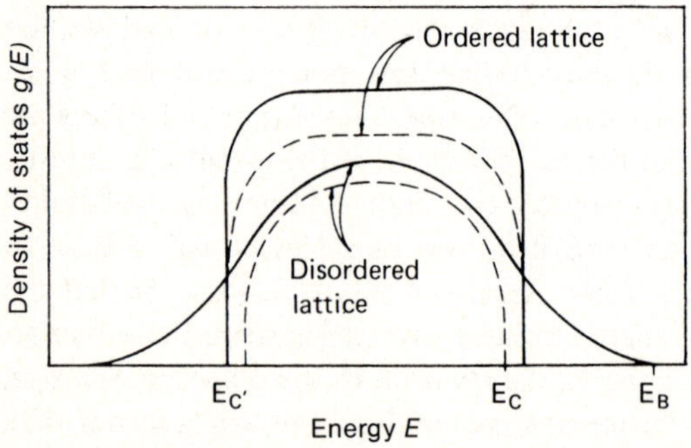

Fig. 10.4. Density of states (solid curves) and distribution of the extended states (dashed curves) for an energy band of an ordered and a disordered lattice (schematic). E_B—band edge; E_c, $E_{c'}$—limits between localized and extended states [from E. N. Economou, M. H. Cohen: Phys. Rev. Lett. **24**, 218 (1970)].

(8.86). Accordingly, E_c is referred to in the literature as the *mobility edge.* It should be emphasized, however, that there are also arguments against a *sharp* mobility edge. This question is still unclear.

Change in the parameters of a solid which influence the disorder can lead to a shift of the mobility edge relative to the Fermi energy. When E_c changes from values $< E_F$ to values $> E_F$, we arrive at a metal-insulator transition, called the *Anderson transition.* We have already referred to it in Section 8.3.3.

10.1.3 Density of States

In energy ranges where the one-electron states E_i are localized, $\boldsymbol{k}$ is no longer a good quantum number. A band structure function $E(\boldsymbol{k})$ can then not be introduced, nor can the derived concepts like effective mass, crystal momentum, etc. (Sec. 2.2.8). A concept which is valid as long as one-electron states can be defined, however, is the *density of states* $g(E)$. Its general definition is

$$g(E) = \frac{1}{V_g} \sum_i \delta(E - E_i) \,. \tag{10.4}$$

We begin with a qualitative discussion of the density of states in a disordered solid. Both experiment and theory show that, as in the crystalline case, bands of delocalized states can exist. Tails with localized states are attached to their edges. Between the tails of two adjacent bands a region without states, a *gap*, can occur, or the tails can overlap (*pseudogap*). We can distinguish three groups of solids, in which a different region of the density of states is of interest in each case.

In *alloys*, as in fluids and amorphous metals, regions in the middle of the band are of interest, since there the Fermi level separates occupied and unoccupied states. In these regions the density of states is often rather structureless. Structural features which occur at critical points in the crystalline case are smeared out in the amorphous phase. Disorder scattering limits the mean free path of the electrons. States can be localized by Anderson localization.

In *impurity band semiconductors* the density of states is of interest in the vicinity of the boundary between localized and extended states. Let us follow the origin of an impurity band with the help of Fig. 10.5. At low defect concentrations, the defects in the (ordered) host lattice form a disordered lattice with large interatomic separation. With increasing impurity concentration, the wave functions of the valence electrons of the different impurity atoms begin to overlap. The impurity levels split up into a band. If the impurity concentration increases further, a metal-insulator transition takes place—this is analogous to the case discussed in Section 8.3.2. Fig. 10.5b corresponds to an impurity concentration which is just below that required for the transition.

The impurity band splits up in accordance with Fig. 8.11 into two subbands (Hubbard model). The lower band is full, the higher empty. The disorder of the

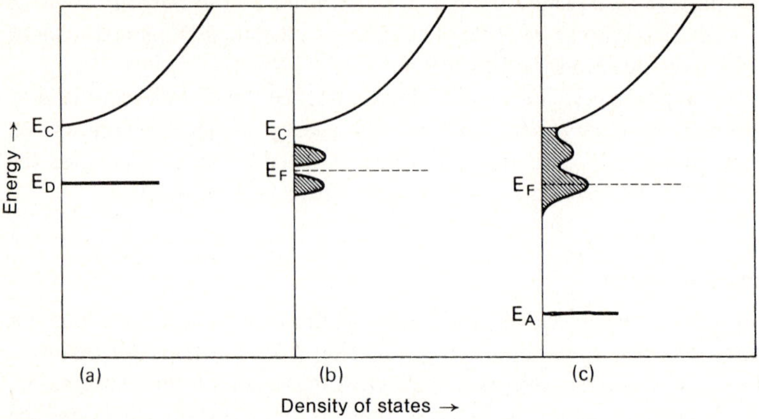

Fig. 10.5a-c. Density of states for a parabolic conduction band with adjacent donor states (schematic): (a) discrete impurity levles, (b) impurity band split into two subbands, (c) overlap of the impurity bands with each other and with the conduction band of the crystal.

impurity lattice, however, is responsible for some important differences to the case treated in Section 8.3.2. One is that the width of the impurity band is not given just by the overlap of the wave functions of the impurity states. A disorder broadening can be involved. Furthermore a portion of the states, or if the Anderson criterion is fulfilled, all the states, will be localized.

Impurity band semiconductors are thus of interest because, besides the ordered crystalline lattice, they contain a disordered lattice of impurity atoms, whose lattice constant can be changed, which can show Anderson transition, etc. However, another parameter is also important here, one which is not given in Hubbard bands in crystalline solids. By additional insertion of acceptors, some of the electrons in the impurity band of Fig. 10.5 are transferred into the deeper acceptor states. In this way it is possible to displace the Fermi energy at will. The disordered distribution of the (charged) acceptors represents a fluctuating potential, which shifts locally the position of the donor levels. The impurity band is further broadened thereby. These, and other factors, can lead to impurity bands overlapping with the conduction band of the host lattice (Fig. 10.5c). The conduction band then has a tail of localized states with a structure which is characteristic of the original impurity band.

We can construct a similar model for the density of states in *amorphous semiconductors* (Fig. 10.6). Following Fig. 10.4, the band edges of the conduction and valence bands contain localized states, which are separated from the extended states by mobility edges. The band edges can be abrupt (Fig. 10.6a), or gradual. They can be separated by a gap, or overlap (pseudogap). Further local defects (dangling bonds, etc.) can cause impurity bands or levels to appear in the gap or pseudogap (Fig. 10.6c). These give the density of states a characteristic structure. The properties of amorphous semiconductors are to a large extent deter-

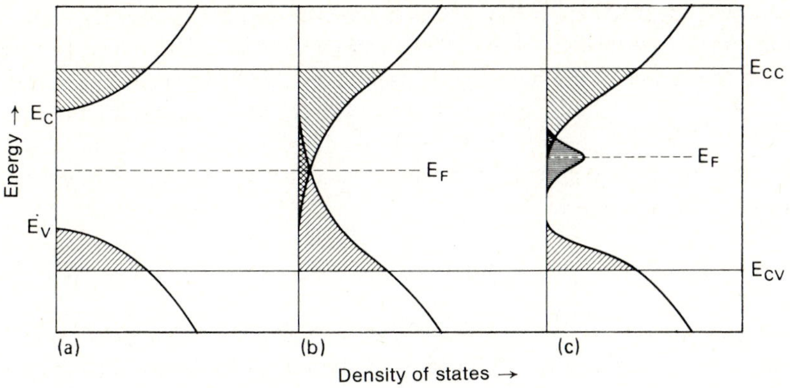

Fig. 10.6a-c. Possible structure of the density of states in amorphous semiconductors (a) localized states at the edges of the valence and conduction bands, (b) overlap of the tails of the localized states, (c) additional impurity band of local defects.

mined by such states. Although the extended states above the mobility gap are also important in deciding the optical properties of amorphous semiconductors, transport mainly occurs between states in the pseudogap. The transport properties of an amorphous semiconductor (at low temperature) are thus often determined by the states of the local defects (gap states) rather than by the localized states of the amorphous matrix (tail states).

Although the densities of states shown in Figs. 10.5 and 10.6 are certainly typical in qualitative terms for the classes of solids mentioned, they cannot answer the question about the characteristics of the density of states which determine the properties of a noncrystalline solid. For this, more sophisticated theoretical models are necessary. The methods to be used, moreover, depend on the group of substances considered. For alloys, other methods of calculation will be applied, for example, than for amorphous semiconductors.

The real difficulty in quantitative calculations lies in the necessity to simplify the theoretical ansatz. All the methods are rather cumbersome since the simplifications which stemmed from the invariance properties of the crystalline lattice are not valid when we deal with the disordered solid. A further complication is added. Even if it were possible to calculate the one-electron energies for a given disordered configuration of atoms, the result would not even be representative for a given noncrystalline solid. A *configurational average* over the different configurations possible must also be involved.

The methods used so far to calculate the energy spectra of disordered lattices can be classified according to the extent to which the long-range order and short-range order are regarded as disturbed (*Ioannopoulos* and *Cohen* [101.31]). The absence of a periodic potential is, of course, a fundamental characteristic of every disorder. This does not mean, however, that the loss of long-range order is the reason for particular properties of a noncrystalline phase. Small changes in the short-range order can be much more important here.

The simplest methods are those in which short-range order is strictly maintained, but long-range order is absent. They will be most successful if one is not interested in the details of the states in the pseudogap but in the states above the mobility edge. In amorphous semiconductors this region is important for the interpretation of optical spectra. Along with the density of states one therefore calculates the imaginary part of the dielectric constant. Corresponding to (10.4) one can define the latter by

$$\varepsilon_2(E) \propto \sum_{if} |M_{if}|^2 \delta(E - E_f + E_i)\,. \tag{10.5}$$

The simplest ansatz for a comparison between crystalline and noncrystalline modifications of a substance is obtained by retaining the crystalline band structure for both modifications, but in the noncrystalline case omitting the $\boldsymbol{k}$-selection rules in the computation of the matrix elements M_{if}. In an improved method, the loss of long-range order is described by correlation functions, which still give the positions of the crystalline case for the nearest neighbours of a particular atom, but for more distant atoms introduce a given uncertainty in the position, thus representing the gradual loss of long-range order for example, (see, *Kramer* [103.XII]). Such methods allow one to describe the optical spectra of amorphous semiconductors quite well. On the other hand, the density of states is often reproduced poorly. Since these methods start out from the change in the crystalline band structure due to loss of long-range order, they cannot describe details at the mobility edge and give no information on the localization of states.

On these points one can only use methods which make more exact assumptions about the deviation of short-range order from the ideal. One starts in general with the calculation of the energy eigenvalues for a cluster which consists of atoms arranged in a particular way, and subsequently averages over the possible configurations of the disturbed short-range order. There are two main difficulties. First, the cluster must be sufficiently large to give reasonable results, i.e., the localization length of states to be calculated should be smaller than the diameter of the cluster. On the other hand, the cluster must be small enough to allow numerical computations. Second, it is difficult to define the correct boundary conditions at the surface of the cluster. The simplest case is the periodic continuation of the cluster. The amorphous solid is then approximated by a periodic lattice with an extremely large number of basis atoms in the Wigner-Seitz cell. In other work the potential in the outer space is set to zero or equal to the mean cluster potential. Finally the cluster can be continued outward by a simplified lattice (Bethe lattice).

Quite different methods have been found successful for alloys. We mention here just two simple approaches which concern compositional disorder. The first is the rigid band model already discussed in Section 3.4.5 (Fig. 3.16). In it, one not only assumes a given band structure, and hence density of states, for the different transition metals, but also for alloys derived from them. The second possibility is the use of a mean potential $V = xV^{\mathrm{A}} + (1 - x)V^{\mathrm{B}}$ for a binary alloy $\mathrm{A}_x\mathrm{B}_{1-x}$. Both methods lead to a definite band structure and thus take no account of the statistical disorder. In spite of this they are often successfully applied. More detailed theories of alloys—as of other disordered phases—make use in each case of multiple scattering theory. We refer to an article by *Ehrenreich* and *Schwartz* in [101.31] as an introduction to such methods; for dilute magnetic alloys see also *Kondo* [101.23] and the literature cited in Section 10.1.1.

10.2 Transport in Disordered Lattices

10.2.1 Transport in Extended States

To close this chapter we want in the following sections to examine electrical conductivity in disordered lattices. In doing so, we want in particular to look at the different conduction mechanisms involved for extended states and localized states.

The contribution of the extended states to the conductivity is always dominant when there is a sufficiently large number of electrons in these states, hence when the Fermi energy lies above the mobility edge E_c or within a few $k_B T$ of this limit. The first case occurs in *liquid and amorphous metals* and in *alloys*. The second is important in *impurity band semiconductors* and in *amorphous semiconductors* at not too low a temperature. We want to look briefly at both cases in this section. The more important case—hopping conductivity in the localized states—is then the subject of discussion in the following sections.

Following Section 10.1.2, the extended states above E_c are characterized by a reduced mean free path. The disorder itself acts as a scattering mechanism for the charge carriers. As long as the mean free path is large compared with the lattice constant, the Boltzmann equation formalism can be used. The scattering at the disordered lattice then takes place similarly to scattering at randomly distributed defects. An example is *alloy scattering* which is an additional scattering mechanism in alloys. In general this approximation is not good enough for the calculation of resistivity in fluids and amorphous metals. A theoretical basis is provided here by multiple scattering theory and the Kubo formula (Sec. 8.3.4). These go beyond the scope intended for this book (see texts such as *Jones* and March [15], for example). For estimates of the conductivity directly at the mobility edge (mean free path comparable with the lattice constant), we can interpret the motion of a charge carrier as Brownian motion: the charge carrier diffuses through the lattice by transitions between equivalent states of nearest neighbours. With this, one arrives at the following expression for the mobility directly above E_c:

$$\mu_{E_c} = \frac{1}{6}\frac{ea^2}{k_B T}\nu\,. \tag{10.6}$$

a is the lattice constant and ν a "nearest-neighbour hopping frequency". This parameter depends on the transfer integral, i.e., on the overlap of the wave functions of pairs of nearest neighbours. Together with the density of states, (10.6) yields the conductivity σ_{E_c}.

If the Fermi energy lies in the region of localized states, the contribution of extended states to conductivity can be readily given via the Kubo-Greenwood formula. From (8.85) we find

$$\sigma = \frac{1}{k_B T} \int \sigma_E f_0 dE \tag{10.7}$$

with $f_0 \approx \exp[(E_F - E)/k_B T]$. The integral must run over all the extended states, i.e., from E_c to ∞. If finally we make the approximation $\sigma_E = \sigma_{E_c}$, we find

$$\sigma = \sigma_{E_c} \exp[-(E_c - E_F)/k_B T]\,. \tag{10.8}$$

The conductivity is thus activated, with an activation energy which corresponds to the distance between Fermi energy and mobility edge. The temperature dependence of this conductivity is the same as for semiconductors in which the Fermi energy lies in the energy gap. This equivalence stems of course from the fact that we have assumed that the localized states below E_c make no contribution to conductivity. Eq. (10.8) has therefore to be extended to include this contribution. It is fundamentally different in that, for $T = 0$, σ_E vanishes below E_c. It is not possible for a charge carrier to diffuse from one localized state to another without any expenditure of energy. Transitions are only made possible by the thermal vibration of the lattice. Below E_c we therefore find the transport mechanism to be *phonon assisted hopping*. It is to this that we now turn.

10.2.2 The Hopping Probability

We already encountered a hopping conduction by charge carrier transitions between localized states in Section 8.3.6. The localization there came from the formation of small polarons. A transition of an electron from a self-induced potential well to a neighbouring lattice site was made possible by a suitable local distortion of the lattice. The transition took place by involving a phonon.

In disordered lattices, the energies of the localized states are spread over a wide range. Adjacent localized states can have widely different energies. In any transition the energy difference must be supplied by a phonon. The transition probability can become so small that transition to further removed states, which involves the expenditure of a lesser amount of energy, becomes more likely. Thus we have a new hopping possibility to add to the *nearest-neighbour hopping*, namely, *variable-range hopping*. We consider such processes in the following,

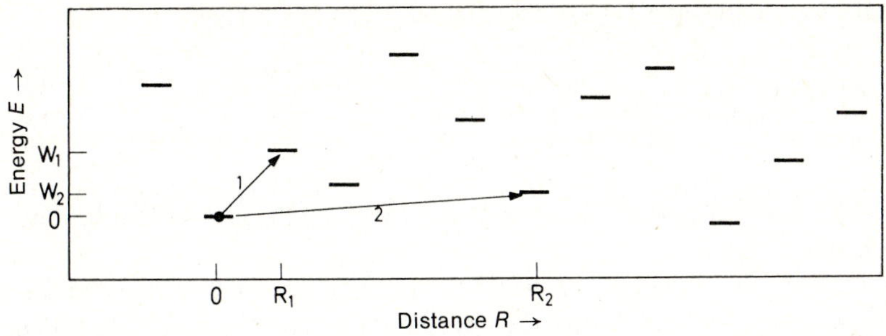

Fig. 10.7. Hopping processes between localized states with statistically distributed position and energy. The hopping probability is determined by the spatial distance and the energy difference between the two states.

with the assumptions that polaron effects can be neglected and only *one* phonon takes part in the transition. This is a simplification which is not necessarily realized in practice.

Figure 10.7 illustrates hopping processes between states with statistical distributions in space and energy. Let us look in particular at two states at locations $\boldsymbol{R}_i$ and $\boldsymbol{R}_j$. Let them have energies E_i, E_j ($E_j > E_i$). The dominant terms in the transition probability w_{ij} can be readily deduced.

The electron crosses the *distance* $R = |\boldsymbol{R}_j - \boldsymbol{R}_i|$ by tunneling. The factor which determines the tunneling probability is the overlap in the wave functions of the two states. The simplest ansatz for the wave function of a localized state is an exponential decay from the centre $\psi \propto \exp(-|\boldsymbol{r} - \boldsymbol{R}_i|/\lambda)$, where λ is a measure of the extent of the state (localization length). If λ is the same for the two states, the tunneling probability is proportional to $\exp(-2R/\lambda)$.

The *energy difference* $W = E_j - E_i$ must (for positive W) be provided by a phonon. W cannot therefore be greater than the maximum energy in the phonon spectrum. Herein lies the difficulty with our simplified model. In view of the restriction to nonpolar solids, we can use the Debye approximation (Sec. 3.3.4). The condition for one-phonon processes is then $W \leq k_B\Theta_D = \hbar s q_D$. Θ_D is the Debye temperature, q_D the maximum wave number in the Debye spectrum, and s the velocity of sound. The number of phonons of energy W in thermal equilibrium goes into the transition probability. For sufficiently low temperature ($k_BT \ll W$) it is given by the Boltzmann factor $\exp(-W/k_BT)$. Altogether we thus have for the transition probability

$$w_{ij} = w_0 \exp\left(-\frac{2R}{\lambda} - \frac{W}{k_B T}\right) \quad \text{for} \quad W > 0\,. \tag{10.9}$$

The factor w_0 can only be calculated when we make further assumptions about the localized states and the electron-phonon interaction. In a fundamental paper, Miller and Abrahams found (for a list of references on variable range hopping see, for example, the review by *Overhof* [103.XVI]) that

$$w_{ij} = \frac{E_1^2 A_0^2 W}{\pi \rho_0 s^5 \hbar^4} \exp\left(-\frac{2R}{\lambda} - \frac{W}{k_B T}\right) \qquad (W > 0)\,. \tag{10.10}$$

Here E_1 is the deformation potential constant (Sec. 4.1.2), ρ_0 the density, and A_0 an exchange integral.

For a fixed W, the transition probability w_{ij} diminishes with increasing distance and falling temperature. However, since W is a function of R (cf. Fig. 10.7), W will have a maximum value at a certain R. We return to this below.

So far we have only looked at the transition probability into a state of higher energy. For the reverse case, a jump from E_j back to E_i, corresponding equations are found for w_{ji} in which only the factor $\exp(-W/k_B T)$ is missing. One sees this most simply from the requirement that, in equilibrium, the transition rates in the two directions must be the same. The transition rates are defined as the product of transition probability, occupation probability of the initial state, and nonoccupation probability of the final state. Thus in equilibrium

$$\Gamma_{ij}^0 \equiv f_i(1 - f_j) w_{ij} = f_j(1 - f_i) w_{ji} \equiv \Gamma_{ji}^0 \tag{10.11}$$

from which, with $f_i^{-1} = 1 + \exp[(E_i - E_F)/k_B T]$, it immediately follows that

$$w_{ji} = w_{ij} \exp\left(\frac{E_j - E_i}{k_B T}\right) = w_{ij} \exp\left(\frac{W}{k_B T}\right). \tag{10.12}$$

Equation (10.11) can be simplified if we take all energies to be large compared with $k_B T$. The occupation probability $f_i = [1 + \exp(x/k_B T)]^{-1}$ is then 1 for negative x, and equal to $\exp(-x/k_B T)$ for positive x. We find

$$\Gamma_{ij}^0 = \gamma_0 \exp[-2R/\lambda - (|E_i - E_F| + |E_j - E_F| + |E_i - E_j|)/2k_B T]\,. \tag{10.13}$$

The factor γ_0 here is only weakly dependent on W.

To determine the conductivity and its dependence on temperature and other parameters, we must now consider the transition rate $\Gamma_{ij} - \Gamma_{ji}$ in an electric field. We then have to add a factor $-e\boldsymbol{E}\cdot\boldsymbol{R}_{ij}$ to the energy difference between the states $W = E_j - E_i$. At the same time we have to take account of the fact that the occupation probability of the localized states changes. We do this by replacing the Fermi energy E_F (equilibrium value of the chemical potential) by a space-dependent electrochemical potential $\mu_i = E_F + \delta\mu_i$

$$f_i^{-1} = 1 + \exp[(E_i - E_F - \delta\mu_i)/k_B T]\,. \tag{10.14}$$

The contribution which transitions between two states i and j make to the electric current is proportional to the difference $\Gamma_{ij} - \Gamma_{ji}$. If we put

$$w_{ij}(E) = w_{ij}^0 + \delta w_{ij}, \qquad f_i = f_i^0 + \delta f_i \tag{10.15}$$

it follows that

$$\Gamma_{ij} - \Gamma_{ji} = \Gamma_{ij}^0 \left[\frac{\delta f_i}{f_i^0(1 - f_i^0)} - \frac{\delta f_j}{f_j^0(1 - f_j^0)} + \frac{\delta w_{ij}}{w_{ij}^0} - \frac{\delta w_{ji}}{w_{ji}^0} \right] \tag{10.16}$$

with Γ_{ij}^0 from (10.13). For weak fields and small deviations of the chemical potential from equilibrium, one can expand the exponential and obtains

$$\Gamma_{ij} - \Gamma_{ji} = \frac{\Gamma_{ij}^0}{k_B T}(e\boldsymbol{E} \cdot \boldsymbol{R}_{ij} + \delta\mu_i - \delta\mu_j)\,. \tag{10.17}$$

The quantity in parentheses on the right side of (10.17) is the total potential difference between points $\boldsymbol{R}_i$ and $\boldsymbol{R}_j$.

10.2.3 Fixed Range and Variable Range Hopping

For an estimate of the temperature dependence of the hopping conductivity, we consider a model of the type illustrated in Fig. 10.7. Let the energies of statistically distributed localized states be distributed over a *finite* energy range. We are thus concerned with an impurity band, which has been broadened by the potential fluctuations of compensated impurities. Hopping processes between adjacent states require greater energy than between more distant states. We inquire about the most probable jump distance $\bar{R}$, and the associated energy difference $\bar{W}$. We use these in (10.17) and (10.13) to determine the temperature dependence of the individual jump, and put this equal to the temperature dependence of the hopping conductivity itself.

At high temperatures there are enough phonons of energy W^0 (mean energy difference between adjacent states) available to allow hopping processes between nearest neighbours to occur. In the transition rate, $\bar{R}$ becomes equal to R^0 (mean separation of nearest neighbours). Two possibilities have to be considered for the temperature-dependent factor in the transition rate: If the transitions occur at the Fermi energy, i.e., from a state $E_i < E_F$ into a state $E_j > E_F$, (10.13) leads to the factor $\exp[-(E_j - E_i)/k_B T] = \exp(-W^0/k_B T)$. If the transition occurs between two E_i, E_j which both lie above E_F, a factor $\exp[-(E_i - E_F + W^0)/k_B T]$ follows. We return to this difference in the next section. For the moment we state only that at high temperature this estimate leads to an *activated conductivity*.

At low temperature, no phonons are available with energy W^0. The electron has to tunnel to reach more distant states ($R > R^0$, but $W < W^0$). The most likely jump distance $\bar{R}$, and energy difference $\bar{W}$, can be readily evaluated. To do this we consider a state (R_i, E_i) at the Fermi energy and ask what radius R a sphere around R_i must have in order to find one state with $E_j = E_i + W$ within it. If we assume the density of states g to be constant over the range of energies considered, the number of states with energies between E_i and $E_i + W$ in a sphere of radius R is equal to $(4\pi/3)R^3 g W$. One finds *one* state for $W = 3/4\pi R^3 g$. One can use this dependence $W(R)$ to determine the extremum of the exponent $-2R_{ij}/\lambda - W_{ij}/k_B T$ in the transition probability. It occurs at $\bar{R} = (9\lambda/8\pi k_B T g)^{1/4}$, $\bar{W} = 3/4\pi\bar{R}^3 g$. If one inserts these values into the transition probability, one finds for W_{ij} and hence for the conductivity σ the temperature dependence

$$\sigma \propto \exp\left(-[T_0/T]^{1/4}\right), \qquad T_0 = \frac{5/2}{9\pi\lambda^3 k_B g}\,. \tag{10.18}$$

This is *Mott's $T^{1/4}$-law*.

The assumption that all jumps are over a fixed distance $\bar{R}$ (*fixed range hopping*) is only justified for hops between nearest neighbours. For $\bar{R} > R^0$, hops of different distances will follow one another (*variable range hopping*). In our model above of an impurity band of limited width, variable range hopping will therefore go over to fixed range hopping as the temperature increases.

The difference between fixed and variable range hopping cannot be just simply carried across to the case where the density of states is as in Fig. 10.6. The mean energy difference between nearest neighbours W^0 corresponds to the largest energy difference in the spectrum given, e.g., for an impurity band it will be its width. In the pseudogap of an amorphous semiconductor, the "largest energy difference" is the separation between the initial level and the mobility edge. A conduction process which then competes with the fixed range hopping—if this can be defined at all—is that due to the extended states above the mobility edge. At high temperatures in this case the variable range hopping will be replaced by the activated conductivity in the extended states.

The estimate of the temperature dependence of the conductivity via the assumption of hops of mean distance is too coarse. For a quantitative theory we make use of the form of (10.17). It gives the transition rate for a hopping process in an electric field in the form of a product of a *conductance* G_{ij} and the *voltage* V_{ij} between the points $\boldsymbol{R}_i$ and $\boldsymbol{R}_j$

$$\begin{aligned} &\Gamma_{ij} - \Gamma_{ji} = G_{ij}V_{ij} \\ &\qquad \text{with} \quad G_{ij} \propto \exp\left(-\frac{2R_{ij}}{\lambda} - \frac{|E_i - E_F| + |E_j - E_F| + |E_i - E_j|}{2k_B T}\right). \end{aligned} \tag{10.19}$$

In principle the conductivity can thus be obtained from a model in which the solid is described as a network of conductances, which link all lattice points in pairs. Kirchhoff's laws then apply to each lattice point. One can now see immediately why the assumption of a mean hopping distance is not valid: The total resistance of a chain of widely differing resistances is not determined by the mean individual resistance, but by the highest individual resistance. In describing the lattice as a network of resistances we therefore have to seek the *most favourable current paths* in order to find the conductivity. Within the network we shall find regions which are practically short circuited by low resistances. On the other hand, there will be regions which make no contribution to current flow, since there are preferred current paths which go round them.

Percolation theory is often used for the solution of problems such as this (for a review see, for example, *Shante* and *Kirkpatrick* [111a.20]). By percolation one means the motion of a classical particle in a scattering medium made up of randomly distributed scattering centres with given properties. The motion is thus determined by the properties of the medium. This is different from diffusion, where each scattering process is statistically independent of the previous history of the scattering and the scattered particle. The medium in which the percolation takes place is defined in abstract terms by *sites* which are connected in a given manner by *bonds*. One distinguishes two possibilities.

1) *Bond percolation*—for each bond there is a probability p that it will be open (for a particle to pass). Correspondingly $q = 1 - p$ is the probability that the passage is blocked. We can also have rules here that individual bonds can only be passed in a given direction, etc.
2) *Site percolation*—corresponding probabilities p and q are defined for the opening or blocking of sites.

The *percolation probability* $P(p)$ is defined as the probability that a particle can percolate an infinite distance from any site. It can be shown that there is a value p_c such that $P(p < p_c) = 0$. For $p = p_c$, the first *percolation path* traces out the lattice. The magnitude of p_c depends on the lattice connections given.

The similarity between this problem and that of the conductance network is clear. We first eliminate all G_{ij} from our model. Then we build them in again in order, beginning with the largest G_{ij}. When all $G_{ij} > G_1$ (with any value of G_1 initially) are built in, some bonds are already connected into clusters. For a smaller G_2 the clusters will be larger until, at a given G_c, the first continuous path (for example, from one electrode to the electrode opposite) traverses the network. From (10.19) the G_{ij} vary over orders of magnitude. One can consequently assume that the last conductance to be built in, G_c, determines the resistance of the entire current path, for all other G_{ij} are much larger than G_c. The *critical conductance* G_c, rather than the mean conductance, determines the conductivity.

We do not want to pursue further here the detailed application of percolation theory to variable range hopping. One finds the same temperature dependence as in the simplified model considered above. The conclusions about the hopping conductivity at the start of this section are thus still valid. Likewise, we do not want here to go into attempts to evaluate networks with limited numbers of lattice sites by direct numerical solution of Kirchhoff's laws. On these questions we refer the reader to the review article by *Overhof* [103.XVI].

The motion of an electron in the potential model of Fig. 10.2 is also a percolation problem. The critical path appears at the energy E_c. It is identical to the lowest extended state.

10.2.4 Conductivity in Impurity Bands and in Amorphous Semiconductors

We collect the most important results on the temperature dependence of conductivity from the last section.

a) Extended states, $E_F \ll E_c$

$$\sigma = \sigma_{E_c} \exp[-(E_c - E_F)/k_B T], \tag{10.20}$$

b) localized states near the mobility edge, $E_c > E_i \gg E_F$ (fixed range hopping),

$$\sigma = \sigma_1 \exp[-(E_i - E_F + \overline{W})/k_B T], \tag{10.21}$$

c) localized states near the Fermi energy, high temperature (fixed range hopping),

$$\sigma = \sigma_2 \exp(-\overline{W}/k_B T), \tag{10.22}$$

d) localized states near the Fermi energy, low temperature (variable range hopping),

$$\sigma = \sigma_3 \exp[-(T_0/T)^{1/4}]. \tag{10.23}$$

σ_{E_c} is the conductivity at the mobility edge, and $\overline{W}$ a mean hopping energy.

All these contributions can be seen in the temperature behaviour of the conductivity of impurity band semiconductors or of amorphous semiconductors. Fig. 10.8 shows the electrical conductivity of highly doped *p*-Ge, which has an acceptor impurity band adjacent to its valence band. Thus, compared to the density of states shown in Fig. 10.5 acceptors and donors, holes and electrons, exchange roles. The experimental curves can be reproduced by three sum terms with different activation energies. The magnitudes of the individual contributions strongly depend on the acceptor concentration (formation of the impurity band) and the donor concentration (compensation, position of the Fermi energy). Based on the density of states shown in Fig. 10.5c, we can explain the high-temperature contribution (straight sections of the conductivity curves in this logarithmic presentation) as the contribution of the holes in the extended states to the valence band, according to (10.20). The second contribution—not distinct in all curves—arises from the unoccupied Hubbard subband of the impurity

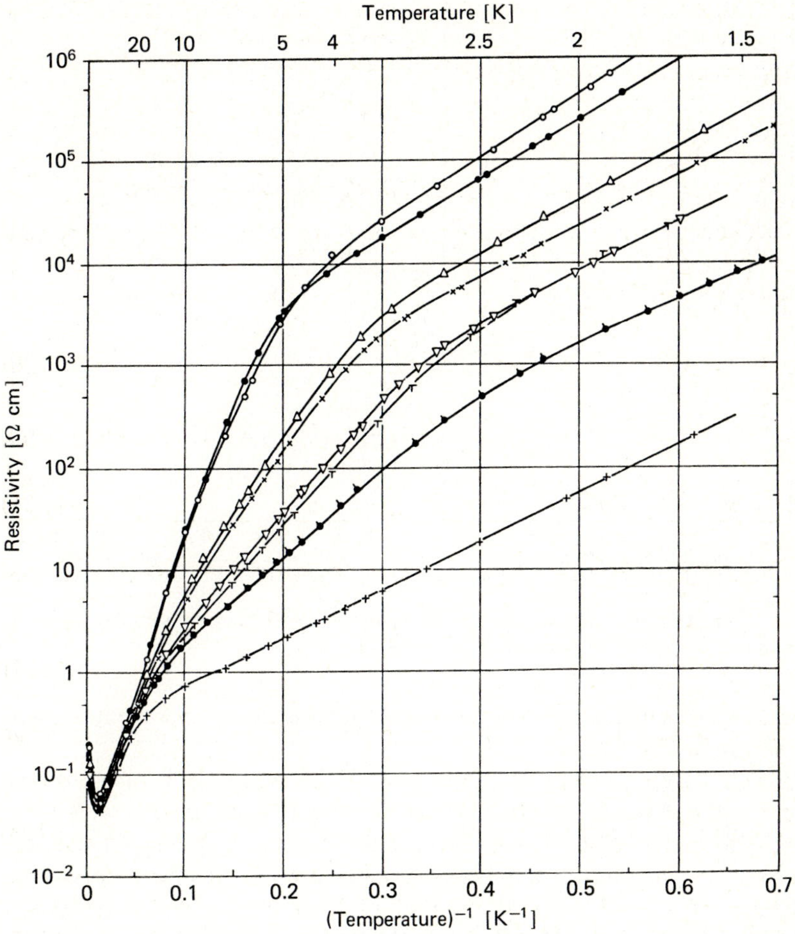

Fig. 10.8. Appearance of three regions with different activation energies for impurity band conduction in p-Ge, according to (10.20–22). The individual curves are for samples with different impurity concentrations ($n_A = 2 \cdot 10^{16}$ cm^{-3} for the uppermost curve, $= 7.3 \cdot 10^{16}$ cm^{-3} for the lowest curve) [from H. Fritzsche: Phys. Rev. **99**, 406 (1955)].

band, in accordance with (10.21). The low-temperature contribution then stems from (10.22) and concerns hopping processes in the vicinity of the Fermi energy in the lower part of the impurity band. From these and other measurements, we can obtain further details on the dependence of hopping energies on the impurity concentrations, and on metal-insulator transitions in the impurity bands.

The low-temperature conductivity in n-type germanium can also be represented by contributions with different activation energies. The intermediate range given by contributions from (10.21) is not so apparent, however, in the

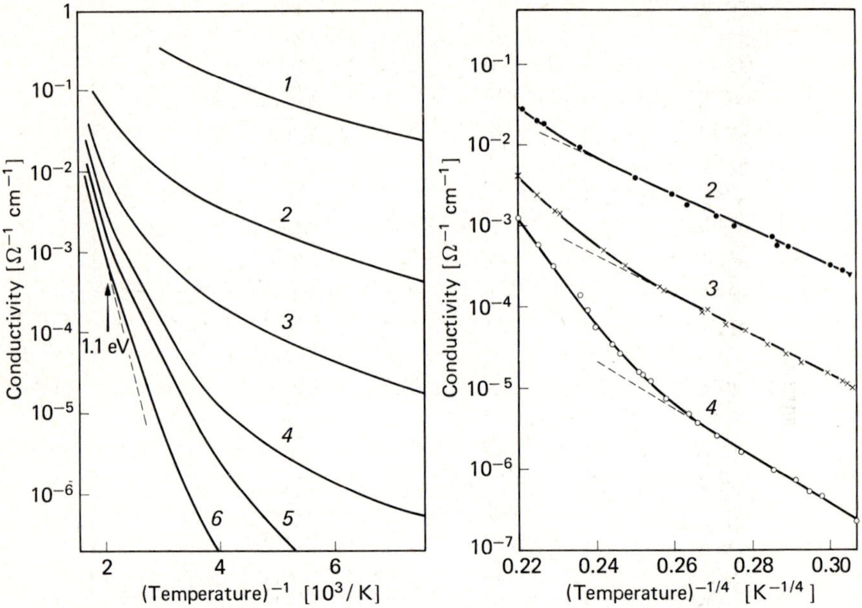

Fig. 10.9. Conductivity of amorphous Ge-films for differently treated samples [from W. Beyer, J. Stuke: J. Non-Cryst. Solids **8–10**, 321 (1972)].

published experimental curves. We have therefore presented experimental results for *p*-type Ge in Fig. 10.8.

Figure 10.9 gives an example of the Mott $T^{1/4}$-law [10.23]. It refers to conduction in amorphous Ge layers. One can see clearly from the T^{-1} and $T^{-1/4}$ plots the activated conductivity of the extended states at high temperature, and the $T^{1/4}$-law for hopping in the localized states near E_F at low temperatures, respectively.

Experimental results such as shown in the last two figures cannot, of course, explain the mechanism of electrical conduction in noncrystalline semiconductors quantitatively. Other characteristics have to be measured before one can make unambiguous statements. Moreover, only a few individual results on individual substances can be as clearly analyzed as the examples above. The cases we have discussed are thus intended only to illustrate the basic validity of the hopping concept outlined. They represent only a very small part of the wide field of electrical transport in disordered solids. Many problems in this area are still in need of a fundamental explanation.

Appendix

The Occupation Number Representation

The quantum mechanical description of elementary excitations becomes much clearer when one employs the occupation number representation. We give here a short review of the method. More detailed descriptions are to be found in, for example, *Pines* [27], *Taylor* [22], *Ziman* [42], and in general form also in many textbooks on quantum mechanics.

In this representation we have to distinguish between bosons and fermions. We start with *bosons*.
Let the Hamiltonian of a boson gas be

$$H = \frac{1}{2}\sum_{\boldsymbol{k}} (P_{\boldsymbol{k}}^* P_{\boldsymbol{k}} + \omega_{\boldsymbol{k}}^2 Q_{\boldsymbol{k}}^* Q_{\boldsymbol{k}}), \quad P_{-\boldsymbol{k}} = P_{\boldsymbol{k}}^*, \quad Q_{-\boldsymbol{k}} = Q_{\boldsymbol{k}}^*, \quad \omega_{-\boldsymbol{k}} = \omega_{\boldsymbol{k}}. \tag{A.1}$$

We found an operator like this, which consists of a sum of terms that are formally identical to the Hamiltonian of the harmonic oscillator, when introducing plasmons [(3.23)] or phonons [(3.128)], for example.

The operators $P_{\boldsymbol{k}}$, $Q_{\boldsymbol{k}}$ in (A.1) obey the commutation relations

$$[Q_{\boldsymbol{k}}, P_{\boldsymbol{k}'}] = \mathrm{i}\hbar\delta_{\boldsymbol{k}\boldsymbol{k}'}. \tag{A.2}$$

As is well known, the energy of the boson gas follows from (A.1) as the sum of the individual energies $E_{\boldsymbol{k}} = \hbar\omega_{\boldsymbol{k}}(n_{\boldsymbol{k}} + \frac{1}{2})$ of harmonic oscillators. We now introduce new operators $a_{\boldsymbol{k}}^+$ and $a_{\boldsymbol{k}}$ by

$$a_{\boldsymbol{k}}^+ = (2\hbar\omega_{\boldsymbol{k}})^{-1/2}(\omega_{\boldsymbol{k}} Q_{\boldsymbol{k}}^* - \mathrm{i}P_{\boldsymbol{k}}), \quad a_{\boldsymbol{k}} = (2\hbar\omega_{\boldsymbol{k}})^{-1/2}(\omega_{\boldsymbol{k}} Q_{\boldsymbol{k}} + \mathrm{i}P_{\boldsymbol{k}}^*). \tag{A.3}$$

Hence

$$H = \sum_{\boldsymbol{k}} \hbar\omega_{\boldsymbol{k}}\left(a_{\boldsymbol{k}}^+ a_{\boldsymbol{k}} + \frac{1}{2}\right) + \sum_{\boldsymbol{k}} \mathrm{i}\omega_{\boldsymbol{k}}(Q_{\boldsymbol{k}} P_{\boldsymbol{k}} - Q_{-\boldsymbol{k}} P_{-\boldsymbol{k}}). \tag{A.4}$$

In the summation over $\boldsymbol{k}$, where there is a term with $-\boldsymbol{k}$ for every term with $\boldsymbol{k}$, the last term disappears. It follows that

$$H = \sum_{\boldsymbol{k}} \hbar\omega_{\boldsymbol{k}}\left(a_{\boldsymbol{k}}^+ a_{\boldsymbol{k}} + \frac{1}{2}\right). \tag{A.5}$$

The Schrödinger equation then becomes

$$\sum_k \hbar\omega_k\left(a_k^+ a_k + \frac{1}{2}\right)\psi = \sum_k \hbar\omega_k\left(n_k + \frac{1}{2}\right)\psi \tag{A.6}$$

hence we also have

$$a_k^+ a_k \psi = n_k \psi \,. \tag{A.7}$$

Thus $a_k^+ a_k$ can be interpreted as a *particle number operator*. If we denote the eigenvector ψ of this operator by giving the occupation numbers n_k of the states defined by $\boldsymbol{k}$, (A.7) becomes

$$a_k^+ a_k |n_1 \ldots n_k \ldots\rangle = n_k |n_1 \ldots n_k \ldots\rangle \,. \tag{A.8}$$

From (A.2) we find the following commutation relations for the operators (A.3):

$$[a_k, a_{k'}^+] = \delta_{kk'} \,. \tag{A.9}$$

Eqs. (A.8) and (A.9) immediately reveal the meaning of these operators. We find

$$a_k^+ a_k a_k^+ |n_1 \ldots n_k \ldots\rangle = (n_k + 1) a_k^+ |n_1 \ldots n_k \ldots\rangle \,. \tag{A.10}$$

Thus

$$a_k^+ |n_1 \ldots n_k \ldots\rangle = N^+(n_k) |n_1 \ldots n_k + 1 \ldots\rangle \tag{A.11}$$

where N^+ is a normalization factor, which is still to be determined. Correspondingly

$$a_k^+ a_k a_k |n_1 \ldots n_k \ldots\rangle = (n_k - 1) a_k |n_1 \ldots n_k \ldots\rangle \tag{A.12}$$

hence

$$a_k |n_1 \ldots n_k \ldots\rangle = N^-(n_k) |n_1 \ldots n_k - 1 \ldots\rangle \,. \tag{A.13}$$

Thus a_k^+ *creates* and a_k *annihilates* a quantum in state $\boldsymbol{k}$ (*creation and annihilation operators*). One determines the normalization factors from

$$\begin{aligned} &\langle n_k | n_k \rangle = 1 \\ &\langle n_k | a_k^+ a_k | n_k \rangle = n_k = N^-(n_k) N^+(n_k - 1) \\ &\langle n_k | a_k a_k^+ | n_k \rangle = n_k + 1 = N^+(n_k) N^-(n_k + 1) \,. \end{aligned} \tag{A.14}$$

Here we have used the abbreviation $|n_k\rangle$ for $|n_1 \ldots n_k \ldots\rangle$. Eqs. (A.14) are satisfied for $N^+(n_k) = \sqrt{n_k + 1}$, $N^-(n_k) = \sqrt{n_k}$; hence

$$\begin{aligned} a_k^+|n_1 \ldots n_k \ldots\rangle &= \sqrt{n_k + 1}|n_1 \ldots n_k + 1 \ldots\rangle, \\ a_k|n_1 \ldots n_k \ldots\rangle &= \sqrt{n_k}|n_1 \ldots n_k - 1 \ldots\rangle. \end{aligned} \tag{A.15}$$

One can clearly obtain each eigenvector $|n_1\, n_2 \ldots . n_k \ldots\rangle$ by repeated application of creation operators a_k^+ on the *vacuum state* $|0\,0 \ldots . 0 \ldots\rangle$

$$|n_1 n_2 \ldots n_k \ldots\rangle = (a_k^+)^{n_k} \ldots (a_1^+)^{n_1}|00 \ldots 0 \ldots\rangle. \tag{A.16}$$

We can introduce a corresponding representation for *fermions*. According to the Pauli principle, each state can only be occupied by one fermion. The n_k in the eigenvectors can therefore only take the values 0 and 1. Thus we find for the corresponding creation and annihilation operators, which we call c_k^+ and c_k

$$c_k^+|0\rangle = |1\rangle, \qquad c_k^+|1\rangle = 0, \qquad c_k|0\rangle = 0, \qquad c_k|1\rangle = |0\rangle. \tag{A.17}$$

We have yet to satisfy the Pauli principle, i.e., the condition that the eigenvectors are antisymmetric. In an exchange of two particles they have to change sign. To define such an exchange, we arrange the n_k, for example, according to the magnitudes of the eigenvalues associated with them. The corresponding relation to (A.16) for constructing an eigenvalue from the vacuum state

$$|0_1 1_2 1_3 \ldots 1_k \ldots\rangle = (-1)^{n+1} \ldots c_k^+ \ldots c_3^+ c_2^+ |0_1 0_2 0_3 \ldots 0_k \ldots\rangle \tag{A.18}$$

is then to be read in such a way that particles are created in succession in the states 2, 3. . . . $\boldsymbol{k}$. See (A.23) for the sign. The order of the states is thus now of importance. The exchange of two particles means the exchange of two n_k, $n_{k'}$ in the eigenvector, hence from (A.18) the exchange of two c_k on the right side. In order that the exchange shall be associated with a change of sign,

$$c_k^+ c_{k'}^+ = -c_{k'}^+ c_k^+ \tag{A.19}$$

and correspondingly

$$c_k c_{k'} = -c_{k'} c_k, \qquad c_k c_{k'}^+ = -c_{k'}^+ c_k \qquad (\boldsymbol{k} \neq \boldsymbol{k}'). \tag{A.20}$$

For $\boldsymbol{k} = \boldsymbol{k}'$ (A.19) and the first equation (A.20) are likewise valid, since from (A.17) the products $c_k^+ c_k^+$ and $c_k c_k$ give zero. From (A.17) it follows that for the second equation (A.20)

$$\begin{aligned} c_k c_k^+|0\rangle &= |0\rangle \qquad & c_k^+ c_k|0\rangle &= 0 \\ c_k c_k^+|1\rangle &= 0 \qquad & c_k^+ c_k|1\rangle &= |1\rangle, \end{aligned} \tag{A.21}$$

hence

$$c_k c_k^+ + c_k^+ c_k = 1 \quad \text{or} \quad [c_k c_{k'}^+]_+ = \delta_{kk'} . \tag{A.22}$$

The same commutation relations apply for the c_k as for the a_k, except that commutators are replaced by anticommutators.

The relations (A.17), together with the commutation relations, lead to an equation analogous to (A.15)

$$\begin{aligned} c_k^+|\ldots n_k \ldots\rangle &= \sqrt{1-n_k}(-1)^{\nu_k}|\ldots n_k + 1 \ldots\rangle \\ c_k|\ldots n_k \ldots\rangle &= \sqrt{n_k}(-1)^{\nu_k}|\ldots n_k - 1 \ldots\rangle \end{aligned} \tag{A.23}$$

with $\nu_k = \sum_{i<k} n_i$. Here we have a positive sign when there is an even number of states in the eigenvector to the left of n_k, and a negative sign when the number is odd.

We now have to examine how to convert quantum mechanical equations from the $\boldsymbol{r}$-representation into the occupation number representation. To do this we look at the simplest case of an operator H which is made up by adding together a number of one-particle operators $h(\boldsymbol{r}_i)$: $H = \sum_i h(\boldsymbol{r}_i)$. Let it operate on a wave function which describes bosons. Φ is then given by the combination which is invariant to particle exchange

$$\Phi = \frac{1}{\sqrt{N!n_1!n_2!\ldots}} \sum_P P\varphi_\alpha(r_1)\ldots\varphi_\alpha(r_{n_1})\varphi_\beta(r_{n_1+1})\ldots\varphi_\lambda(r_i)\ldots\varphi_\omega(r_N) \tag{A.24}$$

where the sum is over all permutations of the indices $\alpha, \beta \ldots$, and n_1 factors have the index α, n_2 factors the index β, etc. Thus

$$\sum_\lambda n_\lambda = \sum_i 1 = N . \tag{A.25}$$

Application of H to Φ yields

$$H\Phi = \frac{1}{\sqrt{\ldots}} \sum_i \sum_P P\varphi_\alpha(\boldsymbol{r}_1)\ldots h\varphi_\lambda(\boldsymbol{r}_i)\ldots\varphi_\omega(\boldsymbol{r}_N) . \tag{A.26}$$

Since $h(\boldsymbol{r}_i)\varphi_\lambda(\boldsymbol{r}_i) = \sum_{\lambda'} \varphi_{\lambda'}(\boldsymbol{r}_i)\langle\lambda'|h|\lambda\rangle$ (with i-independent matrix element!), we find

$$H\Phi = \frac{1}{\sqrt{\ldots}} \sum_{i=1}^{N} \sum_{\lambda'} \sum_P P\varphi_\alpha(\boldsymbol{r}_1)\ldots\varphi_{\lambda'}(\boldsymbol{r}_i)\ldots\varphi_\omega(\boldsymbol{r}_N)\langle\lambda'|h|\lambda\rangle . \tag{A.27}$$

If we now write Φ in the form $|n_1 \ldots n_\lambda \ldots\rangle$, we have to distinguish in the sum between two cases for

$$\varphi_\lambda = \varphi_{\lambda'}: \sum_i \langle\lambda|h|\lambda\rangle|n_1 \ldots n_\lambda \ldots\rangle = \sum_\lambda n_\lambda\langle\lambda|h|\lambda\rangle|n_1 \ldots n_\lambda \ldots\rangle \tag{A.28}$$

and for

$$\varphi_\lambda \neq \varphi_{\lambda'}: \sum_i \sum_{\lambda'}{}' \sqrt{\frac{n_{\lambda'}+1}{n_\lambda}}\langle\lambda'|h|\lambda\rangle|n_1 \ldots n_{\lambda'}+1 \ldots n_\lambda - 1 \ldots\rangle$$
$$= \sum_{\substack{\lambda\lambda' \\ (\lambda\neq\lambda')}} \sqrt{n_\lambda}\sqrt{n_{\lambda'}+1}\,\langle\lambda'|h|\lambda\rangle|n_1 \ldots n_{\lambda'}+1 \ldots n_\lambda - 1 \ldots\rangle. \tag{A.29}$$

In view of (A.8) and (A.15), we can combine (A.28) and (A.29) to

$$H|n_1 \ldots n_\lambda \ldots\rangle = \sum_{\lambda'\lambda} \langle\lambda'|h|\lambda\rangle a^+_{\lambda'} a_\lambda |n_1 \ldots n_{\lambda'} \ldots n_\lambda \ldots\rangle . \tag{A.30}$$

Hence in the occupation number representation the operator H becomes

$$H = \sum_i h^{(1)}(r_i) = \sum_{\lambda'\lambda} \langle\lambda'|h^{(1)}|\lambda\rangle a^+_{\lambda'} a_\lambda \tag{A.31}$$

where the index (1) denotes that $h^{(1)}$ is an *one-particle operator* and where

$$\langle\lambda'|h^{(1)}|\lambda\rangle = \int \varphi^*_{\lambda'}(\boldsymbol{r}_1) h^{(1)}(\boldsymbol{r}_1)\varphi_\lambda(\boldsymbol{r}_1)d\tau_1 .$$

Equation (A.31) is also valid (with c_λ instead of a_λ) for *fermions*. This follows from the antisymmetry of the wave function (1.19), which can also be written as

$$\Phi = \frac{1}{\sqrt{N!}} \sum_P P(-1)^P \varphi_\alpha(\boldsymbol{q}_1)\varphi_\beta(\boldsymbol{q}_2) \ldots \varphi_\omega(\boldsymbol{q}_N) . \tag{A.32}$$

The changed factors on the individual sum terms do not affect the line of argument in any way. We do of course have to take account of the fact that λ in (A.31) counts *states*. The spin summation is thus also contained in the sum.

Correspondingly, one can show that for *two-particle operators* $H = \sum_{ij} h^{(2)}(\boldsymbol{r}_i, \boldsymbol{r}_j)$ it follows for bosons that

$$H = \sum_{ij} h^{(2)}(\boldsymbol{r}_i, \boldsymbol{r}_j) = \sum_{\lambda\lambda'\mu\mu'} \langle\lambda'\mu'|h^{(2)}|\lambda\mu\rangle a^+_{\lambda'} a^+_{\mu'} a_\mu a_\lambda \tag{A.33}$$

with

$$\langle\lambda'\mu'|h^{(2)}|\lambda\mu\rangle = \int \varphi^*_{\lambda'}(\boldsymbol{r}_1)\varphi^*_{\mu'}(\boldsymbol{r}_2) h^{(2)}(\boldsymbol{r}_1, \boldsymbol{r}_2)\varphi_\lambda(\boldsymbol{r}_1)\varphi_\mu(\boldsymbol{r}_2)d\tau_1 d\tau_2 . \tag{A.34}$$

For fermions the $a^{(+)}$ have again to be replaced by the $c^{(+)}$. The derivation of this equation is similar to the derivation of (A.31). We shall not repeat the rather tedious calculations here.

Problems to Chapters 1–9

1.1

a) Show that the variation $\langle \delta\Phi | H | \Phi \rangle = 0$, where

$$H = \sum_{\lambda\lambda'} \langle \lambda' | h^{(1)} | \lambda \rangle c^+_{\lambda'} c_\lambda + \frac{1}{2} \sum_{\lambda\lambda'\mu\mu'} \langle \lambda'\mu' | h^{(2)} | \lambda\mu \rangle c^+_{\lambda'} c^+_{\mu'} c_\mu c_\lambda$$

[(A.31) and (A.33) for fermions] leads to the equation

$$\langle \lambda | h^{(1)} | \mu \rangle + \sum_\nu (\langle \lambda\nu | h^{(2)} | \mu\nu \rangle - \langle \lambda\nu | h^{(2)} | \nu\mu \rangle) = 0$$

where $|\mu\rangle$ is an occupied and $|\lambda\rangle$ an unoccupied state, and where the sum is over all occupied states.

b) Discuss the properties of the "self-consistent one-particle Hamiltonian"

$$H^{\rm sc} = \sum_{\lambda\mu} (\langle \lambda | h^{(1)} | \mu \rangle + \sum_\nu (\langle \lambda\nu | h^{(2)} | \mu\nu \rangle - \langle \lambda\nu | h^{(2)} | \nu\mu \rangle)$$

where there is no restriction on the $|\mu\rangle$ and $|\lambda\rangle$. In particular, show that all matrix elements between occupied and unoccupied states are zero.

c) Use the results from a) and b) to derive the Hartree-Fock equations in occupation number representation from (1.13).

1.2

a) Discuss the exchange interaction in the simple case of a system consisting of two identical particles. The Hamiltonian is $H = H_1 + H_2 + H_{12}$, $H_1 = h(\boldsymbol{r}_1)$, $H_2 = h(\boldsymbol{r}_2)$, $H_{12}(\boldsymbol{r}_1, \boldsymbol{r}_2) = H_{12}(\boldsymbol{r}_2, \boldsymbol{r}_1)$. Treat the interaction H_{12} as a perturbation.

b) At time $t = 0$ let the two particles be in state $\psi(\boldsymbol{r}_1, \boldsymbol{r}_2, 0) = \psi_i(\boldsymbol{r}_1)\psi_k(\boldsymbol{r}_2)$ where ψ_n is given by $H_n\psi_n = E_n\psi_n$. Discuss the time dependence of $\psi(\boldsymbol{r}_1, \boldsymbol{r}_2, t)$.

c) Discuss the three-particle Hamiltonian $H = H_1 + H_2 + H_3 + H_{12} + H_{23} + H_{31}$. To what extent can the calculation be carried out analogously to a)? What difficulties arise?

2.1

a) How are the free energy F, the thermodynamic potential Ω, and the entropy S of the free electron gas related to its total energy U?

b) Calculate the following free electron gas parameters as a function of the electron concentration and the volume at $T = 0$: $k_{\rm F}$, $E_{\rm F}$, $v_{\rm F}$, U, F, Ω, S.

c) For $T = 0$, the Fermi distribution function is a step function. It therefore follows that

$$\int_0^\infty F(E)f_0(E)g(E)dE = \int_0^{E_F} F(E)g(E)dE\,.$$

Calculate the first-order correction to this relation for $T \neq 0$ and use the result to determine the temperature dependence of μ, $U = N\bar{E}$ [(2.30)], F, Ω, and S at low temperatures.

2.2

Determine the density of states $g(E)\,dE$, the Fermi energy E_F, and the mean energy $\bar{E}(T = 0)$ for a one-dimensional and a two-dimensional electron gas.

2.3

The expression for the magnetization given in (2.66) follows from (2.65) after a rather tedious calculation. To reproduce the derivation, make use of *Wilson* [34]. The derivation of the first two terms in (2.66) is relatively straightforward. Pauli paramagnetism (first term): Use Fig. 2.9 to calculate the electron contribution to the magnetization.
Landau diamagnetism (second term): Calculate the mean energy $\bar{E}$ using the density of states (2.63), and from it the magnetization.
(Hint: For small magnetic fields the following approximation applies:

$$\sum_{n=0}^{\infty} F\left(n + \frac{1}{2}\right) = \int_0^\infty F(x)dx - \frac{1}{24}\left.\frac{dF(x)}{dx}\right|_0^\infty + \cdots\Bigg)\,.$$

How is the factor in front of the brackets in (2.66) related to the density of states at the Fermi surface $g(E_F)$?

2.4

Symmetries of the simple cubic lattice (Wigner-Seitz cell: Fig. 2.12a, Brillouin zone: Fig. 2.22a).

a) Determine the elements of the space group and of the point group.
b) Into which classes can the point group (the group of the vector $\boldsymbol{k} = 0$) be resolved? What dimensions do the irreducible representations have?
c) What elements, classes, and irreducible representations has the group of the vector $\boldsymbol{k} = (k_x, 0, 0)$ (point on the Δ-axis of the Brillouin zone)?
d) What elements, classes, and irreducible representations has the group of the vector $\boldsymbol{k} = (\pi/a, 0, 0)$ (point X on the surface of the Brillouin zone)?
e) From the results of a)–d), what statements can be made about possible degeneracies and connections between bands along the Δ-axis?

2.5

a) Calculate the five lowest "free electron" bands along the Δ-axis in the Brillouin zone for the simple cubic lattice (analogous to Fig. 2.35). Calculate the associated eigenfunctions.
b) For bands which are degenerate along the Δ-axis (or at Γ or at X), construct linear combinations of the wave functions in such a way that the new func-

tions are either invariant to the group operations, or reduce to sets of functions which transform into one another. Such sets form the basis functions for the irreducible representations of the group of the $\boldsymbol{k}$-vector.

2.6

In the vicinity of extrema and saddle points of a band, the band-structure function $E_n(\boldsymbol{k})$ can be described in the form

$$E_n(\boldsymbol{k}) = E_0(\boldsymbol{k}_0) + \sum_i a_i(k_i - k_{0i})^2 \,.$$

There is a singularity in the density of states integrand in (2.152) at such *critical points*.

a) Show that there is a discontinuity in dg/dE ("kink" in the density of states) at these points.
b) What types of critical points arise in the one-, the two-, and the three-dimensional case? How does $g(E)$ behave in the neighbourhood of $g(E_0)$? (cf. Fig. 6.2).

The critical points in the density of states give rise to a characteristic structure in the absorption spectrum of semiconductors (see Sec. 6.2.2).

2.7

Let the functions $E_n(\boldsymbol{k})$ and $\psi_n(\boldsymbol{k}, \boldsymbol{r})$ be known for $\boldsymbol{k} = \boldsymbol{k}_0$. By using the Schrödinger equation for $u_n(\boldsymbol{k}, \boldsymbol{r})$, a perturbation method can be developed to determine $E_n(\boldsymbol{k})$ in the vicinity of $\boldsymbol{k}_0$ ($\boldsymbol{k}\cdot\boldsymbol{p}$-method).

a) Let $\boldsymbol{k}_0$ be a band extremum ($\nabla_k E_n(\boldsymbol{k})|_{\boldsymbol{k}_0} = 0$). Calculate $E(\boldsymbol{k}_0 + \boldsymbol{\kappa})$ in second-order perturbation approximation. Distinguish between the cases of a non-degenerate and a degenerate band extremum.
b) How does the result a) simplify for a two-band model (n takes only the values 1, 2)?
c) Using the nearly free electron approximation, calculate $E(\boldsymbol{k})$ in the neighbourhood of a Bragg reflection and compare the result with that obtained in b).

2.8

For semiconductors with diamond or zinc blende structure, the $\boldsymbol{k}\cdot\boldsymbol{p}$ method, mentioned in Problem 2.7, gives in the vicinity of the point Γ (Fig. 2.33) the relationship

$$E'\{E'(E' - E_G)(E' + \Delta) - k^2P^2[E' + (2/3)\Delta]\} = 0$$

with $E' = E(k) - \hbar^2k^2/2m$ and given parameters and matrix elements E_G, P, and Δ. From this find $E(k)$ for small k for the conduction and valence bands. What effective masses emerge for the electrons in the conduction band and the holes in the valence bands? (The uppermost valence band is poorly described in this approximation since only the mutual interaction of the four bands has been taken into account). What is the structure of the conduction band of InSb

for larger values of k? (For InSb one can put $\Delta \gg E_G, kP$).
It will be useful to derive the equation above by following the original treatment by *Kane* [E. O. Kane: J. Phys. Chem. Solids *1*, 249 (1957)].

3.1

Find a canonical transformation whereby the electron-plasmon interaction can be removed from the Hamiltonian (3.23). How do the effective electron mass and plasmon frequency change?

3.2

Discuss the renormalization terms in (3.44) and (3.45) and show in particular that

a) the renormalization of the fermion mass near the Fermi surface ($k \approx k_F$) leads to a reduction in the electron velocity (mass-enhancement) and
b) that the renormalization of the boson frequency leads to a logarithmic divergence in the $\omega(\boldsymbol{q})$ spectrum at $q = 2k_F$ (Kohn anomaly).

3.3

Calculate the 1*s*-, 2*s*-, and 2*p*-states of a Wannier exciton in a semiconductor with anisotropic effective electron mass ($m_\parallel, m_\perp$) and anisotropic dielectric constant ($\varepsilon_\parallel, \varepsilon_\perp$) (example: CdS).

3.4

Let a boson gas have concentration $n = N/V_g$. The distribution of bosons among the possible quantum states E_q is given by the Bose distribution function $f_B = (\exp[(E_q - \mu)/k_B T] - 1)^{-1}$.

a) Determine the temperature dependence of the chemical potential $\mu(T)$ and compare it with that for the chemical potential of a fermion gas.
b) Why is $\mu = 0$ for phonons for all temperatures?

3.5

Dispersion curves for phonons in diamond (Fig. 3.11).

a) Why are the LO- and TO-branches degenerate at $\boldsymbol{q} = 0$?
b) Use the results of Problem 2.4 to discuss the dispersion curves along the Δ-axis.

Diamond has a face-centred cubic lattice with diatomic basis.

3.6

a) For the linear chain with one and with two atoms in its basis, calculate the density of states shown in Fig. 3.12a.
b) For the monoatomic linear chain, calculate the specific heat and discuss its temperature dependence at low and high temperatures.

3.7

a) Calculate the magnetization $M(T)$ at low temperatures for a two-dimensional square lattice and for a simple cubic lattice.
b) What difficulties arise in the two-dimensional case? What are the reasons for them and what consequences do they have?
c) What is the effect of a (weak) magnetic field?

3.8

a) Calculate the magnetization of a free electron gas at low temperature using the Stoner collective electron model with given electron concentration n and exchange energy V.

b) For what parameter values is there no spontaneous magnetization at $T = 0$· for what values is $M(0)$ less than the saturation magnetization?

4.1

Use the deformation potential approximation to calculate the relaxation time $\tau(E)$ for the interaction between electrons and LA-phonons in a nondegenerate semiconductor with parabolic conduction band. How large is the exponent r in $\tau(E) = \tau_0 E^r$?

4.2

Find a canonical transformation which eliminates the electron coordinates from the Hamiltonian (4.22). This transformation is the first step to an improved theory of the large polaron (Lee-Low-Pines method). The next step is a further transformation of the Hamiltonian with the unitary operator $U = \exp\{\sum_q [a_q^+ f(q) - a_q f^*(q)]\}$. The function $f(q)$ has to be chosen in such a way that the expectation value $E = \langle 0|H_{\text{trans}}|0\rangle$ becomes an extremum ($|0\rangle$ = ground state, no phonons). What is the meaning of the second transformation? What differences to (4.29) result for small $\boldsymbol{q}$?

4.3

In an anisotropic solid the general relationship between the components of the current densty and of the electric and magnetic fields applied is

$$i_i = a_{ij}E_j + a_{ijk}E_jB_k + a_{ijkl}E_jB_kB_l + \cdots .$$

a) Which components of the tensors a_{ij}, a_{ijk}, a_{ijkl} vanish for cubic symmetry? What relationships occur between the remaining components?

b) What are the deviations from the isotropic case for conductivity, Hall coefficient, and magnetoresistance?

4.4

Assume a nondegenerate mixed semiconductor. Electrons and holes thus take part simultaneously in the transport phenomena.

a) Using the relaxation time approximation, compute the transport coefficients of (4.85) and from these the electrical conductivity, the thermoelectric power, the Peltier coefficient, and the thermal conductivity.

b) Why does the Peltier coefficient have a different sign in n-type and p-type semiconductors? Under what conditions does the Peltier coefficient vanish in an intrinsic semiconductor ($n = p$)?

c) In addition to an electron and a hole component, the expression for the thermal conductivity contains an additional term which only appears in the case of mixed conduction. What is the physical explanation for this "ambipolar" term?

d) Use the relaxation time approximation to calculate the coefficient for the isothermal Nernst effect. How is the result to be interpreted? In what circumstances does the Nernst coefficient change sign with change of conductivity type?

4.5

From (4.90) determine how the electric current density depends on all possible "forces". In the relaxation time approximation, what are the expressions for the *diffusion coefficient* (constant of proportionality between $\boldsymbol{i}$ and ∇n) and the coefficient of *thermo-diffusion* (constant of proportionality between $\boldsymbol{i}$ and ∇T)? Which gradient of space-dependent quantities can lead to further contributions to $\boldsymbol{i}$?

4.6

The relationship between diffusion coefficient and mobility given in most textbooks is the *Einstein relation* $D_n = (k_B T)\mu_n$. This relation, however, is valid only for a nondegenerate electron gas. Derive the general relation. How does the diffusion coefficient depend on the electron concentration for strong degeneracy?

4.7

In high magnetic fields one finds in the degenerate electron gas an oscillatory magnetoresistance similar to the de Haas-van Alphen effect treated in Section 2.1.4 (*de Haas-Shubnikov effect*). Discuss the origin of these oscillations qualitatively for the simplest case of longitudinal magnetoresistance ($\boldsymbol{B} \parallel \boldsymbol{E}$).

5.1

Use (5.18) and (5.36) to calculate the electronic specific heat of a superconductor. What is the temperature dependence of this quantity for $T \to 0$?

5.2

a) Introduce creation and annihilation operators $b_{\boldsymbol{k}}^+$, $b_{\boldsymbol{k}}$ for Cooper pairs and occupation probabilities $p_{\boldsymbol{k}}$ for pairs of states $(\boldsymbol{k}, \sigma; -\boldsymbol{k}, -\sigma)$ into the Hamiltonian (5.9) and the wave function (5.31). What commutation relations hold for the $b_{\boldsymbol{k}}^+$, $b_{\boldsymbol{k}}$?
b) Calculate the expectation value of the energy and minimize it with respect to the $p_{\boldsymbol{k}}$.
c) Discuss the meaning of this procedure and verify the main results derived in another way in Section 5.3.

5.3

a) Consider a superconductor which for $T \neq 0$ contains Cooper pairs and single excited electrons. Introduce occupation probabilities for both excitations. Calculate the expectation value of the energy, derived for zero temperature in Problem 5.2b, for $T \neq 0$.
b) Use this result to write down the free energy of a superconductor as a function of T and minimize it as in Problem 5.2.

5.4

a) Discuss the flux quantization by using the Ginzburg-Landau approach [(5.66, 70)].

b) Discuss the Josephson effect by using a time-dependent Schrödinger equation for the order parameter, where $\partial\Psi/\partial t$ on one side of the junction is coupled linearly to Ψ on the other side and vice versa. Show that the *dc* current is proportional to $\sin(\delta - 2eVt/\hbar)$, where V is the applied voltage and δ the phase difference of the order parameter on both sides of the junction [(5.70)].

6.1

Discuss the "magnon-polariton". What equations replace (6.14)? What are the dispersion relations?

6.2

In many semiconductors (e.g., Ge, Si, III–V compounds) the valence band is degenerate at $\boldsymbol{k} = 0$, i.e., two (approximately isotropic and parabolic) bands with effective masses m_1, m_2 together form the upper edge of the valence band. A third valence subband has its maximum at an energy Δ, the spin-orbit-splitting energy, below this edge (see Problem 2.8). Direct optical transitions can take place between these three subbands (intraband transitions). How should the results in Section 6.2.2 be modified for this case?

6.3

Equation (6.60) for absorption due to direct optical transitions is often written in another form by introducing *oscillator strengths* $f_{j'j}$

$$f_{j'j} = (2/m\hbar\omega_{j'j})|p_{j'j}|^2 ,$$

where $p_{j'j}$ is the momentum matrix element and $\omega_{j'j}$ the transition frequency. Prove the sum rule

$$\sum_{j'} f_{j'j} = 1 - (m/m_j^*) .$$

6.4

Derive (6.64) by a semi-classical method analogous to that employed in Section 6.2.7. Describe the valence electrons by classical oscillators with the transition frequencies as eigenfrequencies. Introduce the oscillator strengths $f_{j'j}$ defined in Problem 6.3. Show that the contribution from the direct valence-conduction band transitions to the real part of the dielectric constant is given by

$$\varepsilon_1 \propto \omega^{-2}(2\sqrt{E_G} - \sqrt{E_G + \hbar\omega}) .$$

6.5

Extend the treatment in Problem 6.4 by including an external magnetic field (interband magneto-absorption). Confirm (6.116) and find the corresponding expression for ε_1.

6.6

a) Using the relaxation time approximation, calculate the imaginary part of the electrical current density at high frequencies. By equating the result with the imaginary part of (6.101), define an "optical effective mass" m_{op}. Compare m_{op} with the effective mass m^* defined by the second derivative of energy with respect to the components of the wave vector

1) for an isotropic parabolic band $E = \hbar^2 k^2/2m$,
2) for an anisotropic parabolic band $E = (\hbar^2/2)(k_1^2/m_1 + k_2^2/m_2 + k_3^2/m_3)$,
3) for two bands with effective masses m_1, m_2 degenerate at $\mathbf{k} = 0$,
4) for an isotropic band with complete degeneracy of the electron gas.

b) In cases a2) and a3), what combinations of effective masses appear in the density of states $g(E)$? (Definition of a *density of states mass*).

c) In case a2), what combination of effective masses appears in the cyclotron resonance frequency ω_c? Note that the combination depends on the orientation of the magnetic field.

6.7

As a result of the interaction conduction band—spin-orbit split valence band in InSb, not only is the effective electron mass $m^*(k)$ significantly reduced [$m^*(0) \approx 0.015$ m] but the g-factor of the electrons is also greatly changed [$g^*(0) \approx -50$ compared with $g = +2$ for free electrons]. Calculate the g-factor using the equation for the band structure of InSb given in Problem 2.8. In a magnetic field the equation should be extended to include a term $\Delta P^2 eB/3\hbar$ and the energy parameters modified in accordance with Section 2.1.6. Show that in the limit $\Delta \gg E_G$ the formula becomes simply $g^*(0) = 1 - m/m^*(0)$.

7.1

Derive the linear thermal expansion coefficient $\alpha = (1/l)(dl/dT)$ of an ionic crystal (NaCl structure). The temperature should be so high that Boltzmann statistics can be used for the calculation of the mean deviation of the ions from their equilibrium position at $T = 0$. Describe the ion-ion interaction by both possibilities treated in Section 8.2.4 (exponential and power law for the repulsive part of the interaction).

7.2

In metals phonons are scattered at the free electrons in addition to the scattering mechanisms treated in Section 7.4. In this case the Boltzmann equation of the phonons can be solved similarly to the reverse case of the electron scattering at phonons (Sec. 4.4.1). Discuss the calculation procedure under the assumption that the electron system remains in equilibrium.

8.1

What combinations of s-, p-, and d-orbitals lead to hybrid-orbitals having the following symmetries:

a) identical bond to two neighbours in a linear chain,
b) identical bond to three neighbours in a plane,
c) identical bond to six neighbours arranged octahedrally.

8.2

Derive a general quantum-mechanical expression for the resonance energy in (8.28).

8.3

The expansion of the Madelung number, in which the contributions from nearest, next-nearest . . . neighbours are summed, converges very slowly. The following method is more suitable. One combines cells, each of which has zero net charge, e.g., a negative central ion ($-e_i$), and $+e_i/v$ charges of each of the v nearest neighbours. To the contribution of such a cell one then adds the contributions from the nearest, next-nearest . . . identical neighbouring cells.

a) Demonstrate the convergence of this method by comparing both expansions for the linear chain with alternating positive and negative charges. The Madelung number for this case is $A = 2 \ln 2 = 1.3863$.

b) Calculate the Madelung number for the NaCl-lattice [three-dimensional extension of a)].

8.4

The similarity of many properties between the III–V compounds and the group IV elements can be traced to the fact that, in both cases, on average four valence electrons are available to form covalent or mixed covalent/ionic bonds with the four nearest neighbours.

a) Continue the scheme

IV_2
/ \
III V

Which groups of other binary and ternary compounds with coordination number four can be formed? What qualitative criteria can be set up for the stability of such compounds?

b) III–VI compounds such as In_2Te_3 can also be included in this series. The space group of "ordered" In_2Te_3 is the same as that of the III–V compounds. What does the lattice look like? How large is the unit cell?

c) Other criteria for solids with localized bonds are

$$n_e/n_a + b_a = 8 \quad \text{or} \quad n_e/n_a + b_a - b_c = 8$$

where n_e is the number of valence electrons per structural unit, n_a the number of anions, and b_a and b_c the number of anion-anion and cation-cation bonds. Look for examples of such solids.
Rules of this kind have often been used to predict semiconductor properties in binary and ternary compounds [see, for example, E. Mooser, W. B. Pearson: *Progress in Semiconductors*, Vol. 5 (Heywood, London 1960) p. 105].

9.1

Consider a one-dimensional chain of potentials with one defect. Transfer the method used in connection with Fig. 9.9 to show the splitting-off of an energy level from an energy band as in Fig. 9.1. Use the LCAO (tight binding) method treated in Section 2.2.7.

9.2

Calculate the linear and quadratic Zeeman effect for an impurity electron, assuming an isotropic and parabolic conduction band. Use the effective mass approximation and neglect spin (orbital Zeeman effect).

9.3

How do the $(2l + 1)$-fold degenerate energy levels of a free atom split up in a crystal field invariant to all proper rotations which transform a cube into itself? The free atom is invariant to operations of the (infinite) rotation group. The characters of the irreducible representations of this group are: $\lambda^{(l)}(\varphi) = \sin(l + \frac{1}{2})\varphi/\sin\varphi/2$. The point group of the crystal field has 24 elements in five classes and hence also five irreducible representations. Set up the character table for this group first.

9.4

Consider a semiconductor with isotropic parabolic conduction band. Discuss the temperature dependence of the chemical potential and of the electron concentration

a) if shallow donors are present (concentration n_D, energy levels E_D),
b) if additionally recombination centres of acceptor type are present (concentration $n_R < n_D$, energy levels E_R in the middle of the energy gap).

9.5

Calculate the lifetimes of electrons and holes in a semiconductor with recombination centres (acceptors with levels E_R in the energy gap). Treat explicitly the limits of large and small defect concentration n_R. Discuss the recombination mechanism in both cases. Compare the two possible definitions: $\delta n(t) \sim \exp(-t/\tau)$ (decay time) and $\delta n = G\tau$ (steady state).

9.6

Let an n-type semiconductor (electron concentration n_0) contain recombination centres (n_R, E_R) and traps (n_T, E_T), which differ from the recombination centres in that only transitions between trap and valence band are possible.

a) G electron-hole pairs per unit time are created homogeneously within the semiconductor by optical irradiation. It can be assumed that $\delta n, \delta p \ll n_0$. Calculate the photoconductivity $\delta\sigma = \sigma - \sigma_0$ produced.
b) Suppose that at time $t = 0$ the light is switched off. Discuss the decay of photoconductivity. Introduce τ_t and τ_f for the lifetime of a hole in the valence band before capture by a trap, and the time the hole spends in the trap, respectively.

9.7

When discussing the galvanomagnetic effects in Chapter V, a zero lifetime of electron-hole pairs was implicitly assumed. What are the consequences of a nonzero lifetime? What additional effects appear when one irradiates a surface parallel or perpendicular to the magnetic field?

9.8

A steady concentration of electron-hole pairs is formed on the surface of a semiconductor ($x = 0$) by optical excitation. The pairs diffuse through the semiconductor to the opposite surface at $x = a$. Within the semiconductor the pair concentration decays by band-to-band recombination. Pairs which reach the surface recombine there via surface states. What is the recombination law in the volume and what at the surface. How does $\delta n(x)$ depend on τ and on the corresponding parameter associated with the surface states? What is the effect of a different mobility of the electrons and the holes?

Bibliography

Introductions to Solid-State Physics

Excellent and very complete representations of solid-state physics are

1. N. W. Ashcroft, N. D. Mermin: *Solid State Physics* (Holt, Rhinehart & Winston Inc., New York 1976)
2. Ch. Kittel: *Introduction to Solid State Physics*, 5th ed. (J. Wiley & Sons, New York 1976)
3. M. A. Omar: *Elementary Solid State Physics: Principles and Applications* (Addison-Wesley, London 1975)

Other standard introductions are

4. J. S. Blakemore: *Solid State Physics*, 2nd ed. (W. B. Saunders Co., Philadelphia 1973)
5. R. J. Elliott, A. F. Gibson: *An Introduction to Solid State Physics and its Applications* (Barnes & Noble, New York 1974)
6. H. M. Rosenberg: *The Solid State: An Introduction to the Physics of Crystals for Students of Physics, Material Sciences and Engineering* (Oxford University Press, Oxford 1975)
7. Ch. A. Wert, R. M. Thomson: *Physics of Solids*, 2nd ed. (McGraw-Hill, New York 1970)

Monographs on Solid-State Theory

The most modern and complete texts are

8. A. Animalu: *Intermediate Quantum Theory of Crystalline Solids* (Prentice Hall, Englewood Cliffs, N.J. 1977)
9. J. Callaway: *Quantum Theory of the Solid State*, 2 vols. (Academic Press, New York 1974), students edition 1976
10. M. M. Cohen: *Introduction to the Quantum Theory of Semiconductors* (Gordon & Breach Ltd., London 1972)
11. E. E. Hall: *Solid State Physics* (J. Wiley & Sons, New York 1974)
12. H. Haken: *Quantum Field Theory of Solids* (North Holland Publ. Co., Amsterdam—New York 1976)
13. W. A. Harrison: *Solid State Theory* (McGraw-Hill, New York 1969)
14. A. Haug: *Theoretical Solid State Physics*, 2 vols. (Pergamon Press, New York 1972)
15. W. Jones, N. H. March: *Theoretical Solid State Physics*, 2 vols. (J. Wiley & Sons, New York 1973)
16. Ch. Kittel: *Quantum Theory of Solids* (J. Wiley & Sons, New York 1963)
17. R. Kubo, T. Nagamiya: *Solid State Physics* (McGraw-Hill, New York 1969)
18. P. T. Landsberg: *Solid State Physics, Methods and Applications* (J. Wiley & Sons, New York 1969)
19. J. D. Patterson: *Introduction to the Theory of Solid State Physics* (Addison-Wesley, London 1971)
20. J. C. Slater: *Quantum Theory of Molecules and Solids*, 4 vols. (McGraw-Hill, New York 1964—74)
21. R. A. Smith: *Wave Mechanics of Crystalline Solids*, 2nd ed. (Chapman & Hall, London 1969)

22. P. L. Taylor: *A Quantum Approach to the Solid State* (Prentice Hall, Englewood Cliffs, N.J. 1970)
23. J. H. Ziman: *Principles of the Theory of Solids*, 2nd ed. (Cambridge University Press, Cambridge 1972)

as well as

24. P. W. Anderson: *Concepts in Solids* (W. A. Benjamin, New York 1963)
25. H. Clark: *Solid State Physics* (Macmillan-St. Martin's Press, London 1968)
26. H. J. Goldsmid: *Problems in Solid State Physics* (Academic Press, New York 1968)
27. D. Pines: *Elementary Excitations in Solids* (W. A. Benjamin, New York 1963)
28. M. Sachs: *Solid State Theory* (McGraw-Hill, New York 1963)
29. G. Weinreich: *Solids: Elementary Theory for Advanced Students* (J. Wiley & Sons, New York 1965)

The following older books will also be found very readable:

30. N. Mott, W. Jones: *The Theory of Properties of Metals and Alloys* (Oxford University Press, Oxford 1958)
31. R. E. Peierls: *Quantum Theory of Solids* (Oxford University Press, Oxford 1955)
32. F. Seitz: *The Modern Theory of Solids* (McGraw-Hill, New York 1940)
33. G. H. Wannier: *Elements of Solid State Theory* (Cambridge University Press, Cambridge 1959)
34. A. H. Wilson: *The Theory of Metals* (Cambridge University Press, Cambridge 1958)
35. J. H. Ziman: *Electrons and Phonons* (Oxford University Press, Oxford 1960)

Monographs on Special Topics in Solid-State Theory

In the following only the most important newer books as well as some older standard texts are listed.

Mathematical Tools of Solid-State Theory, especially Many-Body Theory and Group Theory

36. A. A. Abrikosov, L. P. Gor'kov, I. Ye. Dzyaloshinski: *Quantum Field Theoretical Methods in Statistical Physics* (Pergamon Student Edition, New York 1965)
37. A. L. Fetter, J. D. Walecka: *Quantum Theory of Many Particle Systems* (McGraw-Hill, New York 1971)
38. R. D. Mattuck: *A Guide to Feynman Diagrams in the Many Body Problem* (McGraw-Hill, New York 1967)
39. Ph. Nozières: *Theory of Interacting Fermi Systems* (W. A. Benjamin, New York 1964)
40. D. Pines, Ph. Nozières: *The Theory of Quantum Liquids* (W. A. Benjamin, New York 1966)
41. S. Raimes: *Many Electron Theory* (North Holland Publishing Company, Amsterdam 1972)
42. J. M. Ziman: *Elements of Advanced Quantum Theory* (Cambidge University Press, Cambridge 1969)
43. J. F. Cornwell: *Group Theory and Energy Bands in Solids* (North Holland Publ. Co., Amsterdam 1969)
44. A. P. Cracknell: *Group Theory in Solid State Physics* (Taylor & Francis, London 1975)
45. M. Hammermesh: *Group Theory and its Applications to Physical Problems* (Addison-Wesley/Pergamon, New York 1962)
46. V. Heine: *Group Theory in Quantum Mechanics* (Pergamon Press, New York 1960)
47. G. F. Koster, J. O. Dimmock, R. G. Wheeler, H. Statz: *Properties of the 42 Point Groups* (MIT Press, Cambridge/Mass. 1963)
48. M. Lax: *Symmetry Principles in Solid State and Molecular Physics* (J. Wiley & Sons, New York 1974)
49. M. Tinkham: *Group Theory and Quantum Mechanics* (McGraw-Hill, New York 1964)

Electronic Properties, Energy Bands, Transport

50. F. Bassani, G. P. Parravicini: *Electronic States and Optical Transitions in Solids* (Pergamon Press, New York 1975)
51. W. R. Beam: *Electronics of Solids* (McGraw-Hill, New York 1965)
52. F. J. Blatt: *Physics of Electronic Conduction in Solids* (McGraw-Hill, New York 1968)
53. L. Brillouin: *Wave Propagation in Periodic Structures* (Academic Press, New York 1960)
54. R. H. Bube: *Electronic Properties of Crystalline Solids* (Academic Press, New York 1974)
55. J. Callaway: *Energy Band Theory* (Academic Press, New York 1964)
56. A. P. Cracknell, K. C. Wong: *The Fermi Surface, Its Concept, Determination, and Use in the Physics of Metals* (Oxford University Press, Oxford 1973)
57. G. C. Fletcher: *The Electron Band Theory of Solids* (Elsevier, New York 1971)
58. W. A. Harrison: *Pseudopotentials in the Theory of Metals* (W. A. Benjamin, New York 1966)
59. C. M. Hurd: *Electrons in Metals: An Introduction to Modern Topics* (J. Wiley & Sons, New York 1975)
60. H. Jones: *The Theory of Brillouin Zones and Electronic States in Crystals*, 2nd ed. (North Holland Publ. Co., Amsterdam 1975)
61. T. L. Loucks: *Augmented Plane Wave Method* (W. A. Benjamin, New York 1967)
62. L. Pincherle: *Electronic Energy Bands in Solids* (MacDonald, London 1971)
63. S. Raimes: *The Wave Mechanics of Electrons in Metals* (North Holland Publishing Company, Amsterdam 1970)
64. A. C. Smith, J. F. Janak, R. B. Adler: *Electronic Conduction in Solids* (McGraw-Hill, New York 1967)
65. J. M. Ziman (ed.): *The Physics of Metals*, Vol. 1: Electrons (Cambridge University Press, Cambridge 1969) (Vol. 2: see [96])

Lattice Dynamics, Superconductivity, Magnetism, Optics

66. T. A. Bak: *Phonons and Phonon Interactions* (W. A. Benjamin, New York 1964)
67. M. Born, K. H. Huang: *Dynamical Theory of Crystal Lattices* (Oxford University Press, Oxford 1954)
68. B. DiBartolo, R. C. Powell: *Phonons and Resonances in Solids* (John Wiley and Sons, New York 1976)
69. G. K. Morton, A. A. Maradudin: *Dynamical Properties of Solids*, 3 vols. (North Holland Publ. Co., Amsterdam 1974—76)
70. G. Venkataraman, L. A. Feldkamp, V. C. Sahni: *Dynamics of Perfect Crystals* (MIT Press, Cambridge, Mass. 1975)
71. J. M. Blatt: *Theory of Superconductivity* (Academic Press, New York 1964)
72. C. G. Kuper: *Introduction to the Theory of Superconductivity* (Oxford University Press, Oxford 1968)
73. E. A. Lynton: *Superconductivity* (Methuen & Co. Ltd., London 1964)
74. G. Rickayzen: *Superconductivity* (Interscience Publishers, New York 1965)
75. A. C. Rose-Innes, E. H. Rhoderick: *Introduction to Superconductivity* 2nd ed. (Pergamon Press, New York 1978)
76. J. R. Schrieffer: *Theory of Superconductivity*, 2nd ed. (W. A. Benjamin, New York 1971)
77. M. Tinkham: *Introduction to the Theory of Superconductivity* (McGraw-Hill, New York 1975)
78. A. H. Morrish: *The Physical Principles of Magnetism* (J. Wiley & Sons, New York 1965)
79. D. C. Mattis: *The Theory of Magnetism* (Harper & Row, New York 1965)
80. G. T. Rado, H. Suhl (eds.): *Magnetism*, 5 vols. (Academic Press, New York 1963—73)
81. S. V. Vonsovskii: *Magnetism* (Halsted, New York 1975)
82. R. M. White: *Quantum Theory of Magnetism* (McGraw-Hill, New York 1970)
83. H. J. Zeiger, G. W. Pratt: *Magnetic Interactions in Solids* (Oxford University Press, Oxford 1973)
84. F. Abelès (ed.): *Optical Properties of Solids* (North Holland Publ. Co., Amsterdam 1972)

85. D. L. Greeneway, G. Harbeke: *Optical Properties and Band Structure of Semiconductors* (Pergamon Press, New York 1968)
86. B. DiBartolo: *Optical Interactions in Solids* (J. Wiley & Sons, New York 1968)
87. B. Seraphin (ed.): *Optical Properties of Solids: New Developments* (North Holland Publ. Co., Amsterdam 1975)
88. F. Wooten: *Optical Properties of Solids* (Academic Press, New York 1972)

Chemical Bond, Metal-Insulator Transitions, Surfaces, Nonperfect Solids

89. C. A. Coulson: *Valence* (Oxford University Press, Oxford 1961)
90. L. Pauling: *The Nature of the Chemical Bond and the Structure of Molecules and Crystals*, 3rd ed. (Cornell University Press, Ithaka, N.Y. 1960)
91. J. C. Phillips: *Covalent Bonding in Crystals, Molecules and Polymers* (The University of Chicago Press, Chicago 1969), and: *Bonds and Bands in Semiconductors* (Academic Press, New York 1973)
92. W. A. Harrison: *Electronic Structure and the Properties of Solids* (W. H. Freeman and Co. Ltd., Oxford 1979)
93. N. F. Mott: *Metal-Insulator Transitions* (Taylor & Francis Ltd., London 1974)
94. N. F. Mott, E. A. Davis: *Electronic Processes in Non-Crystalline Materials* 2nd ed. (Oxford University Press, Oxford 1979)
95. N. B. Hannay (ed.): *Treatise on Solid State Chemistry*, 6 vols. (Plenum Press, New York 1973—76)
96. P. B. Hirsch (ed.): *The Physics of Metals*, Vol. 2: Defects (Cambrige University Press, Cambridge 1975) (Vol. 1: see [65])
97. C. G. Scott, C. E. Reed (eds.): *Surface Physics of Phosphors and Semiconductors* (Academic Press, New York 1975)
98. A. M. Stoneham: *Theory of Defects in Solids: Electronic Structure and Defects in Insulators and Semiconductors* (Oxford University Press, Oxford 1975)
99. J. Tauc (ed.): *Amorphous and Liquid Semiconductors* (Plenum Press, New York 1974)
100. P. T. Townsend, J. C. Kelly: *Color Centres and Imperfections in Insulators and Semiconductors* (Crane & Russack, New York 1973)

Series and Journals with Review Articles

Citations out of this group are made by the number of the series followed by the number of the volume, e.g., [101.17] means: *Solid State Physics*, Vol. 17.

101. *Solid State Physics, Advances and Applications*, ed. by H. Ehrenreich, F. Seitz, D. Turnbull (Academic Press, New York, since 1954) (1981: 36 vols.)
102. Supplements to [101]:
   1) T. P. Das, E. L. Hahn: *Nuclear Quadrupole Spectroscopy*
   2) W. Low: *Paramagnetic Resonance in Solids*
   3) A. A. Maradudin et al: *Theory of Lattice Dynamics in the Harmonic Approximation*
   4) A. C. Beer: *Galvanomagnetic Effects in Semiconductors*
   5) R. S. Knox: *Theory of Excitons*
   6) S. Amelinckx: *The Direct Observation of Dislocations*
   7) J. W. Corbett: *Electron Radiation Damage in Semiconductors*
   8) J. J. Markham: *F-Centres in Alkali Halides*
   9) E. Conwell: *High Field Transport in Semiconductors*
   10) C. B. Duke: *Tunneling in Solids*
   11) M. Cardona: *Modulation Spectroscopy*
   12) A. A. Abrikosov: *Introduction to the Theory of Normal Metals*
   13) P. M. Platzman, P. A. Wolff: *Waves and Interactions in Solid State Plasmas*
   14) L. Liebert (ed.): *Liquid Crystals*
   15) R. M. White, T. H. Geballe: *Long Range Order in Solids*

103. *Advances in Solid State Physics (Festkörperprobleme)*, ed. by F. Sauter, O. Madelung, H. J. Queisser, J. Treusch (F. Vieweg & Sohn, Braunschweig, since 1962, (1981: 21 vols.), since vol. 8 partly in English, since vol. 13 only in English)
104. *Semiconductors and Semimetals*, ed. by R. K. Willardson, A. C. Beer (Academic Press, New York, since 1966) (1981: 14 vols.)
105. *Handbook of Semiconductors*, series editor: T. S. Moss, 4 vols. (North Holland Publ. Co., Amsterdam 1980/81)
106. *Encyclopedia of Physics/Handbuch der Physik*, ed. by S. Flügge (Springer, Berlin, Heidelberg, New York) (partly in English)
107. *Springer Series in Solid State Sciences*, ed. by M. Cardona, P. Fulde, H. J. Queisser (Springer, Berlin, Heidelberg, New York) (1981: 16 vols.) especially:
Vol. 7: E. N. Economou: *Green's Functions in Quantum Physics*
Vol. 10: H. Bilz, W. Kress: *Phonon Dispersion Relations in Insulators*
Vol. 12: S. Nakajima, Y. Toyozawa, R. Abe: *The Physics of Elementary Excitations*
108. *Dynamical Properties of Solids*, ed. by G. K. Horton, A. A. Maradudin (North Holland Publ. Co., Amsterdam, since 1974) (1981: 4 vols.)
109. *Current Topics in Material Sciences*, ed. by E. Kaldis (North Holland Publ. Co., Amsterdam, since 1978) (1981: 7 vols.)
110. *Critical Reviews in Solid State Sciences* (CRC Press, Cleveland, Ohio)
111. Review papers on solid state problems may also be found in:
(a) *Advances in Physics* (Taylor & Francis, London)
(b) *Physics Reports* (part C of *Physics Letters*) (North Holland Publ. Co., Amsterdam)
(c) *Reports on Progress in Physics* (The Institute of Physics and the Physical Society, London)
(d) *Springer Tracts in Physics* (English continuation of *Ergebnisse der Exakten Naturwissenschaften*) (Springer, Berlin, Heidelberg, New York)
(e) *Topics in Applied Physics* (Springer, Berlin, Heidelberg, New York)
(f) *Topics in Current Physics* (Springer, Berlin, Heidelberg, New York)

Conferences, Summer Schools

112. Proceedings of the International School of Physics, Varenna/Italy (Academic Press, New York)
Vol. 22: *Semiconductors*, ed. by R. A. Smith
Vol. 31: *Quantum Electronics and Coherent Light*, ed. by C. H. Townes
Vol. 34: *Optical Properties of Solids*, ed. by J. Tauc
Vol. 37: *Theory of Magnetism of Transition Metals*, ed. by W. Marshall
Vol. 42: *Quantum Optics*, ed. by R. J. Glauber
Vol. 52: *Atomic Structure and Properties of Solids*, ed. by E. Burstein
113. *Trieste Lectures* (International Centre for Theoretical Physics, Trieste/Italy, International Atomic Energy Agency, Vienna)
a) *Theory of Condensed Matter* (1968)
b) *Theory of Imperfect Crystalline Solids* (1970)
c) *Electrons in Crystalline Solids* (1972)
d) *Surface Science* (2 vols.) (1974)
114. *NATO Advance Study Institutes Series* (Plenum Press, New York)
a) *Tunneling Phenomena in Solids*, ed. by E. Burstein, S. Lundqvist (1967)
b) *Far-Infrared Properties of Solids*, ed. by S. S. Mitra, S. Nudelman (1968)
c) *Optical Properties of Solids*, ed. by S. Nudelman, S. S. Mitra (1969)
d) *Elementary Excitations in Solids, Molecules and Atoms*, ed. by J. T. Devreese, A. B. Kunz, T. C. Collins (1973)
e) *Optical Properties of Ions in Solids*, ed. by B. DiBartolo, D. Pacheco (1974)
f) *Physics of Structurally Disordered Solids*, ed. by S. S. Mitra (1976)

115. *IBM Research Symposia Series* (Plenum Press, New York)
New York 1970: *Computational Methods in Band Theory*, ed. by P. M. Marcus, J. F. Janak
Wildbad 1971: *Computational Solid State Physics*, ed. by F. Herman, N. W. Dalton, T. R. Koehler
San Jose 1972: *Computational Methods for Large Molecules and Localized States in Solids*, ed. by F. Herman, A. D. McLean, R. K. Nesbet
116. *Amorphous and Liquid Semiconductors*, Conference Proceedings:
Cambridge 1969, ed. by N. F. Mott (North Holland Publ. Co., Amsterdam 1970)
Ann Arbor 1971, ed. by M. H. Cohen, G. Lucovsky (North Holland Publ. Co., Amsterdam 1972)
Garmisch 1973, ed. by W. Brenig, J. Stuke (Taylor & Francis, London 1974)
Leningrad 1975, ed. by B. T. Kolomiets (Isdatelstvo Nauka, Leningrad 1976)
Edinburgh 1977, ed. by W. E. Spear (Centre for Industrial Consultancy and Liaison, University of Edinburgh 1977)
Cambridge/Mass. 1979, ed. by W. Paul, M. Kastner (North Holland Publ. Co., Amsterdam 1979)
117. *Elementary Excitations in Solids* (Cortina Lectures 1966), ed. by A. A. Maradudin, G. F. Nardelli (Plenum Press, New York 1969)
118. *Polarons and Excitons* (Proceedings of the Scottish Universities Summer School, Edinburgh), ed. by C. G. Kuper, G. D. Whitfield (Oliver & Boyd, London 1963)
119. *Polaritons* (Int. Conference Taormina, Italy 1972), ed. by E. Burstein, F. DeMartini (Pergamon Press, New York 1975)
120. *Lattice Dynamics* (International Conference, Kopenhagen 1964), ed. by R. F. Wallis (Plenum Press, New York 1965)
121. *Phonons in Perfect Lattices and in Lattices with Point Imperfections* (Proceedings of the Scottish Universities Summer School, Edinburgh), ed. by R. W. H. Stevenson (Oliver & Boyd, London 1965)
122. *Phonon Scattering in Solids* (International Conference, Nottingham 1975), ed. by L. J. Challis, V. W. Rampton, A. F. G. Wyatt (Plenum Press, New York 1976)
123. *Light Scattering in Solids* (International Conference Paris 1971), ed. by M. Balkanski (Flammarion, Paris 1971)
124. *Localized Excitations in Solids* (International Conference Irvine/Calif. 1964), ed. by R. F. Wallis (Plenum Press, New York 1968)
125. *Proceedings of the International Conference on Metal-Nonmetal Transitions* (San Francisco 1968): Rev. Mod. Phys. **40,** 673—844 (1968)
126. *Solid Surfaces* (International Conference Providence, R. I. 1964), ed. by H. Gatos (North Holland Publ. Co., Amsterdam 1964)
127. *Semiconducting Surfaces*, ed. by A. Many, Y. Goldstein, N. B. Grover (North Holland Publ. Co., Amsterdam 1965)
128. *The Structure and Chemistry of Solid Surfaces* (International Conference Berkeley 1968), ed. by G. A. Somorjai (J. Wiley & Sons, New York 1969)
129. *Amorphous Materials* (International Conference Sheffield 1970), ed. by R. W. Douglas, B. Ellis (J. Wiley & Sons, New York 1972)
130. *Conduction in Low Mobility Materials* (International Conference Eilat, Israel 1971), ed. by N. Klein, M. Pollak, D. S. Tannhauser (Taylor & Francis, London 1972)
131. *Electronic and Structural Properties of Amorphous Semiconductors* (Proceedings of the Scottish Universities Summer School, Edinburgh), ed. by P. G. LeComber, J. Mort (Academic Press, New York 1973)
132. *Tetrahedrally Bonded Amorphous Semiconductors* (International Conference Yorktown Heights 1974), ed. by M. H. Brodsky, S. Kirkpatrick, D. Weaire (American Institute of Physics 1975)
133. *Modern Solid State Physics*, Vol. I: *Electrons in Metals*, ed. by J. F. Cochran, R. R. Haering, Vol. II: *Phonons and Their Interactions*, ed. by R. H. Enns, R. R. Haering, Simon Frazer University Summer Schools (Gordon & Breach, London 1968/69)

Literature to the Problems

In the following we give some literature either containing a discussion of the more advanced problems or being helpful in solving them. Since most of these problems are discussed in several monographs or review papers, the literature cited is only one of many sources of information.

1.1: D. J. Thouless: *Quantum Mechanics of Many Body Problems* (Academic Press, New York 1961), Sec. III
2.4,5: [60] Chap. 3
2.6: [50] p. 159
2.7: [15] p. 77
3.1: D. Pines [101.1]
3.2: [22] Sec. 5.2, 5.3
3.7: [106.XVIII/2] p. 90
4.2: T. D. Lee, F. E. Low, D. Pines: Phys. Rev. **90**, 297 (1953)
4.3—6: [106.XX] Chap. C
5.2./3: T. Bardeen, L. N. Cooper, T. R. Schrieffer: Phys. Rev. **108**, 1175 (1957)
6.2: [113a] p. 917
6.4/5: B. Lax [112.22]
6.7: W. Zawadzki: Phys. Lett. **4**, 190 (1963)
7.2: [34] p. 292
8.1: [90] p. 146
8.2: [90] App. V
8.3: [26] Chap. 3
9.2: B. Lax [112.22]
9.4: [54] p. 337
9.5—8: [106.XX] Chap. D

Subject Index